2950 Em

INVERTEBRATE ZOOLOGY
VOLUME III

INVERTEBRATE

Crustacea

ZOOLOGY

VOLUME III

by Alfred Kaestner

Professor University of Munich Emeritus
Former Director of the Natural History
Collections of Bavaria

translated and adapted from the
second German edition by

HERBERT W. LEVI

Museum of Comparative Zoology
Harvard University

and
LORNA R. LEVI

INTERSCIENCE PUBLISHERS

A Division of John Wiley & Sons, New York · London · Sydney · Toronto

translated and adapted from
Lehrbuch der Speziellen Zoologie, Band 1, 2 Teil
Copyright 1959, 1967 by VEB Gustav Fischer Verlag, Jena

Copyright © 1970, by John Wiley & Sons, Inc.

Library of Congress Catalogue Card Number: 67-13947

SBN 471   45417   6

Printed in the United States of America

10 9 8 7 6 5 4 3 2 1

Preface to the English Edition

The third volume of *Invertebrate Zoology* contains the second part of Volume 2 of the *Lehrbuch der Speziellen Zoologie*, 2nd ed., except for the introduction to the mandibulates which is included in *Invertebrate Zoology*, Volume 2, Chapter 16. As with the previous two volumes, this volume was freely translated and brought up-to-date. The emphasis has been shifted from species of Central Europe to the world fauna with an emphasis on North America and Europe. Some illustrations were added, some were replaced. Certain parts of the original German edition (the cephalocaridans, copepods, and cirripeds) were partly rewritten with the help of specialists.

As the title of the German version implies, the volume started as a *Lehrbuch,* i.e., a textbook, rather than as a treatise. Although the volume has been expanded and may now go far beyond textbook level, it still has some of the shortcomings of a textbook. For instance, there are generalizations based on the study of only one or few species of a group, while all classes and genera contain species that depart from the typical anatomy. For this reason almost every sentence would have to be modified with the words "usually" or "almost always." In the interest of readability this was not done. Some of the facts may have been simplified and aberrant forms have been mentioned only if they are of anatomical or biological importance. Because the volume started as a textbook, literature references are listed at the end of each chapter. The few citations given in the text are recent works added in the English adaptation; they appear only in the longer chapters where it would be difficult to find the source of the information among the many literature references. References generally emphasize recent work and only rarely include the classical citations before Kükenthal's *Handbuch der Zoologie*. The bibliographic titles, especially those in French and German, had to be shortened.

This volume, it is hoped, will fill a need for an up-to-date book on Crustacea in English. Although several volumes have been published recently,* the most

---

* Citations are given at the end of Chapter 1.

important being Waterman's *The Physiology of Crustacea*, there is at present no textbook of Crustacea comparable to the numerous textbooks on insects. The most used volume in English is still Calman's, published in 1909 and recently reprinted because of demand. Not only have numerous crustacean groups been discovered since Calman, but also a huge amount of research has been published, widely scattered throughout biological literature. More important, since the time of Calman, physiology has become an integral part of comparative and adaptive anatomy and systematics.

Again, as in Volume 2, in the hope of avoiding unnecessary future name changes, we have starred the names of those families, genera, and, in the decapods, species on which the Commission on Zoological Nomenclature has acted. These names are listed in the *Official List of Family Group Names in Zoology* and the corresponding lists of genera and species, published in 1958 and 1966 by the International Trust for Zoological Nomenclature, London.

We are grateful for the advice and assistance of the numerous colleagues whose help has made possible the English version of this volume. Dr. W.D.I. Rolfe, Hunterian Museum, University of Glasgow carefully read the entire volume, made suggestions on style, removed "American barbarisms" and added information on fossil crustaceans. Drs. W.A. Newman and R.R. Hessler freely gave advice and assistance. Dr. A.G. Humes took time to suggest new illustrations for the copepods, Dr. F.A. Chace Jr. for the decapods. Dr. V. Petriconi and Dr. K.E. Lauterbach supplied new figures. Dr. R. Benesch offered unpublished information on the development of *Artemia*, Dr. J.R. Grindley on *Spelaeogriphus*. Dr. G. Hartmann made suggestions for the ostracod part of the original German version. We are grateful for the numerous suggestions and changes made by specialists who read chapters or parts of chapters: Dr. E.L. Bousfield, National Museums of Canada (Amphipoda); Dr. T.E. Bowman, Smithsonian Institution, U.S. National Museum (Euphausiacea and Mysidacea); Dr. F.A. Chace Jr., Smithsonian Institution, U.S. National Museum (Eucarida); Dr. I. Efford, University of British Columbia (parts of the introduction and Eucarida); Dr. G. Fryer, Freshwater Biological Association, England (Branchiura); Dr. L.F. Gardiner, Woods Hole Oceanographic Institution (Tanaidacea); Dr. R.R. Given, Marine Science Center, Avalon, California (Cumacea); Dr. J.R. Grindley, Port Elizabeth Museum, South Africa (Copepoda and Spelaeogriphacea); Dr. C.E. Goulden, Academy of Natural Sciences, Philadelphia (Branchiopoda); Dr. B.A. Hazlett, University of Michigan (Eucarida); Dr. R.R. Hessler, Woods Hole Oceanographic Institution and Scripps Institution of Oceanography (Cephalocarida, Mystacocarida, Pancarida and Isopoda); Dr. H.H. Hobbs, Smithsonian Instituttion, U.S. National Museum (Astacidae); Dr. L.B. Holthuis, Natural History Museum, Leiden (Eucarida); Dr. A.G. Humes, Boston University (Copepoda); Dr. R.V. Kesling, Museum of Paleontology, University of Michigan (Ostracoda); Dr. R.B. Manning, Smithsonian Institution, U.S. National Museum (Hoplocarida);

Dr. R.J. Menzies, Florida State University (Isopoda); Dr. W.A. Newman, Scripps Institution of Oceanography (introduction and Cirripedia); Dr. W. Noodt, University of Kiel (Syncarida); Dr. L. Pardi, University of Pisa (orientation of Amphipoda); Dr. A.J. Provenzano Jr., University of Miami (Paguroidea and Coenobitoidea); Dr. W.D.I. Rolfe, Hunterian Museum, University of Glasgow (Phyllocarida); Prof. S.M. Shiino, Faculty of Fisheries, Prefecture University of Mie, Japan (Isopoda); Dr. P. Tongiorgi, University of Pisa (orientation of Isopoda and Amphipoda) and Dr. A.A. Weaver, College of Wooster (Copepoda). Many others have supplied information by correspondence. The helpful suggestions supplied by the specialists were for the most part adopted but once in a while, usually to maintain consistency of terminology or for brevity, one or another could not be followed. Although accuracy has been a major goal, we take personal responsibility for the inevitable errors.

Here we should also thank Mr. R.E. Truelsen, managing editor and Miss Lola Peters, staff editor, for their patience and efficient, friendly handling of the manuscripts.

A. Kaestner
H.W. Levi
L.R. Levi

December 1969

Preface to the Second German Edition

The Crustacea (Part 4 of the first edition) have been completely reworked for the second edition using the literature available to me up to fall 1966. The parts on Mandibulata,* Cephalocarida, and Ostracoda were completely rewritten; the part on habits of the Cladocera, Copepoda, Decapoda, and Isopoda were partially rewritten. Despite this I have tried to increase the size of the volume only slightly, but at the same time give information usable for the increasing number of skin divers. In order to increase the illustrations, or clarify older figures, 25 new figures have been prepared and seven illustrations used from other volumes published by VEB G. Fischer.

Dr. H.-E. Gruner and Prof. L.B. Holthuis have answered questions on nomenclature. Dr. G. Kolb and Prof. F. Seidel, Dr. H. Altner and Dr. H. Löffler offered advice for the new edition from their teaching experience. I would like here to thank these colleagues. Also I would like to thank Mrs. E. Puhonny for making the new illustrations and the director and editor of the publisher for their helpfulness.

Alfred Kaestner

Munich
November 1966

* This chapter is printed in Volume 2 as Chapter 16 of the English edition.

Contents

INVERTEBRATE ZOOLOGY

VOLUME III

1.

Class Crustacea: Anatomy

The Crustacea include about 30,500 known Recent species,* among them the largest living arthropods: the spiny lobster, *Jasus huegeli*, 60 cm long; the American lobster *Homarus americanus*, to 60 cm long; the giant crab of Australia, *Pseudocarcinus gigas*, having the carapace 40 cm wide; and the giant spider crab of Japan, *Macrocheira kaempferi*, with a leg span of 3.6 m.

Crustacea are mandibulate arthropods. There are two pairs of antennae on the head. They usually respire through gills derived from leg appendages or through the inner carapace wall.

Crustacea are adapted for life in water. They exist in great diversity in the ocean and fresh water, as nekton and plankton as well as on a variety of benthic substrates from the shore to 10,000 m.† Members of four subclasses have become parasitic. Despite diversity in structure only very few terrestrial forms have evolved, and of these only the wood lice have become completely terrestrial, including adaptations for terrestrial development.

The unknown ancestral crustaceans probably had a trunk consisting of a series of homonomous metameres, each with a pair of appendages. This idea is supported by many Recent malacostracans, each metamere of which bears a pair of limbs. The homonomy of somites is also preserved in several non-malacostracans. In others some thoracomeres are fused to the head. From these generalized types diverse forms have evolved: minute, soft dwarf forms as well as huge decapods with strong calcareous armor, often strikingly colored. The ancestrally elongate body often has become compact through narrowing of the abdomen, by folding the abdomen under the venter of the thorax as in crabs, and by reduction

---

* This figure is only a rough approximation. More exact information will be available after the completion of the planned *Catalogus Crustaceorum*.

† For comparison the greatest depth known is 11,034 m from the Mariana Islands in the Pacific Ocean, and the greatest penetrated by man was 10,916 by Piccard and Walsh in 1960; the Galathea expedition collected animals from these depths.

of the somite number (Figs. 4-10, 13-32). By similar modifications spherical or ovoid species have evolved, in which the trunk, as in mollusks, is enclosed between two valves formed from folds of the integument (Fig. 4-7). Many parasitic forms, among Copepoda, Cirripedia, and Isopoda, with many organs reduced, have lost all superficial resemblance to crustaceans or even arthropods; their relationship can be recognized only in their larval forms.

Anatomy

In this introduction the parasitic forms are ignored, as these have lost many of the structures characteristic of the class.

METAMERES.   The body of a crustacean consists of the acron, the telson, and a variable number of metameres in between (Figs. 1-1, 1-3). The largest number of metameres, more than 50, is found in the Notostraca, the smallest number in the Ostracoda. The ostracod trunk, like that of mites, is secondarily reduced, an adaptation to small size (Fig. 5-3). Only within the subclass Malacostraca is the number of metameres constant: some of the primitive ones have 20 (Phyllocarida, Fig. 10-6; Lophogastrida, Fig. 15-3), while in all others the last metamere, the 20th, becomes reduced during ontogeny so that only 19 remain, 5 in the head and 14 in the trunk (Figs. 1-1, 1-4). The metameres, except in the very short forms, make their appearance in three different forms: as head somites (cephalomeres), thoracic somites (thoracomeres), and abdominal somites (pleomeres).* While each cephalic and thoracic somite has one pair of appendages, the abdominal somites in most subclasses other than Malacostraca form no appendages, not even in the embryo and larva (Fig. 2-16). When abdominal somites have appendages, as in the Malacostraca, the pleopods differ from the thoracic appendages, the thoracopods, by being shorter and having a different shape (Figs. 1-3, 1-4). The three tagmata (groups of metameres) therefore differ from each other by their appendages.

TAGMATA.   The variability in somite number precludes designation of a tagma border, as is possible in the chelicerates in which the 7th metamere or pregenital somite is always the first abdominal one, regardless of its appendages. Thus the thorax and abdomen (pleon) of the different suborders of Crustacea are not homologous to each other as they are in arachnids or insects, but are only analogous. It is not even worthwhile to invent special names that would distinguish the crustacean tagmata from those of insects, as to be strictly homologous the tagmata would have to have a different name in almost every subclass, or each order among the Branchiopoda.

---

* Recently T. Wolff (1956, *Galathea Rep.* 2: 187–253) suggested that the metameres of the pereon be called pereonites, those of the abdomen, pleonites. This has been widely followed although the meaning of the older terms pereomere and pleomere is much easier to understand for those unacquainted with the vocabulary of the carcinologist. We have followed here the older German version in terminology.

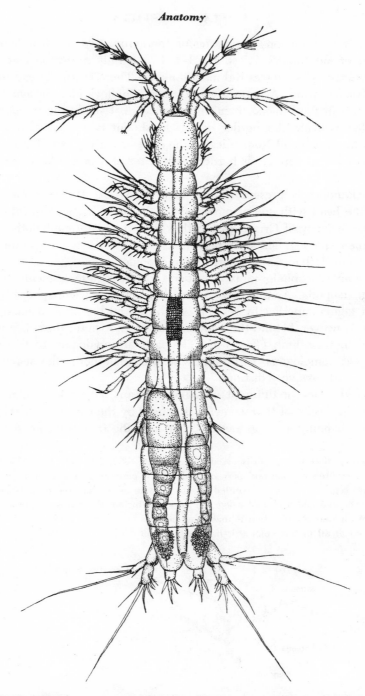

Fig. 1–1. *Bathynella natans*, a crustacean with similar somites; up to 2 mm long. There are 8 thoracomeres and 6 pleomeres, the last fused to the telson. Female viewed from above: the 8th thoracopod and 1st pleopod are very short, the 6th pleopod (uropod) is long. In the middle the gut can be seen through the integument, on the sides are the ovaries. (After Jacobi.)

A true head (cephalon) consisting of the acron, two pairs of antennae and three pairs of mouthparts (Figs. 1-1, 1-2, 4-7) is only rarely present. This is in marked contrast to the terrestrial mandibulates. There is still a head in Cephalocarida, Mystacocarida, Anostraca, Notostraca, Phyllocarida, Bathynellacea, and less distinctly in the Conchostraca and Cladocera. Usually the integument lacks grooves that indicate the border of the head, but two transverse grooves are present in Anostraca and Notostraca, at the edge of slightly raised areas. However, these cannot be interpreted as borders of tergites of a particular metamere (Fig. 1-2).

A cephalothorax is present in the majority of Crustacea, formed from the fusion of the head with one or more trunk (thoracic) metameres into a uniform tagma. The majority of Crustacea has thus three tagmata: cephalothorax, pereon and abdomen (pleon). The term pereon[*] is used for the tagma comprising those thoracic somites[†] that are not part of the cephalothorax.

In the simplest cephalothorax only one thoracic somite is combined with the head (e.g., Isopoda, Amphipoda, *Anaspides*, and many copepods) (Fig. 1-3). This short tagma superficially resembles a head and has often been so called by specialists. However, behind the two maxillae on the venter there is a maxilliped, a modified thoracic limb (Fig. 1-3). The parts are analogous to those of a real head but not homologous. In groups with a cephalothorax the segmentation of the pereon is always very distinct.

The cephalothorax is different in malacostracans with a large carapace (Fig. 1-4). Here the fusion of thoracomeres is caused by the carapace growing to their tergites and hanging as a protective cover over the free posterior thoracomeres,

[*] The spelling of this term is discussed in T. Wolff (1962, *Galathea Rep.* 6:1–320). Pereon has also been written pereion and peraeon and with it pereiopods and peraeopods.

[†] The term segment has been avoided in this volume as in the preceding because it refers to both a metamere and a part of a leg. To avoid confusion, the terms somite and article are used here. An added complication is that some authors use joint in the sense of article, others hinge. If used at all in this volume it has the sense of hinge.

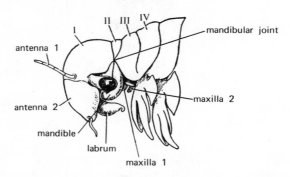

Fig. 1–2.   Head of *Eubranchipus vernalis* in lateral view; II–IV, parts of head separated by grooves; first thoracic tergite follows IV. (After Snodgrass.)

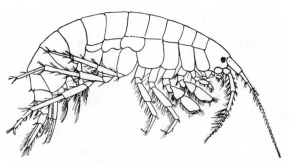

Fig. 1–3. *Gammarus lacustris,* **a crustacean with a very short cephalothorax; 2 cm long. (After Sars from Schellenberg.)**

the pereomeres (Fig. 1-7). In the Mysidacea the carapace fuses with thoracomeres 1–3, in the Tanaidacea 2, in the Cumacea 4–6, and in the Eucarida usually all thoracomeres. In malacostracans at most three pairs of thoracopods of fused thoracomeres differentiate into maxillipeds. The other thoracopods do not change and remain locomotory limbs called pereopods, although they are inserted on the cephalothorax. This avoids profusion in terminology. If the carapace is fused to and encloses more than three thoracomeres, including thoracomeres bearing walking or swimming limbs, the cephalothorax is analogous to the prosoma (cephlothorax) of chelicerates. As illustrated by Fig. 1-6 this condition develops in primitive decapods only during postembryological growth. Even though the adults do not have free thoracomeres anymore, their limbs are still considered pereopods.

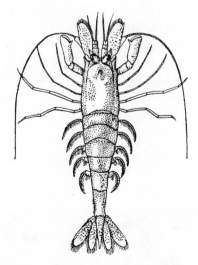

Fig. 1–4. **The shrimp *Crangon crangon*; cephalothorax here includes all thoracomeres; 5 cm long. (After Calman from Schellenberg.)**

In many malacostracans, however, the carapace is not completely fused with the dorsum of the head anteriorly. It is free and forms a roof for a short distance anteriorly. The short, covered apical part underneath is usually a plate on each side of which are the eyestalks (Fig. 1-5), suggesting that it belongs to the acron. In the Stomatopoda this anterior section is unusually long and consists of two skeletal parts connected by membranes; the anterior part bears the eyestalks, the posterior part the two pairs of antennae. These anterior, free plates, roofed over by the carapace, can hardly be considered to be of comparative anatomical significance, as they are completely fused to the carapace in the larvae of these animals (Fig. 1-6, 2-20b).

The abdomen (pleon) is made up of pleomeres, abdominal somites, which are fused in some groups. Commonly the last somite is fused with the telson and forms a pleotelson. In some isopods other somites are fused to the telson; in Asellota, all abdominal tergites are fused (Fig. 17-28).

The telson, which bears the anus, in contrast to true metameres has neither ganglia nor appendages. In many subclasses, however, it bears the furca, a pair of processes (Figs. 1-1, 4-7), each ramus often consisting of several articles.

As in the chelicerates, some crustacean groups have a much shortened trunk, lacking segmentation (Cladocera, Ostracoda, and Cirripedia). While in Cladocera a thorax can still be distinguished from the abdomen, the appendageless pleon has disappeared in many Cirripedia (Fig. 9-6). The ostracods have only two thoracic somites and posteriorly, a short and saclike pleon (Fig. 5-3).

CARAPACE.   It is characteristic of Crustacea that on each side of some or all somites there are dorsolateral, flat overhanging keels, called epimeres (Fig. 17-1). (Unfortunately the term pleura is at times used for these keels; pleura are the lateral walls of the arthropod body.) These epimeres may be more or less horizontal projections from the body or may hang down ventrally (abdomen in Fig. 1-3). The posterior epimeres of the head, called carapace, may be very large, and may grow into a long uniform piece sometimes considered a shell. The carapace often extends posteriorly far beyond the head (Fig. 4-7). In the Notostraca and Branchiura it forms a more or less dorsoventrally flattened shield; in the Clado-

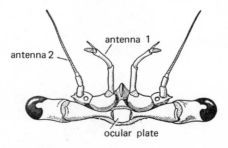

Fig. 1–5.   Anterior end of head of the crab *Callinectes sapidus*, normally hidden by the overhanging carapace, dorsal view. (After Snodgrass.)

cera, Conchostraca, Ostracoda, Cirripedia, and Phyllocarida it forms a pair of lateral, domed valves. These clamshell-like valves may enclose the whole body, the trunk, or only the thorax and part of the abdomen (Fig. 1-6). But in the Malacostraca (except the Phyllocarida) the carapace never extends beyond the thorax, and always hangs down on the sides, enclosing more or less the cephalo-thorax (Fig. 1-4). Between the carapace and the lateral wall of the cephalothorax is enclosed a space, the branchial or respiratory chamber (gill chamber) that functions in various ways in respiration. The carapace of the Conchostraca and many Ostracoda, less distinctly of the Cladocera, originates from the somite bearing the second maxillae (Figs. 4-7, 4-10, 5-3). In other subclasses, however, the carapace is fused without a seam with the lateral parts of the more anterior head somites. In all non-malacostracans and also the phyllocarids the fusion of the carapace with the body posteriorly ends at the posterior border of the somite of the second maxillae. In the malacostracans alone (except for the Phyllocarida), the carapace is also fused with thoracic somites at their dorsal surface, the extent of the fusion varying in different orders and genera. In Eucarida the carapace is almost always fused to 8, that is all thoracic somites (Fig. 1-4). In this case the median region of the carapace forms a dorsal cover for the whole cephalothorax, its sides (called branchiostegites) forming the wall of the respiratory chamber (Fig. 13-8). The process of fusion can be observed in the postembryonal develop-ment of the Penaeidae and Hoplocarida. The carapace first appears in the protozoea. In its early stage it connects only to the somite of the second maxillae but forms a roof over all thoracic somites. As always in the presence of a carapace, the posterior thoracic somites do not develop any other epimeres (Figs. 1-6, 11-5c). Only later does the initially free shield fuse to the dorsum of the thoracomeres.

Thus it can be concluded that the carapace of the Eucarida and Hoplocarida is a cephalic outgrowth and is not produced by fusion of the successive thoracic epimeres. At the same time one can recognize that the formation of a cephalo-thorax in many malacostracans is closely connected with the carapace. The anterior of the carapace often forms a median overhanging rostrum (Fig. 1-19).

The ability to form large plates is not a property of the somite of the second maxillae alone, but is also possessed by the metameres anterior to it, as can be observed in the nauplius of ostracods. Such larvae have a pair of large valves even though the second maxillary somite has not yet been formed (Fig. 5-5).

The widespread occurrence of carapace and epimeres in the Crustacea as well as in trilobites and in groups extinct since the Carboniferous (the Merostomoida, Marrellomorpha, and Pseudocrustacea; of epimeres only in Arthropleurida) makes it probable that both are ancestral characters that have been preserved in aquatic Xiphosura and Crustacea, but lost in terrestrial arthropods (arachnids and ter-restrial mandibulates). Some biologists assume therefore that Crustacea without a carapace have lost it secondarily, an assumption supported by paleontological

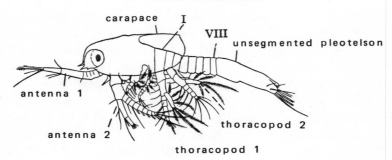

Fig. 1–6. Protozoea of *Penaeopsis*; lateral view shows carapace extending freely above thoracic somites; 0.7 mm long. I, first thoracomere; VIII, eighth thoracomere. (After Gurney.)

evidence. But some carcinologists believe that since some of the most primitive-appearing crustaceans (e.g., Cephalocarida) lack a carapace it is a more recent acquisition.

The biological significance of the carapace in free-living species is probably protective; and it also functions in respiration. Its dorsal and lateral walls are strongly cuticularized, often mineralized. The carapace covers the back in most Malacostraca, the gills, and in Notostraca, all appendages. Even more protective in the Cirripedia, Ostracoda, and Conchostraca, the right and left valves can enclose the whole body within a capsule. The closing is accomplished by contraction of the large transverse adductor muscle connecting the valves, a muscle that arises also in the somite of the second maxillae (Fig. 4-7).

Another carapace function is respiration. Its mesal (inner side) wall is thin, permitting exchange of respiratory gases with the hemolymph present within the lumen of the carapace wall. Connective tissue pillars and islands within the walls separate the blood current, directing it into a network of ducts that presents a large surface for O_2 uptake. This gas exchange device is especially well developed in many Malacostraca and Ostracoda, Conchostraca, Cladocera, and Cirripedia (Fig. 15-4). Usually the lumen within the wall is narrow, especially in the Malacostraca. But in other subclasses and in brachyurans visceral organs extend into the spaces within the wall (Fig. 4-10).

APPENDAGES. The diverse crustacean appendages are structurally of three different kinds: biramous appendages, cylindrical walking legs, and leaflike limbs.

The biramous appendage consists of a protopodite bearing two branches (rami), an outer exopodite, and an inner endopodite (Fig. 1-7). Lateral exites or endites may also extend from the protopodite. The endites function in transporting or chewing food material and are therefore always well developed in mouthparts. The exites, however, may have a very thin cuticle, and are then called epipodites; they may become simple or branched gills.

The most generalized biramous appendages are found in the Cephalocarida (Fig. 1-7), and from them probably all crustacean limbs can be derived. They

have the unsegmented protopodite, the two-articled exopodite and its pseudepipodite all are anterioposteriorly flattened. The endopodite has five cylindrical
articles and functions as a walking limb. The integument of the whole appendage
is soft; the articles are kept stiff by the hydrostatic pressure of the hemolymph.
There are no distinct hinges between the articles.

In other crustaceans the biramous appendages have a stiff exoskeleton and
their articles are separated by well-developed articulations. The protopodite is
usually segmented: there are usually a basal coxa and a distal basis (Figs. 1-1,
1-6, 7-4).

The endopodites, if they are used for swimming or to filter currents, consist of
similar articles (e.g., swimming legs of many copepods and the pleopods of many

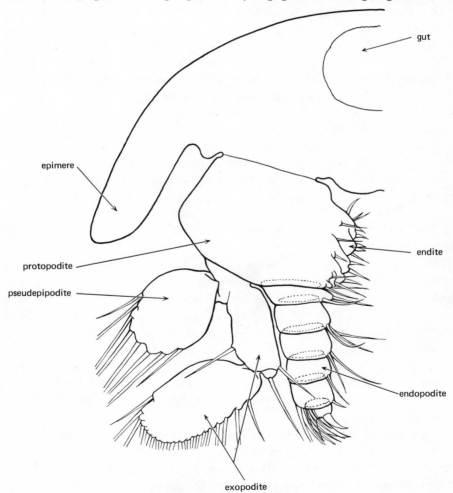

**Fig. 1–7. Thoracic leg of the cephalocarid *Hutchinsoniella macracantha*, with cross
section of left half of trunk somite; 0.16 mm wide. (After Hessler.)**

malacostracans, Fig. 1-19). If they are used for walking or handling prey, the articles are specialized and the endopodite resembles an insect or spider walking leg (e.g., pereopods of Malacostraca, Fig. 1-19). Five articles are usually named: ischium, merus, carpus, propodus, and dactyl, by no means homologous among the different subclasses and orders, as diverse branching or fusions occur.

The exopodite, in contrast to most endopodites, is differently segmented in most orders, often only annulate, antenna-like, and the articles have not been named (Fig. 15-2). If they are not mouthparts, they usually are swimming limbs. Long, dense setae increase their surface. In swimming crustaceans the exopodites of the thoracopods are always well developed (Fig. 15-2); in walking malacostracans (except Amphipoda), they can only be found on the pleopods (Fig. 1-19).

The cylindrical walking legs which resemble the legs of other classes of arthropods, consist of a single series of articles belonging to both the protopodite and the endopodite. While the exopodite is always absent (Fig. 1-4), epipodites or endites may be present. Examples are the thoracopods of decapods, isopods, and amphipods all of which are used as walking legs or as grasping appendages, only rarely is the dactylus flattened for swimming (in the Portunidae and Munnopsidae). The temporary presence of exopodites on the thoracic legs of decapod larvae and during the embryonic development of the isopods indicates that the primitive leg of Crustaceans is biramous (Figs. 1-6, 2-4, 2-17). This view is supported by the presence of biramous appendages where known in all classes of early Paleozoic marine arthropods (Marellomorpha, Pseudocrustacea). The legs of trilobites differ from the biramous appendages of crustaceans in the insertion of the long lateral branch: in trilobites it is inserted very near the insertion of the first article, the precoxa, not on the tip of a 2- or 3-articled protopodite, as in Crustacea.

Foliaceous thoracic legs are found only in the Branchiopoda and the Phyllo-carida. Typically (not one of the first 10 limbs of Notostraca) they have a thin colorless integument, in which the hinge membranes can barely be seen and their cross section is flattened. Each leg then forms a transverse pocket, stiffened partly by internal blood pressure. The segmentation of the leaflike limbs is indicated mainly by deep grooves on both sides, separating a row of endites toward the middle, and by a row of exites toward the outside (Fig. 1-8). The protopodite, according to the view of many authors, takes up most of the leg length (Fig. 1-8); its two rami are said to be the apical branches. But these are doubtful homologies.

The activity of the foliaceous limbs is made possible by a complicated inter-action of successive pairs. Three simultaneous functions are combined: swimming, respiration, and filter current production. Because of complexity of functions, this leg type cannot be considered ancestral to the biramous appendage or walking legs. Furthermore, during ontogeny of the other legs there is never a stage resembling the foliaceous limbs (Figs. 1-6, 2-17). However, these leaflike limbs

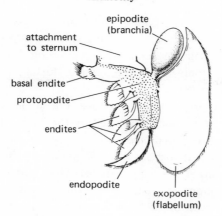

Fig. 1–8. Lamellate limb of *Triops cancriformis*; the homologies of the articles are uncertain; exopodite 0.5 cm long. (After Seifert.)

can readily be derived from the biramous, turgor appendages of the Cephalo-carida.

The appendages are arranged in three groups in each individual: the preoral two pairs of antennae, the mouthparts, and the trunk appendages. There are two kinds of trunk appendages: thoracopods and pleopods. The one to three anterior thoracopods are often transformed into accessory mouthparts, with strong endites and strong basal articles and less-developed branches. If there are several pairs these modifications are less distinct in the posterior ones (Fig. 13-6). Such specialized thoracopods are called maxillipeds, the others, walking legs or pereopods. Sometimes the grouping of the trunk somites corresponds to the distribution of these appendages. In this case there are as many thoracopods transformed into maxillipeds as thoracomeres fused with the head into a cephalo-thorax (e.g., Isopoda, Amphipoda, many Mysidacea, and Cumacea). Here the pereon bears only pereopods. But this is not always the case; in decapods, which usually have all thoracic somites within the cephalothorax, there is no pereon (Fig. 1-4, 1-19).

The last pair (or pairs) of pleopods, if different from the others, are called uropods (Fig. 1-4).

ANTENNAE. The first antennae (antennules) in non-malacostracans, and in some malacostracans (Isopoda, Tanaidacea, Amphipoda, many Cumacea, and the decapod genus *Lucifer*) consist of a single series of articles (Fig. 1-3). The other malacostracans have usually an antennule shaft that bears two flagella; rarely in Stomatopoda and some Decapoda (Natantia) it bears three flagella at its tip (Fig. 1-4). In the Notostraca the first antennae become stronger during the course of the postembryonic development, while in Cirripedia they become reduced. In Conchostraca and Cladocera and in terrestrial isopods they are very small (Fig. 4-7).

The antennules function as organs of touch and olfactory organs. In Decapoda and Anaspidacea the first shaft article encloses a statocyst. Nauplius larvae and adult ostracods use their first antennae for swimming (never in Branchiopoda). In many copepods the male holds the female during mating with antennules modified as a grasping organ.

The second antennae are usually biramous, but some representatives of most orders have unbranched second antennae. In the Stomatopoda, most Eucarida and Mysidacea, and some Tanaidacea and Anaspidacea, the exopodite has the shape of a flat scale and is almost always unsegmented (Fig. 1-4). In the adult the second antennae are mainly sensory organs. They are used for swimming by nauplii, by later developmental stages of the Anostraca, Euphausiacea, and Penaeidae, and by the adults of the Conchostraca, Cladocera, and Ostracoda (Figs. 1-6, 4-7, 4-10, 5-4). The Cirripedia, however, lose the second antennae after the metanauplius stage.

MOUTHPARTS.   The mouthparts have the protopodite and its endites well developed, while the branches usually are vestigial or lost.

The mandibles in nauplii are almost always swimming appendages, but near their base they may have large hooklike or feathered setae that aid in food uptake (Fig. 2-22). In adults the mandible is only rarely biramous, for instance in some copepods that filter food particles (Fig. 1-9). Usually the exopodite is

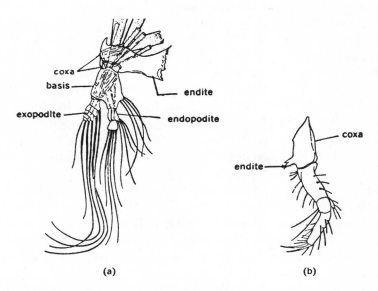

(a)   (b)

Fig. 1–9. Leglike mandibles. The article proximal to the exopodite and endopodite bearing the basis is the coxa. The coxa may consist of two parts (a), probably a secondary fragmentation as the distal part does not have its own muscles. The proximal part is called the precoxa. All three articles have endites. (a) Mandible of the 5th copepod stage of *Calanus cristatus*; (b) mandible of the ostracod *Philomedes globosa*. (After Snodgrass.)

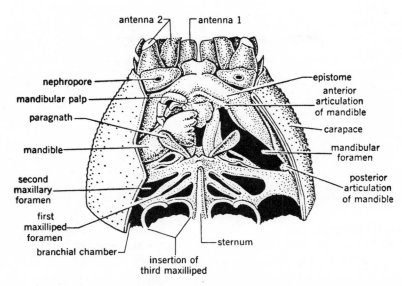

Fig. 1–10. **Mouth region of the crayfish *Cambarus longulus* in ventral view; about 2.5 cm wide. All appendages except both pairs of antennae and right mandible have been removed, showing insertion, on the border of which the appendages are attached to the sternum. (After Snodgrass.)**

completely lost and the distal part of the protopodite, together with the reduced endopodite, forms a palp (Fig. 1-10). The basal article, however, considered by most authors to be the coxa or precoxa, forms a huge, sclerotized endite (Fig. 1-9), which may develop into a cutting incisor process or a molar process bearing humps and grooves (Fig. 1-11). Or it may become stylet-shaped in parasites or in groups that suck up plant material (Fig. 5-6). As in terrestrial mandibulates, the palp sometimes is lost so that only the endite-bearing basal article remains (Fig. 1-2). In adults the mandible is used exclusively for feeding, except in the filter-feeding copepods. In these copepods, as in nauplii and metanauplii, the mandibles take part in locomotion and in producing the filter current (Fig. 7-16).

The first maxillae (maxillulae) and second maxillae are close to each other and are closely appressed to the mandibles (Figs. 1-12, 1-13, 15-6); both are flattened. Each maxilla may, on each of its two basal articles, form an endite, usually platelike and, unlike that of the mandibles, inserted on the basal article so as to permit movement. The two branches are never well developed, and one or the other may be absent. Both pairs of maxillae are used in feeding, for transporting food to the mandibles, or in filter-feeding. One of the branches of the second maxilla or its epipodite, in those malacostracans in which the carapace forms a respiratory chamber, may extend posteriorly into this chamber and serve as a cleaning brush or to produce a respiratory current (Fig. 13-11).

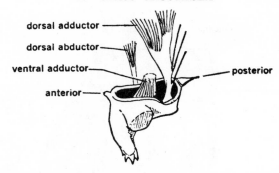

Fig. 1-11. Right mandible of the amphipod *Talorchestia longicornis*, mesal view showing chewing surface (lined); the mandible moves around the anterioposterior axis. The adductor muscles used in biting are stronger than the abductors. (After Snodgrass.)

THORACIC APPENDAGES. As already mentioned, the first thoracic appendages are often transformed into maxillipeds in the Malacostraca and are used in feeding (Fig. 17-5). Often their articles (in Peracarida the endites) bear setae or thorns on the inner surface, which coordinate the movements of the two members of the pair of protopodites or endites. They are analagous to the labium in insects.

In most malacostracans the thoracic appendages that are not modified, the pereopods, have a strong endopodite resembling the walking leg of terrestrial arthropods. In other crustaceans, depending on the presence of a large exopodite, they may be used for swimming (Fig. 1-1). The terminal articles of the endopodite may work against each other to form a grasping structure, the subchela (Figs. 1-3, 1-4), or may form a chela of large size (Fig. 1-19). In non-malacostracans, the thoracic appendages are foliaceous (Branchiopoda) or biramous with two

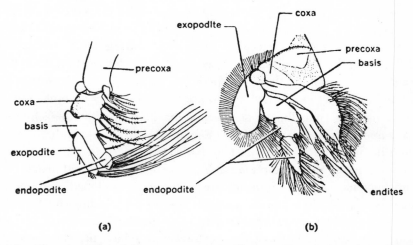

(a)   (b)

Fig. 1-12. First maxilla of Crustacea: (a) leglike maxilla of the ostracod *Polycope*, lacking endites; (b) first maxilla of the mysid *Gnathophausia*, having two endites. (After Hanssen from Vandel.)

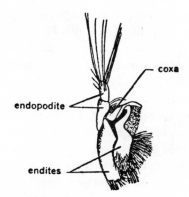

coxa

endopodite

endites

Fig. 1–13.   The second maxilla of *Gnathophausia*. (After Hanssen from Vandel.)

similar branches (Figs. 4-7, 7-4). Ostracods have only two pairs of legs, and these are very specialized.

ABDOMINAL APPENDAGES.   Only Malacostraca have pleopods; unlike malacostracan thoracopods, pleopods never have the appearance of walking legs. They are primarily swimming limbs (Fig. 1-4), and therefore are of medium length, often covered by setae (Fig. 1-4), and their two branches may be similar (Fig. 1-19). They also may produce respiratory currents and, much flattened, serve as gills (Isopoda). In males the first two pairs may be transformed into copulatory appendages (gonopods) (Fig. 1-19). The last pair of pleopods, the uropods, are usually flattened in the Malacostraca and together with the telson form a tailfan. This tailfan considerably increases the propellant action of the downward beat of the abdomen (Fig. 1-4).

INTEGUMENT.   The integument consists of several layers. The hypodermis secretes a many-layered chitinous cuticle. An epicuticle above the cuticle protects it. Only in Decapoda has this epicuticle been carefully examined. In contrast to the other layers it can be stained by acid fuchsin and contains lipids; it is not a continuous layer of wax, as is the epicuticle of insects.

In large species (e.g., decapods, stomatopods, and isopods), in sessile barnacles, and in some ostracods the hypodermis secretes calcium into the cuticle, making it strong and rigid. The integument also serves as a skeleton. In large malacostracans especially it forms an internal ventral skeleton by growth of apodemes into the body.

Resilin, an elastic structural protein, has been demonstrated only in the articulation between ischium and merus of thoracopods of the crayfish *Astacus*. The elastic hinge membrane acts antagonistically to the flexor muscles.

The coloration of the body in many species is due to soluble or granular pigments in the cuticle or hypodermis. In malacostracans with thin cuticle (e.g., prawns and many isopods) the body color depends largely on starshaped pigment-containing cells (chromatophores) in the connective tissue below the hypodermis. The contents of the gut and especially blue or red lipid spheres in the body cavity may give color to minute planktonic crustaceans.

CONNECTIVE TISSUE.   Cellular connective tissue is found in all parts of the crustacean body. Cords or groups called fatbodies lie near the gut or its ceca; they take up proteins, lipids, and glycogen. Their reserves are important during periods of fasting, for example, during molting. As in other arthropods, connective tissue layers and membranes may cover the internal organs as a peritoneum, or may govern the circulation of body fluids by dividing the body cavity.

SENSE ORGANS.   Simple innervated setae are tactile in function; the more complicated jointed setae of the decapods sense currents or waves of water pressure. Long tubelike processes, esthetascs, however, are chemoreceptors. These are usually found on the second antennae, rarely on the mouthparts. Proprioceptors are discussed in the chapter on decapods. Statocysts are known from many malacostracans. In decapods statocysts function not only as equilibrium organs but also give an awareness of movement.

There are two fundamentally different kinds of eyes: compound lateral eyes and simple median eyes, as in many other arthropods (Xiphosura; most insects.) The median or frontal eyes are different in the adults, but in the nauplius stage there are three (Fig. 1-14b) forming a group. Because they are the only optic organs of the nauplius, this group is called the nauplius eye.

The frontal eyes even if vestigial connect to the median optic center in the midline on the anterior edge of the protocerebrum. Their number and degree of differentiation differ very much in different subclasses and groups. In general the nauplius eye consists of three pigment cups, two directed dorsolaterally and one ventrally (Fig. 1-14a). Each cup contains retinal cells which form rhabdoms on surfaces facing each other (Fig. 1-15). Only ostracods and the nonsessile Maxil-

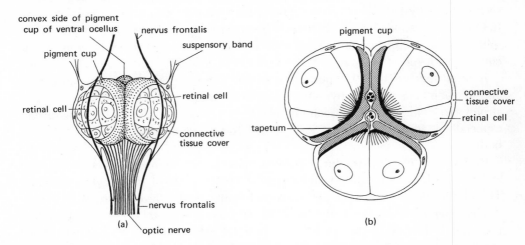

Fig. 1–14.   Median eye of *Eucalanus elongatus*: (a) in dorsal view; (b) cross section; 0.85 mm diameter. (After Vaissière.)

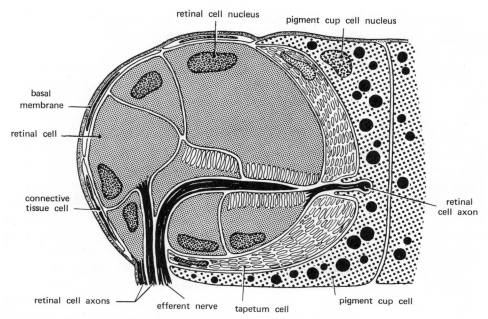

Fig. 1–15. **Longitudinal section through a dorsolateral ocellus of the median eye of** *Macrocyclops albidus*; **0.05 mm diameter. The nerve that ends between the rhabdom microvilli is probably efferent; microvilli very diagrammatic. (After Fahrenbach.)**

lopoda (Mystacocarida, Copepoda, and Branchiura) have a tapetum. The retinal cells differ in regard to the origin of their axons. The axon originates from the tip of the retinal cell touching the bottom of the pigment cup penetrating the cup in Malacostraca, in other crustaceans it originates from the opposite, the opening of the cup and the nerve going around the pigment cup (Fig. 1-15).

Generally the Malacostraca have seven frontal eyes, a median group of three nauplius cups, one dorsal and one ventral pair. In Stomatopoda and some Decapoda (e.g., Pandalidae, Hoplophoridae, Hippolytidae, Palaemonidae, and Processidae) a pair of cups each with 9–30 retinal cells lies posteriodorsally against the three nauplius cups so the functional part of the frontal eyes is made up of five parts. The paired ventral frontal eyes are rudimentary. In all other Malacostraca not only the ventral but also the dorsal pair is vestigial. Both pairs consist often of tubes filled with cells which at times may still contain rhabdoms or pigment. Each of the seven frontal eyes may independently be more or less reduced. For example in *Astacus* the only indication of the three nauplius eye cups is pigment spots with groups of undifferentiated cells in the gangliar cortex of the median optic center. The dorsal and ventral pairs of frontal eyes however form tubes filled with cells originating in the optic center.

The many combinations and degrees of reduction, or their complete absence in a group or species has to be considered secondary. Their absence is most common in crustaceans which lack free nauplius or metanauplius stages. Thus Leptostraca and almost all Brachyura and Peracarida have no vestiges of frontal eyes left (Table 1-1).

Cephalocarida and Mystacocarida lack eyes. Branchiopoda (except Anostraca) have a nauplius eye the lateral cups of which touch a median 4th cup during larval development. In large species with many somites the lateral cups contain many cells, in Notostraca each contains several hundred and in Conchostraca, 50–150. The small *Lynceus* has fewer than 10 cells in each lateral cup. The Cladocera have likewise few retinal cells in their nauplius eye, e.g., *Daphnia,* four in each cup. The better the development of the compound eyes of Cladocera, the smaller the nauplius eyes. *Sida* and *Leptodora* have in place of the naupliar eye only an accumulation of cells with rhabdoms, while *Evadne, Podon,* and *Bythotrephes* lack a nauplius eye and other vestigial frontal eyes. Species with a functional nauplius eye have usually median or paired tube-shaped vestigial frontal eyes differing in different groups of branchiopods and presumably not homologous with each other. The Anostraca have a naupliar eye with three cups, each with 25–75 cells and also a paired rhabdom forming a tube-shaped vestigial frontal eye.

Ostracods, copepods, and branchiurans have only the typical nauplius eye with three cups, which in free living species contain retinal, pigment, tapetum, and sometimes corneal cells (Figs. 1-14, 1-15). In the Branchiura the number of retinal cells of the lateral cups is large, each with 60 to about 200. In some ostracods there are altogether about 100 retinal cells. But usually each eye cup of the Ostracoda has only 10–15 retinal cells and each eye cup of the copepods has hardly more than 10. Vestigial frontal eyes are absent in Ostracoda and Maxillopoda. The Podocopida, unlike the Myodocopida, have a pair of transparent corneal cells in front of each pigment cup. Their ventral eye cup may be directed anteriorly.

Sometimes the three cups of the nauplius eye are separated and in the Cytheridae the lateral cups push in at the dorsal margin of the valves, in the Balanidae into the mantle (carapace) between scutum, rostral, and lateral plates, or at the articular membrane.

Table 1–1. The Presence of Frontal Eyes in Malacostraca

	Natantia	Stoma-topoda	Euphausi-acea	Anaspi-dacea	Mysidacea
Nauplius eye	+	+	+	+	−
Vestigial ventral frontal eyes	+	+	+	−	+
Vestigial dorsal frontal eyes	sometimes	−	−	+	−
Functional dorsal frontal eyes	sometimes	+	−	−	−

In some copepods the nauplius eye may be specialized, and especially the lateral cups of the Corycaeidae and Sapphirinidae. In these families a septum consisting of pigment cells growing from the bottom of the cup divides each lateral cup into an upper and a lower one. This is found most specialized in some marine Poecilostoma. Their large upper section of each lateral cup lies parallel to the long axis of the body and its dioptric apparatus stretches to one-fifth of the body length (Fig. 1-15). In front of the two to three cylindrical retinal cells the enveloping connective membrane extends as a conical tube to the tip of the body so that each eye has the shape of a telescope. In front of the retinal cells lies a diaphanous body, in *Corycaeus* a second lenslike body close to the hypodermis. In *Copilia, Corycaeus,* and *Sapphirina* the cuticle forms a lens (Fig. 1-16).

Since the number of retinal cells of each eye cup is usually small, the eye cannot perceive images. Also larvae and adults that have nauplius eyes only cannot use the plane of polarized light for orientation. But the eyes are of importance to the swimming position of planktonic nauplii and to some planktonic copepods. These animals position their body normally in the direction of the light and swim either toward the light or away from it. In ponds they maintain a vertical

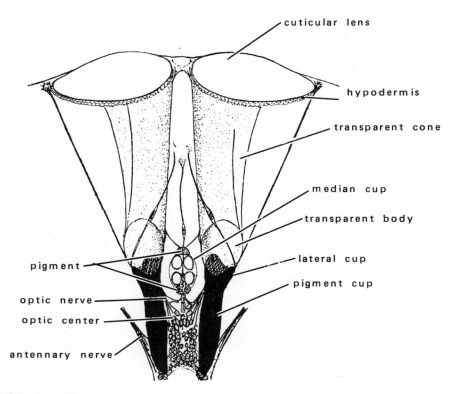

Fig. 1–16. Specialized nauplius eye of the copepod *Sapphirina maculosa*; diagram **0.3 mm wide at anterior lenses. (After Vaissière.)**

position, and the three eye cups are thereby diagonal to the light direction. If one suddenly illuminates the water from the side, the crustacean will turn its body in this direction so that all three eye sections have equal illumination.

Compound eyes are found in most subclasses and in their presence the nauplius eyes may be reduced or completely lost. During the development the compound eyes appear only after the metanauplius stage. There are no compound eyes at all in cephalocarids, mystacocarids, and copepods. Cirripeds and myodocopid ostracods have the compound eyes only temporarily in the last larval stage. From their wide distribution within the class Crustacea it can be concluded that compound eyes are ancestral structures. They may be present even in animals enclosed by valves so that direct light reaches them only through a slit (Conchostraca, Myodocopina, and cypris larvae of Cirripedia). Sometimes the compound eyes move from the sides toward the middle. In Notostraca and Cumacea they lie on the dorsum of the carapace close to each other (Fig. 4-5) and in Cladocera they are fused into a single sphere (Fig. 4-16).

EYESTALK. During larval development, the area between compound and nauplius eye underlain by the paired cylindrical optic centers may become a raised transverse swelling, and finally form a stalk. In the adult the compound eyes each perch on the end of this cylindrical stalk, the cavity of which contains the optic center (Fig. 1-18). This occurs postembryonically in the Anostraca, Hoplocarida, and many Malacostraca. The eyestalk can be moved by muscles.

The ontogeny of the eyestalk is completely different from that of the appendages which develop as saclike evaginations (Fig. 2-4). That the eyestalk is derived from ectoderm of the cephalic lobes, as are the optic masses below, can be demonstrated by sectioning zoea larvae. Thus the eyestalks cannot be considered appendages. It is most surprising, however, that after tearing off the whole eyestalk from a decapod or *Squilla*, experimentally and also in nature, rather than regenerating, in its place grows a structure resembling the first antenna. This antenna-like structure is sensitive to chemicals and its stimulation may cause cleaning movements by the first antennae. In the place of the torn optic centers a new nerve forms that goes to the protocerebrum. But Crustacea that have the optic masses in the cephalothorax rather than in the eyestalk (e.g., *Porcellana*) will regenerate a new eyestalk in the place of one that is lost. The regeneration may be influenced by the neighboring ganglia.

The compound eyes are sometimes apposition eyes (e.g., Anostraca, Notostraca and Cladocera, Hoplocarida, Isopoda and Amphipoda), sometimes superposition eyes (e.g., many Decapoda, Mysidacea, and Euphausiacea).* The number of ommatidia varies between 4 (some isopods) and 14,000 (some decapods). The number increases during postembryonic development; new ommatidia appear on

---

* See Chapter 16 and especially Figs. 16–20, 16–21, pp. 317 and 318, Vol. 2 of this treatise, for a discussion on mandibulate eyes and their function.

the mesal border of each eye. The dioptric apparatus differs among the various groups.

Swimming predators generally have much larger eyes than filter feeders: filter-feeding *Daphnia* has only 22 ommatidia altogether (in both eyes), while *Leptodora* has 500. Deep sea crustaceans, if they are swimmers, have the eyes often differentiated into especially sensitive superposition eyes. Deep-sea-bottom dwellers often have the eyes much reduced, but there are exceptions: the deep sea isopod *Bathynomus* has 3,000 ommatidia.

In darkness the distal pigment cells move their pigment in the direction of the cornea by drawing the cell body together or by protoplasmic currents without any change of the cell shape. These movements depend on hormones. In decapods a pigment-moving hormone is found in the eyestalk and in some neurosecretory cells of the brain and some ventral ganglia, which causes expansion of the pigment. The antagonist is the hormone of the postcommissural organ and also additional substances from the eyestalk and neurosecretions of the central nervous system. Decapod groups differ from each other in details in the function of different eyestalk organs so that there still exist many uncertainties. Many species of *Hemigrapsus* and *Orconectes* also have a rhythmic daily movement of the pigment, as do some Natantia, even if they are kept in constant darkness.

The compound eyes also function in position orientation. Once they have developed in postembryonic ontogeny, the swimming crustacean turns to a horizontal position, its transverse phototaxis becomes dominant, and the longitudinal phototaxis disappears. The unstable horizontal swimming position which causes predominately horizontal swimming, is the result not only of the light position but also of a simultaneous geotaxis. Some species turn their back to the light (*Triops, Argulus,* and *Palaemon*), others their venter (Anostraca, *Simocephalus,* and *Euphausia*).

Frontal organs of unknown function are found near the nauplius eyes mainly in nonmalacostracans. These are groups of a few sensory cells whose axons go into the median area of the protocerebrum, the optic center of the nauplius eyes (Fig. 1-14a). The dendrites, however, enter the hypodermis; in some non-malacostracans they enter rod or coneshaped cuticular extensions. In the Myodocopida these extensions may be noticeably elongated, in the Conchostraca and juvenile stages of the Notostraca they are much shorter. In mature Notostraca the frontal-organ dendrites lie in the hypodermis of the eye chamber. In other branchiopods there is generally a median ventral and a pair of dorsal organs; cladocerans have both sets in pairs. The dorsal frontal organ of *Daphnia* consists of a paired row of about five sense cell groups lying in the dorsolateral margins of the head, extending from the region dorsal of the brain to the region dorsal to the insertion of the 2nd antennae. In the Myodocopina the frontal organ is usually median and outside on the median pigment cup of the nauplius eye. Many copepods have paired frontal organs. Groups of sensory cells, with axons going to the middle of the protocerebrum, have also been found in some Malacostraca and therefore are probably frontal organs. For instance, such organs are found in the Natantia, in which there is still a nauplius eye, in stomatopods, in *Eucopia,* and also in *Gammarus.*

   NERVOUS SYSTEM.   The central nervous system consists of a supraesophageal ganglion and a ventral chain of ganglia that usually traverses the entire length of the body. In most Branchiopoda, and in *Anaspides,* all ganglia, even those of the three pairs of mouthparts, are separated by connectives (Vol. 2, Fig. 4-9, p. 47). In other groups at least the mandibular and the two pairs of maxillar ganglia have fused to a subesophageal ganglion (Vol. 2, Fig. 4-10, p. 48). But often other trunk ganglia move anteriorly and join to form a large ventral ganglion which in the Branchiura, Balanomorpha, and most Brachyura takes up all the ganglia of the trunk (Fig. 1-17b). There is an obvious relationship between shortening (oligomery) of the trunk and fusion of ganglia. In other crustaceans the long nerve

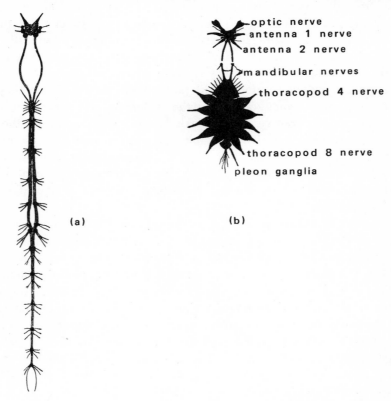

optic nerve
antenna 1 nerve
antenna 2 nerve
mandibular nerves
thoracopod 4 nerve
thoracopod 8 nerve
pleon ganglia

(a)   (b)

Fig. 1–17.   Central nervous system: (a) crayfish; subesophageal ganglion is behind long circumesophageal connectives, posterior to it is ventral nerve chain, 10 cm long. (After Matthes.) (b) Brachyuran; all trunk ganglia have moved forward and have joined ganglia of the mouthparts, forming a large subesophageal ganglion about 5 cm long. (After Giesbrecht.)

chain persists but the paired ganglia and often also their connectives have fused in the midline of the body (Fig. 1-17a in the abdomen).

In most Malacostraca there are 17 ganglion pairs, of which 3 belong to the mouthpart somites, 8 to the thoracomeres, and 6 to the pleomeres, so that each somite is served by one pair of ganglia. Fusions can be recognized by comparing the number of appendage nerves. Other subclasses, corresponding to their variable number of metameres, have a variable number of ganglia.

In the crayfish (*Cambarus*) two giant fibers starting at the circumesophageal commissure run the whole length of the nerve chain without break or transverse connections. They rapidly (15–20 m/sec at +20°C) transmit a stimulation from the brain to the abdomen, causing a strong beat of the posterior end of the pleon, causing the animal to shoot backward. The importance of the giant fibers, then, is the same as in polychaetes. The lateral fibers have a similar function, but these are interrupted in each ganglion by cells and also are connected to each other. They transmit more slowly (10–15 m/sec) and are connected behind the fourth thoracomere with the segmental ganglia. Their stimulation causes the same flight response.

In those animals having large compound eyes the optic masses extend stalklike from the brain wall and adjoin the back of the eye (Fig. 1-18). In Malacostraca, as in insects, there are three optic masses on each side (lamina ganglionaris, medulla externa, and medulla interna); the other subclasses have only two. Adjoining the optic masses is the medulla terminalis, the part of the protocerebrum

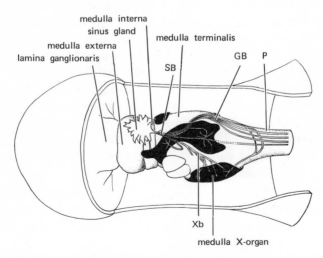

Fig. 1–18. Neurosecretory cell groups and neurohemal organs in eyestalk of the cray-fish *Cambarus virilis*; neurosecretory areas are black. Medullae are optic areas; medulla terminalis is an optic association center of the duct from protocerebrum; SB, neuro-secretory duct to sinus gland; GB, neurosecretory duct from supraesophageal ganglion to sinus gland; P, connection to median area of cerebrum; Xb, neurosecretory ducts of X-organ to sinus gland. (After Bliss, Durand and Welsh.)

that bears the globuli. The deutocerebrum contains the olfactory centers, glomeruli which receive nerve fibers from the chemical sense organs of the first
antennae. In terrestrial isopods and some terrestrial amphipods, which have only
very small first antennae, the deutocerebrum lacks an antennal glomerulus; instead
a glomerulus in the tritocerebrum connects to the large second antennae.

The sympathetic nervous system connects to the gut, heart, and musculature. In
all Crustacea the part belonging to the gut is divided into an anterior and a
posterior portion. The anterior part comes from the circumesophageal (stomodeal)
connectives, the tritocerebrum. In *Astacus* the paired branch unites on the
anterior surface of the esophagus to form the ganglion oesophageale. From the
latter the unpaired stomatogastric nerve arises and follows the midline of the
chewing stomach, giving off a dorsal branch (nervus cardiacus) to the heart and
ending in the ganglion ventriculi superior. This ganglion is the motor center for
the stomach musculature, and all of its nerves supply stomach muscles except for
the branching hepatic nerve, which goes to the gut ceca. The posterior part of the
intestinal nervous system arises (in *Astacus*) from the very last ganglion of the
ventral chain as a median branch. This branch divides into three, of which the
median passes under the hind gut to the anus; the remaining two pass to either
side of the gut anterior to the midgut. All three nerves send off branches forming
a motor plexus on the gut musculature.

NEUROHORMONES AND ENDOCRINE GLANDS.   The nervous system, as in
other arthropods, functions not only to transmit impulses but also to coordinate
organs and metabolism by means of secretions, neurohormones. Neurohormones
can be demonstrated morphologically by staining, unlike other neurosecretions
such as acetylcholine and adrenalin. This is an active field of research and our
knowledge is fairly good only in decapods in relation to color change and molting.
Only the physiology of the decapods is considered here.

Neurosecretory cells have been demonstrated in various regions of the protocerebrum, tritocerebrum and the ganglion chain. Because of their positions in
most decapods in the eyestalk, neurosecretory cells of the optic ganglion and
those of the postcommissural organ are ideal for experimental purposes. These
areas can be homologized within the decapods, even within the Malacostraca,
and have been given special names.

1.   The medulla X-organ (medulla terminalis ganglionic X-organ, MTGX)
consists of a group of neurosecretory cells in the medulla terminalis and is the
main source of eyestalk hormones (Fig. 1-18). Its axons go mainly to the sinus
gland, a few to the papilla X-organ.

2.   The sinus gland has club-shaped swellings of axons. Here the hormones
are stored until secreted into the hemolymph. The sinus gland is thus a storage
organ; its walls are formed from the covers of the optic masses and it borders a
blood sinus in the region of the medulla externa and medulla interna of the optic
ganglia (Fig. 1-18). Besides axons of the medulla X-organ there are axons from

neurosecretory cells of other optic ganglia and of the protocerebrum. The neurohormones are secreted into the blood sinus.

3. The papilla X-organ (Sensory papilla X-organ, SPX) is made up from
material of a reduced sensory papilla of the eyestalk wall. Often two parts are
recognizable. Distally there are bipolar sensory cells, the axons of which enter
the medulla terminalis. Below is a neurohemal organ, which probably contains
hormones from the medulla X-organ, perhaps also its own. (The organ is difficult
to study histologically.) In the Brachyura and Stomatopoda the organ clings
closely to the medulla X-organ. Its function is not known but it may give off a
hormone furthering molting.

4. The postcommissural organ arises from the tritocerebral commissure which
clings to the posterior wall of the esophagus. The commissure contains several
axons carrying neurosecretions entering from the stomodeal connectives. Their
ganglion cells are probably in the tritocerebrum and perhaps also in the stomodeal
connectives. The axons arise combined with motor nerves to two cords from the
commissure and terminate in a network of fibers on the surface of a blood sinus.
The neurohormones function in physiological color change and retinal pigment
movement.

5. The pericardial organ consists of a nerve plexus, the ends of which contain
droplets of neurosecretions that are secreted into the blood. The networks are
either on the pericardial membrane on the side toward the heart, or they are
suspended in the pericardial sinus, covered by a sheath. Its ganglion cells (at
least in Brachyura and the Stomatopoda studied) lie in a ventral nerve cord. In
nine somites of *Squilla* median connections arise from the median nerve cord
between the segmental ganglia and run toward the pericardial organ. The
neurohormones stimulate the heart, as can be demonstrated with extracts from
the plexus which have been injected into *Cancer* or *Maja*.

The neurohormonal organs of the eyestalk of decapods are formed only late in
ontogeny. The five zoea stages of *Palaemonetes* lack the sinus gland, and the
medulla X-organ is represented by a single large cell, the secretion of which
probably reaches the saclike neurohemal organ of the papilla X-organ. Removal
of eye ganglia together with the neurohemal organ often causes a strong contraction of the chromatophore pigment but does not influence the next molt.
(Larvae of this size molt every 2 days.)

The study of the function of these secretory organs is difficult. Only those of
the eyestalk can be easily removed by an experimenter. Also the hormones may
function indirectly via a neighboring endocrine gland. In addition neurohemal
organs like the sinus gland contain several different neurosecretions, the separation
of which has been successful only a few times, using techniques of precipitation
or electrophoresis. Extirpations of the organs are therefore not conclusive. Also
not all neurohormones are secreted in all seasons and for the same period before
each molt or reproductive period so that similar experiments with the same species

have often been difficult to repeat. Eyestalk neurosecretory function and especially pigment changes and molting are discussed in Chapter 13 with the decapods. Also in nonreproductive periods the glands inhibit gonads; the gonads start enlarging if the eyestalks are removed during this period. The neurohormones also appear to be involved with the daily rhythm of activity (in *Gecarcinus* and *Astacus*). Their effect on metabolism has been observed in crustaceans during the period of molt. The hormones control formation of gastrolith, calcium and glucose level in the blood and at the same time lower glucose in the midgut ceca, water take up in the hemolymph and tissues, protein metabolism, nitrogen excretion, and oxygen uptake, as well as size of respiratory quotient. Little is known about neurosecretions which are not carried to the neurohemal organs of the eyestalk.

A very important antagonist of some eyestalk neurohormones is the secretion of a pair of endocrine glands that arise from an ectodermal proliferation of the wall of the cephalothorax, the Y-organ or molting organ. This gland is innervated from the subesophageal ganglion and its cells secrete a hormone in cycles which initiates molting and facilitates regeneration. In *Orconectes* the secretions in this gland disappear 3–5 days before the molt. The Y-organ also has an influence on the oocyte number (Chapter 13). In contrast to the analogous thoracic gland of insects, the Y-organ often does not degenerate at maturity and most crustaceans continue to molt when adult. In all Malacostraca this organ is found near the second maxillae or second antennae depending upon whether the space is already taken by excretory organs. In Decapoda, Euphausiacea, Mysidacea, and Amphipoda it is in the metamere of the second maxillae; in other orders (Cumacea, Isopoda, Thermosbaenacea) as in the Hoplocarida, it is in the metamere of the second antennae.

DIGESTIVE SYSTEM.   The median labrum covers the mouth from the anterior. The labrum with its base, the epistome, is derived from the cephalic lobe of the embryo (Fig. 16-15, p. 312, Vol. 2). At the posterior border of the mouth, the metastomal area, there may be a projection from the sternal region (the labium and hypopharynx) which may have a pair of lobes, the paragnaths. In *Astacus* (Fig. 1-10) the paragnaths are very long. The projection is homologous to (part of) the hypopharynx of insects and analogous to the labium of araneids and serves the same function.

The gut of crustaceans differs from that of other mandibulates in having large paired midgut ceca (Fig. 1-19). The ceca are never metamerically arranged as they are in the chelicerates but branch out from the anterior of the midgut. Only rarely are there additional ceca at the end of the midgut, and more rarely in the middle. The size of the ceca and the extent of their branching depends on the size of the animal. (Compare Figs. 4-10 and 13-9.) The length of the straight midgut, which is almost always endodermal, varies. In contrast to the midgut of chelicerates, it may sometimes be so small that the whole midgut is only the section where the anterior ceca open. Posteriorly the long gut, as is indicated by its

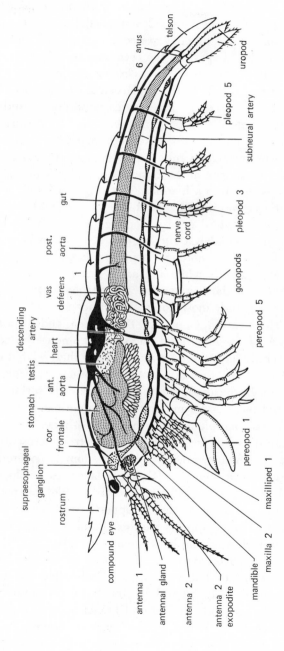

Fig. 1-19. Decapod (Astacura) anatomy. (Modified from Huxley and Siewing.)

27

cuticularized lining, is of ectodermal origin and must be considered the hindgut (crayfish, terrestrial isopods, *Limnoria, Idotea*). In all crustaceans that feed on coarse particles the ectodermal foregut has a chewing and a filtering mechanism that lets only minute particles pass. Such fine particles enter the lumen of the ceca without plugging them. Coarse particles, which cannot be broken down by chewing surfaces or by enzymes coming from the midgut ceca into the stomach, pass into the gut, bypassing the cecal entrance. Because most non-malacostracans feed on microscopic sized particles or are parasites, the chewing or filtering stomach is found only in the Malacostraca (Fig. 10-1).

EXCRETORY SYSTEM.   Phagocytes, nephrocytes, and the epithelium of midgut glands and nephridia function in excretion. The nephridia are comparable to the coxal glands of chelicerates. Like these they have a sacculus, a coelomic remnant, which opens through a funnel into a more or less coiled excretory canal. The excretory canals open at the base of appendages (Fig. 1-19). Before opening the duct may widen into a bladder. This organ is found in adult crustaceans in the base of the second antennae or that of the second maxillae. Only rarely do both somites have nephridia (e.g., lophogastrids and Mysidacea). In the Cephalocarida, some Tanaidacea and *Idotea* there are large maxillar nephridia but also very small persistent antennal nephridia have been observed. But the larva often has antennal glands while the adult has maxillary glands or the reverse. The adult Myodocopida, Euphausiacea, Decapoda, Mysidacea, and Amphipoda have an antennal gland; most non-malacostracans (Ostracoda except Myodocopida) and the Hoplocarida, Cumacea, Tanaidacea, and Isopoda have maxillary glands. The length of the nephridial ducts varies among the genera. Freshwater inhabitants generally have the duct much longer than related marine species, which can be explained by the fact that the organ controls water as well as salt metabolism. In decapods the sacculus and duct surfaces are mostly increased by folding and the bladder may be large, sometimes having a number of diverticula taking up considerable space in the cephalothorax.

The sacculus has been examined electronmicroscopically in *Orconectes* and *Artemia*. Its one-layered, folded epithelium is covered by a complete layer of connective tissue. Between the cell layers is a blood sinus supplied by the antennal artery. The epithelial cells resemble the podocytes of the vertebrate kidney. In both they stand on supports on a basal membrane which faces the blood sinus. These supports are their basal processes from which extend laterally toothed branches on the basal membrane (Fig. 1-20). The areas between these branches are covered toward the basal membrane by a thin membrane. The interstices continue between the cell bodies into the lumen of the sacculus. The blood is most likely filtered by the basal membrane and the thin membrane between the supports and lateral branches of the epithelial cells. The filtered liquid reaches the lumen of the sacculus flowing in the spaces between the epithelial cells. Filled vacuoles in epithelial cells indicate simultaneous secretion into the lumen.

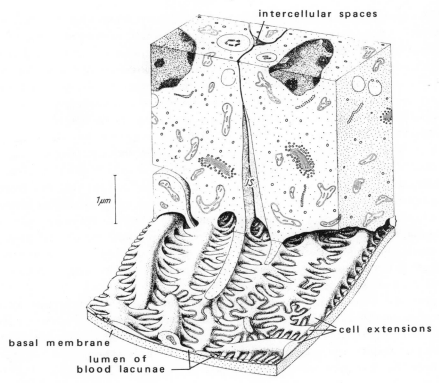

intercellular spaces

1μm

basal membrane

lumen of
blood lacunae

cell extensions

Fig. 1–20. Diagram of three epithelial cells of coelomic sacculus of the antennal gland of the crayfish *Orconectes limosus*. (After Kümmel.)

Gills and other areas whose integument is thin also probably take part in excretion, letting nitrogenous products in the form of ammonium salts escape into the water. Also uptake and release of salts through the integument has been demonstrated in decapods (Chapter 13). This explains why large crustaceans may have only one pair of relatively small excretory organs.

RESPIRATORY ORGANS. Appendages of the limbs, epipodites, but rarely pleopods, and the thin internal carapace wall, serve as respiratory organs, and have a differentiated hypodermis. Often more than one organ is functional in one crustacean. The minute copepods and mystacocarids do not have any special respiratory surfaces, but the whole thin integument takes part in gas exchange.

The inner wall of the carapace, facing the trunk, contains blood in narrow channels having a large surface and may be the most important or the only respiratory organ in many groups (Conchostraca, Cladocera, Ostracoda, Branchiura, Cirripedia, as well as many Malacostraca: Leptostraca, Mysidacea, Cumacea, Tanaidacea, some terrestrial decapods, and zoea larvae). The hypodermis of the abdomen may have also respiratory function in ostracods. In most decapods, however, the gills are the main respiratory organs.

The current of fresh respiratory water into the branchial chamber, the space between carapace and trunk, is produced by the steady movement of all walking legs in Conchostraca, Cladocera, Cirripedia, and Leptostraca. At the same time the current provides the food supply (Fig. 4-7). In other crustaceans the epipodite of a single appendage causes the respiratory current; in ostracods that of the first or second maxillae (Fig. 5-4); in decapods that of the second maxillae; in Mysidacea, Cumacea, and Tanaidacea, that of the first thoracopods.

The epipodites, which function as gills, are found on all or nearly all thoracopods of Branchiopoda, Anaspidacea, Leptostraca, Stomatopoda, Lophogastrida, and Eucarida. Usually they are lamellate, being branched only in the Lophogastrida and Eucarida. The Cumacea have an epipodite on each of the first thoracopods; their respiratory area is increased by the formation of many tubular or lamellate extensions (Fig. 16-7). The position of the gills near the appendages is very favorable for receiving an uninterrupted flow of respiratory water.

Only in the Stomatopoda and Isopoda do the pleopods function as respiratory organs. In the isopods the pleopods lack epipodites but by becoming wide and lamellate their surface has been increased.

Terrestrial Crustacea continue to use the aquatic respiratory organs without important modification. This is especially true for the decapods, whose carapace protects the gills from desiccation. However, most terrestrial isopods develop additional respiratory organs while retaining the gills. These new organs are tracheae-like invaginations in the pleopod exopodites (Fig. 17-23).

CIRCULATORY SYSTEM.   A respiratory pigment, hemocyanin, is present in the blood of most crustaceans. Its presence may be masked by other pigments dissolved in the blood (e.g., Cladocera). There are blood corpuscles, various kinds of leucocytes that function in clotting and phagocytosis. In *Astacus* these may take up experimentally injected glass powder. The blood of decapods forms clots by agglutination of blood corpuscles and this may be followed by later gelatinization of the blood plasma. The active formation of new blood corpuscles takes place in cells that are distributed in the fibrillar connective tissue. In decapods lobes of such tissue are found near the anterior aorta or at the base of the rostrum. The young lymphocytes migrate through the walls of the lobe into the surrounding lacunae, not directly into the artery.

The heart lies within a pericardial sinus. Primitively it extended as a tube with spiral muscles in its walls the complete length of the trunk, developing a pair of ostia in each somite. But only rarely does the heart retain its entire initial length; in the Anostraca (*Branchipus* and *Artemia*) there are 18 pairs of ostia; in Hoplocarida the heart is reduced only in the 14th trunk metamere (Fig. 11-4). In most crustaceans the spiral musculature is concentrated in that part of the heart that passes near the respiratory organs (Table 1-2). In front of and behind that region it lacks thick muscles, and forms an anterior and posterior aorta (Fig. 1-19).

The heart is well adapted to rapid transport of the O_2 enriched blood. The ostia closest to the opening of the podopericardial channels into the pericardium

Table 1–2. Length of the heart

Genus	Anterior end	Posterior end	Number of ostia pairs
Squilla	Posterior of head	13th somite = 5th pleomere	13
Anaspides	1st thoracomere	12th somite = 4th pleomere	1
Nebalia	Head	12th somite = 4th pleomere	7
Triops	Posterior of head	11th somite	11
Eucopia (Mysidacea)	2nd thoracomere	8th somite = 8th thoracomere	3
Gammarus (Amphipoda)	2nd thoracomere	7th somite = 7th thoracomere	3

are never reduced (Fig. 15-4). The position of the heart is in the cephalothorax or thorax in those orders in which the carapace forms a respiratory chamber (Eucarida, Mysidacea, Cumacea, and Tanaidacea), and in the Eucarida and Amphipoda in which the gills are on the thoracic legs (Figs. 1-19, 15-3). However, the heart is mainly in the abdomen, extending only a little into the thorax, in the isopods, in which the pleopods are respiratory organs (except in *Jaera* where it extends into the thorax).

In the Eucarida the heart is short and wide, although many representatives have a long trunk (Fig. 1-19). The shortening is an adaptation, together with the more elaborate vessels, to the large size of the animals. But the shortening of the heart in the subclasses and orders with few metameres (some Ostracoda, Branchiura, Cladocera) must be a secondary adaptation to the small size of the animals. In many ostracods and copepods the heart has disappeared.

The pericardial membrane is attached to the ventral heart wall. If there are transverse (alar) muscles in the pericardial membrane, they may take part in the contraction of diastole, as has been demonstrated by cutting on one side or both sides in a somite of *Phronima*. Even the very long heart of *Artemia*, which traverses 18 somites, contracts almost simultaneously in all parts by a rapid wave. In all Crustacea the heart is neurogenic.

In small crustaceans the heart beats faster than in large ones; the beat slows in proportion to the weight ($g^{-0.12}$). A typical crustacean weighing 1 g has 160 heartbeats per minute at 20°C; *Daphnia* has 380–450/min; *Gammarus pulex*, 260–340/min; crayfish, 100/min. The rhythm of the heart is not correlated with that of the appendages. Cold-adapted animals have a faster heartbeat at low temperatures than warm-adapted animals of the same species. To ascertain the time required for circulation of the blood, $Na_4[Fe(CN)_6]$ has been injected into the most posterior thoracic leg and a puncture made in the joint membrane of an anterior leg. For some crabs (*Dromia vulgaris* and *Maja verrucosa*) with a body weight of 75–375 g the shortest circulation time was 40–65 sec at 22–23°C. A mammal of corresponding weight would have an average circulation time of 5 sec at 37°C.

Non-malacostracans usually lack arteries; the blood goes directly from the heart into the body, usually at the head, which is favored in the distribution of oxygenated blood (Fig. 4-7). Only in those crustaceans in which the heart does not reach the posterior of the head (Calanoida and Branchiura) is there an anterior aorta which supplies the brain directly with blood.

Malacostracans, however, have various arteries: always an anterior extension of the heart (anterior aorta) and a posterior extension (posterior aorta). Also there are paired lateral arteries (Fig. 1-19). Like the heart, the anterior and posterior aorta they arise from walls of coelomic pouches in the midline of the body and in the head the preantennal coelomic pouch may take part, as has been demonstrated for *Anaspides* and *Nebalia* (Figs. 2-11, 2-12). Furthermore, at the origin of each metameric branch of the posterior artery there can be found valves that correspond to those of arteries leaving the heart.

The anterior aorta splits into several forks of which some are much-branched and supply the subenteric ganglia. The increased friction is overcome by extension of the heart into the head in the Stomatopoda and Leptostraca. In those groups in which the heart ends in the thorax there is a special pulsating organ just before the branching of the artery (except in the Cumacea and some Tanaidacea). This so-called cor frontale (Fig. 1-19) arises from a widening of the aorta on the dorsal side of the stomach. This organ lacks its own wall musculature; contraction is caused by outer tangential muscles or internal transverse muscles derived from muscles that had other functions.

The lateral arteries in relatively generalized members of the Malacostraca resemble those of chelicerates, having arteries in all somites. This can be seen in stomatopods (Fig. 11-4). The common single artery which supplies the mouthparts (arteria lateralis anterior) indicates that there were also such vessels in the posterior head metameres. But most lateral arteries have been lost with the shortening of the heart. A large number and regularity of their distribution is more common in generalized species, while the branching seems to vary with body volume. The largest number of lateral arteries and the most pronounced metameric arrangement is found in the stomatopods and leptostracans (Figs. 10-7, 11-4), while the greatest reduction in number is found in the short heart of the decapods, with only three pairs of lateral arteries (Fig. 1-19). But decapods have metameric branches of the posterior aorta.

Each lateral artery originally forked into an outer (appendage) and an inner (visceral) branch. Sometimes arteries close to the heart or away from it are branched (Fig. 1-19). The visceral branch is usually short and goes to the gut, midgut glands, gonads, and often to the ventral nerves (Fig. 1-19). The outer branch enters the appendages. In the Stomatopoda, all appendages are supplied by arteries; in Isopoda and Cumacea, only the thoracopods; in *Anaspides,* the pleopods (Figs. 1-19, 12-1). But in the primitive leptostracans the branches are so short that they reach only to the trunk musculature of the appendages; in the thoracic area only the visceral branches remain (Fig. 10-7). The appendage arteries have completely disappeared in the thorax of the lophogastrids and eucarids, and in the abdomen on the isopods. These limbs are then supplied with blood by branches of the subneural artery (Figs. 1-19, 1-21).

The subneural artery in some isopods is a short, isolated, medioventral longitudinal metameric vessel. Starting from a visceral branch of a lateral artery it supplies each ganglion with blood from the ventral side. In most isopods and most other malacostracans the different isolated branches join into a continuous ventral vessel, which curiously only in *Anaspides,* comes to lie on the dorsum of the nerve (Fig. 1-21). While in stomatopods and isopods several lateral arteries supply the ventral artery, in Decapoda and Mysidacea it is supplied by a single artery, one of the most posterior laterals of the heart (arteria descendens) (Figs. 1-19, 1-21), whose partner is so short that it ends just below the heart, or may disappear during development (Fig. 1-21).

As in other arthropods the arteries are open at the end and blood flows through spaces between organs, the mixocoel, finally reaching a large ventral space. Flowing from the head posteriorly, from the tail anteriorly, the path of the blood is determined by muscles or membranes and often is difficult to follow (Fig. 4-1).

In Malacostraca the pericardial membrane between the transverse muscles extends far ventrally and may even continue into the appendages (Fig. 1-21). It thus divides the central body cavity from a lateral sinus, just below the lateral walls of the body. The blood from the gill-bearing limbs rises into the pericardial sinus without mixing with the hemolymph of the viscera (branchiopericardial or podopericardial veins). In other crustaceans (Cladocera), there is a horizontal membrane just above the ventral wall, that separates two lateroventral sinuses from the visceral space. In addition, on the sides a vertical membrane extends into the legs, forcing the blood entering from the venter to flow to the leg tips before returning. In the region of the carapace attachment a vertical membrane extends dorsally to the entrance of the lumen within the carapace. This membrane guides the blood into the carapace lumen. There the lumen is divided into many small branches where gas exchange takes place.

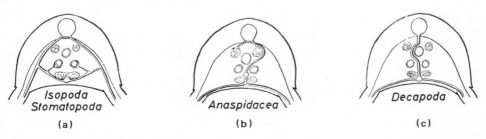

Fig. 1-21. Diagrammatic cross section through the thorax showing sub- and supraneural arteries and descending arteries. Line below heart toward legs indicates border of the pericardial sinuses. In the center, from dorsal to ventral, are heart, pair of ovaries, gut, pair of midgut glands, and ventral nerve cord. (After Siewing.)

References

Crustacea—General References

Calman, W.T. 1909. Crustacea *in* Lankester, R. *A Treatise on Zoology*. Black, London, vol. 7. (Reprinted 1964, Stechert-Haffner, New York.)

Giesbrecht, W. 1921. Crustacea *in* Lang, A. and Hescheler, K. *Handbuch der Morphologie der Wirbellosen Tiere*, Fischer, Jena, vol. 4.

Green, J. 1961. *A Biology of Crustacea*. Witherby Ltd., London.

Kükenthal, W. 1926–1927. *Handbuch der Zoologie*, vol. 3, "Crustacea," De Gruyter, Berlin.

Lockwood, A.P.M. 1967. *Aspects of the Physiology of Crustacea*. Aberdeen Univ. Press.

Moore, R.C. ed. 1969. *Treatise on Invertebrate Paleontology*. Univ. Kansas Press, Geol. Soc. Amer., Manhattan. Arthropoda 4 (1, 2) part R.

Schmitt, L.W. 1965. *Crustaceans*. Univ. Michigan Press, Ann Arbor.

Snodgrass, R.E. 1952. *A Textbook of Arthropod Anatomy*. Comstock Publ. Assoc., Ithaca.

———. 1956. Crustacean Metamorphosis. *Smithsonian Misc. Coll.* 131(10): 1–78.

Waterman, T.H. 1960–1961. ed. *The Physiology of Crustacea*. 2 vol. Academic Press, New York.

Whittington, H.B. and Rolfe, W.D.I. ed. 1963. *Phylogeny and Evolution of Crustacea*. Mus. Comp. Zool., Cambridge, Mass.

Chapter 1 (See also References to Chpt. 16, p. 335 of vol. 2)

Anderson, S.O. and Weis-Fogh, T. 1964. Resilin, a rubberlike protein. in *Advances in Insect Physiology* 2: 1–65.

Cooke, I. 1964. Electrical activity and release of neurosecretory material in crab pericardial organs. *Comp. Biochem. Physiol.* 13: 353–366.

Dahl, E. 1953. Frontal organs in free living copepods. *K. Fysiogr. Sällsk. Lund Förh.* 23: 32–38.

———. 1956. On the differentiation of the topography of the crustacean head. *Acta Zool.* 37: 123–192.

———. 1965. Frontal organs and protocerebral neurosecretory systems in Crustacea and Insecta. *Comp. Endocr.* 5: 614–617.

Dall, W. 1965. Studies on the physiology of a shrimp. *Australian J. Marine Freshwater Res.* 16: 1–23.

Elofsson, R. 1963. The nauplius eye and frontal organs in Decapoda. *Sarsia* 12: 1–68.

———. 1965. Nauplius eye and frontal organs in Malacostraca. *Ibid.* 19: 1–54.

———. 1966. Nauplius eye and frontal organs in Nonmalacostraca. *Ibid.* 25: 1–128.

Fahrenbach, W.H. 1964. The fine structure of a nauplius eye. *Z. Zellforsch.* 62: 182–197.

Gräber, H. 1933. Gehirne der Amphipoden und Isopoden. *Z. Morphol. Ökol.* 26: 335–371.

Hanström, B. 1924. Die Sehganglien der Crustaceen. *Ark. Zool.* 16: 1–119.

———. 1931, 1933. Sinnesorgane und Nervensystem der Crustaceen. *Z. Morphol. Ökol.* 23: 80–236; *Zool. Jahrb. Abt. Anat.* 56: 388–520.

Helm, F. 1928. Das Gehirn der Dekapoden. *Z. Morphol. Ökol.* 12: 70–134.

Kümmel, G. 1964. Das Cölomsäckchen der Antennendrüse von *Cambarus*. *Zool. Beitr.* 10: 227–252.

Siewing, R. 1956. Zur Morphologie der Malacostraca. *Zool. Jahrb. Abt. Anat.* 75: 39–176.

Snodgrass, R.E. 1950. The jaws of mandibulate arthropods. *Smithsonian Misc. Coll.* 116(1): 1–85.

Travis, D. 1965. The deposition of skeletal structures in the Crustacea. *Acta Histochem.* 20: 193–222.

Vaissière, R. 1961. Les yeux des crustacées copépodes. *Arch. Zool. Exp. Gén.* 100: 1–125.

2.

Class Crustacea: Reproduction, Development, and Relationships

Reproduction

Crustacea are dioecious. Only a few are hermaphrodites: most freeliving and parasitic Cirripedia, some Natantia, Tanaidacea and a few parasitic isopods. Except for the Cirripedia these hermaphrodites are protandric. Parthenogenetic development is found in most filter-feeding Cladocera, many ostracods, and in *Artemia*. But in all of these, males may be present in the population under certain conditions. For instance, males are usually present in populations of Notostraca and Ostracoda in more southern, warmer, areas of their distribution. But there is no such relationship between presence of males and climate in *Artemia*. In filter-feeding Cladocera, parthenogenetic and bisexual generations may alternate (heterogony) or the same female may first produce parthenogenetic diploid female eggs, parthenogenetic diploid male eggs or haploid eggs.

GONADS. The gonadial cavities are derived from coelomic pouches as in other arthropods. This has been demonstrated in *Artemia*, *Estheria* and some Malacostraca (e.g., *Anaspides*, *Nebalia*, and *Panulirus*). In many Peracarida the germ cells become surrounded by the dorsal border of the mesoderm somites. Testes and ovaries are usually paired but may fuse in the midline. As in other mandibulates, but unlike chelicerates, the gonads are dorsal or dorsolateral to the gut (Fig. 1-19). The gonads may extend through most of the trunk or be limited to thorax or abdomen (Figs. 1-1, 1-19). In sessile barnacles the ovaries have shifted to the anterior of the head (Fig. 9-13). Although originally tube-shaped, there may be modifications. The efferent ducts may be partially differentiated by having glandular epithelium, sometimes in the shape of a diverticulum. In the vas deferens there may be material for a spermatophore, and in the oviduct, material to keep the eggs agglutinated.

The position of the gonopores varies in different subclasses. In the Anostraca and Copepoda they are on the first two abdominal somites, which have fused. In Notostraca, they are on the 11th pair of limbs. In the order Thoracica (of the Cirripedia) the gonopores of the female are on the base of the first pair of thoracopods, those of the male at the posterior end of the body. In all Malacostraca the gonopores of the male are on the 8th, those of the female on the 6th, thoracic somite, either on the sternite or its coxae (Fig. 1-19). As the seminal receptacles usually open with a median pore between the gonopores, the copulatory pore and birth pore are separate in copepods.

HORMONES.   Hormonal control of maturation of the gonads and development of secondary sexual characters has been studied only in Decapoda, Amphipoda, and Isopoda.

1.   Inhibition of maturation of the ovaries by the neurohormones of the eyestalk can be demonstrated by extirpation of the eyestalk. *Pandalus, Lysmata,* and *Carcinus* produce eggs earlier than expected after the operation. The inhibiting hormone is produced in the medulla X-organ, and is stored in the sinus gland. Probably hormones from the neurosecretory cells of the brain and thoracic ganglia also influence the ovary.

2.   Androgenic glands have been found in several orders of Malacostraca (Decapoda, Mysidacea, Cumacea, Amphipoda, and Isopoda). They are paired, mesodermal, solid, cellular bodies of conical, egg, or in the isopods ribbon shape. Usually they lie within the last thoracic somite, close to the male efferent vessel. In isopods they extend far anteriorly parallel to the testes. The hormone they secrete induces formation of secondary male sexual characters and development of gonad primordia to testes. It is not the same as testosterone proprionate, injection of which into adult females transforms only the ovary into a testis. In contrast, females implanted with androgenic glands transform externally and internally, after several molts, into males (experiments in *Pandalus, Ocypode, Armadillidium, Porcellio, Asellus, Sphaeroma, Talitrus,* and *Orchestia*).

Protandry has been observed in nature in *Pandalus, Lysmata,* and *Hippolyte,* and is always accompanied by degeneration of the androgenic glands. In *Cyathura* it is the reverse: the androgenic glands function only in older animals. In *Pandalus,* during the male phase, there is a clear-cut inhibition of ovary development by the eyestalk hormones. If these are injected, the transformation of a male into a female can be stopped. In the protandric isopod *Anilocra,* the same result can be obtained by transplanting optic lobes of young males into older males. If these lobes are transplanted into females, there is renewed activity of the androgenic gland and the animals become male.

Antagonists to the androgenic gland appear to be the neurosecretory cells of the central region of the protocerebrum of *Porcellio.* If these are extirpated, the androgenic gland and all male reproductive organs hypertrophy; after several weeks in juvenile males, the operation results in early maturation of sperm.

3. Hormonal activity of the ovary has been demonstrated in terrestrial wood lice, *Asellus* and *Orchestia*. After extirpation of ovaries and oviducts, the oostegites and other parts of the brood chambers did not form. The same effect could be obtained in *Porcellio* after castration with x-rays. If testes of *Talitrus* are transplanted into a male of *Orchestia* after its androgenic glands have been extirpated, the testes are transformed into ovaries, and after a while the genetically male animal forms oostegites.

All results make it probable that the sex-determining genes in Malacostraca determine the formation or suppression of the androgenic gland only. If developed, the embryo becomes male, if not, it becomes female.

TRANSFER OF SPERM. In Anostraca, Ostracoda, and Thoracica (Cirripedia), the sperm is transferred by a penis, in many copepods by the last limb, in Malacostraca usually by the branches of the first two pairs of pleopods (Fig. 1-19). To hold the female, diverse structures are used, structures that otherwise have a completely different function: in copepods, one or both first antennae, in Anostraca the second antennae, in many Conchostraca and Cladocera the first thoracic legs.

OVIPOSITION. Females rarely abandon their eggs. Usually the eggs are carried in one or two clumps attached to the genital somite in the Copepoda. They may be cemented into a ball and held by the maxillipeds (Stomatopoda) or fastened to the thoracopods (Euphausiacea) or pleopods (Decapoda). Or they may be carried loose in special brood pouches. In Conchostraca and Cladocera the dorsal part of the cavity under the carapace is a marsupium (Fig. 4-10); in Peracarida there is a ventral basket, the walls of which are medially directed flat epipodites called oostegites (Fig. 17-7).

Although there are usually only small differences between males and females, in many parasitic crustaceans and sessile Cirripedia there is an extreme sexual dimorphism. Besides some parasitic females of giant size, some are so much transformed that they can hardly be recognized as crustaceans (Figs. 7-13d, 17-2). Males of parasites retain the typical crustacean shape, but often remain larva-like and small. In sessile Cirripedia the internal organization may retrogress in dwarf males (Fig. 9-19).

Development

The developmental patterns among Crustacea are very diverse. Some species produce all metameres within the eggs; others spread metameric growth over a longer period, extending it into embryonal or postembryonal periods.

Cleavage is usually superficial as in most arthropods, but it is of great interest that some representatives of all subclasses have total cleavage. The total cleavage is sometimes determinate thus resembling that of annelids. In addition there are other similarities to annelid embryos that are not seen in terrestrial arthropods. The crustacean embryology is thus of great importance for an understanding of the relationship between arthropods and annelids. Even though some of these similarities are not found in the majority of crustaceans, they are of phylogenetic interest and therefore are mentioned here.

1. Total cleavage occurs in some or in all Branchiopoda, Ostracoda, Copepoda, Cirripedia, Anaspidacea, Euphausiacea, and rarely in Decapoda (e.g., *Lucifer*). It occurs not only in eggs in which the embryo hatches in the nauplius stage (e.g., *Artemia, Cyprideis, Cyclops, Balanus, Ibla*, and *Euphausia*) but also in species with direct development, some of them with few metameres (*Polyphemus*) as well as some with many metameres (*Anaspides*). The occurrence of total cleavage does not result merely from the small amount of yolk but must be considered an ancestral character as is shown by the transition of total cleavage to a mixed total and superficial cleavage in related cladocerans. The spherical egg of *Polyphemus* has total cleavage (Fig. 2-1). The yolk-rich oval eggs of *Moina, Holopedium* and the summer eggs of *Daphnia pulex* form cell membranes, limited to the surface of the egg, first in the eight-cell stage. Later the cell membranes traverse for a period the whole embryo, in *Holopedium* during the 16 to 62 cell stage, in the summer egg of *Daphnia pulex* during the 32 to about

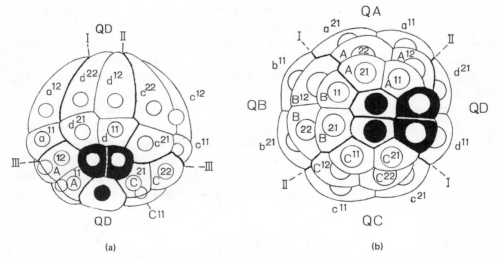

(a)   (b)

Fig. 2–1. Total and determinant cleavage of egg of *Polyphemus pediculosus*, much enlarged. (After Kühn.) (a) 31-cell stage, viewed from side. D-quartet is toward viewer. First 3 cleavage lines are darker (I, II, meridional; III, equatorial). Cells are the result of 5 successive divisions of the zygote, except the primordial germ cell (black nucleus), which did not take part in the last division; (b) 32-cell stage, view toward vegetal pole, after primordial germ cell (black nucleus) has divided; first and second cleavage lines are thick. Letters a-d, cells of animal half; A-D, cells of vegetal half below the 3 equatorial cleavage planes, in the center of which gastrulation will begin; Q, quartet separated by first two meridional cleavage planes. Arabic numbers give origin of each cell (cell a divides into daughter cells a1-a2, cell a1 into a1.1 and a1.2, etc.). In (b) the primordial germ cell (black nucleus) has divided into D1.1 and D1.2. The primordial endoderm cells are black.

128 cell stage. The cytoplasm is in the peripheral part of the cells and the yolk is pushed into the central part from which the cell membranes disappear. One can readily see when a large amount of yolk complicates cleavage. Such intermediates of total and superficial cleavage can also be observed in other orders. For instance in *Gammarus* the first divisions are total up to the eight-cell stage, whereas later divisions are superficial.

2.   The total cleavage and even the transition of total to mixed total-superficial cleavage may be accompanied by early determination of cells which is also characteristic of annelids. This is clearly seen in some Cladocera (in the spherical egg of *Polyphemus* and the oval eggs of *Moina*). It is also indicated in eggs which have the same composition of cells at their vegetal pole after the fourth to fifth cleavage (*Polyphemus.*) In such eggs one cell shifts to the center of the vegetal pole; the divisions of that cell and its descendants are retarded compared to those of the surrounding cells (Fig. 2-1). This occurs in the eggs of some cladocerans (*Holopedium,* summer eggs of *Daphnia pulex*) and *Cypris.* The cleavage of these are total at least when the cell pattern is the same as in Fig 2-1b. In *Cyclops,* the Cirripedia and in *Euphausia* the cell pattern of the vegetal pole during this period is similar.

A careful investigation of cell lineage has been made in *Polyphemus.* The mature egg contains the remainder of at least one of three nutrient cells of the ovary. In the cell divisions this ectosome* is always shifted to that descendant cell which continues the cell lineage. After the fourth cleavage a primordial germ cell is differentiated from the other blastoderm cells. Its descendants are all marked by the ectosome and can be followed until they form the gonad. In the eight-cell stage the animal pole has four large cells (labeled a, b, c, d in Fig. 2-1) which produce ectoderm cells. The vegetal pole consists of smaller cells A, B, C, D. A to C in the course of development produce ectoderm and mesoderm. D, however, which contains the ectosome, gives rise to the germ and endoderm cells. In the 16-cell stage the germ and endoderm cells are already completely separated. Cell D has divided into D1 and D2. The spindle fibres of the mitotic division lie in the meridional plane in contrast to that of other cells. Thus the D1, the primordial germ cell, containing the ectosomes is shifted to the center of the vegetal pole (Fig. 2-1). Cell D2 without ectosomes has become the primary endoderm cell which produces only endoderm cells. The axes of cells D1 and D2 mark the presumptive mediosagittal axis of the embryo.

During the next cleavages the mitoses of these two cells and their descendants are retarded. Therefore after the 7th cleavage there are only 118 cells, instead of 128 cells as might be expected. Now there are already three germ layers and the germ cells. We can separate 64 ectoderm cells from a, b, c, d; 42 ectoderm cells derived from A, B, C; 6 primary mesoderm cells from A, B, C; 4 endoderm cells

---

* Certain cells have special bodies which since these cells later become germ cells **are** called germ cell bodies or ectosomes (Siewing, 1969).

from D2, and 2 primary germ cells from D1. Each of these cells produces descendants for its own germ layer only. Gastrulation begins at the ninth cleavage indicating the fate of various cells. The mesodermal, germ and endodermal cells are separated from the surrounding ectoderm cells by sinking into the blastocoel. In the next stage, the youngest embryo, the 12 mesoderm cells which in the blastula had encircled the endoderm cells have grown into a U-shaped plate. The lateral walls of the plate are directed dorsally, parallel to the lateral walls of the embryo (compare with Fig. 2-12a, a later stage). A group of endoderm cells rests ventromedially on the mesoderm plate. The germ cells forming the gonad lie more dorsally.

The cell lineage of the eight long cells which surround the animal pole in the 32-cell stage has been studied up to the forming of the organs (Fig. 2-1). The descendants of cells a1, a2, b1, b2, c1, c2 give rise to the compound eyes, the optical ganglia and the supraesophageal ganglion. Only a few neighboring ectoderm cells take part (cell a1 and a2 are daughter cells of a; cell b1 and b2 are daughter cells of b, etc.). The cells d1 and d2 give rise to the subesophageal ganglion, very few other ectoderm cells taking part.

In other cladocerans gastrulation is similar. But the long ventral part of the oval yolk-rich egg causes differences. In the moderately long egg of *Moina*, first the germ and mesoderm cells sink in, the endoderm soon follows. But the process of gastrulation is prolonged. In many crustaceans the gastrulation of the germ cells and mesoderm is separated by a temporary interval from the later sinking in of the endoderm cells. The gastrulation thus has two phases. For instance, in the summer eggs of *Holopedium* the first phase occurs during the 10th, the second during the 11th cleavage. The same is true for the long, oval summer eggs of *Daphnia pulex* whose ventral side is very long. Therefore not only the time but even the location of the second gastrulation differs from the first. The presumptive mesoderm and germ cells sink in at the posterior region, the presumptive endoderm cells at the anterior region of the egg in the region into which later the foregut is invaginated. It is of interest that in the winter eggs of *Holopedium* and *Daphnia pulex* the first gastrulation occurs after the cleavages, the second much later in the season, in spring.

3. Only little experimental work has been done on the determination of crustacean eggs. In the summer eggs of *Daphnia pulex* at the beginning of development the polarity and the bilateral position are already determined. Shifting of the egg contents by centrifugation does not change this determination. Thus the distribution of plasma, nucleus, yolk grains and the oildrop does not determine the normal cleavage pattern. Structures in the stable egg cortex underneath the membrane may have caused the cytoplasmic currents which still make possible the normal course of events after centrifugation. It has been shown in *Cyclops* that the two blastomeres of the two-cell stage have different potentials. If one is killed by X-rays, the other one will develop as in an uninjured embryo

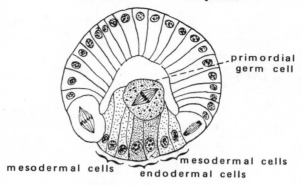

Fig. 2–2. Section through zygote of *Megacyclops viridis*, beginning gastrulation after 8th cleavage; much magnified. (After Fuchs.)

and gives rise to the expected number of cells; however, in addition to ectoderm and mesoderm there may be germ cells or not.

However, in no crustacean has there been found an arrangement of the blastomeres as in spiral cleavage; there are never the quartets so typical of annelids. It is probably of no significance that in some Cladocera and Ostracoda the mitotic spindles are diagonal and alternate in a number of divisions; this feature is not always the same in all egg quartets.

4. In many crustaceans with total equal cleavage or total unequal cleavage, the formation of the internal germ layers, mesoderm and endoderm, shows resemblances to the development of annelids. In some Cirripedia, for instance, epiboly takes place. In other groups, however, endoderm, mesoderm, and germ cells push into the blastocoel at the vegetal pole and are covered in gastrulation by invagination (Fig. 2-2). A true blastopore may develop (*Moina, Cyprideis, Cyclops, Anaspides, Euphausia,* and *Lucifer*) or there may hardly be a depression on the surface (*Polyphemus* and *Daphnia*). Only rarely is there a real archenteron (*Lucifer*); the cells that have moved into the blastocoel clump together and later proliferate. It seems strange that in decapod eggs with superficial cleavage there may be an archenteron extending into the central mass of yolk (Fig. 2-3). This archenteron is especially distinct in *Astacus* in which yolk is later brought into its lumen after the mesoderm has separated off anteriorly. The endoderm cells take up yolk with the end turned toward the yolk and become huge pyramidal cells (Fig. 2-3c). The nucleus later migrates with cytoplasm toward the peripheral end of the cell and the yolk alone fills the long inner part. The central sections of the cell soon disintegrate and the yolk fills the lumen of the archenteron. The peripheral cell sections alone form the wall of the gut enclosing the yolk. The embryo in this way transports the large mass of yolk into the archenteron while forming endoderm by invagination; otherwise in superficial cleavage this would come about by polar growth into the endoderm and the migration of endoderm cells toward the periphery of the yolk.

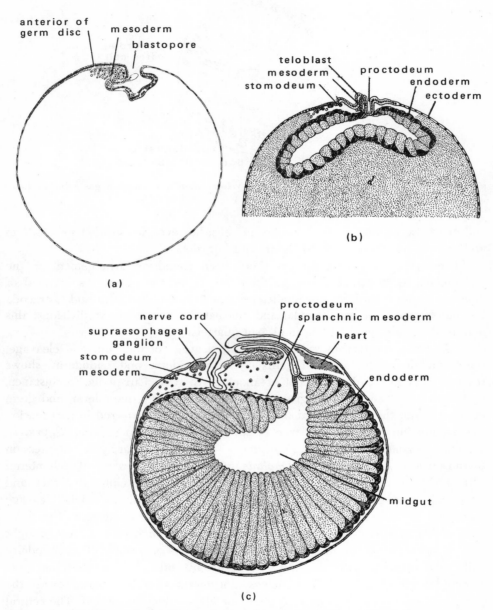

Fig. 2–3. **Endoderm formation in the crayfish *Astacus astacus*, shown by longitudinal sections through embryo; venter uppermost. (After Reichenbach from Korschelt.) (a) Invagination of endoderm in the egg already surrounded by blastoderm; yolk fills all the area shown white; diameter 2.4 mm; (b) endodermal cells begin to take up yolk (stippled); diameter 2.6 mm; (c) endodermal cells have taken up all yolk (stippled) and have stretched; their plasma (hatched) and nuclei have moved to the periphery. The preanal growth zone has formed somites, of which the most posterior ones form an anteriorly projecting cone; diameter 2.8 mm.**

This phenomenon can best be explained as the survival of an ancient mode of endoderm formation. Most likely the ancestors of malacostracans had eggs with little yolk, total cleavage and gastrulation by invagination. Further evidence is seen in primitive decapods (Penaeidae) which leave the egg as small nauplii.

5. Crustaceans, like polychaetes, have at first a short germ band, formed of the cephalic lobes and three appendage-bearing metameres which appear at the same time (first and second antennae, mandible). Often, especially in non-malacostracans, the egg hatches at this stage as a nauplius, and in the swimming larva the still missing parts grow on the posterior preanal (teloblastic) budding zone in front of the telson. One after another the metameres of the first maxillae, and other metameres are produced, slowly increasing the distance between mouth and anus. This process also resembles the ontogeny of annelids. While such growth can take place in free-living larvae, it can also, as in many malacostracans, occur within the egg. In Malacostraca the budding zone consists of one row each of ectoderm and mesoderm teloblast cells, a further resemblance to polychaete development.

6. In polychaete eggs that have spiral cleavage, all regions of the egg participate in forming a trochophore. In arthropods, including many crustaceans, this development usually differs: first only the future venter contributes to a germ band on which all development (nervous system, appendages, mesoderm) takes place (Figs. 2-4, 2-8, 2-9). The sides and dorsum of the egg undergo no differentiation and only late in development does the germ band become a cylindrical body by growth of the coelomic pouches dorsally to meet in the dorsal midline (Figs. 2-12, 2-13). Crustaceans are the only arthropods in which the embryo often, as in polychaetes, has a cylindrical body form from the start. This form of development may be found in crustaceans that hatch as nauplius larvae with few metameres (*Cyclops, Lepas, Balanus,* and *Euphausia*) as well as in some non-malacostracans with few metameres that hatch after they have their final body form (most Cladocera; Fig. 2-5).

7. Also the fact that the larva often leaves the egg after a short period of development and can live freely resembles annelids. This is an adaptation to freshwater and marine environments in which the minute larvae can move about and find microscopic food.

8. The formation of mesoderm in many crustaceans also resembles that of annelids. At first the differences in origin of the naupliar and imaginal mesoderm are much more distinct than in other arthropods. In *Anaspides* and *Nebalia*, as well as in several peracarids and decapods, the nauplius mesoderm is derived from the immigration of mesoderm in the middle or more posterior region of the germ disc. The small mesoderm cells migrate from this blastopore region anteriorly to come to lie under the V-shaped cephalic part of the germ disc

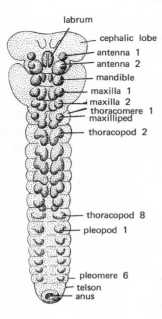

labrum
cephalic lobe
antenna 1
antenna 2
mandible
maxilla 1
maxilla 2
thoracomere 1
maxilliped
thoracopod 2

thoracopod 8
pleopod 1

pleomere 6
telson
anus

Fig. 2–4. Germ band of the isopod *Ligia italica*, which produces all somites within the egg. The primordia of the paragnaths can be seen between the mandibles; 1.5 mm long. (After Silvestri.)

arranging themselves under the ectoderm as V-shaped, paired unsegmented mesodermal bands (Figs. 2-6, 2-7). In marked contrast, the imaginal mesoderm of the postmandibular metameres arises later in a second phase, one after another, in the preanal zone from large mesoderm cells. The naupliar mesoderm can be compared with the deutometameres, the imaginal mesoderm with the tritometamere* of annelids.

The naupliar mesoderm (deutometameres) does not segment. Nevertheless, it later forms coelomic pouches in the metamere of the second antennae and sometimes (as in *Panulirus* and *Artemia*) in the areas of all three pairs of naupliar appendages (Fig. 2-13). In *Nebalia, Hemimysis, Gammarus,* some isopods (e.g., *Asellus, Idotea, Limnoria, Cirolana*), and several Natantia (*Palaemonetes*) a paired immigration of cells from the surface in front of the metamere of the first antennae is observed which transforms into a preantennal mesoderm and can develop a pair of temporary coelomic pouches. Its cells become muscles of the labrum, the stomodeum and the walls of the anterior aorta (Fig. 2-13).

The production of imaginal mesoderm (tritometameres) by preanal budding is especially interesting and annelid-like in those malacostracans in which the budding zone consists only of two transverse rows of unusually large cells, the ectoderm teloblast and the mesoderm teloblast cells (Fig. 2-9). While the

* Deutometameres are the original metameres formed during larval development from primordial mesoderm cells, tritometameres those formed by later teloblast growth and budding anterior to the telson (pygidium) (see also Siewing, 1969).

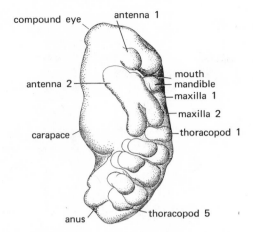

Fig. 2–5. Embryo of the cladoceran *Moina brachiata*; 0.2 mm long. (Modified after Grobben.)

number of the ectoteloblast cells may vary, in Crustacea there are always eight mesoteloblast cells.*

Each row of mesoteloblasts, by an anteriorly directed unequal, synchronous transverse division, first produces small daughter cells which lie directly behind the mandibular metamere. This first row in *Nebalia, Neomysis,* and *Ligia* proliferates to produce all mesodermal cells of the first maxillar metamere. Simultaneously with the division of the mesoteloblast cells the ectoteloblasts above it divide. Sometimes (e.g., *Ligia* and *Limnoria*) this happens twice, so that for each transverse row of mesoteloblast cells there are two rows of ectoteloblast cells (Fig. 2-9). Another division of the teloblast cells soon follows.

As these teloblast cells have been described only in those crustaceans in which development shows very little determination, we cannot derive them from particular blastomeres. But in constancy of number and in their activity, they agree well with the teloblasts of the oligochaete *Tubifex,* which originate from cells 2d (ectoderm) and 4d (mesoderm); the resemblance justifies comparison with annelids, but it should not be assumed that crustaceans originally had teloblast cells.

---

* Also in nonmalacostracans typical ecto- and mesoteloblast cells forming a single transverse row of cells synchronously splitting off anteriorly a row of daughter cells. But these observations have not been repeated and are doubtful except in the ectoderm of *Ibla*. Benesch (1968) has demonstrated that in *Artemia* the germinative complex consists of a rather long zone of ectodermal and mesodermal cells which are not arranged in transverse rows. The mitoses are scattered irregularly. Each newly formed somite consists of a long complex of irregularly arranged cells which are fused to the budding zone for a considerable time and which have begun to differentiate before detaching: it does not consist, as might be expected, of two transverse rows of ectodermal and mesodermal cells.

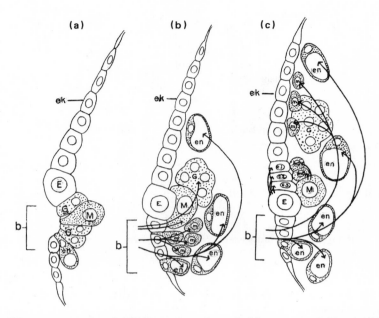

Fig. 2–6. Diagrammatic longitudinal section through blastopore of *Hemimysis*; zygote has superficial cleavage and development resembles determinant type. (After Manton.) (a) Beginning of gastrulation; (b) formation of m, naupliar and M, metanaupliar mesoderm; (c) completion of gastrulation, with migration of naupliar mesoderm to the anterior and beginning of teloblastic somite formation. b, blastopore region; E, ectoteloblast, which by transverse divisions produces ectodermal primordia of the postmandibular somites: e1, e2, e3, etc. (there are 15 adjacent teloblast cells vertical to the plane of the illustration); ek, naupliar (larval) ectoderm derived from the blastoderm; en, endodermal cell; G, germ cells; M, metanaupliar mesoderm (consisting of 8 cells in a row vertical to the plane of illustration), which gives rise by transverse cleavage to a row of 8 cells each forming the primordium of a postmandibular somite; m1, mesoderm band, from which metamere of first maxilla is derived; m, mesodermal cells moving into the naupliar region and there forming the mesoderm (deutometameres).

In crustaceans in which the telson and preanal budding zone are flat on the yolk (e.g., Peracarida), the teloblasts produce only the venter of the metameres, a germ band on the yolk (Fig. 2-4). In others, however, (*Anaspides, Nebalia*, and many decapods) the preanal budding zone arises as a hump after the formation of a few metameres (tritometameres) (Fig. 2-8). The teloblasts are arranged in a ring and produce a two-layered, segmented, yolk-filled tube, which grows anteriorly under the egg membrane, resembling the growth of the scorpion postabdomen and the abdomen of primitive spiders (Figs. 2-3, 2-10, and compare to Figs. 4-11, p. 50, 4-14, p. 53, and 4-15, p. 54 of Vol. 2).

While the posterior metameres are produced, the cells of the earlier anteriorly formed metameres divide and differentiate into organs as in annelids with equal

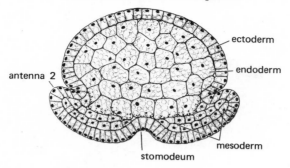

Fig. 2–7. Cross section through zygote of *Anaspides tasmaniae* in the naupliar stage. (After Hickman.)

amounts of yolk, often the embryo shows a regular gradient of the degree of development from anterior to posterior.

The imaginal mesoderm, the tritometameres, contain only small coelomic pouches, unlike those of annelids, chelicerates, and chilopods. They are so small that they never meet on the dorsum (Figs. 2-11, 2-12, 2-13). Because of their small size and brief existence these coelomic pouches have been missed by some embryologists who looked for them, resulting in some controversy about their existence. But recent observations leave no doubt that they do develop in nearly all thoracic and abdominal somites of *Nebalia, Anaspides, Hemimysis, Idotea, Limnoria, Astacus,* and *Panulirus,* as well as in the thoracic somites of *Gammarus, Artemia, Chirocephalus,* and some Conchostraca.

The coelomic pouches appear in the dorsal part of the mesoderm only (Fig. 2-12), and these cavities are rapidly obliterated.

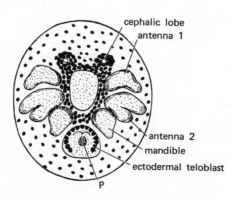

Fig. 2–8. Germ disc of *Anaspides tasmaniae* in naupliar stage, 13 weeks old; on the caudal papilla (P), ectodermal teloblast cells begin to reorganize themselves into rings. (After Hickman.)

central band

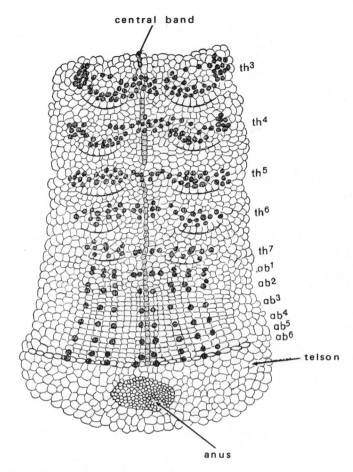

Fig. 2–9. Germ band of the isopod *Ligia*; posterior somites (ab4–ab6), recently developed from teloblast cells, each consists of two rows of ectodermal cells and one row of 8 mesodermal cells (dark). In the older metameres the mesodermal cells have doubled (th7–ab2) or multiplied (th3–th6); ab, abdominal somite primordium; th, thoracic somite primordium. (After McMurrich from Korschelt and Heider.)

Which of two pathways is followed in the differentiation of the trunk mesoderm depends on whether two bands or two half tubes are formed. The mesoderm bands of *Hemimysis,* lying flat on the shrinking yolk, segment immediately, the oldest and most anterior metameres growing dorsally, up the sides of the egg between ectoderm and yolk (Figs. 2-12b, 2-13). At the same time coelomic pouches appear in the dorsally growing parts. On further shrinking of the yolk, a dorsolateral space appears between the ectoderm and the outer (somatic) wall of the coelomic pouch. This space becomes the pericardial sinus.

Also a more ventral space appears and becomes a blood lacuna (perivisceral lacuna) between the inner (splanchnic) wall and the yolk (the future gut, Fig. 2-12b,c). The later development also resembles that of other arthropods (Figs. 2-11 to 2-3 and compare with Vol. 2, p. 52). Finally the somites dissolve. Most of their cells transform into longitudinal muscles or other cells shift to the surface of the gut and its ceca forming their muscle layer (Fig. 2-12b,c).

The development of the mesodermal organs is the same in species whose mesodermal somites are at first ring or half-ring shaped, filling completely all space between ectoderm and yolk or the gut except the hemocoel (Fig. 2-12a).

The embryonal development takes a different part in the ontogeny of different groups of Crustacea. We can thus distinguish two forms of development which are termed epimorphic and anamorphic. In epimorphic development all metameres are produced within the egg. In anamorphic development, a larva with few metameres hatches from the egg and only during the course of postembryonal development forms the missing metameres and appendages.

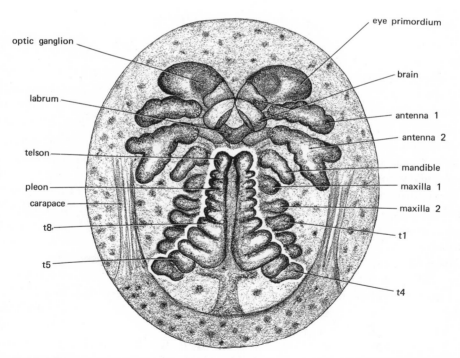

Fig. 2–10.   Late embryonic stage of the crayfish *Astacus astacus*; 2.6 mm diameter. The germ band is folded; its posterior end from the 4th thoracomere on, has been pushed toward the venter of the anterior; t, thoracopod primordium. (After Reichenbach from Korschelt.)

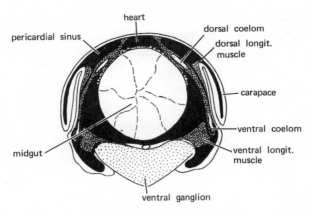

Fig. 2–11. Cross section of a thoracic somite of *Hemimysis* embryo whose mesoderm has proliferated dorsally. Along the dorsal midline the somite borders grow toward each other and produce at first the floor of the heart. Primary body cavity containing circulating blood is indicated in black; mesoderm, densely stippled; primordia of ganglia, lightly stippled. (After Manton.)

Epimorphic Development

Complete development within the egg, which leaves only final growth and formation of reproductive organs to the postembryonal period, is common. Unlike chelicerates and insects, crustaceans probably acquired the epimorphic development as a secondary condition, in association with brood care. Epimorphic development is rare in crustaceans with few somites, but common in those with many. It is found in the following groups, though sometimes limited to certain genera: Branchiopoda (in most Cladocera), Anaspidacea, a few Decapoda; all Leptostraca, and the many species of Peracarida.

In many invertebrates the freshwater species develop epimorphically, unlike their marine relatives. This may be an adaptation to the hazard of having larvae carried downstream by currents. This adaptation does not show up as distinctly in Crustacea. Departures are seen, for instance in the Peracarida, mostly marine, which carry their young about in many crustaceans which have free-living larvae. The explanation is that the adults of freshwater crustaceans that have free-living larvae are small species that can live only in places where currents do not carry them off, as ponds, lakes or slow rivers among plants. In such places the nauplii, having three pairs of appendages, can also survive (Anostraca, Conchostraca, Ostracoda, and Copepoda).

Young that hatch with all metameres usually do not differ markedly from adults. In young lobsters the pleopods are only stumps, the pereopods biramous (mysis stage, Fig. 2-17e). At times appendages may be missing: many Peracarida lack the seventh pereopod, crayfish lack uropods. Pronounced transformation is seen only in species that become parasites after a free existence (female Epicaridea, Fig, 17-19) or that live first as parasites and on reaching sexual maturity become free-living, subsisting on stored nutrients (Gnathiidae, Fig. 17-24). Only in the Gnathiidae can comparison be made with insect metamorphosis, not in the Epicarida. Epicarid females do not display the typical structure of the order; they undergo a reduction of organ systems that represents an adaptation to their parasitic habits.

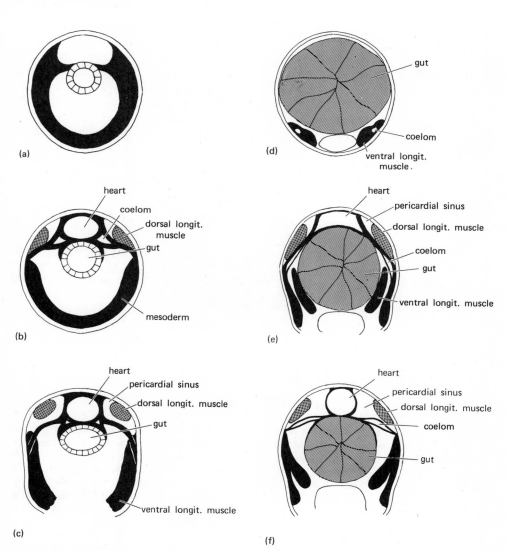

Fig. 2–12.   Mesodermal organ formation in embryo having ring arrangement of teloblasts (*Estheria*, a–c), and in embryo having linear arrangement of teloblasts (*Hemimysis*, d–f); (a,d) early stages, *Hemimysis* has mesoderm limited to venter; (b,e) pericardial sinus and heart formation; (c,f) degeneration of the coelom. The mesoderm is indicated in black, and dorsal longitudinal muscle is stippled. (After Manton.)

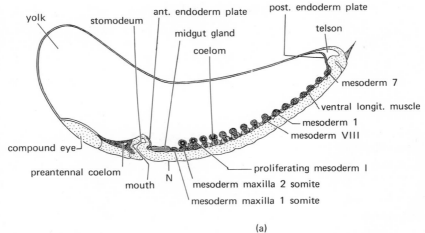

(a)

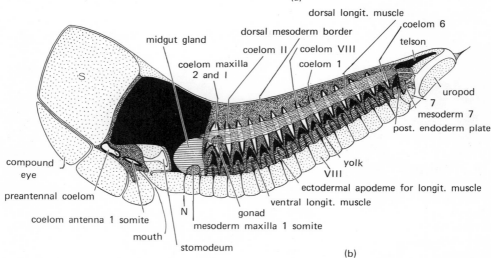

(b)

Fig. 2–13. Coelom formation and dorsal growth of mesodermal somites in *Hemimysis lamornae.* Reconstruction of lateral view of embryos illustrated as though they were transparent. Ectoderm, lightly stippled; mesoderm, densely stippled; endoderm, with horizontal lines. (After Manton.) (a) Mesodermal somites begin to grow dorsally in the oldest (= most anterior) somites along the wall of the embryo; yolk indicated by white areas. (b) Older embryo in which mesodermal somites have grown up on the sides and have fused dorsally; yolk is indicated in black. The coelom appears in the dorsolateral section (the later formed ventral coelomic pouches are visible in Fig. 2-12d). The venter of each somite is narrowed because its anterior and posterior walls have transformed into longitudinal muscles above which the yolk (black) is visible. N, border between naupliar part of body (deutometameres) and teloblastic trunk (tritometameres). Roman numerals indicate thoracomeres I–VIII; Arabic numerals, pleomeres 1–7. S, vertical ectodermal median septum which enters the yolk from anterior.

Anamorphic Development

The anamorphic development of some Crustacea is unlike the development in most other groups of arthropods. In anamorphic development the larvae hatch with few metameres and the postembryonal appearance of additional metameres by teloblastic budding increases the somites to the final number characteristic of the species. This form of development is otherwise known only for trilobites, pycnogonids, many myriapods, and Protura.

In many crustaceans the embryonal development is interrupted when there are three pairs of appendage-bearing somites, those of the first and second antennae and the mandibles. The embryo hatches as a spherical or oval nauplius larva (Fig. 2-14) typical for Crustacea. That this is an ancient larval form is indicated by the following characteristics: the constancy of the three metameres, the appearance of nauplii in many subclasses and orders (in some families or all Branchiopoda, Copepoda, Ostracoda, Cirripedia, Euphausiacea, and Decapoda), and the appearance of an additional cuticle under the egg membrane in some species having epimorphic direct development, after the three naupliar limbs are formed (*Anaspides,* some Decapoda including *Astacus*).*

The ancestral crustaceans probably did not resemble nauplii. The few segments and small size of the nauplii are adaptations made possible by the production of many eggs with little yolk. Although this requires a long and hazardous planktonic life, it expedites the distribution of the species by currents, as in many mollusks and polychaetes.

Not all anamorphic crustaceans hatch from the egg as nauplii. By extending their embryonal development, many hatch after a larger number of metameres

---

* The cuticle indicates that this is the conclusion of a definite developmental stage, even though the embryo does not leave the egg but continues its development within two membranes, the egg membrane and the naupliar cuticle. In such direct development, after the formation of all metameres the embryo secretes another cuticle after which the naupliar cuticle and egg membranes are broken and the animal hatches. The production of the naupliar cuticle appears to be without function; it can only be interpreted as a vestige.

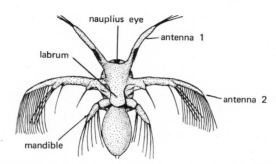

Fig. 2-14. Freshly hatched nauplius of *Branchinecta occidentalis*; 0.4 mm long. (After Heath.)

has been formed, thereby jumping several stages of postembryonal development and molts.

Some stages may be cut out. *Hutchinsoniella,* for instance, has primordia of both pairs of maxillae in the first free-living stage which normally arise on later metameres. *Chirocephalus* on hatching has the primordia of the two pairs of maxillae and two pairs of limbs and the succeeding trunk metameres. Many more metameres are produced in the eggs of the Branchiura, the majority of decapods, and the Stomatopoda. Many decapods (Natantia and Brachyura) hatch only after having developed many somites (Fig. 2-19).

Larvae hatching with many somites need less time to develop to adults than comparable larvae hatching as nauplii, reducing the hazardous period of planktonic existence. As might be expected, many-segmented larvae appear in those species producing fewer, more "valuable," yolk-rich eggs. The Chinese mitten crab *Eriocheir* produces many hundred thousand eggs, 0.2–0.3 mm in diameter, which give rise to larvae of an intermediate stage, the zoea. Lobsters produce 80–90 thousand eggs, 1–1.5 mm in diameter, which hatch in the mysis stage, late in larval development. The crayfish *Astacus* produces only 60–150 eggs, 2–3 mm in diameter, which hatch as decapodids and acquire their adult form in only 10–15 days.

Many crustaceans have a regular anamorphic development, which produces in regular procession one posterior somite after another with their limb primordia. The first larval stage slowly progresses to the adult form in small steps, separated by molts, with corresponding changes in habits. This is found especially in crustaceans that swim with their second antennae until reaching a late stage. Adults of Cephalocarida, Mystacocarida, and Ostracoda still paddle with the second antennae; the Anostraca (except Polyartemiidae) do so until all 11 pairs of limbs are present, the last still undifferentiated (Fig. 1-16): the Euphausiacea do so until the furcilia larval stage.

One might expect the stepwise development to take place with the appearance, after each molt, of a new metamere with the primordia of its pair of appendages, behind the previously formed one, while the previous one develops further. Usually there are slight modifications. The development is never quite regular. The smallest modifications are found in *Hutchinsoniella* (Cephalocarida). It hatches with two trunk somites and by the 11th stage has formed all 19 trunk somites, though the posterior five of the trunk appendages are functional only in the adult, in the 19th stage. The postembryonal development of copepods (*Cyclops*) and ostracods is more irregular. Significant shortcuts are made in the Anostraca where *Artemia,* after five molts,* has all metameres as well as 11 limb pairs (the last ones not well developed). *Chirocephalus* before each molt forms

* Some embryological and growth data on *Artemia* of various authors conflict. This may be due to observations on different species—all called *A. salina* because of morphological similarities (see also taxonomic section of Anostraca, Chapter 4).

several somites and appendage primordia (Fig. 2-15). There is a larger short-cut from the metanauplius to the cypris larva in Cirripedia, which at least six pair of thoracopod primordia become functional after a single molt (Fig. 9-14).

Irregular anamorphic development[*] is common in Crustacea in many decapods and stomatopods. Here the somites are not formed one after another with their appendages, but some younger metameres, more posterior in position, develop earlier than older, more anterior ones. In decapods and *Squilla* this phenomenon also occurs in embryonal development, so that the first free larval stage, the zoea or pseudozoea, has the abdomen more developed than the cephalothorax (Fig. 2-24). As the larvae in both cases use the abdomen in locomotion, the advanced development appears to be an adaptation.

The steady increase in number of appendages corresponds to the change in function of the limbs, of minor importance in Cephalocarida and Mystacocarida. The change is seen clearly (Fig. 2-17) in the primitive decapod family Penaeidae. The swimming function of the antennae (metanauplii) is first transferred to the two anterior thoracopods (protozoea), later to be taken over entirely by the thoracopods (mysis larva), and finally by the pleopods. Similarly in copepods the thoracic limbs take up the locomotory function of the second antennae after a single molt from the last metanauplius to the copepodid stage; the thoracic appendage primordia of the metanauplius have no apparent function. This occurs also in barnacles in the transformation from the metanauplius to the cypris larva. In some ostracods that have a regular anamorphic growth, the large walking claw is retained and is present on three different pairs of appendages at different times (see under Development in Chapter 5).

The anamorphic transformation of crustacean larvae differs from the metamorphosis of insects. If considered from the naupliar stage it is also different from the anamorphosis of Protura and the myriapods. Insect larvae have all metameres and appendages from the start, while Protura and myriapods have, if not quite the adult number, at least all functional types of metameres and appendages. However, the nauplius hatches with only three appendage pairs that have to perform all the functions of more than 10 pairs in the adult. If a shrimp hatches as a zoea it lacks at least five pairs of thoracopods and all pleopods, which are the locomotory appendages. The swimming function of the three pairs of thoracic appendages is temporary, later to be taken over by the pleopods, a change unknown among insects and myriapods. The crustacean larval appendages, then, always appear more unspecialized than those of the adult, perhaps representing passage through ancestral forms, but corresponding with their function in the larva. The second antennae and the mandibles of the nauplius, which function in swimming and feeding, are much more leglike than the more

---

[*] Waterman and Chace (1960) in Waterman, T.H. ed. *The Physiology of Crustacea*, call this metamorphic development, a terminology which does not correspond to usage in other arthropods.

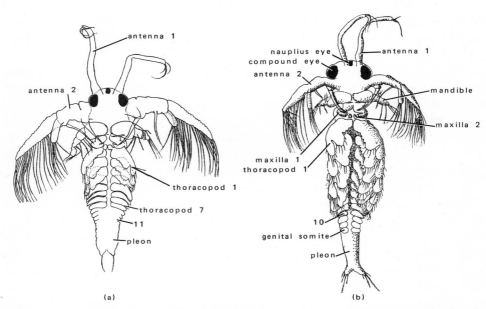

Fig. 2-15. **Two stages in the anamorphic development of** *Chirocephalus grubei***; labrum omitted. (After Oehmichen.) (a) After the first molt of the metanauplius that hatched from the egg; 1 mm long. (b) After the second molt; 13 trunk somites are visible, of which 9 carry foliaceous limbs, more differentiated anteriorly than posteriorly; 1.6 mm long.**

specialized homologs of the adult (Figs. 2-14, 2-17a). This is also true for the second antennae of the stages of Anostraca and Euphausiacea which have many somites, groups in which the locomotory organs differentiate only late (Figs. 2-15, 2-16). The maxillipeds of decapods and stomatopods are first unspecialized swimmerets, at the time when the larva (protozoea, zoea, and antizoea) has no other limb pairs (Figs. 2-17c,d, 2-19). They differentiate into maxillipeds only after the posterior thoracopods (mysis stage) or pleopods appear.

The differential number and form of somites and appendages and the faster development of younger, more posterior metameres results in great diversity of crustacean larvae. In addition to these are cenogenic peculiarities reflecting the diverse habits differing from those of the adults. Examples are the noticeable, long spines of the carapace of cirriped nauplii and zoea larvae, as well as the long pinnate hairs of their appendages which increase their resistance to sinking (Fig. 2-20).

The many different crustacean larvae can be arranged into the larval stages discussed below. But it has to be considered that in each stage there are specific (or generic) differences in the number of molts passed. (The Cephalocarida, Mystacocarida, *Bathynella*, and Branchiura are not considered here.)

Larval Stages of Regular Anamorphic Development

NAUPLIUS (Figs. 2-14, 2-17a, 2-22). There are only three pairs of metameres with their appendages and no other somites. Ostracods already have a carapace (Fig. 5-5).

METANAUPLIUS (Fig. 2-17b). There are only three functional pairs of appendages (both pairs of antennae and mandibles) as well as additional metameres whose appendages are nonfunctional primordia. In Notostraca, Conchostraca, and Anostraca the metanaupliar stages smoothly transform into the postlarval stage. The end of the metanaupliar stage is arbitrarily set at the appearance of functional maxillae (but the antennae still function as locomotory organs; Fig. 2-16). This postmetanaupliar stage ends when the second antennae get their final form and are no longer locomotory. Then the postlarval period starts.

COPEPODID. The copepodid arises from the metanauplius of Copepoda and swims with some of the thoracic appendages.

CYPRIS. The cypris larva of the Cirripedia forms from the metanauplius in which after a single molt up to six pairs of appendage primordia become functional and at which the previously flat carapace bends ventrally and encloses the larva (Fig. 9-14).

PROTOZOEA. The protozoea of the primitive family Penaeidae (Decapoda, Fig. 2-17c) comes from a metanauplius and continues to use both pairs of antennae for swimming. In addition the biramous first two appendages of the thorax are locomotory. All thoracic somites are formed, some covered by the overhanging

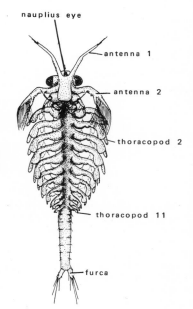

Fig. 2–16. *Branchinecta occidentalis,* 7th stage; the 10 anterior thoracopods have endites, the second antenna is still used for locomotion.

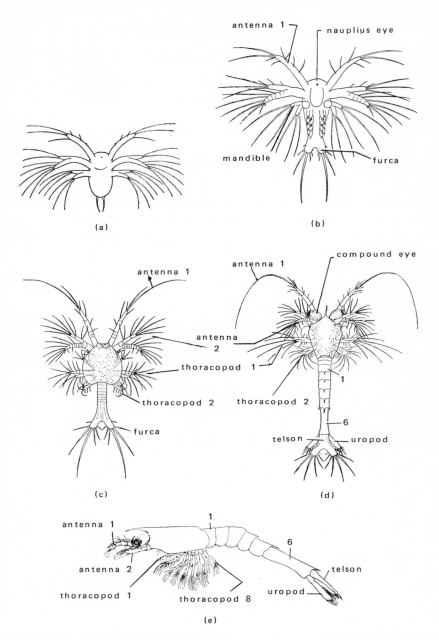

Fig. 2–17. Several stages from the four periods of anamorphic development of the primitive decapod *Trachypenaeus constrictus.* **Development from hatching to decapodid takes 2–3 weeks. (After Pearson.) (a) Nauplius, venter; 0.26 mm long. (b) Metanauplius, 4th (last) stage, venter; 0.42 mm long. (c) Protozoea, first stage, dorsum; 0.7 mm long. The thorax becomes segmented, a short carapace and the first two thoracopods are present; pleon is unsegmented. Both antennae and thoracopods are used for locomotion; the mouthparts are used only in feeding. (d) Protozoea, third (last) stage; 1.9 mm. Thoracomeres 3–8 have limb primordia; the pleon is segmented; uropods are present and compound eyes are stalked. (e) Mysis, first stage; 2.8 mm long. Antennae are reduced and are no longer used for swimming; all thoracopds are present as biramous appendages and are used in locomotion. (Pleopods appear in the second mysis stage.)**

carapace which is attached to the somite of the second maxillae (Fig. 1-6). Besides the two anterior biramous thoracic appendages, stumps of the third are present. The unsegmented abdomen lacks appendages and the telson bears a furca. The compound eyes are covered by the carapace.

In the molt to the second protozoea stage the carapace narrows and forms an anterior rostrum, the compound eyes become free and stalked. At the same time the borders of the five pleomeres appear. The third (and last) protozoea stage has the primordia of all thoracic appendages although only the first two and the precocious uropods are functional. Further molts transform the protozoea via the mysis stage into a postlarva.

MYSIS STAGE. The mysis stage (Fig. 2-17e) is a transformed protozoea, having all its appendages formed, the swimming function transferred from the antennae first to the thoracopods, later to the pleopods.

ANTIZOEA. Many Stomatopoda have an antizoea, which unlike the zoea, hatches with a well-developed thorax covered by a carapace. The biramous legs of the thorax, often five pairs, function as locomotory organs (Fig. 11-5). The pleon is shorter than the thorax, still unsegmented and lacks limbs. The compound eyes are not stalked.

CALYPTOPIS. The calyptopis of the Euphausiacea probably belongs here. It is formed from the metanauplius, has a very short but well-segmented thorax, and a large, long abdomen (Fig. 2-18). The carapace attached to the somite of the second maxillae hangs over the posterior of the thorax and covers the compound eyes. Both pairs of maxillae as well as the first thoracopods are formed, the 2nd to 6th only as unsegmented buds. The antennae alone are locomotory. In its second stage, the first five pleomeres form and in the third stage, the 6th. Also

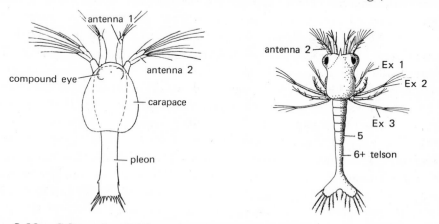

Fig. 2–18.   Calyptopis of *Thysanoessa inermis* (Euphausiacea); dorsal view of first stage which developed from a metanauplius; 1 mm long. (After Lebour from Zimmer.)

Fig. 2–19.   Zoea of the prawn *Pandalina brevirostris* (Decapoda, Natantia), dorsum of first stage just hatched from egg; 1.8 mm long. Ex, exopodites of thoracopods, which are used for swimming. (After Lebour.)

the uropods now appear, early, before the 2nd–8th thoracopods and 1st–5th pleopods.

Larval Stages of Irregular Anamorphic Development

In irregular anamorphic development, the abdomen is ahead in development, and all three pairs of mouthparts are functional (zoea types).

ZOEA.   The zoea, often just called larva by specialists, appears in most decapods and hatches directly from the egg. The first two or three pairs of thoracopods, which later become maxillipeds, are locomotory (Fig. 2-19). The pleon is very long and segmented. The thorax is very short, only partly segmented, and the entire length of its dorsum is fused to the carapace. The compound eyes are sessile in the first stage, slightly stalked later. The anomuran and brachyuran zoeae have long spined projections on the carapace (Figs. 2-20, 2-24). The musculature and internal organs resemble those of the adult.

PSEUDOZOEA.   The pseudozoea is found in many stomatopods (e.g., *Squilla*). Its abdomen is segmented and each metamere bears a pair of biramous appendages (Fig. 11-5). The thorax is also segmented, but only the first two metameres have appendages, the second pair of which is already specialized as raptorial limbs.

Anamorphic Larval Stages in which the Advanced Development of the Pleon Is Balanced

METAZOEA.   The metazoea, found only in Anomura and Brachyura, usually slowly transforms from a zoea by budding of simple uniramous limbs on the posterior thoracomeres. These appendages usually are shorter than the two anterior swimming limbs. Also pleopods 1–5 appear usually all at the same time. The eyes are stalked (Fig. 2-20b). Many authors call this the late zoea stage.

MYSIS STAGE.   The mysis stage of Natantia and some Reptantia (Decapoda) develops usually from a zoea, rarely as in lobsters (Fig. 2-17e), it hatches directly from the egg. It has biramous thoracopods, almost all of which have exopodites, providing numerous swimmerets with which the larva swims like the Mysidacea (giving rise to the old, phylogenetically unjustified name). Later stump primordia of pleopods 1–5 appear simultaneously. A rostrum forms and the eyestalks grow.

FURCILIA.   The furcilia is found only in the Euphausiacea. The branches of the second antennae are unsegmented, as in the calyptopis, and are locomotory. The compound eyes are stalked. In the course of numerous molts (up to 12) the thoracomeres grow and differentiate appendages, from anterior to posterior, only the 7th and 8th thoracopods remain as stumps. At the same time, pleopods appear. In late stages simple gills as well as bioluminescent organs appear (Fig. 2-21). The eyestalks grow extending laterally from below the carapace. This stage changes gradually into the cyrtopia.

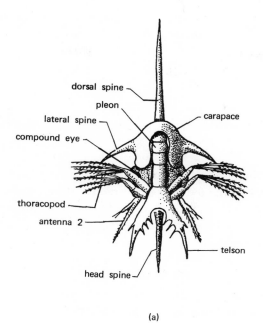

(a)

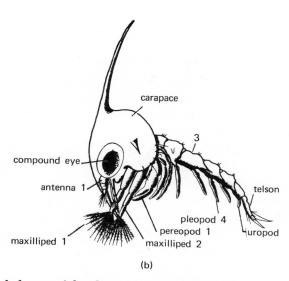

(b)

Fig. 2–20. Larval forms of brachyuran decapod *Atelecyclus septemdentatus.* (After Lebour.) (a) Zoea, just hatched, viewed from behind; 1.6 mm long. (b) Metazoea, last (5th postembryonal) stage, which will molt later to a decapodid; 4 mm long. The maxillipeds are used for locomotion; pereopods 1 already have a chela; the pleomeres are more advanced than the thoracomeres.

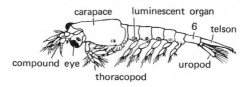

Fig. 2–21. Furcilia of *Nyctiphanes couchi* (Euphausiacea), last stage; 4.3 mm long. (After Lebour from Zimmer.)

CYRTOPIA.   The cyrtopia stage of Euphausiacea has the inner branch of the second antennae segmented, the outer one differentiated to a scale. The antennae have lost all locomotory function. During the course of numerous molts the final shape of the gills, appendages, and telson is gradually attained and, with them, the form of the adult Euphausiacea.

Postlarval Period

Young animals possessing the full complement of metameres and appendages have all the characteristics of their order. This stage then may best be named by adding the syllable -id to the subclass or order name (e.g., copepodid, decapodid). If the adult is specialized and deviates from the usual structure of the group, as in the Brachyura and Anomura, the first postlarval stage does not resemble the adult, and usually has a special name. This stage transforms into the adult in a single molt.

Habits of Larvae

NAUPLII AND METANAUPLII: LOCOMOTION AND FEEDING.   The nauplii and metanauplii, with few exceptions (e.g., harpacticoids), are probably permanent swimmers of the plankton and have been studied carefully only in some branchiopods, copepods, and in barnacles.

Except in branchiopods having minute first antennae, all three pairs of appendages are used in locomotion, beating in a metachronal rhythm, one after another, mandibles first. In the ocean or fresh water with the light from above, the swimming direction is always vertical. In the posterior and medioventral paddling beat the appendages are extended. In forward stroke the appendages are bent and their setae appressed, reducing resistance in the water. Although the forward momentum is kept up, a jerky movement results.

Food is collected by the setal fans of the two posterior appendages, or in *Artemia* by the second antennae alone. Small floating material is thrown behind the labrum by backward movement of the appendages. Long basal spines of the second antennae working with ventral mandibular setae near the mouth push the particles under the mucus-secreting labrum (Figs. 2-15, 2-22). The *Artemia* nauplius uses only the second antennae for collecting food; the distal articles of the antennae are combed by the mandibles. The mandibles then push the particles obtained along the median of the venter to the basal antennal spines.

The efficiency of the second antennae and mandibles as a collecting apparatus can be demonstrated by presenting the nauplius with a suspension of carmine particles. A *Balanus* nauplius fills its gut completely with the particles within 15 min. In water containing particles 4–9 μm in diameter, the species examined took only particles 4–6 μm in diameter, diatoms, green algae, and protozoa, swallowing them whole.

The nauplii of *Calanus* and *Diaptomus* usually swim jerkily but can also move slowly at a steady rate by rapid metachronal vibrations of their two posterior pairs of appendages.

On each side of the metanauplius of *Calanus* a vortex appears, the current close to the sides moving posteriorly and carrying floating particles. The mandibles quickly move their basketlike arranged setae diagonally across this current, toward the venter, throwing planktonic organisms toward the sternites close to the mouth. The tips of the setae sweep some distance along the venter and with this movement carry the planktonic animals toward the labrum aided by the feathered

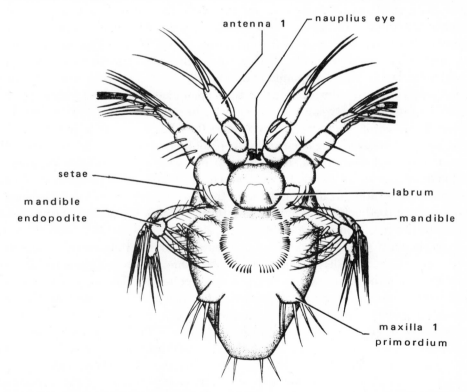

Fig. 2–22. Metanauplius (3rd stage) of *Cyclops strenuus* with 2nd antennae not yet preoral; 0.2 mm long. (After Storch.)

protopodite setae of the mandibles. The particles are swept under the labrum by strong hooked spines at the base of the second antennae, in successive back-stroke movements of this appendage (Fig. 2-22). Unlike the adults, the nauplius does not have its mandibular setae close to each other; their spacing of 15–20 μm does not permit filtering. The second metanauplius stage takes up algae to 20 μm long, but does not take small flagellates 2–4 μm in diameter. The first two naupliar stages of *Calanus,* as of many other crustaceans, subsist on the remaining yolk and do not feed otherwise.

The *Diaptomus* larvae produce similar currents which can be made visible by use of carmine powder. One current arises between the second antennae and the mandibles and moves toward the labrum; the other arises between the mandibles and moves toward a hump behind the labrum on the ventral wall. The floating particles are whipped under the labrum with feathered setae on the basal articles of the two posterior appendages.

ZOEA AND OTHER STAGES: LOCOMOTION AND FEEDING.   The habits of zoea and other larvae have been well studied in only a few species. The zoea stages swim up or down, as do the nauplii, and also have daily vertical migrations. The limbs of most observed zoea larvae of the Natantia and Anomura are strongly concave anteriorly, and they paddle in the opposite direction to those of the nauplius. The zoea thus move with the telson ahead, up or down in the water. The body is extended vertically, its shape providing little resistance to the water. As the zoea larvae have compound eyes, their light orientation is contrary to the usual observations that crustaceans with compound eyes have transverse photo-taxis. A compromise between this movement and posterior orientation is found in the mysis larvae of *Nephrops.* These extend the abdomen vertically toward the light above, but bend the eye-bearing cephalothorax horizontally, so that both orientations are found at the same time. In other zoeae if the light direction is changed there may be momentary transverse phototaxis. If a vertical swimming *Alpheus* zoea is suddenly illuminated from the side, it turns around its longitudinal axis, exposing the back to the light before turning its body so that its longitudinal axis is in the direction of the light.

As is shown in Fig. 2-23, the current produced by the thoracopod exopodites moves anteriorly, parallel to the sides, in the same direction as the respiratory currents within the carapace produced by the second maxillae. Such observations have been made on the zoeae of *Palaemonetes, Leander, Galathea, Munida* and *Pagurus;* however, the zoeae of *Crangon* and *Porcellana,* which swim with their heads forward, are exceptions. Those zoeae larvae of Brachyura that have been observed thus far swim vertically, with the abdomen bent down (Fig. 2-24). With the dorsal spine ahead, they move up or down, but the eye-bearing cephalothorax is kept horizontal, corresponding with the transverse phototaxis of the compound eyes. Infrequently, the posterior part of the body takes part in movement, by a ventrally directed stroke.

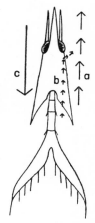

Fig. 2–23. Zoea (first stage) of *Munida bamffica*, swimming. The arrows, a, give direction of beat of the first two thoracopods; b, respiratory current; c, direction of movement. (After Foxon.)

At 16–17°C zoeae of *Galathea* swim 1 m up or down, in 45–56 sec. *Porcellana* with their long extensions take 65–92 sec. How long the animals can maintain these speeds in nature is unknown. But it seems sufficient to provide daily vertical migrations. While some species (*Munida*) have a resting (extended horizontal) position that provides maximum resistance to sinking, brachyurans have to continuously move the thoracic appendages to avoid sinking.

Except for the Brachyura, the zoeae have a characteristic method for turning in water. Either they change the direction of beating the thoracopods, or they produce a strong beat with the pleon and tail fan. In *Galathea, Porcellana,* and *Pagurus* the anterioventral beat serves the purpose; in *Munida,* however, the subsequent extension of the trunk is used for turning.

Little is known about the long carapace spines of the brachyuran zoea, but the following information may be pertinent. The spines are related to negative geotaxis as can be shown with the zoeae of *Cancer* and *Portunus.* Although in red light they are unable to recognize the direction to the surface of the water visually, the larvae nevertheless swim in a straight line. But a larva with the long carapace spines removed swims without direction in spirals, and soon ends up on the bottom, unable to rise again. Only after a white light is turned on above the surface do the operated animals resume their normal swimming behavior.

The spines also reduce the rate of sinking. This is of importance during molting, when the animal cannot paddle. During molting the larva takes a position, as though anesthetized, determined by its center of gravity, the cephalothorax down and the two large spines (Fig. 2-24) horizontal, parallel to the substrate. In this position an anesthetized brachyuran zoea sinks 1 m in 83 sec at 17°C; with the dorsal spine removed it sinks this distance in 66 sec. Anesthetized zoeae of *Porcellana,* with their long appendages, require 350 sec to sink to the same depth.

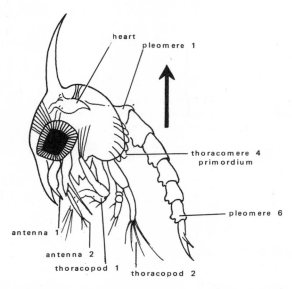

Fig. 2–24.   Body position of a crab zoea of *Maja* swimming up and down; the arrow indicates the direction of this movement; about 1.5 mm long. (Combined after Claus and Foxon.)

The spines provide protection only against small mouthed predators, such as pipefish (Syngnathidae). Plankton feeders such as ctenophores, medusae, and lobster larvae regularly take zoeae.

The spines are of no mechanical importance in maintaining body position or directional movement. Cutting the spines off alters the body position in no way, as the appendages are not influenced. That the appendages are critical can be seen in an anesthetized zoea. Its body turns 90° toward the center of gravity, so that the lateral spines are horizontal and the abdomen up.

When the thoracopods determine locomotion (zoea, mysis), many decapod larvae move telson first through the water. But if the thoracic exopodites become reduced and the pleopods are used in locomotion, the body will move in the opposite direction, anterior first. This happens in the postlarval stage when the exopodites become reduced, in *Penaeus* from the 4th mysis stage and in Brachyura in the megalopa stage. In the Euphausiacea the pleopods function just before or at the cyrtopia stage.

Feeding has been observed only in the genus *Palaemonetes;* taking zooplankton (e.g., nauplii) with their maxillipeds, they chew them apart with maxillae and mandibles. They do not filter-feed and therefore cannot be fed cultures of unicellular algae.

Molting

The different larval stages are separated by molts, during which changes of internal and external structure take place. But internal changes may be surprisingly small even with a large external change, as for instance in transformation of a decapod larva into a young adult. The physiology of molting has been studied only in half-grown or adult decapods, not in larval stages. The course of the molt includes several stages, described in detail for *Cancer pagurus* and *Leander*.

The molt commences with proecdysis, during which the animal stops feeding and its integument thins and softens. Simultaneously a new epicuticle with a thin cuticular layer forms on the hypodermis under the old integument. Feeding ceases and metabolism depends upon nutrient reserves (lipids, fatty acids, glycerol, and glycogen) periodically stored in the digestive gland. Ecdysis takes place after the body has swollen by taking in water. The shed cuticle includes the cuticularized part of fore- and hindgut. The hardening of the new exoskeleton, metecdysis, by mineralization with calcareous materials and the formation of an endocuticle layer follows. During the periods between molts the animal feeds again and its internal organs function. The intermolt periods may be either diecdysis, lasting only several weeks in animals that molt several times each year, or anecdysis in animals that molt once a year. A long molt-free period that terminates in the natural death of the animal is called terminal anecdysis.

PROECDYSIS. Preparation for breaking the old exoskeleton begins with the secretion of a chitinase from the digestive gland. This enzyme dissolves the chitinous portions of the old cuticle from inside on the surface of the hypodermis. At the same time the mineralizations of the exoskeleton are dissolved and resorbed by the body surface, raising the blood calcium level. In *Cambarus* the weight of the old exoskeleton is reduced to one-fourth its previous weight, the calcium compounds to one-third the normal amount. While the old material dissolves, the hypodermis below secretes a new cuticular layer including new spines, setae, etc. Between the new cuticle and the old lie the mucilaginous remains of the liquified endocuticle. The lubricating effect of this liquid later facilitates the slipping off of the old exuviae. The new cuticle, wider and longer than the old, is corrugated and thus can expand after shedding the old skin.

ECDYSIS. The swelling of the body, now of great importance at this time for tearing the cuticle, results from drinking seawater, which reaches the blood through foregut and digestive glands. *Cancer* and *Carcinus* increase their weight by 66% in a few hours, *Maja squinado* doubles its weight (from 100 to 207 g). The blood volume of *Cancer* increases 9 times, of *Maja* 10 times (from 8.7 to 82 cm³). One-third of the water swallowed by *Carcinus* is taken up by the cells.

The cuticle tears along distinct sutures that run diagonally through the joint membrane of the carapace and the first abdominal tergite in crayfish and lobsters. The details are described later in Chapter 13 on decapods. The preformed sutures of Brachyura start on the above mentioned joint membrane and follow the sides of the carapace anteriorly. The carapace opens up as if hinged along the anterior edge. The crab then arches dorsally and pulls the abdomen out of the old exoskeleton.

METECDYSIS. An adult Chinese mitten crab, as it emerges from its exuvia, has its final body size, while a crayfish continues to grow. The freshly emerged crab is completely soft, but soon thickening and mineralization of the new exoskeleton start. The mineral material, consisting of calcium and glycogen, comes from the digestive gland where it is periodically stored. (Chitin is made up of amino sugars.) In *Carcinus* during molting the Ca level of the blood increases from 126 to 164% that of the Ca contents of sea water. The calcium reserves in the digestive gland, of both *Astacus* and *Homarus,* are very small. The sclerotized gastroliths under the intima of the cardiac stomach also cannot provide sufficient minerals for hardening, but must be supplemented by Ca intake through the gills from the surrounding water. This can be demonstrated in the following experiment, which has also been made with brachyurans. If a freshly molted *Orconectes* is kept in running hard water without food, the exoskeleton hardens in less than a month. If the animal is kept in soft standing water, the exoskeleton remains paperlike and elastic. If the *Orconectes* in soft standing water is fed calcium-rich food, (e.g., *Chara*), the exoskeleton nevertheless remains soft. The Ca is taken up from the water, as radioactive tracers have shown in *Metapenaeus,* 90% of it through the gill epithelium. In the digestive gland of a 100 g crab, *Maja,* only 0.2 g Ca is present; in the exoskeleton there is 40 times as much, or 9 g. (There are no gastroliths in *Maja.*)

During molting the gastroliths of the lobster fall into the lumen of the gut, are dissolved and resorbed; they contain, however, only 1/130 of the calcium necessary for hardening the exoskeleton.

DIECDYSIS AND ANECDYSIS. The swelling produced by the intake of water persists but the water is replaced by tissues, either by enlargement or by division of individual cells. While the water content decreases, the protein content of the animal increases.

CONTROL OF MOLTING CYCLE. Although many neurohormones and environmental factors act on the molt, many experiments with Astacura and Brachyura have clearly demonstrated that there are two antagonistic hormones which chiefly control the molt. No crustaceans other than developing decapods have been studied.

The molt inhibiting neurohormone is secreted by the medulla X-organ, axons of which take the secretions to the sinus gland, a storage organ. The hormone acts on the molting glands. If, surgically, one breaks the axon connection between the medulla X-organ and sinus gland, the molt inhibition remains. The neurohormones, then, must come from the X-organ alone without any contribution from the sinus gland. The same neurohormone also inhibits Ca storage in the gastroliths and secretions of Ca from the digestive gland via the gut into the hemolymph by influencing the Y-organ (carapace organ). The molting hormone in Reptantia and Brachyura is produced by the Y-organ.

The following experiments are a few examples of the many experiments made which indicate the importance of these hormones. Proof of the molt inhibiting function of the medulla X-organ and sinus gland has been obtained in the crab *Carcinus maenas,* which after reaching adult size does not molt again. If both eyestalks are removed or if only the medulla X-organ and sinus gland are removed from an adult crab, the animal immediately enters the initial stage of an additional molt: the cuticle changes, the blood calcium level goes up, and finally the molt occurs. During the course of the year, three additional molts may follow, so that a giant crab, 132 mm in carapace width (normal width is 86 mm) is produced. Injecting sinus gland extracts will stop the molts immediately, proof that the sinus gland governs the cessation of molting. By injecting sinus gland extracts into young prawns, *Leander serratus,* it can be shown that the extract from adult *Carcinus* is 2.6 times as strong as that from young crabs; the procedure indicates that large amounts of hormone are present during terminal anecdysis.

The action of the molting hormones of the Y-organ in *Carcinus maenas* can be demonstrated by extirpation of the Y-organ. After its extirpation in young or medium-aged crabs, further molts are stopped for a period of a year. If one injects an extract of the Y-organ of young crabs repeatedly into old crabs in terminal anecdysis, the animal will molt within a month. The Y-organs of old crabs are relatively smaller than those of young animals and their extracts are less active if injected into a test prawn (*Leander serratus*). If both the eyestalks with molt inhibiting glands and the Y-organ are extirpated no molt takes place; the presence of the Y-organ hormone is necessary for molting. A simple injection of Y-organ extract in *Carcinus maenas* caused a threefold increase of the blood calcium level.

Initiation of molt, by arrest of the molt inhibiting neurohormone, is caused in *Carcinus maenas* by a certain increment of growth. In a well-fed animal, the molts follow each other faster than if the animal is starved. That this will influence the hormone is indicated by the fact that *Carcinus maenas* with its eyestalks removed will molt in a constant rhythm regardless of feeding or growth.

Other environmental determinants of molt are the following: Low temperatures delay the next molt by many months. North Sea *Carcinus maenas* molt only at water temperatures over 8°C; subtropical and tropical crabs have higher minimum temperatures. *Pachygrapsus* molts only at temperatures higher than 14°C, *Gecarcinus* at temperatures above 17°C. Continuous light or continuous darkness delay the molt in the few crustaceans examined.

Several internal determinants may influence the molt. Regenerating legs initiate molt early in *Carcinus maenas;* probably this is an adaptation, as the regenerated limbs become functional only after a molt.

The presence of larger members of its species, recognized in the dark by touch or in light visually (even through glass panes) delays molting in *Carcinus maenas*. This may be an adaptation preventing attack on vulnerable soft crabs by larger members of the species.

Most insects and many arachnids molt only in larval stages, rarely as sexually mature animals, this is very different in many crustaceans. In many crustaceans the molt to maturity is also the last one (e.g., in many parasites, in *Maja squinado*, and probably in all oxyrhynch Brachyura). In *Maja* it can be shown that cessation of molting is related to the pronounced reduction of the Y-organ, the weight of which sinks to 1/18 of the previous weight; the injection of extracts from the reduced Y-organ appears to have no effect on test crustaceans.

Crustaceans that continue to molt as adults belong to two groups. In one, there is an unlimited number of molts which is only stopped by the death of the animal. The resultant adults vary in size and giant individuals are found. Examples are lobsters, *Cancer pagurus*, many or all Natantia, probably also some Cirripedia (*Lepas*) and, to a lesser extent, *Astacus*.

In others, the number of molts of mature individuals is limited and between the last molt and death the animal spends several months in terminal anecdysis, never growing beyond a certain size. Thus in *Carcinus maenas*, unlike *Maja*, further molting is inhibited by very strong secretions of the molt inhibiting neurohormone. This is probably the case for all Portunidae, perhaps for all brachyrhynch Brachyura. To this group also belong many Peracarida and Cladocera although we know nothing about their hormones.

Crustaceans may get very old. European crayfish mature in 3–4 years, but may reach 20 years of age. Mature American lobsters have a body length of 25 cm at 6 years of age, and may have a body length of more than 60 cm after 25 years.

Habits

Crustacea, like many other marine invertebrates, have modified and adapted their anatomy to equip them for the various habitats in the ocean. There are filter-feeding, sand cleaning, mud-, plant-, carrion feeding, predatory, as well as parasitic Crustacea. There are permanent swimmers, slow moving bottom-dwellers, interstitial inhabitants, semisessile burrowers, and permanently sessile crustaceans.

Relationships and Classification

The crustaceans appear more primitive than other mandibulates. Primitive features, the presence of the second antennae, and the mandibular palp, show up in their development. Paleontology supplies little information on the phylogeny of the eight subclasses which have Recent representatives. The crustaceans with a body consisting of a large number of similar somites with similar appendages appear most primitive. But as Branchiopoda indicate, shortening of the body may have evolved several times within a subclass and one suspects for instance that the reduction of somite numbers of the ostracods has evolved secondarily after the group separated from other crustaceans.

The subclasses can be arranged in five groups, each of which presumably had a common ancestor: Cephalocarida, Branchiopoda, the Maxillopoda, which include the Mystacocarida, Copepoda, Branchiura, and Cirripedia, the Ostracoda, and the Malacostraca. Except for the Ostracoda these groups all have some representatives that have the trunk well segmented and some with many metameres that may be considered an ancestral character. Especially primitive are the head segmentation, trunk musculature, and appendages of the Cephalocarida (Fig. 3–4). In these characters they may actually resemble the ancestors of the Crustacea, but their lack of eyes and carapace may be secondary. Also from the turgor-extended, biramous appendages of the Cephalocarida, the appendages of the Branchiopoda, Maxillopoda and Malacostraca may be derived. But absolute proof for relationships has not been found.

The non-malacostracans used to be combined as Entomostraca. While this seems very practical, it is obvious that the Entomostraca are not a monophyletic group; their representatives belong to diverging evolutionary lines.

The relationship of crustaceans and trilobites is still unexplained; no known crustacean has a leg branch near the trunk articulation like the preepipodite of trilobites (except the arthrobranchs and pleurobranchs of decapods, which are obviously secondary). Also the trilobite preepipodite is more differentiated than the epipodites of primitive crustaceans.

The classification adopted here is the following:

Class Crustacea
    Subclass Cephalocarida
    Subclass Branchiopoda
        Order Anostraca, fairy shrimps
        Order Notostraca, tadpole shrimps
        Order Conchostraca, clam shrimps
        Order Cladocera, water fleas
    Subclass Ostracoda, seed shrimps
        Order Myodocopa
        Order Cladocopa
        Order Podocopa
        Order Platycopa
    Subclass Mystacocarida
    Subclass Copepoda, copepods
        Order Calanoida
        Order Harpacticoida
        Order Cyclopoida
    Subclass Branchiura, fish lice

Subclass Cirripedia, barnacles
    Order Ascothoracica
    Order Thoracica
    Order Acrothoracica
    Order Rhizocephala
Subclass Malacostraca
    Superorder Phyllocarida
    Superorder Hoplocarida
        Order Stomatopoda
    Superorder Syncarida
        Order Anaspidacea
        Order Stygocaridacea
        Order Bathynellacea
    Superorder Eucarida
        Order Euphausiacea
        Order Decapoda
    Superorder Pancarida
        Order Thermosbaenacea
    Superorder Peracarida
        Order Mysidacea
        Order Cumacea
        Order Spelaeogriphacea
        Order Tanaidacea
        Order Isopoda
        Order Amphipoda

References

Adelung, D. and Bückmann, D. 1965. Innersekretorische Organe und Häutungsrhythmus von *Carcinus maenas*. *Verhandl. Deutschen Zool. Ges.* 28: 131–136.

Aiken, D.E. 1969. Photoperiod, endocrinology, and the crustacean molt cycle. *Science* 164: 149–155.

Aoto, T. and Nishida, H. 1956. Effect of removal of eyestalks on growth and maturation of oocytes in a hermaphroditic prawn, *Pandalus kessleri*. *J. Fac. Sci. Hokkaido Univ.* 12: 412–424.

Arvy, L. *et al.* 1956. Organe Y et gonade chez *Carcinides*. *Ann. Sci. Natur. Zool.* (11)18: 263–267.

Baldass, F. von. 1941. Entwicklung von *Daphnia pulex*. *Zool. Jahrb. Abt. Anat.* 67: 1–60.

Balesdent-Marguet, M.–L. 1960. Disposition, structure et mode d'action de la glande androgène d'*Asellus*. *Compt. Rend. Acad. Sci. Paris* 251(5): 803–805.

———— 1963. Identification et rôle d'une glande annexe de l'appareil génital femelle chez les Crustacé Isopode *Asellus*. *Ibid.* 257: 4053–4056.

Bauchau, A.G. 1949. Intensité du metabolisme et glande sinusaire chez *Eriocheir*. *Ann. Soc. Zool. Belgique* 79: 73–86.

Benesch, R. 1969. Ontogenie und Morphologie von *Artemia salina*. *Zool. Jahrb. Abt. Anat.* 86:(in press).

Berreur-Bonnenfant, J. and Charniaux-Cotton, H. 1966. Hermaphroditism protéandrique et fonctionnement de la zone germinative chez *Pandalus borealis*. *Bull. Soc. Zool. France* 90: 243–259.

Bliss, D.E. *et al.* 1954. Some decapod crustacean neurosecretory systems. *Pubbl. Staz. Zool. Napoli* 24 (*Suppl.*): 68–69.

Broad, A.C. 1957. Larval development of *Palaeomonetes*. *Biol. Bull.* 112 144–161.

Burgers, A.C.J. 1959. The actions of hormones and light on the erythromelanosomes of the swimming crab, *Macropipus*. *Pubbl. Staz. Zool. Napoli* 31: 139–145.

Cannon, H.G. 1926. On the postembryonic development of the fairy shrimp, *Chirocephalus*. *J. Linnean Soc. London Zool.* 36: 401–416.

Carlisle, D.B. 1955. Endocrinology of moulting in *Ligia*. *J. Marine Biol. Ass.* 35: 515–520.

———— 1955. Hormonal control of water balance in *Carcinus*. *Pubbl. Staz. Zool. Napoli* 27: 227–231.

———— 1957. Hormonal inhibition of moulting in decapod Crustacea. II. The terminal anecdysis in crabs. *J. Marine Biol. Ass.* 36: 291–307.

———— 1959. Moulting hormones in *Palaemon* and sexual biology of *Pandalus*. *Ibid.* 38: 351–359, 381–394, 481–506.

———— and Butler, C.G. 1956. The "queen substance" of honeybees and the ovary-inhibiting hormone of crustaceans. *Nature* 177: 276–277.

———— and Knowles, F.G.W. 1959. Endocrine control in crustaceans. *Cambridge Monogr. Exp. Biol.* 10: 1–120.

Charniaux-Cotton, H. 1956. Organe analogue à la "glande androgène" chez un pagure et un crabe. *Compt. Rend. Acad. Sci. Paris* 243: 1168–1169.

———— 1958. La glande androgène de quelques Crustacés Décapodes. *Ibid.* 6: 2814–2817.

———— 1959. Développement post-embryonnaire de l'appareil génital et de la glande androgène chez *Orchestia*. *Bull. Soc. Zool. France* 84: 105–115.

———— 1963. La sécrétion d'hormone femelle par le testicule inversé en ovaire de *Talitrus*. *Compt. Rend. Acad. Sci. Paris* 256: 4088–4091.

Dahl, E. 1957. Embryology of X organs in *Crangon*. *Nature* 179: 482.

Dall, W. 1965. On the physiology of freshwater shrimp. *Australian J. Marine Freshwater Res.* 16: 1–23.

Demeusy, N. 1964. La croissance somatique et la vittellogenèse du crabe *Carcinus*. *Compt. Rend. Acad. Sci. Paris* 260: 323–326; 2925–2928.

———— and Veillet, A. 1958. L'ablation des pédoncules oculaires sur la glande androgène de *Carcinus*. *Ibid.* 246: 1104–1107.

Durand, J.B. 1960. Limb regeneration and endocrine activity in crayfish. *Biol. Bull.* 118: 250–261.

Fingerman, M. 1956. Phase difference in the tidal rhythms of color change of fiddler crabs. *Ibid.* 110: 274–290.

———— 1956. Physiology of the black and red chromatophores of *Callinectes*. *J. Exp. Zool.* 133: 87–105.

———— 1957. Endocrine control of the red and white chromatophores of the dwarf crawfish, *Cambarellus*. *Tulane Stud. Zool.* 5(6): 140–148.

———— 1958. The chromatophore system of the crawfish *Orconectes*. *Amer. Midland Natur.* 60: 71–83.

———— 1966. Neurosecretory control of pigmentary effectors in Crustaceans. *Amer. Zool.* 6: 169–179.

———— and Aoto, T. 1958. Chromatophorotropins in the crayfish *Orconectes* and their relationship to long-term background adaptation. *Physiol. Zool.* 31: 193–208.

———— and Lowe, M.E. 1957. Influence of time on background upon the chromatophore systems of two crustaceans. *Ibid.* 30: 216–231.

————— and Yamamoto, Y. 1965. Neuroendocrine control of the crustacean hepatopancreas. *Biol. Bull.* 129: 389–390.

————— et al. 1959. The red chromatophore system of the prawn *Palaemonetes*. *Physiol. Zool.* 32: 128–129.

Foxon, G.E.H. 1934. Swimming methods and habits of certain crustacean larvae. *J. Marine Biol. Ass.* 19: 829–849.

Fuchs, K. 1914. Die Keimblätterentwicklung von *Cyclops*. *Zool. Jahrb. Abt. Anat.* 38: 103–156.

Gabe, M. 1956. Histologie comparée de la glande de mue (organe Y) des Crustacés Malacostracés. *Ann. Sci. Natur. Zool.* (11)18: 145–152.

Gauld, D.T. 1959. Swimming and feeding in crustacean larvae: the nauplius larva. *Proc. Zool. Soc. London* 132: 31–50.

Gersch, M. 1964. *Vergleichende Endokrinologie der wirbellosen Tiere*, Geest & Portig, Leipzig.

Green, J. 1965. Chemical embryology of the Crustacea. *Biol. Rev.* 40: 580–600.

Gurney, R. 1942. *Larvae of Decapod Crustacea*. Ray Soc., London.

Haffner, K. von. 1934. Der Brutkreislauf von *Phronima sedentaria*. *Z. Wiss. Zool.* 146: 283–328.

Heath, H. 1924. The external development of certain phyllopods. *J. Morphol.* 38: 453–483.

Hickmann, V.V. 1937. The embryology of the Syncarid crustacean, *Anaspides tasmaniae*. *Papers Proc. Roy. Soc. Tasmania* 1936: 1–36.

Hodge, M.H. and Chapman, G.B. 1958. The fine structure of the sinus gland of a land crab, *Gecarcinus*. *J. Biophys. Biochem. Cytol.* 4: 571–574.

Hubschman, J.H. 1963. Development and function of neurosecretory sites in the eyestalks of larval *Palaemonetes*. *Biol. Bull.* 125: 96–113.

Humbert, C. 1965. Rôle de l'organe X dans les changements de couleur et la mue de *Palaemon serratus*. *Trav. Inst. Scient. Chérif.* (*Zool.*) 32: 1–86.

Jacobs, M. 1925. Entwicklungsphysiologische Untersuchungen am Copepodenei, *Cyclops*. *Z. Wiss. Zool.* 124: 487–541.

Juchault, P. 1967. La différentation sexuelle mâle chez les isopodes. *Ann. Biol. Paris* (4)6: 191–212.

————— and Legrand, J.J. 1965. L'intervention des neurohormones dans le changement de sexe d'*Anilocra*. *Compt. Rend. Acad. Sci. Paris* 260: 1491–1494; 1783–1786.

————— et al. 1965. Inhibition protocérébrale de la glande androgène. *Ibid.* 261: 1116–1118.

Katakura, U. 1960. Transformation of ovary into testis following implantation of androgenous glands in *Armadillidium*. *Annot. Zool. Japan* 33(4): 241–244.

Kandewitz, F. 1950. Entwicklungsphysiologie von *Daphnia pulex*. *Roux Arch. Entwicklungsmech.* 144: 410–447.

Keller, R. 1965. Hormonale Kontrolle des Polysaccharidstoffwechsels bei *Cambarus*. *Z. Vergl. Physiol.* 51: 49–59.

Kleinholz, L.H. 1966. Purification of crustacean eyestalk hormones. *Amer. Zool.* 6: 161–167.

————— et al. 1962. Neurosecretion and crustacean retinal pigment hormone: distribution of the light-adapting hormone. *Biol. Bull.* 122: 73–85.

Knowles, F.G.W. 1953. Neurosecretory pathways in the prawn *Leander serratus*. *Nature* 171 (4342): 131–132.

————— 1954. Neurosecretion in the tritocerebral complex of crustaceans. *Pubbl. Staz. Napoli Zool.* 24 (*Suppl.*): 74–78.

————— and Carlisle, D.B. 1956. Endocrine control in the Crustacea. *Biol. Rev.* 31: 396–473.

Kühn, A. 1913. Die Sonderung der Keimesbezirke in der Entwicklung der Sommereier von *Polyphemus*. *Zool. Jahrb. Abt. Anat.* 35: 243–340.

Kühnemund, E. 1929. Die Entwicklung der Scheitel-platte von *Polyphemus pediculus* von der Gastrula bis zur Differenzierung. *Zool. Jahrb. Abt. Anat.* 50: 385–432.

Legrand, J.-J. 1955. Rôle endocrinien de l'ovaire dans la differenciation des oostégites chez les crustacés isopodes terrestres. *Compt. Rend. Acad. Sci. Paris* 241: 1083–1085.

————— 1959. Modifications du cycle sexuel chez les femelles des Oniscoides superieurs ayant reçu une implantat testiculaire. *Ibid.* 248: 1043–1046, 1240–1243.

————— and Juchault, P. 1963. Un type d'intersexualité chez l'Oniscoide *Armadillidium*. *Ibid.* 256: 1606–1660, 2931–2933.

Lehmann, R.L. and Scheer, B.T. 1956. Effect of eyestalk removal on phosphate uptake by *Hemigrapsus nudus*. *Physiol. Comp. Oecol., den Haag* 4: 164–171.

Lochhead, J.H. 1936. Feeding mechanism of the nauplius of *Balanus*. *J. Linnean Soc. London* 39: 429–442.

Manton, S.M. 1928. Embryology of a Mysid crustacean *Hemimysis*. *Philos. Trans. Roy. Soc. London* (B) 216: 363–463.

————— 1934. Embryology of the crustacean *Nebalia bipes*. *Ibid.* (B) 223: 163–238.

Maynard, D.M. 1961. Thoracic neurosecretory structures in Brachyura. *Biol. Bull.* 121: 316–329.

Miyawaki, M. 1956. Histological observations on the incretory elements in the eyestalk of a Brachyura, *Telemessus*. *J. Fac. Sci. Hokkaido Imp. Univ.* 12: 325–332, 516–520.

Nair, S.G. 1956. Embryology of the isopod *Irona*. *J. Embryol. Exp. Morphol.* 4: 1–33.

Patané, L. and DeLuca, V. 1957. Sul determinismo di alcuni caratteri sessuali secondari nell' oniscoide *Porcellio*. *Bull. Zool.* 24: 259–264.

Pearson, J.C. 1939. The early life histories of some American Penaeidae. *U.S. Dept. Commerce Bull. Bureau Fish.* 49(30): 1–73.

Rangneker, P.V. *et al.* 1961. A hypoglycemic factor in the eyestalks of freshwater crabs *Paratelphusa*. *J. Anim. Morphol. Physiol.* 8(2): 137–144.

Sanders, H. L. 1957. The Cephalocarida and crustacean phylogeny. *Syst. Zool.* 6: 112–128.

Schwartzkopff, J. 1953. Pulsfrequenzen von Garneelen. *Naturwissenschaften* 40(23): 609.

————— 1955. Die Grössenabhängigkeit der Herzfrequenz von Krebsen im Vergleich zu anderen Tiergruppen. *Experientia* 11: 323–325.

————— 1955. Vergleichende Untersuchungen der Herzfrequenz bei Krebsen. *Biol. Zentralbl.* 74: 480–497.

Siewing, R. 1969. *Lehrbuch der vergleichenden Entwicklungsgeschichte der Tiere.* P. Parey, Hamburg.

Smith, R.I. 1948. Sinus glands in retinal pigment migration in grapsoid crabs. *Biol. Bull.* 95: 169–185.

Snodgrass, R.E. 1956. Crustacean metamorphosis. *Smithsonian Misc. Coll.* 131(10): 1–78.

Stephens, G.J. 1952. Mechanisms regulating the reproductive cycle in the crayfish, *Cambarus*. *Physiol. Zool.* 25: 70–84.

Taube, E. 1909, 1915. Beiträge zur Entwicklungsgeschichte der Euphausiden. *Z. Wiss. Zool.* 92: 427–464; 114: 577–656.

Tchernigoutzeff, C. 1965. Multiplication cellulaire et régénération aux cours du cycle intermue des Crustacés Décapodes. *Arch. Zool. Exp. Paris* 106: 377–497.

Terao, A. 1929. The embryonic development of the spiny lobster *Panulirus*. *Japan. J. Zool.* 2: 387–449.

Teyan, B.S. *et al.* 1959. Oxygen consumption in *Uca* before and after eyestalk removal. *Biol. Bull.* 117: 429.

Travis, D.F. 1954. The molting cycle of the spiny lobster, *Panulirus argus*. I. Molting and growth in laboratory maintained individuals. *Ibid.* 107: 433–450.

————— 1955. II. Pre-ecdysial histological and histochemical changes in the hepatopancreas and integumental tissues. *Ibid.* 108: 88–112.

————— 1955. III. Physiological changes in the blood and urine during the normal molting cycle. *Ibid.* 109: 484–503.

_____ 1957. IV. Post-ecdysial histological and histochemical changes in the hepatopancreas and integumental tissues. *Ibid.* 113: 451–479.

Trilles, J.-P. 1964. L'évolution régressive de l'appendix masculina chez les Isopodes Cymothoidea. *Compt. Rend. Acad. Sci. Paris* 258: 5739, 5989, 6248, 6545.

Waterman, T.H. 1960–1961. *The Physiology of Crustacea,* I, II. Academic Press, New York.

Webb, H.M. *et al.* 1954. A persistent diurnal rhythm of chromatophoric response in eyestalkless *Uca. Biol. Bull.* 106: 371–377.

Weygoldt, P. 1958. Die Embryonalentwicklung des Amphipoden *Gammarus pulex. Zool. Jahrb. Abt. Anat.* 77: 51–110.

Wotzel, F. 1937. Zur Entwicklung des Sommereies von *Daphnia. Ibid.* 63: 455–470.

Yamamoto, Y. 1961. Relation of formation of gastrolith to ovarian development and spawning in the crayfish, *Procambarus. Annot. Zool. Japan* 34: 38–42, 117–127.

Young, J. H. 1956. Anatomy of the eyestalk of the White Shrimp *Penaeus. Tulane Stud. Zool.* 3: 171–190.

Zehnder, H. 1934. Die Embryonalentwicklung des Flusskrebses. *Acta. Zool.* 15: 261–408.

Zimmer, C. 1927. Crustacea: Algemeine Einleitung in die Naturgeschichte der Crustacea *in* Kükenthal, W. and Krumbach, T. *Handbuch der Zoologie,* de Gruyter, Berlin, 3(1): 277–304.

3.

Subclass Cephalocarida

Only four species of Cephalocarida are known, the largest of which is up to 3.6 mm long.

The trunk of cephalocaridans is composed of head, thorax, and abdomen (Fig. 3-3). Limbs are present on all anterior but not the abdominal somites. The telson bears a caudal furca. There are neither eyes nor carapace. The thoracic append-ages are similar to each other and to the second maxilla in both external appearance and musculature. They are turgor appendages with the proto-, and exo-, and pseudepipodites flattened and leaflike; the many-articled endopodites, however, are tubular.

The subclass was discovered in 1955. Its members are common where they occur, and are remarkable for their primitive appearance.

Anatomy

The fairly large head and each of the eight thoracomeres bear a pair of lateral folds of the tergites (epimeres) (Fig. 3-1). The abdomen consists of 11 ringlike somites and a telson with a furca, each branch of which has a seta sometimes half as long as the animal. Both pairs of antennae are multiarticulate, the second antennae are biramous. The mandibles lack a palpus. All other appendages are similar, except for the first maxillae which have especially long basal endites and lack pseudepipodites. The eighth thoracic limbs have lost their endopodites. There are also limb rudiments on the first pleomere, used for carrying eggs. The second maxillae and the following seven pairs of thoracic limbs have the same general plan. They are turgor appendages whose anteroposteriorly flattened protopodite bears a leaflike exopodite and pseudepipodite laterally, a many-articled tubular endopodite distally, and a row of setose endites medially (Fig. 3-2). The exopodite consists of two articles.

The labrum is ventral and extends posteriorly to cover the mandibular endites and the posteriorly directed mouth (Figs. 3-2, 3-3). A pair of cephalic digestive glands extends posteriorly to open into the anterior end of the midgut. An antennal

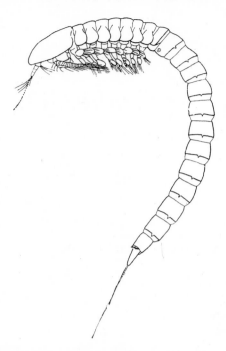

Fig. 3–1. *Lightiella incisa*, lateral view; 2 mm long, not including furca. (After Gooding.)

gland is active in the larva. It persists in the adult, but here the maxillary gland is the dominant excretory organ.

The Cephalocarida are hermaphroditic. Each individual bears paired testes and ovaries. The common gonoduct exits on the sixth thoracic limb. The egg sacs, carried on the first abdominal somite, each contains only one embryo.

Development

The first free-living stage is a metanauplius with the usual complement of uniramous first antennae, biramous second antennae, mandibles, and with rudimentary maxillae. The second antennae are armed with the typical basal endite called the naupliar process. All stages are benthic. In the naupliar stages the animal uses the exopodites of the second antennae and mandible to bring detritus to the mouth, and endites of the second antennae and mandibles push the particles under the labrum.

Hutchinsoniella passes through 13 metanaupliar stages. By stage 11 all somites are present. The molt to the first juvenile stage (stage 14) results from loss of the naupliar process and mandibular palp, and reorientation of the first maxillae. This change marks the beginning of the adult mode of feeding. In juvenile stages 14 through 18, thoracopod pairs 5 to 8 appear.

Relationships

The serial similarity of the somites, appendages and internal anatomy marks the Cephalocarida as a truly primitive group. Not only do all thoracopods resemble each other and resemble the second maxillae, but paired dorsal and ventral longitudinal muscles extend through the trunk (Fig. 3-4) in as regular a pattern as in the Notostraca. The gradual pattern of development is also very primitive.

The trunk musculature of all other subclasses can be derived from the simple cephalocaridan type. Even the complex abdominal muscles of the Malacostraca can be derived from that of the Cephalocarida, with the Hoplocarida and Leptostraca being at intermediate levels. The dorsoventral muscles of the thoracomeres are quite regular, resembling those of the Notostraca and Leptostraca. The diverse limb morphologies of many crustacean subclasses can easily be derived from the generalized cephalocaridan type.

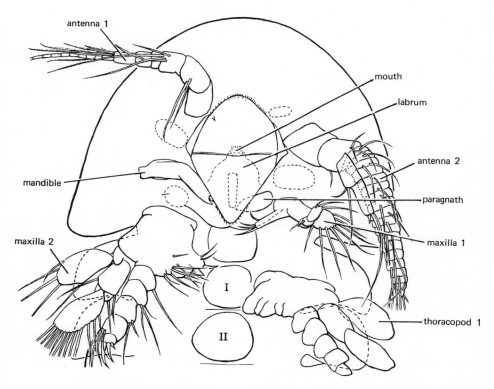

Fig. 3–2.   Ventral view of head and first two thoracomeres of *Lightiella incisa*, somewhat diagrammatic; 0.4 mm wide. The appendages are shown alternately on one or the other side, the opposite ones shown only by the outline of the area to which they are attached. The mandible, as in many other particle feeders, lacks a palp; the mouth is seen below the labrum. (After Gooding.)

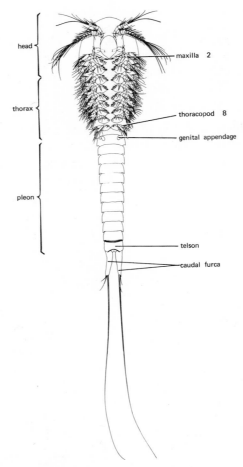

head

maxilla   2

thorax

thoracopod   8

genital appendage

pleon

telson

caudal furca

Fig. 3–3. *Hutchinsoniella macracantha* in ventral view; 3 mm long not including furca. (From Sanders.)

In general, the Cephalocarida are most similar to the most primitive members of other subclasses, such as the Notostraca and the Devonian *Lepidocaris* within the lipostracan Branchiopoda, and the Leptostraca within the Malacostraca. The cephalocaridan method of feeding is such that the primitive feeding types in both the branchiopods and malacostracans can be derived from it. Thus do the cephalocarids form a key link, in showing how very diverse crustacean groups could have evolved from a common ancestor.

Habits

Cephalocarida are bottom-dwellers, found on both coasts of North America, in the West Indies, and Japan. *Lightiella* was found from the intertidal to 2.5 m deep, and *Hutchinsoniella* from 1 to 300 m. *Hutchinsoniella* lives on soft sediments in densities as high as 177 individuals/m².

LOCOMOTION.   The limbs of *Hutchinsoniella* beat in metachronal waves, starting with the last. The locomotory backstroke is 6 to 8 times as fast as the forward movement. Maxillae and antennae take part in the movements. Unlike *Artemia,* cephalocaridans move venter downward.

FEEDING.   Feeding is tied to locomotory movements that create water currents. The exopodites of the second antennae and the claws of the thoracic endopodites whirl up detritus. This detritus is sucked up into the midventral space between the paired rows of thoracic appendages. As in the Anostraca, two pairs of limbs following each other form an interlimb suction chamber. The

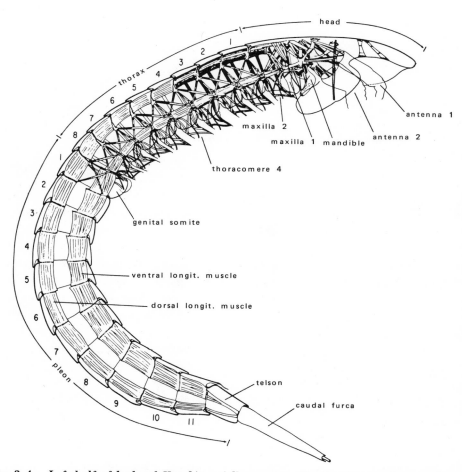

Fig. 3–4.   Left half of body of *Hutchinsoniella macracantha* showing trunk musculature and proximal portion of extrinsic limb musculature; about 3 mm long. (From Hessler.)

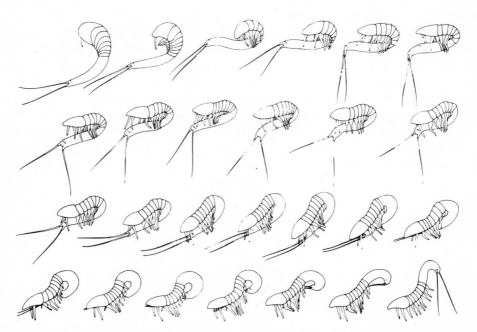

Fig. 3–5.   Grooming movements of *Hutchinsoniella*; about 3 mm long without furca. (From Sanders.)

chamber is then closed on the sides by the posterior bending of the pseudepipodites and exopodites (Fig. 3-1). As those two branches move laterally, slits open
and water is expelled when the backstroke of both limbs diminishes the interlimb
space. Floating material sticks to feathered setae on the mesal edge of the protopodites and endopodites. A real filtering fence of setae is not present. With
subsequent limb beats the detritus is combed up by the neighboring limbs and
gradually transferred to the food groove that lies along the venter medially
between the legs (Fig. 3-2). The basal endites of the limbs pass food forward to
the mouth. The food consists of diatoms, detritus, etc.

GROOMING.   To clean the appendages the flexible abdomen is bent under the
head and thorax and moved through the straightened out appendages (Fig. 3-5).
During this movement the transverse setae of the last pleomere and telson comb
out the legs while the long furcal setae are drawn through the food groove.

Classification

Hutchinsoniellidae. Hutchinsoniella macracantha, up to 3.7 mm long, is quite common in Buzzards Bay, Massachusetts, and Long Island Sound, off the U.S. east coast.
Lightiella has no appendages on the 8th thoracomere, and its postembryonal develop-

ment differs from that of *Hutchinsoniella*. *Lightiella incisa,* 2.6 mm long, is found from 1–3 m deep near Puerto Rico; *L. serendipita,* 3.2 mm long, comes from 1–2 m deep in San Francisco Bay, California. *Sandersiella acuminata,* 2.4 mm long, is found in Japanese waters.

References

Gooding, R.U. 1963. *Lightiella incisa* (Cephalocarida) from the West Indies. *Crustaceana* 5: 293–314.

Hessler, R.R. 1964. Cephalocarida, comparative skeleton musculature. *Mem. Connecticut Acad. Arts Sci.* 16: 1–97.

———— and Sanders, H.L. 1964. Cephalocarida at depth of 300 m. *Crustaceana* 7: 77–78.

Jones, M.L. 1961. *Lightiella serendipita,* a cephalocarid from San Francisco Bay. *Ibid.* 3: 31–46.

Sanders, H.L. 1955. Cephalocarida, a new subclass of Crustacea. *Proc. Nat. Acad. Sci.* 41: 61–66.

———— 1963. Significance of Cephalocarida *in* Whittington, H.B. and Rolfe, W.D.I. *Phylogeny and Evolution of Crustacea.* Mus. Comp. Zool., Cambridge, Mass. pp. 163–179.

———— 1963. Cephalocarida, functional morphology, larval development, comparative external anatomy. *Mem. Connecticut Acad. Arts Sci.* 15: 1–80.

———— and Hessler, R.R. 1964. Larval development of *Lightiella. Crustaceana* 7: 79–97.

Shiino, S.M. 1965. *Sandersiella acuminata,* a cephalocarid from Japanese waters. *Ibid.* 9: 181–191.

4.

Subclass Branchiopoda

There are about 800 species described within the subclass, almost all of them freshwater forms. The largest species is *Branchinecta gigas,* a North American fairy shrimp, which may reach a length of 10 cm.

Branchiopoda are crustaceans having the first antennae and second maxillae vestigial. The large trunk limbs are usually turgor appendages with wide, flat protopodites in the adults. The nauplii, unlike those of other crustaceans, have short, unsegmented first antennae, tipped by setae, and have a mandible consisting of a single branch.

All four orders are also known fossil, and an additional four orders have been described from fossils.

Anatomy

The homonomous metamerism of the trunk, the nervous system and the heart gives the impression of a generalized crustacean. Species with a short body also have distinct segmentation, even in the brain (supraesophageal ganglion). Only the protocerebrum and the deutocerebrum are fused. The tritocerebrum is separated from the deutocerebrum and its position behind or below the mouth is a primitive character (see Vol. 2, Fig. 4-9, p. 47). Nauplius eyes and compound eyes are present. The nauplius eyes consisting usually of 3-4 cups are sometimes reduced.

The mandibles of Branchiopoda lack a palp. The second maxillae are reduced. The trunk limbs, leaflike turgor appendages, are not considered primitive. They are flat only in filter-feeding species, cylindrical in predacious groups (Polyphemoidea and Haplopoda). The nauplii also have cylindrical appendages. Further evidence of the derived nature of leaflike limbs is that the first limbs, which function as palps in the Notostraca, are cylindrical with true joints. Therefore it can be assumed that the leaflike appendages are specialized to function with each other as a complicated filtering mechanism.

Relationships

The Anostraca are most distinct, as they lack a carapace, have free, stalked, compound eyes, the naupliar eye consists of three cups, and the musculature of their legs is unlike that of other branchiopod orders.

The other orders are often combined in the group Phyllopoda, a term that has also been used for the non-cladoceran Branchiopoda. Members of these orders have a large carapace and lack the eyestalk but have the compound eyes in a chamber. A fold covers the compound eyes from behind and forms a transparent lid over them, protecting the cuticle. The chamber between eyes and fold has a pore that opens to the outside, allowing it to be filled with water in the Notostraca and Conchostraca. The nauplius eyes of these groups consist of four cups.

The large dorsally flat carapace is characteristic of the Notostraca, while that of both Conchostraca and Cladocera is folded, laterally compressed and clamlike. Both the Notostraca and Conchostraca have the trunk consisting of numerous metameres while that of the Cladocera has only few somites, doubtless a secondary reduction. The rami of the caudal furca are long and straight in the Notostraca while in both the Conchostraca and Cladocera the furcal rami are clawlike. The clawlike furca is an adaptation for cleaning the inside of the valves and food groove.

ORDER ANOSTRACA, FAIRY SHRIMPS

There are about 175 known species, of which the northwestern North American *Branchinecta gigas*, up to 10 cm long, is the largest.

Anostracans are elongate Branchiopoda, without carapace. The segmentation of the trunk, the nervous system and the heart are almost homonomous and appear primitive. Even the trunk appendages are much like one another.

Anatomy

The head is distinctly set off from the trunk and has one or more transverse grooves, which give the illusion of external segmentation (Fig. 4-1). The trunk is circular in cross section. Each of the anterior 11 (rarely 17–19) metameres bears one pair of leaflike limbs, while the posterior metameres lack limbs. The telson is elongated into a furca and the anus. The first two limbless somites are fused with one another and bear the gonopore. The first antennae are slender and cylindrical; the second are much heavier and in the male are transformed into complicated grasping structures. The grasping structure differs in different species. The long axis of the mandibles is vertical; the grooved chewing surface is bent horizontally toward the midline (Vol. 2, Fig. 16-3, p. 304). There are no palps. The limbs are transverse leaflike appendages, stiffened by turgor pressure rather than by the integument. There is only one functional articulation (Fig. 4-2). Despite this, they are not held in a single plane but are bent at a right angle (Fig. 4-3). The homology of the articles with those of other crustaceans is uncertain.

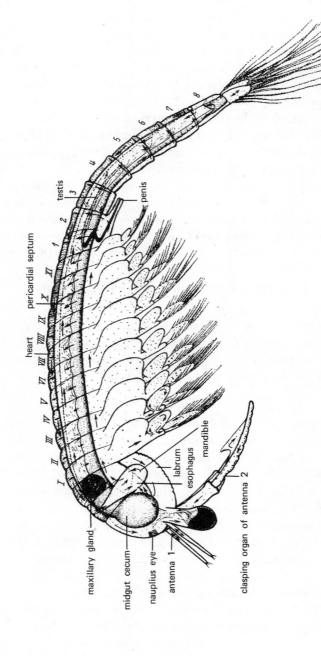

Fig. 4-1. *Artemia salina*, lateral view; 1 cm long. Arrows indicate the direction of blood flow; the ostia of the heart are indicated in black. Roman numerals indicate the appendage-bearing thoracic somites, Arabic numerals the abdominal somites lacking appendages. The animal swims on its back. (After Vehstedt.)

heart

pericardial septum

testis

penis

labrum

esophagus

mandible

maxillary gland

midgut cecum

nauplius eye

antenna 1

clasping organ of antenna 2

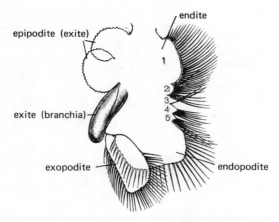

Fig. 4–2. Leaflike appendage of *Chirocephalus diaphanus*, slightly flattened; 2 mm long. As result of comparison with the Devonian *Lepidocaris*, Preuss and recently Benesch have interpreted the endopodite as the 6th endite; the exopodite (flabellum) as endopodite; branchia (exite) as exopodite. However, the homologies of the limb articles have not been settled. (After Smith.)

The exoskeleton is not calcified. The color may be transparent blue, green, orange, or red.

NERVOUS SYSTEM. The postantennal ganglia form a nerve chain with ganglia in each metamere. There are naupliar and compound eyes. During the course of postembryonal development eyestalks develop. Sensory setae and thin cylindrical hairs are organs of touch and chemical sense organs.

DIGESTIVE AND EXCRETORY SYSTEMS. Following the short ectodermal foregut is a very long midgut from the anterior part of which originate a pair of spherical dorsal ceca. Maxillary nephridia that develop during postembryonal development slowly replace the antennal gland.

RESPIRATORY AND CIRCULATORY SYSTEMS. Respiration is mainly through the branchia of the limbs (Fig. 4-2).

The heart extends through the whole trunk. Even though systole takes place anteriorly and posteriorly at the same time, most blood is circulated toward the anterior. There are 18 pairs of ostia, one in each trunk metamere, with an additional median ostium at the posterior end in *Branchipus* and *Artemia*. The ostia conduct blood from the pericardial sinus into the heart. The ventral wall of the pericardial sinus, the pericardial septum, separates the sinus from the body cavity as a dorsal tier. Oval openings of the septum in all metameres permit blood to enter from all lacunae (Fig. 4-1).

REPRODUCTIVE SYSTEM. The gonads are paired; in *Artemia, Chirocephalus* and *Branchipus* they are located in the posterior, limbless region of the trunk. Both oviducts combine into a median uterus which continues as a wide ectodermal

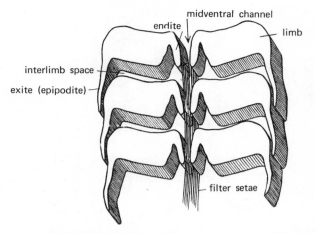

Fig. 4–3.   Horizontal section through basal half of several successive appendages of *Branchinecta gaini.* **The width of each limb is 1.5 mm. (Reconstructed after Cannon.)**

vagina. The vagina fills the two fused reproductive sternites which are swollen into a large saclike brood chamber and opens on their posterior border (in *Branchipus, Chirocephalus* and *Artemia* on the 13th sternite). Males have paired penes.

Reproduction

In mating, the male swims under the female and grasps her from below, from the female's dorsal side, with his 2nd pair of antennae. In males the 2nd antennae are strongly developed, often with spatulate widened or chelate distal articles and often with a more or less long featherlike or antlerlike appendage of the protopodite. This protopodite appendage unfolds while the male follows the female but its function is not completely understood. During mating the male wraps his body around the female so that the penes come to lie opposite the gonopore. *Artemia* pairs may swim about in this manner for hours and copulate every 4 to 5 min. Some populations of *Artemia salina* reproduce parthenogenetically.

Development

The eggs, each encased in a tough shell, pass through total cleavage and blastoderm formation within the brood chamber. The difference between the thin-walled summer eggs and thick-walled resting eggs produced by some species may be only in the coating of the egg. The eggs are released in clutches of 10–250 every 2–6 days. Some *Artemia* are ovoviviparous. The eggs of species inhabiting temporary pools dry up within the mud bottom without becoming damaged and remain viable for 3 to 5 years. But drying and freezing are not essential, and in aquaria these eggs may hatch even if continuously covered by water.

In experiments, soil from dried Wyoming prairie ponds known to be inhabited by several species of anostracans was placed in aquaria. The eggs of some species were found to hatch only within a narrow range of temperatures and salinities, while the adults of the species can endure substantial ranges.

The development within the egg is so advanced that sometimes after flooding of the pools the nauplii of *Chirocephalus* hatch within 2 hr, and in serial sections of the eggs one can already see the primordia of two metameres behind the mandibles. In summer under favorable food and temperature conditions, they may become sexually mature within 8 days, making maximum use of the vernal conditions of the pool. *Chirocephalus*, after the third molt, has all somites, but not all limbs. *Artemia* hatches as a nauplius and has all somites at the 9th postembryonal stage maturing in altogether about 14 instars, or 18–21 days, to live about 4 months. It disappears from the Great Salt Lake if temperatures drop below 4°C. *Chirocephalus grubei* lives about 90 days at 5.2°C, 78 days at 7.9°C, 50 days at 13.5°C. The adult lives only 4–10 days.

Relationships

The Anostraca appear to be primitive and at present are believed to be quite distinct from the other branchiopod crustacean orders. Major differences are the completely different muscle systems of the appendages, the stalked eyes, the transformation of the genital rings into a brood chamber, the position of the gonopore behind the appendages, the median female gonopore, the feeding of the eggs by nutritive tissue, the transformation of the second antennae into clasping organs in the male, and the lack of carapace. Also the leaflike appendages differ in form and are more specialized in function than those of Notostraca. In contrast to these specializations, characteristics in common with the other orders are the relative homonomy of the limbs, nerves, and heart, characteristics that are presumed to be ancestral.

Habits

The Anostraca live almost exclusively in standing fresh- or inland saltwater, most species only in temporary pools that appear as snow melts or after heavy rainfalls. The spring species appear at the time of thaw (e.g., *Chirocephalus grubei* in Europe, and *Eubranchipus vernalis* in eastern U.S.). They may be found from January to May, and the late ones tolerate high water temperatures (25–30°C) unlike those that mature in March at water temperatures of 3–10°C.

Alpine ephemeral ponds observed in the Rocky Mountains, in the Medicine Bow Range at 3000–3500 m, always had only one species (*Branchinecta paludosa, B. coloradensis*, or *Chirocephalopsis bundyi*), while Wyoming prairie ponds had several species in various combinations, and late in the season with conchostracans and notostracans. Early species hatching in May at low temperatures around 10°C were *Branchinecta lindahli* and *B. packardi;* late species, hatching in August above 15°C were *B. lindahli, Streptocephalus texanus,* and *Thamnocephalus platyurus.* The last two seem thermophilic, *B. lindahli* eurythermal. Populations of *B. packardi* from Wyoming, Kansas, and Oklahoma seemed to have different tolerances. *Artemia salina, B. lindahli,* and *S. texanus* were found in saline ponds; *B. lindahli* eggs hatched at the highest salinities tested in aquaria (2996 ppm), while others hatched only at lower concentrations. The adults survived in a pond with 406 mM/l (21572 ppm), though also found at low salinities. If acclimated slowly, all anostracans tried tolerated about isoosmotic concentrations of salt (155 mM/l). Oxygen content and pH appeared of minor importance for hatching and survival. (Prairie ponds tend to be alkaline, alpine ponds acid.) High Mg concentrations were deleterious.

Artemia salina is stenohaline in saltwater, living in brine pools inland as well as in saline basins separated from the ocean by dams and used for commercial salt production, but never in the ocean. *Artemia* can tolerate salt concentrations between 41 and 230 ‰. It also occurs in high carbonate waters and in some rich in potassium. In some areas, especially in the Great Salt Lake of Utah, *Artemia* appears in unbelievably large numbers. It does not survive in freshwater.

In general, Anostraca live all in borderline habitats in which there are no predacious fish or food competitors. They are readily eaten by amphibians inhabiting the pools, caddis flies, and other predacious insects.

LOCOMOTION.   All fairy shrimp swim continuously, slowly and gracefully, with the venter turned up. Only rarely do they remain quiet on the bottom. The locomotory organs are the leaflike limbs, which in adduction (backward beat) push the body forward. By flexing and rapidly extending the trunk they can dart off.

FEEDING.   Most species feed on microscopic particles. They are filtered either from the plankton (Fig. 4-4) or from suspensions covering plants, from currents created by the animals. Some obtain particles from the substrate as the animal slowly pushes itself along the bottom, venter down, scratching off particles with the endopodite claws. Besides fine organic particles diverse planktonic and attached protozoa, unicellular algae, and diatoms are taken. There is no food selection.

The water laden with plankton is sucked in by the limbs, the transverse protopodites of which follow each other in paired rows. Between them are interlimb spaces (Figs. 4-3, 4-4), each with an anterior and posterior wall formed by two successive protopodites. Their lateral walls are the exites which are bent at right angles posteriorly to the protopodites. The floor is the ventral side of the trunk, the ceiling the distal part of the protopodite with the endopodite bent at a right angle, distally. The inner side of the space is closed by a comb of endite setae which are also bent posteriorly. The setae, feathered with little branches parallel to the midline of the body, form a comb (Fig. 4-3, 4-4).

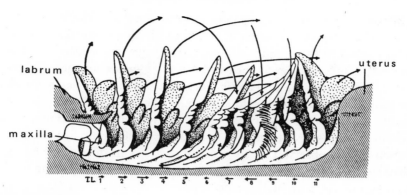

Fig. 4–4. Median section through a swimming *Branchinecta gaini*, in physiological orientation. The channel between setae is cut sagittally and one can see the individual interlimb spaces as the endite setae have been removed (except from appendage 8). Appendage 10 has just completed its forestroke; appendage 9 started the movement later and has not completed it yet; appendage 5 has not yet started its forestroke, having just completed a backstroke.

Interlimb spaces 6–10 are suction chambers; their walls are formed by exites, and their ceilings are the endopodites. Water current flows from the median channel into the chambers. The chambers behind appendages 1–4 expel water. The endopodites are extended and the exites have lost touch with the following limbs. The limbs 1–4 expel water when in backstroke and cause the animal to move. The section is about 1.3 cm long. (After Cannon.)

There is a deep median food channel between the right and left rows of limbs, along the ventral side of the body, walled in on the sides by the vertical setal combs.

Food is obtained by metachronal waves of beating limbs. Depending on temperature and size of the animal, there are about 140–400 beats per min. Both limbs of a pair are in synchronous motion, the most posterior legs starting the forward motion. After completing 1/6–1/10 of their stroke (depending on species), the adjacent anterior legs start their forestroke. Thus a wavelike motion passes along the legs, in *Tanymastix*, for instance, with a difference in phase of 1/10 and with 10 successive leg pairs taking part; in *Branchinella* only 6 leg pairs take part (Fig. 4-4). If a limb moves forward, its adjacent posterior space widens, producing suction and drawing water into the space. Suspended carmine particles show that the water flows exclusively through the midventral channel into which floating particles are carried. From there the water passes through the setal combs into the spaces while the filtered particles accumulate in the channel (Figs. 4-3, 4-4). The lateral walls and ceiling of the spaces close so well that no water enters from other directions; however, as soon as the anterior protopodite begins backstroke the chamber is squeezed, the lateral walls bent outward and the protopodites are stretched (Fig. 4-4, 1 to 5). The water escapes upward or to the sides and at the same time produces sufficient backward thrust to propel the animal forward.

The filtrate accumulated along the combs is carried into the longitudinal food groove of the venter by the water that constantly flows into the channel (Fig. 4-4). This current to the mouth (Eriksson, 1936) may be caused by setae of the basal endites, which, with each backstroke of a leg, move forward. Others (Cannon, 1933) believe that the differential pressures of the spaces alone produce the current toward the mouth due to a curb in the wall near the base of each protopodite. Depending on the phase of limb movement, this curb may produce a low-pressure or high-pressure current in an anterior direction.

Filtration can be observed by suspending particles in the water and by filling individual spaces with methyl blue (not methylene blue) using a fine pipet.

The most anterior part of the food groove is covered by a long, posteriorly directed labrum which secretes a particle-collecting mucus (Vol. 2, Fig. 16-8, p. 307). The forward and backward movements of the first maxillae transport food to the mouth. Their setae move like a hinged door from posterior in an anteromedian direction, throwing the particles enclosed in mucus toward the mandibles. These are also moved so that their inner endites crush the particles and transport them to the mouth at the same time (Vol. 2, Fig. 16-5, p. 306; Fig. 4-11). Unwanted particles are thrown out by lifting the labrum, moving the mouthparts, or by the most anterior limb endites.

The large *Branchinecta gigas* is a predator. As many as five individuals, up to 1 cm long, of related species of fairy shrimps have been found in its gut. Some had a mass of *Diaptomus* in the gut, others had the gut stuffed full with over 100 eggs of their own species. Between the mandibles, a crushed anostracan was found whose posterior end was still within the food groove of the predator. Other individuals had only detritus within the gut, which they must have obtained by filter feeding. Their legs form a long median channel, with walls consisting of stiff setae spaced relatively far apart on endites, that can be closed by crossing over stiff bordered endopodites. The anostracans appearing in this channel are moved anteriorly by the endite setae during swimming motions, taken up by the toothed mandibles, crushed and pushed underneath the labrum.

OSMOREGULATION.   Osmoregulation has been studied in saltwater *Artemia*. The hemolymph even in concentrated saltwater remains hypotonic compared to stenohaline marine animals in normal surroundings. The hemolymph ion concentration is just as independent of the surrounding medium as its osmotic pressure. The high concentration of Cl ions and very low Mg resembles that of freshwater inhabitants. This independence from the medium is made possible by low permeability of the integument (except of the epipodites of the 10 anterior limb pairs) which prevents water loss.

Also because of the permeability of the gut epithelium, swallowed water and NaCl are supplied continuously to the blood, from which the NaCl is continuously removed by the 10 anterior epipodites against the salt concentration of the medium (Table 4-1). In nauplii the dorsal organ has this regulatory function.

Table 4–1. Osmotic concentration of hemolymph of *Artemia salina* and the surrounding medium. Osmotic pressure is expressed as percentage of NaCl. The epipodites are destroyed by placing *Artemia* in a saturated $KMnO_4$ solution for 5 min

	Surrounding medium	Hemolymph
Normal animal	1	0.9
Normal animal	2.5	1.1
Cauterized epipodites	2.5	2.5
Normal animal	4	1.1
Cauterized epipodites	4	4
Normal animal	30	2.7

The functioning of this regulatory mechanism has been demonstrated experimentally in adults. If *Artemia* is placed in seawater (3.24% NaCl) with 15% per volume glycol and phenol red, the gut contents soon become red. This proves that *Artemia* drinks even when it does not feed. In animals that had both head and anus ligated it did not. The osmotic pressure rose more rapidly in unligated animals demonstrating that the glycol penetrated through the gut epithelium but the animals do not become dehydrated. The osmotic pressure of ligated animals however does not rise. The glycol does not enter through the integument.

Artemia kept in seawater and placed in $10^{-2}M$ $AgNO_3$ for 2–5 min, then in photographic developer 1–2 min, had the 10 anterior epipodites turn intensely black. Transport must have been through these 10 appendages, in which Ag precipitated. If *Artemia* was placed in concentrated solution of $KMnO_4$ these same epipodites stained brown and some burned. Although 5 min damaged the epipodites, many animals survived if placed in dilute seawater afterward. But if these animals were later placed in more concentrated salt solution, it was found that they had lost the ability to osmoregulate.

The osmoregulation of *Artemia* resembles that of marine fish. In concentrations of ¼ seawater the osmotic pressure of the hemolymph is hypertonic toward the medium and in freshwater the *Artemia* dies within 24 hr.

CIRCULATION. The blood contains hemoglobin. In concentrated solutions of more than 125 ‰ salt that have little oxygen, *Artemia* has more blood pigment than in less concentrated water. This can be seen readily by the red color of the animal or in physiological experiments by measuring the ability to take up oxygen.

Classification

Most North American species are found in the Great Plains, fewest in the southeastern states.

Branchinectidae. Males of this family have the inner margin of the second antennae serrate, or with a knob or spur, or slender with tips turned up. *Branchinecta gigas,* to 10 cm, is found in western U.S., the smaller *B. coloradensis* is found in small puddles in the Rocky Mountains above the timberline.

Artemiidae. Males of this family have the distal end of the second antennae plate-like. *Artemia salina,* which was thought to be the only species, lives in salt pools all over the world, but never in the ocean. They often appear in unbelievable numbers. In very saline waters the eggs float and may stick to legs of birds and be distributed to other waters. The nauplii hatch, particularly in spring, when the streams are high and the water less saline. The eggs do not have to dry out in order for them to hatch. Of commercial importance, eggs are shipped dry, hatched in saltwater, and used as fish food by fish fanciers. *Artemia salina,* 1.5 cm long, is found, rarely, in central Europe, appearing as diploid, triploid, tetraploid, pentaploid, or octoploid parthenogenetic and diploid or tetraploid bisexual strains. In maturation division of gametes of partheno-genetic races the reduction division may be left out or not (in the same broodpouch) with the polar bodies moving back into the egg. The body proportions vary with the salt concentration. In very saline waters the body is short, the limb-bearing trunk sec-tion is shorter than the section without appendages, the furca is shortened and has fewer setae, but is differently modified in diploid and polyploid clones. The descendants of differently proportioned animals have proportions depending on the saltwater con-centration within which they develop. Thus offspring of the same mother show various modifications depending on the salt concentration, which may be diluted by thaw or rains. The postembryonal development in central Europe takes 6–8 weeks. The pop-ulation in Great Salt Lake is diploid and males are present. The morphological varia-tion of Eurasion populations with salt concentrations is not present in American populations.

Recently it was shown that saltwater ponds in Sardinia and in Hidalgo, Argentina are inhabited by sympatric forms of *Artemia* that differ only slightly in morphology. Because they are sterile in crosses with each other and differ in chromosome number, it is assumed that they belong to different sibling species.

Branchipodidae. *Branchipus stagnalis,* up to 2.3 cm long, is found in open pools of Europe from April to September. *Tanymastix* belongs to this family.

Chirocephalidae. The male's second antennae may have a small blunt appendage, long ribbonlike or long coiled processes. *Chirocephalus diaphanus,* up to 3.5 cm long, has a violet-brown brood pouch. *Chirocephalus grubei,* up to 2.9 cm long, yellowish red, is found all over central Europe in light woods after melting snow fills temporary pools. *Eubranchipus* has seven species in North America, of which the northeastern *E. vernalis* is most widespread and common (Fig. 1-2). *Chirocephalopsis bundyi* is widespread from the northern United States to Alaska.

Streptocephalidae. The second antennae of the male have a long folded outgrowth on the distal end. *Streptocephalus auritus,* to 4 cm long, has a long red brood pouch, which reaches almost to the end of the body. Species are found in eastern U.S., West Indies to Central America. *Streptocephalus proboscideus* and *S. vitreus* to 2.5 cm long are found in North Africa. The eggs of these desert species not only have to dry but may also require exposure to 50°C before hatching. *Streptocephalus proboscideus* appears on the 3rd day, S. *vitreus* on the 4th after the mud was wetted and 5 days after hatching S. *proboscideus* carried egg masses, S. *vitreus* in 10 days. In the Sudan where experiments were made they live together with *Triops granarius*.

Thamnocephalidae. Males of this family have a median branched frontal appendage. Species are found in western and southwestern U.S.

Polyartemiidae. This family, unlike others, has 17 or 19 limb-bearing segments. *Polyartemia* is found in the Old World arctic; *Polyartemiella* has two species in Alaska, Yukon and Northwest Territories.

ORDER NOTOSTRACA, TADPOLE SHRIMPS

There are 9–12 widespread species of tadpole shrimps known. The largest is *Lepidurus lynchi*, found in western U.S.; its trunk may reach 10 cm in length.

The Notostraca have more than 40 body rings, many of them under a wide, domed carapace. A very long furca is attached to the telson.

Anatomy

The large dorsal shield, the carapace, is fused to and grows out from the second maxillary somite (Fig. 4-5). The shield covers most trunk somites but is not attached to them. The trunk, with more than 40 rings, is very flexible. Only

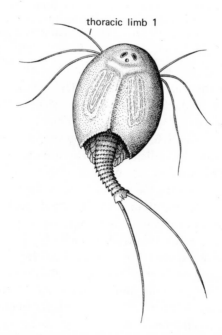

thoracic limb 1

Fig. 4–5. *Triops cancriformis*, dorsal view; 5 cm long. The winding maxillary gland ducts are visible through the carapace. (After Schäffer.)

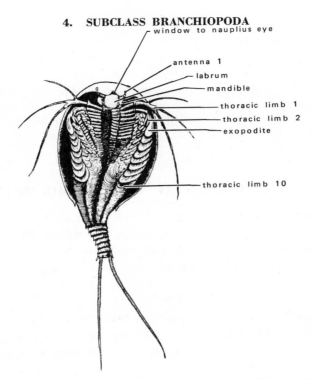

Fig. 4–6. *Triops cancriformis*, ventral view; 6 cm long. The basal endites of the appendages almost touch in the midline and other endites of the anterior limbs are so far apart as to form a wide channel. (After Schäffer from Vollmer.)

the last rings lack appendages. Each of the first 12 trunk somites carries one pair of limbs; each posterior ring has up to 6 pairs. The maximum number is 70 pairs of limbs. The variability in the number of limb pairs, and their excess over the ring number, results from the incomplete segmentation in teloblastic growth. The posterior somites separate completely, only on the ventral side, while on the dorsal side individual somites may be indicated only by annuli, as in the conchostracan *Limnadia* (Fig. 4-7). The number of rings is variable within a species.

The first antennae originate on the ventral side of the carapace as thin slender rods (Fig. 4-6), the second are minute. The first pair of limbs have cylindrical, flagellate endites that in some species extend laterally beyond the carapace (Fig. 4-6), while other appendages are typical, leaflike limbs. The first 11 limb pairs are relatively stiff and have three well-developed articulations (Fig. 1-8). The 11th pair carries a circular egg pouch, with the domed lid formed from the branchia, the bottom from a disclike exopodite and a part from the protopodite. From the 12th pair on, the length and stiffness of the appendages decreases posteriorly.

The compound eyes are situated close together on the top of the carapace just behind a close group of four nauplier eyes. The large labrum covers the mandibular edge which is armed by sharp teeth (Fig. 4-6). Underneath is the mouth which opens into the vertical esophagus. The gut bends to form the long midgut, from which, anteriorly on each side, a cecum originates. The ceca branch elaborately within the carapace.

The maxillary glands, of excretory function, have loops reaching far posteriorly into the carapace and opening laterally on the vestigial second maxillae. The heart extends through the anterior 11 trunk somites and has a pair of ostia in each. The cephalic aorta is branched. The paired ovaries are much branched and, like the testes, extend along both sides of the gut for the entire length of the appendage-bearing trunk, to open on the 11th pair of limbs.

Reproduction

In central and northern Eurpe the Notostraca reproduce only parthenogenetically. Fewer than 10 males are found for each 1000 females. The males are readily recognized by their lack of brood pouches in the 11th pair of limbs. Hermaphroditic gonads have been found at times in *Triops cancriformis, T. longicaudatus* and *Lepidurus arcticus.* However, males are common in the southern and tropical populations.

The eggs remain only briefly within the brood pouch and then are dropped. They survive drying up as well as freezing, and hatch the following year when the pools have refilled with water. They may remain viable for 7–9 years in the soil. That freezing and drying are not necessary for the eggs to hatch has been demonstrated by *Lepidurus arcticus,* found in deep northern waters, and by central Eurpean species kept in aquaria. However eggs of the North African desert inhabitant, *Triops granarius,* have to dry and perhaps to be exposed to heat exceeding 50°C. The eggs remain in mud which may reach 80°C. In experiments the eggs remained viable after being kept in an incubator for 16 hr at 98°C. Boiling killed the eggs. Desiccated eggs survived better than soaked ones.

Development

Nauplii hatch out of the eggs of *Triops cancriformis.* In other species there are probably also metanauplii. Behind the mandibles, the last appendages, the primordia of 5 more somites can be seen as transverse swellings.

Lepidurus apus has molted 17 times by the time it is 1.2 cm long. After there are 13 somites and 7 pairs of limbs, the most anterior ones still have the same shape as in the nauplius; only the mandibles have basal teeth. *Triops cancriformis* molts about 40 times but during midsummer can pass through postembryonal growth within 14 days.

Habits

Notostraca live in standing freshwater. They may appear with anostracans in temporary ponds.

As in the Anostraca there are early season and later season species in Europe. The Eurasian *Lepidurus apus* appears during thaw, between February and late May, rarely until the middle of June; *Triops cancriformis* appears between May and September. *Triops longicaudatus* is found late in the season, in August, in Wyoming. The eggs need warm water (15°C) and low salinities to hatch in prairie pools shared with anostracans.

Lepidurus arcticus, with circumpolar distribution, is also found in lakes to 29 m deep.

LOCOMOTION.   All species move usually on the bottom of pools, ventral side down, so that the appendages, except for the first pair of limbs, are protected by the carapace. In walking they may stretch and bend the trunk, but locomotion depends mainly on the first 11 pairs of limbs. Sometimes they climb along branches that have dropped into the water. They also swim actively, usually dorsal side up.

SENSES.   The position of the body and contact with the ground depend on stimulation received by the limbs. The swimming position is directed by the compound eyes, which are oriented toward the light, and in an aquarium the animal turns upside down if it is illuminated from below. If the compound eyes are painted over, the animal swims in circles up or down. If illuminated from below while on the ground the animal turns over with a rapid swimming movement. The stimulus is probably the light received by the ventrally directed naupliar eyes. From the naupliar eyes a connective tissue channel extends toward the venter to an unpigmented, windowlike area in the integument anterior to the labrum. This shaft admits light from below. If the compound eyes and this window are covered by paint the animal behaves as if completely disoriented.

The long, flagellate endites of the first legs pass over the substrate, serving as chemical sense organs. *Triops*, if its endites touch a chironomid larva suspended near the bottom, will throw itself over the prey. The endites of more posterior legs also seem sensitive to chemicals. They are attracted to pieces of earthworm, but reject those flavored with quinine.

FEEDING.   Notostraca are sediment feeders and also, as suggested by the sharp mandibular teeth, may be predacious, feeding on amphibian eggs, weakened fairy shrimps, smaller tadpole shrimps, ostracods, cladocerans, soft insect larvae, and tadpoles. They throw themselves over the prey, covering it with the whole body. Then they rasp the prey with their sharp endopodites and distal endites of the first nine pairs of legs. The 9th limb pair starts with the forward movement, the 8th starts a moment later, so that waves of motion pass over the legs. The legs tear the pieces from the prey during their backstroke, throwing them posteriorly toward the basal endites along the midline. The basal endites in turn pass the particles from one endite to the next, until they reach the mouthparts where they are chewed by the mandibles and taken into the mouth.

The main source of food is probably sediment stirred up by the anterior endites. The anterior nine limbs form interlimb spaces, as do all limbs of the anostracans, with the differences that the lateral closure is formed by the carapace and that the endites are not bent posteriorly nor have they a comb of setae (Fig. 1-8). The spaces, therefore, open toward the midline. As clouds of sediments rise into the median food channel, only water and fine particles enter the interlimb spaces. Coarse particles cannot enter the narrow spaces, but remain in the food groove and are moved anteriorly. The fine mud that leaves the interlimb spaces collects behind the 10th limb pair and is dispersed by the soft posterior limbs. Some selected particles remain in the food goove and are pushed anteriorly, so that the fecal material contains diatoms, algal filaments, moss thalli, etc. If the posterior legs are removed, the mud collects like a refuse pile behind the 10th pair of legs.

Although there are no combs of setae, *Triops* can feed on algae 15–80 μm diameter, and may even catch some flagellates 3 μm long.

RESPIRATION. After remaining for some time in water only 20% saturated with oxygen at 19°C, *Triops* produces hemoglobin and becomes red.

Classification

*Triopsidae.** There are only two genera. In *Triops** ($=Apus$) the telson is not platelike. *Triops cancriformis* has the carapace 3.5 cm long; it is a late season form of Eurasia and North Africa. *Triops longicaudatus* from western United States, Hawaii, Galapagos, Japan and Argentina is variable. Its preference is for muddy, alkaline waters. In California it creates a nuisance by stirring up mud in rice fields, and may chew leaves. Copper sulfate is used as poison. Similar depredations are known from Spain. In Mexico it is eaten as food. *Triops granarius*, up to 4 cm long, is found from China to South Africa except in forest regions. In the Sudan it survives dry periods as eggs, which have to dry and probably be heated above 50°C to be viable. They hatch within 2 days after the mud is wetted, and mature in 16–20 days at 30–33°C. Reproduction takes place in rain pools that last only 4–6 weeks. There is only one generation a year, all members of which die within 25 days after the first rainfall. *Triops australiensis* is found in Madagascar and dry regions of Australia.

*Lepidurus** has the telson extended into a flat plate posteriorly; *L. arcticus*, of northern Europe, Greenland, Labrador, Alaska and Siberia, is very distinct. There may be five other species, mostly in western North America. *Lepidurus couesii*, found in northern states, Canada to northern Siberia, may be a race of *L. apus* which is almost worldwide. *L. lynchi* is found from Washington to Nevada.

ORDER CONCHOSTRACA, CLAM SHRIMPS

Of about 180 species of conchostracans known, the largest is up to 1.7 cm long.

In the Conchostraca the trunk consists of many metameres with 10 to about 32 pairs of leaflike limbs. The body is completely enclosed by the carapace, a

---

* The family name Triopsidae is on the *Official List of Family Names in Zoology*, *Triops* and *Lepidurus* are on the *Official List of Generic Names*.

bivalved shell (except in *Lynceus*). The furca is clawlike. The conchostracans left
a rich fossil record, ranging from the Silurian. Most of the supposed Conchostraca
from the Cambrian are now regarded as Ostracoda (the Archaeocopida).

Anatomy

The worm-shaped body (Fig. 4-7) and the head, bent ventrally are covered by
brown valves, both attached in the region of the second maxillary somite and
only here fused to the body (Figs. 4-7, 4-8). A transverse adductor muscle in this
region closes the two valves (Fig. 4-7).

In *Limnadia* the carapace is flat and soft; in others it is domed and hard due
to calcification. The carapace is usually not lost in molts, and shows growth rings

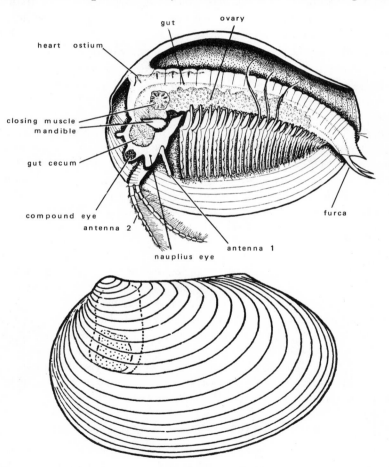

Fig. 4–7. *Limnadia lenticularis*; 1 cm long. The carapace forms a bivalved shell; the
left half and left second antenna have been removed. The head is bent down, permitting
it to be covered by valves. (Modified after Nowikow.)

Fig. 4–8. Lateral view of a conchostracan; 1 cm long. Anterior is on left; the valve
(carapace) shows growth lines. Dotted line indicates area of insertion, stippled area is
muscle attachment. (After Snodgrass.)

(Fig. 4-8). The first antennae are thin, short and rod-like, the second strong with two long branches bearing setae (Fig. 4-7). The 10–28 leaflike limbs become gradually shorter posteriorly. The distal portion of the first, or the first two, are grasping appendages in the male.

The heart extends from the second maxillary somite only to the three or four trunk metameres and therefore has only three to four pairs of ostia (Fig. 4-7). The paired gonads, which are branched, extend through the whole trunk and open on the 11th somite: the vasa deferentia of *Lynceus* on the telson.

Reproduction

In mating, the male, with the strong claws of the modified first limbs, grasps the ventral valve borders on the female and then extends his abdomen into the space between the valves of the female. Some species (e.g., *Limnadia*) reproduce parthenogenetically. The female retains the eggs for some time in a dorsal brood chamber. In most species they stick to dorsal filamentous appendages of the exopodites of several limbs (Fig. 4-7). Later the female molts and at the same time releases the eggs. In *Cyclestheria*, however, the chamber between the back of the trunk and the valve can be closed posteriorly by outgrowths from the 8th posterior body ring, thereby forming the brood chamber (as in the Cladocera), in which summer eggs hatch and the young pass through many stages.

Development

The nauplius has vestigial first antennae. The metanauplius (except in Lynceidae) has a short valve, produced by a pair of folds of the integument, which at first covers only the trunk and its appendages as in the Cladocera. The first stages of *Lynceus* have a large disc-shaped carapace, resembling that of Notostraca. *Limnadia stanleyana* has five nggpliar stages. The third stage begins feeding, and at the fifth molt the animal metamorphoses into a juvenile, which feeds as an adult.

Relationships

Metanauplii of *Cyzicus* have six pairs of leaflike limbs and as the carapace does not yet enclose the head, they strikingly resemble Cladocera, giving the impression that they evolved from a common ancestor. Further evidence for such descent is that in some conchostracan genera there appear characters typical of the Cladocera. For instance in *Lynceus* the head is outside the valves; *Cyclestheria* has a dorsal brood chamber and the compound eyes have fused in the midline of the body.

Habits

Clam shrimps are found only in the littoral zones of lakes, ponds and in vernal pools. They prefer warmer water than fairy shrimp. A number of species of *Eulimnadia* occur in North America, each one known from only one locality, but others are widespread. *Caenestheriella setosa* and *Leptestheria compleximanus* have been found together with anostracans in prairie pools in Wyoming in early and late summer. Most species are found on the bottoms of pools. *Limnadia* lies on its side in the mud. *Cyzicus* burrows with its anterior end so that, on the surface of the substrate the presence of an animal is indicated only by a hole from which emerges a cloud of detritus. The specialized spherical genus *Lynceus,* however, swims through the water back down, with wide open valves. Not only the second antennae but also leaflike limbs take part in locomotion. Other species swim only during the breeding season. They keep their valves closed, so that the second antennae are the only locomotory structures, producing a more or less jerky forward movement during which the dorsum is usually turned up.

FEEDING.   The aberrant, swimming *Lynceus* collects plankton. *Limnadia,* however, while lying between mosses or on its side on the substrate, takes plankton as well as detritus from the surface of plants and substrate, and within its guts have been found *Pediastrum, Closterium, Scenedesmus, Oedogonium,* etc. *Cyzicus* feeds on mud, which it scrapes up with the clawlike furca and sweeps into the current produced by the limbs.

Feeding has been observed in detail only in *Limnadia* and appears to be similar to feeding of Anostraca. The valves open at only two places providing a slit far anterior and another posteriorly extending to the furca. The water enters anteriorly and is drawn in a dorsoposterior direction. Coarse particles are prevented by setae of the basal endites from entering the food groove. Wrapped in a secretion from the limbs, the coarse particles move posteriorly where they leave the valve. The posterior limbs are relatively unimportant for feeding (as in the Notostraca) but aid in removal of unusable particles. Their shortness leaves a ventral space (Fig. 4-7) which is cleaned by the furca.

Classification

Lynceidae. The telson has only rudimentary claws. The head is not enclosed by the carapace. The carapace has no growth rings or few. *Lynceus** *brachyurus,* up to 6.5 mm long, has a strongly domed carapace; it is found in Eurasia and North America.

Limnadiidae. There is a stalked organ behind the eyes. *Limnadia lenticularis,* up to 17 mm, is found in Europe, New England and coastal lakes of the Arctic, *L. stanleyana* in New South Wales, Australia. *Eulimnadia* has a number of localized species in North America.

Leptestheriidae. The rostrum has a spine. *Leptestheria dehalacaensis,* 15 mm long, is found in Europe, *L. compleximanus* in Kansas, Texas, western states and Mexico.

Cyzicidae.* The rostrum lacks a spine and there is no stalked head structure. *Cyzicus** *mexicanus* occurs from Canada to Mexico, other species from Kentucky to Texas and in Europe. *Caenestheriella* species are found from Ohio west to the Pacific coast.

Cyclestheriidae. The compound eyes are fused together dorsally. *Cyclestheria* is found in India, Ceylon, and Australia.

* The names *Lynceus* and *Cyzicus* are on the *Official List of Generic Names in Zoology,* Cyzicidae on the *Official List of Family-Group Names in Zoology.*

ORDER CLADOCERA, WATER FLEAS

There are about 420 species of Cladocera known, the largest being *Leptodora kindtii* of north temperate regions, which may reach 1.8 cm in length.

Cladocera have only a few metameres in the trunk and have five or six pairs of limbs. The furca is clawlike. The bivalved carapace never covers the head but usually covers the trunk and limbs only. Exceptions are the few predacious species having slender limbs; in these the carapace only forms a dorsal brood chamber.

The Cladocera are dwarf forms, usually about 1 mm long. Most appear in huge numbers of individuals and are of great importance in the economy of lakes and in food webs, making available a wealth of food to larger animals.

They differ from the Conchostraca by the reduction in the number of limb pairs, at most six, and in never having the head covered by the carapace.

Anatomy

From their appearance and habits one can easily separate the filter feeders from the predators. Filter feeders, like the Conchostraca, have leaflike limbs enclosed by the carapace. The carapace is folded into two valves; it is not hinged. Within the carapace is also the abdomen, which lacks appendages and is flexed ventrally (Fig. 4-10); the recurved section is called the postabdomen. This very successful group was probably derived from the Conchostraca. The few predators have cylindrical, raptorial legs, with their freedom of movement enabled by the reduction of the carapace. In these the minute carapace is carried on the dorsum like a rucksack, forming the brood chamber (Fig. 4-9). The predators appear very specialized. Filter feeders have the head bent ventrally. The trunk is divided into an appendage-bearing part with 6 metameres at most, and a posterior part lacking appendages. The posterior part still has four metameres in some predators and the same number can also be seen in embryos of Daphniidae (Fig. 4-9).

The first antennae are usually small and slender, long only in *Moina*, Bosminidae, and Macrothricidae. The second antennae are biramous and well developed. In the filter feeders each mandible has a grinding ridge (Fig. 4-11), while in predators they bear long teeth. Two different types of limbs are found among the filter feeders: the superfamily Sidoidea has six similar (except for the last) pairs of leaflike limbs easily derived from the Conchostraca, and the Chydoroidea, has limbs that differ from each other (Fig. 4-15). The cylindrical, jointed limbs of the predators are completely different. *Polyphemus*, Polyphemoidea, still has exopodites, but in the related marine genera *Evadne* and *Bythotrephes*, the exopodites are reduced. These are completely absent in *Leptodora*.

The cuticle of the carapace frequently has a polygonal pattern reflecting the borders of the hypodermal cells below. Most Cladocera are transparent; any coloration present is due to the hemoglobin of the blood and to oil droplets in the

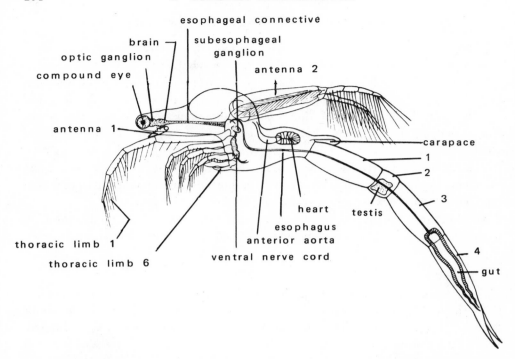

Fig. 4–9. *Leptodora kindtii*, a predacious cladoceran, male; 1 cm long. The head is elongated; numbers indicate posterior trunk segments. (After Saalfeld.)

mixocoel of the trunk and carapace, in which red carotenes and blue reserves have accumulated.

SENSE ORGANS.   While the three to four nauplius eyes are small, may even lack pigment or be absent, the compound eyes are relatively large. These are fused in the midline of the body and are moved by a number of muscles within a completely enclosed eye chamber. There are all together 22 ommatidia in the eyes of *Daphnia*, but many more in predators (*Evadne* 80, *Leptodora* ca. 500), indicating that they might be used to find prey.

DIGESTIVE SYSTEM.   The labrum contains gland cells and extends posteriorly far beyond the mouth, forming a preoral cavity within which the mandibles work. The midgut is usually straight, coiled in most chydorids and macrothricids; often it has an anterior pair of tubelike ceca, sometimes a posterior median cecum (some Chydoridae) or no ceca at all (*Sida, Bosmina*).

CIRCULATORY SYSTEM.   The heart, except in *Sida,* is very short, barrel-shaped, and has only one pair of ostia. Delicate membranes within the body and limbs guide the circulation.

REPRODUCTIVE SYSTEM.   The paired gonads are in the appendage-bearing part of the trunk and extend into the abdomen (Fig. 4-10). They are shorter in *Polyphemus* and *Leptodora*. The ovaries open laterally, through paired short

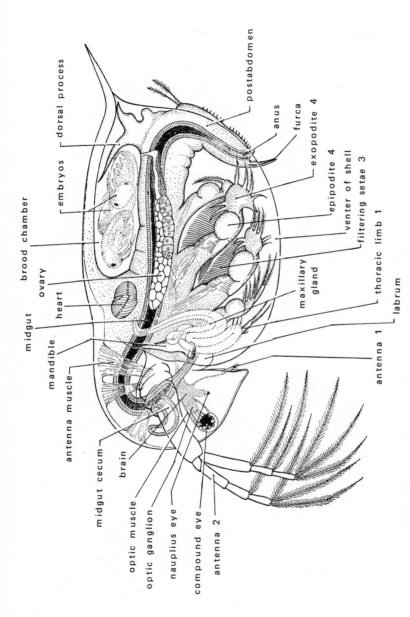

Fig. 4-10. *Daphnia pulex*, a filter feeder, lateral view in morphological orientation; 3.3 mm long. The first antennae hang into and chemically test the water current produced by the limbs. The second are used for swimming. The exopodite marked is that of leg 4, which closes the interlimb chamber ventrally by placing its posterior border against the 5th limb. The filter setae of leg 3 lie in the plane of this page while the main part of the limb is transverse to it and only the edge shows. The brain consists of proto- and deutocerebrum; behind it is the esophageal connective, horizontal due to bending of the head. The dorsal process closes the brood pouch posteriorly. (After Matthes.)

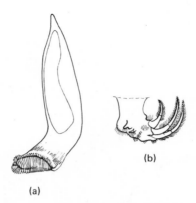

(b)

(a)

Fig. 4–11. Mouthparts. (a) The vertically oriented mandible of *Daphnia pulex*, inner surface; 0.4 mm long. One can see the large area of insertion and the chewing surface with rows of teeth. (b) Maxilla. (After Lilljeborg.)

oviducts, into the space between dorsum and the carapace, the brood chamber (Fig. 4-10). The testes may be connected to a pair of penes.

Reproduction

Males are much smaller than females. Mating has only rarely been carefully observed. During mating the furca with the male gonopores is extended between the valves of the female to the brood chamber. All species produce eggs which initially enter the brood chamber.

In the Sidoidea the brood chamber is closed toward the sides and posteriorly by a horseshoe-shaped ridge of the inner wall of the shell, in Daphniidae it is closed laterally by a paired swelling of the dorsum and posteriorly by fingerlike projections. In other families it is limited by a bend of the posterior of the trunk. In predators the brood pouch is tightly closed and filled with a nutrient fluid that nourishes the eggs, which have little yolk. According to old observations a food organ is present in the predatory Cladocera, and in *Moina* and *Moina-daphnia,* a swelling of the hypodermis, which may secrete nutrients into the fluid. The young, which are completely formed within the brood chamber, escape usually through an opening produced by lowering and bending the posterior part of the trunk.

In reproduction, parthenogenetic and bisexual generations alternate. Within the brood sac the parthenogenetically produced subitaneous eggs (summer eggs) develop without reduction division directly into young cladocerans. The haploid eggs are fertilized; they are winter (resting) eggs, out of which hatch, sometimes after a few days instead of over winter, parthenogenetic, usually amictic, females.

In parthenogenetic reproduction, the offspring are not necessarily all alike. In *Daphnia pulex,* before the only maturation (equatorial) division of the summer eggs there is a kind of meiosis within the nuclear membrane. Bivalents conjugate without separating during anaphase. There may be gene exchange before, which may produce breaks or new linkages. In this case the chromosomes of the eggs have a combination of genes different from that of the mother.

The large resting eggs, rich in yolk, develop in the brood chamber of the mother up to the appearance of the mesoderm. Then there is a diapause lasting days or months, during which time the eggs are resistant to drying and freezing. In Sidoidea, Polyphemoidea, and Haplopoda the double-shelled eggs are released into the water while the mother molts. Chydoroidea, however, do not release the thin-shelled eggs from the brood chamber; instead, the eggs are packed into the exuvia during a molt of the mother. The ovary, during development of the resting eggs, changes the shape of the carapace. Thus the carapace is changed in the molt before the appearance of the resting eggs, but is changed to a different degree in different genera. The change is usually considerable and the whole carapace exuvia may rarely be used to encase the eggs (*Alona*). Usually preformed molt lines appear on the carapace surrounding a saddle-shaped dorsal part which loosens and forms the ephippium, containing the eggs from the brood chamber (Fig. 4-12b). Strong sclerotization and dark pigmentation distinguish the ephippium. In *Daphnia*, in which it always contains two resting eggs, the structure is especially complicated. Here the cuticle of the outer carapace wall is much thickened along the back and to the sides of the eggs. Also hypodermal cells of the outer carapace wall, especially in the area of the wide ventral borders that form the closure of the ephippium, have secreted high, hollow prisms initially filled by fluid instead of solid cuticle. All ephippia sink with their contents to the bottom, but later rise to the surface, especially in Daphniidae, due to the filling of the hollow prisms of their walls with gas (perhaps air). Such floating, unwettable ephippia adhere to the feathers of water birds, which carry the ephippia to other waters. The resting eggs may even survive passage through the bird's digestive tract.

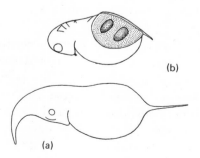

Fig. 4–12. Body outline of *Daphnia cucullata* with curved helmet; (a) June generation; (b) fall generation carrying 2 eggs in the ephippium; 1–2 mm long. (After Wagler.)

Unlike rotifers the same female of many species in different succession can produce parthenogenetic diploid female eggs, parthenogenetic diploid male eggs, or haploid resting eggs. Also ephippia may be formed several times, one after another.

Experiments have been made with *Daphnia magna, D. pulex, D. cucullata; Moina brachiata,* and *Chydorus sphaericus* which demonstrated that unspecific stimuli within the cultures may cause a change in the ovary of the parthenogenetic female. The change does not occur if environmental conditions remain constant. In *Moina brachiata,* lipid materials of the food determined the eggs to be male. In overpopulated cultures fed with normal yeast, 30% of the eggs are male. Populations fed with yeast from which the lipids had been removed produced only females. After adding the fatty extracts to the cultures, 30% of the eggs were again male. The same results were obtained when the lipid extracts were replaced by a mixture of 1 part ergosterin and 1 part olive oil. Only feeding with certain materials will activate the production of males; normally the female factors are more common. The resting eggs also develop as a result of change of food. Fed with normal yeast, 66% of females produce overwintering eggs; fed with lipid-extracted yeast, 5.9% of females. If ergosterin was added to lipid-extracted yeast, 28% of females produced overwintering eggs.

With plenty of food, favorable temperature and aeration, cultures of *Daphnia magna* existed for 4 years, and *Moina brachiata* for 3 years without bisexual reproduction. In contrast to conditions in nature only parthenogenetic generations appeared, in *Moina* 180 of them. Reduction of temperature below 15°C, poor nourishment, and overpopulation, however, cause the production of males and resting eggs. Comparison may be made with similar experiments made with rotifers (Vol. 1, p. 223).

Day length and water temperature influence the determination of eggs of parthenogenetic females of *Daphnia magna.* A culture of amictic females, all offspring of a single amictic mother, reproduced parthenogenetically at 11, 18, and 27°C and a day length of 12 hr. When day length increased to 20 hr, bisexual reproduction was favored, especially if at the same time the temperature was low. At long day lengths and 30°C, besides amictic females only males appeared, at 8°C only ephippial females.

The changes in the ovary are of great importance for the upkeep of cladoceran populations. During steady favorable conditions and parthenogenetic reproduction, in which each individual reproduces, a large population appears rapidly. Each individual of *Daphnia pulex* and *D. magna* can have more than 40 eggs in the brood chamber, and each individual can produce such broods several times. During adverse conditions (e.g., drying out), a new sexual generation makes its appearance and with it the production of resting eggs.

Water fleas that live in the center of a large pond or lake where temperature only slowly follows air temperatures have usually only one sexual generation in the fall when the temperature falls. Thus they are monocyclic (e.g., *Diaphano-*

soma, Holopedium, Leptodora, Bythotrephes, and some *Bosmina*). They overwinter as resting eggs. In very large lakes, *Daphnia longispina, D. cucullata,* and *Bosmina longirostris* may leave out the fall sexual period (acyclic) so that one finds no or very few ephippial females and only a small number of overwintering, brood-carrying females under the ice.

In small ponds the spring and autumn environments may be very different from those during the rest of the year and their cladoceran inhabitants may reproduce sexually twice during the year (dicyclic). But in such waters there may also be changes during other seasons: changes in temperature and dissolved chemicals, etc., and other sexual periods may be present, especially in small shallow vernal pools. *Moina brachiata* may, in the first parthenogenetic generation, hatch out some mictic females from resting eggs. The second generation can already produce resting eggs which, should the water dry up, will prevent the extermination of the population. Each succeeding generation has a larger percentage of males than the preceding, so that the number of resting eggs steadily increases. In the Arctic, there may not be parthenogenetic reproduction at all. However, *Daphnia middendorffiana,* a species almost totally restricted to the Arctic, may form resting eggs in the absence of males (Brooks, 1957).

Development

All Eucladocera develop directly, completely within the egg (Fig. 2-5). Only the resting eggs of Haplopoda (*Leptodora*) have a metanauplius that undergoes a postembryonal anamorphic development. In Europe *Daphnia pulex* has a life span of about 62–85 days during spring, late summer and fall, 51–63 days in late spring and early summer, with abundant food. All cladocerans molt, even females that have had eggs. The young molt frequently, the adults less frequently. In summer, adult *Daphnia* may molt every 2 days.

In molting the cuticle of the posterior border of the head breaks transversely, and the head exuvia folds anteriorly and down. The head emerges first through the opening; valves and trunk follow. During the molt, which takes only 1–3 min, the individual grows in size and changes in proportions (e.g., the brood chambers).

CYCLOMORPHOSIS.* In addition to changes of shape resulting from postembryonic growth, some permanent swimming, limnetic species e.g., *Daphnia longispina, D. cucullata, Bosmina coregoni* and *B. longirostris*) have parthenogenetic generations that differ in appearance. There are differences in average length and body shape, as well as in the number of parthenogenetic eggs in the brood chamber. The changes consist mainly of increased length of the anterior margin of the head (Fig. 4-12) or the tail spine in *Daphnia,* or the height of the dorsal shell hump or length of the first antenna in *Bosmina* (Fig. 4-13). In

---

* An excellent review of the phenomenon appeared recently (Hutchinson, 1967, Chapter 26).

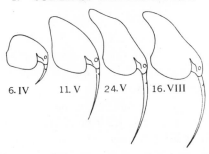

6. IV   11. V   24. V   16. VIII

Fig. 4–13.   Cyclomorphosis of *Bosmina coregoni thersites*; body outline in swimming position; up to 1 mm long. Below, the collecting dates; Roman numerals indicate month collected. (After Lieder.)

Bosmina one can see the differences already in the young within the mother's brood chamber. These changes begin in early summer, increase in the following generations, and disappear completely in fall. All individuals of a population undergo the same changes. Changes in size are mainly due to increased cell size or sometimes to increased cell number. Populations of the same species in neighboring lakes may also have a cyclomorphosis but the resultant appearance of each population might be very different. There may be a straight or a curved helmet spine, or a wide head projection. These morphological differences between races are very distinct in the summer populations, while in winter all local races look the same. These changes, as well as those of postembryology, make the study of cladoceran taxonomy very difficult. Similar changes occur in rotifers.

The dependence of cyclomorphosis on the environment and especially on water temperature, has been studied in field experiments with races of *Daphnia* and with *Bosmina coregoni*. With each favorable increase in temperature there was an increase in the length of the extensions (e.g., the hump of *Bosmina*), and the number of eggs within the brood chamber. If the temperature optimum was passed, the length of extensions and number of eggs decrease, just as if the temperature fell below the optimum. How the environment affects the organ system is not known.

A laboratory study of parthenogenetic females of *Daphnia galeata mendotae* in individual cultures indicated that in addition to temperature, other environmental factors, such as light intensity and photoperiod which increase during the summer, and the amount of food and turbulence of water all affect cyclomorphosis. The various external conditions cause increased growth rate of the upper head or helmet, anterior to the eyes, as compared to the carapace (from the mouth to the tail spine).

The influence of temperature and turbulence is expressed as early as the second embryonal period. The clones from Bantam Lake, Connecticut, proved to be genetically polymorphic in their cyclomorphic reaction. Also the cultures indicated that water temperature produced changes even during the lifespan of an individual in the relative head growth. The offspring of mothers kept at 24.5°C have relatively long helmets. If their postnatal development takes place in water at 7.5°C the positive allometric growth rate of the heads declines so much that as adults they resemble members of clones that passed also their embryonal stages at the lower temperature. The reverse is true for the newborn, short-headed offspring of mothers kept at 7.5°C if they are reared at 24.5°C. In both groups the body proportions approach those of individuals that spent their entire lives at these temperatures. This change supports the view that cyclomorphosis is adaptive to the surroundings, but the function of the adaptation is not known.

Perhaps in predation by fishes, long-helmeted *D. galeata* raised at 25°C, which swim and sink more horizontally, have a selective advantage over the short-headed ones raised at 5°C. Five *Daphnia galeata* at a time and three guppies (*Lebistes reticulatus*) were placed in 2-liter glass cylinders, 15 cm in diameter. In trials lasting 2–15 min, *Daphnia* raised at 25°C were less preyed upon than those having the spine amputated or those raised at 5°C (Table 4-2). However it cannot be assumed that the long spines compensate for the lower viscosity of warmer water and facilitate swimming in upper water layers, because cyclomorphosis occurs in tropical waters at constant temperatures, and in small temperature changes of only 2–3°C accompanied by negligible viscosity changes. Also the extensions may be longer in cold water than in warm in *D. cucullata* and *Bosmina coregoni*.

Habits

Cladocera usually live in standing freshwater. Very few are marine. Though they may inhabit quiet streams, typically they live in shallow ponds, especially among plants close to shore, and swim only short distances, occasionally resting on plants. *Sida* can attach itself to leaf surfaces with an attachment disc of the anterior dorsum; *Simocephalus* attaches with a hooked seta of the second antennae, *Chydorus* with limb spines.

Some Macrothricidae and Chydoridae live on the bottom sediment or within its surface layer. *Daphnia magna* and *D. pulex* live in bays of lakes and ponds but not among plants. *Scapholeberis* (Fig. 4-14) lives below the water surface, a habit for which it is adapted by having straight ventral valve borders armed with

Table 4–2. Results of fish predation experiments on differently shaped *Daphnia galeata*

No. of trials	Temp., °C	Survival of long-helmeted forms, %	Survival of short-helmeted forms, %	Survival of long-helmeted forms whose helmet was cut off, %
28	25	71.4	42.1	
40	15	65	46.5	
	25	48.7		34.4

Fig. 4–14. *Scapholeberis mucronata*, moving along water surface; 1 mm long. (After Wagler.)

laterally directed unwettable hairs. The animal penetrates the surface and hangs itself up on the surface film by the hairs. The second antennae then move the animal along the underside of the surface while it feeds on microorganisms and detritus. *Scapholeberis* can also swim down away from the surface. *Notodromus* has similar habits.

Other species, transparent like many other planktonic organisms, are found far out in the center of lakes where they cannot rest, but must keep permanently in motion; this is probably a secondary habitat for Cladocera (e.g., *Daphnia longispina, D. cucullata, Diaphanosoma, Holopedium,* some *Bosmina,* and *Leptodora*). One rarely finds these species near shore. If at night they drift toward shore, they swim out to open water before sunrise, together with the copepods *Eudiaptomus gracilis* and *Cyclops tatricus.* Experiments with these copepods and *Daphnia longispina* in the Lake of Lunz, Austria, indicated that orientation in these migrations depended on landmarks above the horizon, such as mountains and forest borders. By shifting an artificial horizon, a plate 1.2 m high, the direction of migration could be changed. The migration of the large females of *Acanthodiaptomus denticornis* and *Daphnia longispina,* carrying ephippia in the fall, as in the reverse direction, toward shore and water only 0.1–0.5 m deep.

Vertical migration of *Daphnia longispina* resembles that of the copepods *Eudiaptomus.*

LOCOMOTION.   Most species swim with their second antennae. If the second antennae are large (*Daphnia*), their propelling movements are separated by pauses during which the cladoceran sinks with antennae spread, as if attached to a parachute. If the second antennae are small (*Bosmina*), they are moved very rapidly. Some Chydoridae (*Alona*) also propel themselves with the long posterior end of the body, proceeding by jerky movements through the water.

The swimming position of the trunk differs among genera. Some limnetic species can float, so that they have to paddle only once in a while. The specific gravity of *Diaphanosoma* is extremely low, and in *Holopedium gibberum* it has been reduced by a gelatinous sheath, which gives the very flat animal a spherical appearance. In cultures the sheath was lost within a week. An animal that had foreign bodies stuck to its sheath, forcing the animal down, threw the sheath off and within 13 hr at 15°C formed a new one, almost as thick as the old. The elongated body of *Leptodora* resists sinking (Fig. 4-9).

Some species walk on the substrate, venter down. *Streblocerus serricaudatus* walks on the mud with long stiff setae of the second antennae. *Ophryoxus* walks with the spined tips of the first limbs and the furca of the abdomen. *Ilyocryptus* crawls in the mud holding its large antennae laterally at right angles to the body and moves them like paddle wheels. Beats of the furca help in propulsion. Other species push themselves through the mud with the furca alone. *Chydorus sphaericus* climbs along filamentous algae by loosely wrapping its valves around the filament, the furca posteriorly forced against the thread and the hooked distal spines of the first legs extending and flexing alternately. *Anchistropus* walks similarly on the trunk and tentacles of *Hydra* without the use of the furca.

SENSES. How the chemical nature of the environment is perceived is completely unknown. But *Daphnia* follows increased concentration of O_2, which leads them to areas ideal for its metabolism, with abundant algae and food supply. A strong increase of CO_2 makes Cladocera positively phototactic and they surface.

Temperature differences are perceived by *Daphnia*. Lowering of the temperature causes them to swim up; increased temperature makes them move down.

The naupliar eye is only rarely as large as the compound eye, usually much smaller (except in the Chydoridae). The compound eye itself consists of only a few ommatidia. Among *Daphnia* the three to four cups are much reduced and among predators they are absent.

Daphnia can orient in space with the eyes. It seems to "know" the direction of surface and shore, and can perceive the borders of an area that contrasts with the dark background. Most Cladocera (except some Chydoroidea, but including *Simocephalus* and *Scapholeberis*) swim horizontally with the light above them, a transverse phototaxis indicating that they orient with their compound eyes. *Daphnia*, while maintaining this same orientation of its eyes toward the light, orients its body in any plane. Such freedom of swimming direction is possible because the movable, fused compound eyes of *Daphnia* can be rotated by means of six eye muscles. Even in absolute darkness the top of the eye faces up in all possible swimming positions. The eye rotation probably depends on gravity, and the swimming position on geotaxis, which during the day functions together with phototaxis. The location of the gravity receptors is not known.

The stronger the light above the animal, the more *Daphnia* tips anteriorly, so that in very strong illumination it swims down. In polarized light *Daphnia* swims perpendicular to the plane of light. Many populations, with eyes in intermediate adaptation, move toward the longer wave length, away from blue and violet. Thus they differentiate at least two colors. The biological significance of color vision is not known.

A general light sensitivity of the integument has been demonstrated, which increases between 700 nm (red) and 400 nm (violet). A clear separation of the light sensory function of integument and eyes has not been possible.

FEEDING. The vast majority of Cladocera, all those enclosed within a shell, feed on minute particles (often 8–20 *u*m diameter), nannoplankton forms too small to collect with plankton nets, plankton, bacteria, flagellates (*Euglena, Mallomonas,* and *Chlamydomonas*), *Chlorella,* and tripton (dead detritus particles). Larger algae such as *Scenedesmus* also are taken. Many chydorids enrich the filtered water by scratching up the substrate or whirling up mud. The food depends on the season and location, whether surface or mud. *Anchistropus* is a parasite on *Hydra.*

The plankton feeders filter the water and transport it to the mouth as do the Anostraca; the main difference is that the legs are not used for swimming at the same time. The interlimb spaces differ from those of the Anostraca in that the lateral walls are formed by the valves rather than by the exites (epipodites). The endopodite, if present, is short. Its function of closing the venter has been taken over by a large exopodite (Fig. 4-16). The flexible furca is used to clean the filter and the chambers.

Sidoidea are relatively primitive, Anostraca-like, and have all five anterior legs alike, while the 6th lack filtering setae and only close off the 5th interlimb space. The limbs may make 500 movements per minute. The water sucked through the anterior part of the slit between the shell reaches the median food channel between the legs, and enters the interlimb spaces through filter walls. If the spaces contract, the exopodite extends and the water can escape ventrally into the carapace space and posteriorly between the valves. A basal seta of the first leg moves the filtrate along the food groove to the mouthparts where it is wrapped in mucus secreted by a labral gland.

In Chydoroidea the feeding apparatus is more specialized. The first and second pairs of limbs are specialized in addition to the last (Fig. 4-15). Further, the distances between the limbs varies, and with it the size of the interlimb spaces. Also the shape of the posterior spaces is much different from that of the Sidoidea and Conchostraca because of the anterior bend of the animal's very long postabdomen (Figs. 4-10, 4-16). The legs do not beat in the same rhythm. In the Daphniidae the posterior part of the basal endite setae of the third and fourth limbs transport the filtered food particles toward the mouth parts. Anterior setae of the second leg push the filtrate underneath the long, posteriorly directed, mucus-secreting labrum.

In the Daphniidae the first limb is slender and cylindrical (Fig. 4-15) and does not take part in the rapid movement of the other limbs. Thus there are only three functional interlimb spaces. The first is separated from the food channel by a small basal comb of setae belonging to the second limbs (Fig. 4-15). The much larger setae of the third and fourth leaflike limb pairs border the following two spaces (Figs. 4-15, 4-16) and are the most important in filtering. The 5th and last limb lacks setae and serves only to close posteriorly the large third interlimb space (Figs. 4-10, 4-15, 4-16). While the large second and third interlimb spaces

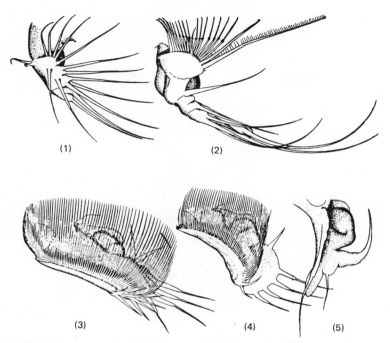

Fig. 4–15. The five right legs of *Daphnia magna* in mesal view seen from the food channel. The endite filters and epipodites are in the plane of the page, the endites in the foreground, the epipodites in the background; the protopodite is vertical to the plane of the page and only the edges are shown. The protopodite of the 3rd and 4th legs hang diagonally posteriorly, and the setae therefore are directed dorsally. The leg tips are to the right (see also Fig. 4–10, showing the legs from the outside). Leg 4 shows the exopodite well. (After Cannon.)

enlarge as in the Anostraca, the small first one enlarges by adduction of the third limbs rather than abduction of the second.

Even more specialized are the limbs in the family Chydoridae and Macrothricidae. Here the three closely spaced anterior limb pairs are stalked and bear distal claw-like curved spines. The first two do not take part in filtering, nor in the filtering movements. They function for climbing and walking, sometimes to pull in large particles or to scrape up particles and sweep them into the current. *Ophryoxus,* using the first limb, moves mud particles toward the valves while the second and third limbs break them up and sweep the fragments toward the ventral food groove. *Chydorus* uses the second limbs to scrape film from algal filaments. *Anchistropus* pinches the epidermis of its host, *Hydra,* between the shell slits and then uses its second and third limb claw to scratch off cells which then float into the food current.

The well developed posterior end and furca, used by substrate inhabitants for locomotion, can reach far anterior, causing the close spacing of the first 3 specialized limb pairs. The filtering current is produced by the constant motion

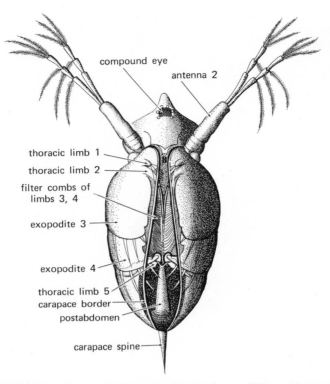

Fig. 4–16. *Daphnia pulex*, **ventral view; 3 mm long. Valves illustrated as though transparent. Limbs 3 and 4 are anterior (abducted), 5 is diagonal to the longitudinal body axis and has closed the posterior interlimb chamber. The exopodites of limbs 3 and 4 close the two large interlimb spaces ventrally. The filter setae of the 3rd and 4th limbs are nearly perpendicular to the plane of this page and form the walls of the median channel. (Modified after Storch.)**

of the fourth and fifth limb pairs (200–300 motions/min). The setae on the endites of legs 3–5 serve as a filter and even in *Bosmina,* a plankton feeder, the limbs are formed as in the Chydoridae, although *Bosmina* never uses them to loosen food particles. Perhaps the explanation for this is that the ancestors of *Bosmina* inhabited the substrate.

The minute captured particles, bound together by mucus of the labral gland are pressed between labrum and mandibles and pushed into the pharynx without being chewed. Large algae, 11 μm diameter, are only rarely taken. The green coloration of the gut fluid of *Daphnia* indicates that enzymes have entered the swallowed but undamaged *Stichococcus, Chlorella, Gonium* and *Chlamydomonas* and have dissolved their contents. Proteins, starch and fat are digested, but not the cellulose and pectin of walls; the emptied cell walls are eliminated.

These Cladocera are of great importance in the food chain, making microscopic particles available to larger aquatic animals. But with this may come the hazard of concentrating man's pollutants. In laboratory experiment with strontium isotopes it was found that the alga *Chlamydomonas* concentrates at a high rate. After 35 days the alga had a concentration 2000 times that of the environment. The alga was eaten by *Daphnia* and *Ceriodaphnia* which in turn were eaten by guppies (*Lebistes*). In the experiments after 35 days the fish had "only" 175 times the strontium concentration of the surroundings, demonstrating that a hazard is present (Bardach, 1964).

The predacious Haplopoda and some Polyphemoidea are armed with slender legs. But predation has been only little studied. *Leptodora* has the venter of the limb-bearing trunk almost at right angles to the head, facing anteriorly and with the grasping limbs covering an anterior area as a trapping basket (Fig. 4-9). Between the limbs of preserved *Bythotrephes* have been found Cladocera and, within the gut, remains of *Diaptomus*. Among the legs of *Polyphemus* (Fig. 4-17) ostracods have been found, but it is uncertain whether this genus is exclusively predacious. Their slender legs have exopodites that enclose a space bordered anteriorly by the head; into this chamber water enters when the head is lifted, and with the water come ciliates and other small animals.

CIRCULATION. The hemoglobin content of the hemolymph is inversely proportional to the oxygen content of the surrounding water, independent of its iron content. Also an increase of metabolism may increase hemoglobin content. Synthesis and destruction of hemoglobin takes place in fat cells and probably in the maxillary gland. In mature females considerable hemoglobin enters the eggs. The different clones of *Daphnia* differ in their ability to synthesize hemoglobin. Various surfaces permit exchange of gases.

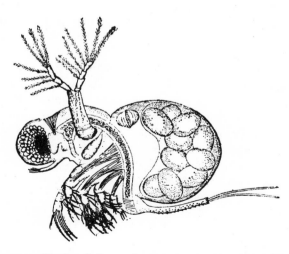

Fig. 4–17. *Polyphemus pediculus*, 1.3 mm long. (After Birge from Ward and Whipple, courtesy J. Wiley and Sons.)

Classification

There are difficulties in diagnosing those species having cyclomorphosis. Also introgressive hybridization has been demonstrated to occur in North American *Daphnia* and European *Bosmina*. Hybrids appear to be fertile, and may make up large populations, there being few physiological isolating mechanisms. The hybrids may be fertile parthenogenetically and may have fertile gametes.

The classification adopted here follows Brooks (1959).

SUBORDER HAPLOPODA

The only species included has 6 pairs of slender legs; the carapace is reduced to a small dorsal brood chamber. There is no nauplius eye; the compound eye is very large. The body and head are elongated. A metanauplius hatches out of winter eggs.

Leptodoridae. *Leptodora kindtii* is completely transparent (Fig. 4-9). This is the largest cladoceran, up to 18 mm long, and is a predator found in northern lakes of America and Eurasia.

SUBORDER EUCLADOCERA

Usually all, but at least the last two pairs of limbs, are leaflike. The carapace usually covers the trunk, and there usually is a nauplius eye. There is never a metanauplius stage; development is always direct. The animals are usually smaller than 6 mm.

SUPERFAMILY SIDOIDEA

This group is also called Ctenopoda. There are six pairs of leaflike limbs, of which the anterior five are alike and equipped with comblike setae on their endites.

Sididae. The carapace lacks a gelatinous mantle. *Sida crystallina*, to 4 mm, lives in ponds and lakes. *Diaphanosoma brachyurum*, up to 1 mm, and *Latona setifer*, up to 3 mm, are both widespread. *Penilia* is marine.

Holopediidae. Females have a spherical gelatinous sheath. *Holopedium gibberum*, up to 2.5 mm long, is found in northern Eurasia, and in North America, west to the Rocky Mountains.

SUPERFAMILY CHYDOROIDEA

This group has also been called Anomopoda. The five or six pairs of limbs are very unlike each other, the first and second slender and cylindrical (Fig. 4-15).

Daphniidae. The small first antennae are attached to the ventral sides of the head. *Daphnia magna*, with females up to 6 mm, males 2 mm, is found in Europe and in North America, in ponds and small lakes. *Daphnia pulex*, up to 4 mm (Figs. 4-10, 4-16), is widespread and variable in Eurasian and North American lakes. *Daphnia longispina*, up to 2.5 mm and planktonic, is variable in Eurasia, absent from North America. *Daphnia cucullata*, up to 2 mm and variable, is found in plankton in the Palearctic. *Scapholeberis*, up to 1.5 mm long, swims on its back below the surface film; *Simocephalus* is up to 4 mm long; *Moina brachiata* (=*rectirostris*), up to 1.6 mm, is widely distributed in small muddy ponds.

Bosminidae. The first antennae of the female are large, immovable, and proboscis-like (Fig. 4-13). *Bosmina longirostris*, 0.6 mm long, is cosmopolitan, and *B. coregoni*, up to 1.2 mm, is holarctic, in open water, with many races. Some *Bosmina* species are marine.

Macrothricidae. The long, movable first antennae of the female are attached to the anterior or venter of the head. *Ilyocryptus,* up to 1 mm long, are particle feeders that crawl in mud using the furca and laterally extended second antennae. The exuviae stay attached to the new integument in *Ilyocryptus,* forming concentric bands. *Streblocerus,* up to 0.6 mm, is a bottom dweller; *Ophryoxus* is widely distributed on weeds.

Chydoridae. The first antennae are covered by the head shield. This is the largest family. *Chydorus sphaericus,* 0.5 mm long and spherical is ubiquitous and cosmopolitan. *Anchistropus emarginatus,* of Europe and *A. minor* of North America, up to 0.46 mm, are ectoparasites on *Hydra. Alona,* up to 1 mm long, has many species.

SUPERFAMILY POLYPHEMOIDEA

Another name for this group is Onychopoda. There are four pairs of slender legs. The shell is a rucksacklike brood chamber, limited to the dorsal side. There is no naupliar eye. The compound eye is very large, the head short. The group superficially resembles the Haplopoda, but is believed to have evolved separately.

Polyphemidae. Polyphemus pediculus, to 2 mm long, is found throughout northern North America and Eurasia in freshwater lakes (Fig. 4–17). *Bythotrephes,* 3–4 mm long without tailspine, has a long horizontal rodlike tail which increases the resistance to sinking. *Podon,* up to 1 mm long, is marine; *Evadne,* 1.2 mm long, lives in shallow ocean water, up to 30 m in depth.

References

Branchiopoda

Anderson, D.T. 1967. Larval development and segment formation in the branchiopod crustaceans. *Australian J. Zool.* 15: 47–91.

Elofsson, R. 1966. Nauplius eye and frontal organs of the non-malacostraca. *Sarsia* 25: 1–128.

Horne, F. 1967. Effects of physical-chemical factors on the distribution and occurrence of some southeastern Wyoming phyllopods. *Ecology* 48: 472–476.

Hutchinson, G.E. 1967. *A Treatise on Limnology,* J. Wiley, New York. vol. 2.

Pennak, R.W. 1953. *Fresh-water Invertebrates of the United States.* Ronald, New York.

Vollmer, C. 1952. Kiemenfuss, Hüpferling und Muschelkrebs. Die Neue Brehm-Bücherei. Akad. Verlagsges. Geest & Portig, Leipzig.

Wagler, E. 1927. Branchiopoda, Phyllopoda in Kükenthal, W. and Krumbach, T. *Handbuch der Zoologie,* de Gruyter, Berlin, 3(1).

Anostraca

Barigozzi, C. 1957. Differentiation des génotypes et distribution géographique d'*Artemia salina. Année Biol.* (3)33: 241–250.

Benesch, R. 1969. Ontogenie und Morphologie von *Artemia salina. Zool. Jahrb. Abt. Anat.* (in press).

Cannon, H.G. 1926. Post-embryonic development of the fairy shrimp *Chirocephalus. J. Linnean Soc. London, Zool.* 36: 401–416.

———— 1929. Feeding mechanism of the fairy shrimp, *Chirocephalus diaphanus. Trans. Roy. Soc. Edinburgh* 55: 807–822.

——————— 1935. The feeding mechanism of *Chirocephalus diaphanus*. *Proc. Roy. Soc. London* 117B: 455–470.

——————— and Leak, F.M.C. 1933. The feeding mechanism of the Branchiopoda with an appendix on the mouthparts of the Branchiopoda. *Philos. Trans. Roy. Soc. London* 222(B): 267–352.

Cole, G.A. and Brown, R.J. 1967. The chemistry of *Artemia* habitats. *Ecology* 48: 858–861.

Croghan, P.C. 1958. The survival of *Artemia* in various media. *J. Exp. Biol.* 35: 213–249, 425–436.

Dexter, R.W. 1959. Anostraca *in* Edmondson, W.T. ed. *Ward and Whipple Freshwater Biology*, Wiley, New York, p. 558–571.

Dornesco, G.T. and Steopoe, J. 1958. Les glandes tegumentaires des Phyllopodes Anostracés. *Ann. Sci. Natur. Zool.* (11)20: 29–69.

Eriksson, S. 1935. Die Fangapparate der Branchiopoden. *Zool. Bidr. Uppsala* 15: 23–287.

Fränsemeier, L. 1940. Zur Frage der Herkunft des metanauplialen Mesoderms und die Segmentbildung bei *Artemia salina*. *Z. Wiss. Zool.* 152: 439–472.

Fryer, G. 1966. *Branchinecta gigas,* a non filter feeding raptatory anostracan. *Proc. Linnean Soc. London* 177: 19–34.

Gilchrist, B.M. 1954. Haemoglobin in *Artemia*. *Proc. Roy. Soc. London* 143(B): 136–146.

Goldschmidt, E. 1953. Chromosome numbers and sex mechanism in euphyllopods. *Experientia* 9:65–66.

Gross, F. 1932. Die Polyploidie und die Variabilität bei *Artemia salina*. *Naturwissenschaften* 20: 962–967.

Halfer-Cervini, A.M. et al. 1967. Fenomeni di isolamento genetico in *Artemia salina*. *Atti ass. Genet. Italiane* 12: 312–327.

——————— 1968. Sibling species in *Artemia*. *Evolution* 22: 373–381.

Hall, R. E. 1961. The natural occurrence of *Chirocephalus diaphanus*. *Hydrobiologia* 17: 205–217.

Heath, H. 1924. The external development of certain phyllopods. *J. Morphol.* 38: 453–483.

Hentschel, E. 1965. Neurosekretion und Neurohämalorgan bei *Chirocephalus grubei* und *Artemia salina*. *Z. Wiss. Zool.* 171: 44–79.

——————— 1967. Experimentelle Untersuchungen zum Häutungsgeschehen geschlechtsreifer Artemien. *Zool. Jahrb. Abt. Physiol.* 73: 336–342.

Hsü, F. 1933. The Anostraca of Nanking and its vicinity. *Contr. Biol. Labor. Sci. Soc. Nanking* (*Zool.*) 9: 329–340.

Kallinowsky, H. 1955. Der Einfluss exogener Faktoren auf Wachstum, Körpergrösse und Lebensdauer von *Chirocephalus grubei*. *Z. Morphol. Ökol.* 44: 196–221.

Krishnan, G. 1958. Cuticular organisations of the branchiopod *Streptocephalus dichotomus*. *Quart. J. Microscop. Sci.* 99: 359–371.

Linder, F. 1941. Morphology and taxonomy of the Branchiopoda Anostraca. *Zool. Bidr. Uppsala* 20: 101–302.

Linder, H.J. 1959. Structure and histochemistry of the ovary and accessory reproductive tissue of *Chirocephalus*. *J. Morphol.* 104: 1–46.

Lochhead, J.H. 1950. Artemia *in* Brown, F.A. ed., *Selected Invertebrate Types*, Wiley, New York, pp. 394–339.

Lynch, J.E. 1964. Packard's and Pearse's species of *Branchinecta*, analysis of nomenclatural involvement. *Amer. Midland Naturalist* 71: 466–488.

Mathias, P. 1937. Biologie des Crustacés Phyllopodes. *Actualités Sci. Ind.* 447: 1–107.

Moore, W.G. 1955. Biology of the spiny-tailed fairy shrimp in Louisiana. *Ecology* 36: 176–184.

Oehmichen, A. 1921. Die Entwicklung der äusseren Form des *Branchipus*. *Zool. Anz.* 53: 241–253.

Preuss, G. 1951. Die Verwandtschaft der Anostraca und Phyllopoda. *Ibid.* 147: 49–64.

———— 1957. Die Muskulatur der Gliedmassen von Phyllopoden und Anostraken. *Mitt. Zool. Mus. Berlin* 33(1): 221–257.

Prophet, C.W. 1963. Physical-chemical characteristics of habits and seasonal occurrence of some Anostraca. *Ecology* 44: 798–801.

Reeve, M.R. 1963. Filter-feeding of *Artemia. J. Exp. Biol.* 40: 195–205, 215–221, 237–249.

Stammer, H.J. 1956. Zeitliches Auftreten von *Chirocephalus. Wiss. Z. Univ. Greifswald* 5: 279–280.

Tyson, G. 1968. Fine structure of the maxillary gland of *Artemia. Z. Zellforsch.* 86: 129–138.

Vehstedt, R. 1940. Bau, Tätigkeit und Entwicklung des Rückengefässes und des lacunären Systems von *Artemia. Z. Wiss. Zool.* 154: 1–39.

Wagler, E. 1927. Branchiopoda, Phyllopoda *in* Kükenthal, W. and Krumbach, T. *Handbuch der Zoologie,* de Gruyter, Berlin, 3(1): 305–398.

Weisz, P.B. 1947. Histological pattern of metameric development in *Artemia salina. J. Exp. Zool.* 81: 45–89.

Notostraca

Carlisle, D.B. 1968. *Triops* eggs killed only by boiling. *Science* 161: 279.

Chaigneau, J. 1959 [1960]. Dessiccation et de la température sur l'éclosion de l'oeuf de *Lepidurus apus. Bull. Soc. Zool. France* 84: 398–407.

Cloudsley-Thompson, J.L. 1966. Orientation responses of *Triops* and *Streptocephalus. Hydrobiologia* 27: 33–38.

Dahl, E. 1958. Protocerebral sense organs in notostracan phyllopods. *Quart. J. Microscop. Sci.* 100: 445–462.

Eriksson, S. 1935. Fangapparate der Branchiopoden. *Zool. Bidr. Uppsala* 15: 23–287.

Fox, H.M. 1949. On *Apus*; its rediscovery in Britain. *Proc. Zool. Soc. London* 119B: 693–702.

Gaschott, O. 1928. Versuche an *Triops cancriformis. Zool. Anz.* 75: 267–280.

Grasser, J. 1933. Die exkretorischen Organe von *Triops. Z. Wiss. Zool.* 144: 317–362.

Hesse, E. 1935. Die Dauer des jährlichen Auftretens von *Lepidurus apus. Zool. Anz.* 112: 80–85.

———— 1937. Welche Höchstemperaturen verträgt *Lepidurus apus. Ibid.* 120: 152–154.

Linder, F. 1952. The morphology and taxonomy of the Branchiopoda Notostraca, with special reference to the North American species. *Proc. U.S. Natl. Mus.* 102(3291): 1–69.

———— 1959. Notostraca *in* Edmondson, W.T. ed., *Ward and Whipple Freshwater Biology,* Wiley, New York, pp. 572–576.

Longhurst, A.R. 1954. Reproduction in Notostraca. *Nature* 173: 781–782.

———— 1955. A review of the Notostraca. *Bull. Brit. Mus. Natur. Hist.* 3: 1–57.

Preuss, G. 1957. Die Muskulatur der Gliedmassen von Phyllopoden und Anostraken. *Mitt. Zool. Mus. Berlin* 33(1): 221–257.

Seifert, R. 1930. Sinnesphysiologische Untersuchungen am Kiemenfuss (*Triops cancriformis*). *Z. Vergl. Physiol.* 11: 386–436.

Thiel, H. 1963. Zur Entwicklung von *Triops cancriformis. Zool. Anz.* 170: 62–68.

Williams, W.D. 1968. Distribution of *Triops* and *Lepidurus* in Australia. *Crustaceana* 14: 119–126.

Conchostraca

Botnaruic, N. 1948. Développement des Phyllopodes Conchostracés. *Bull. Biol. France Belgique* 82: 31–36.

Cannon, H.G. 1924. Development of an estheriid crustacean. *Phil. Trans. Roy. Soc. London* 212(B): 395–430.

Eriksson, W. 1936. Die Fangapparate der Branchiopoden. *Zool. Bidr. Uppsala* 15: 726–732.

Linder, F. 1945. Affinities within the Branchiopoda. *Ark. Zool.* 37: 1–28.

Mathias, P. 1937. Biologie des crustacés phyllopodes. *Actualités Sci. Ind.* 447: 1–107.

Mattox, N.T. 1959. Conchostraca *in* Edmondson, W.T., ed., *Ward and Whipple Freshwater Biology*, Wiley, New York, pp. 577–586.

————— and Velardo, J.T. 1950. Effect of temperature on the development of the eggs of conchostracan. *Ecology* 31: 110–114.

Cladocera

Bacci, G., Cognetti, G. and Vaccari, A.M. 1961. Endomeiosis and sex determination in *Daphnia pulex. Experientia* 17: 505–506.

Bainbridge, V. 1958. Observations on *Evadne nordmanni. J. Marine Biol. Ass.* 37: 349–370.

Baldass, F. von 1937. Entwicklung von *Holopedium gibberum. Zool. Jahrb. Abt. Anat.* 63: 399–454.

————— 1941. Entwicklung von *Daphnia pulex. Ibid.* 67: 1–60.

Banta, A.M. 1939. Physiology, genetics, and evolution of some Cladocera. *Paper Dept. Genetics, Carnegie Inst.*, Washington. no. 39.

Bardach, J. 1964. *Downstream.* Harper and Row, New York.

Baylor, E.R. and Smith, F.E. 1953. Orientation of Cladocera to polarized light. *Amer. Natur.* 87: 97–101.

Borg, F. 1935. Cladoceran-Gattung *Anchistropus. Zool. Bidr. Uppsala* 15: 289–330.

Brooks, J. L. 1946. Cyclomorphosis in *Daphnia. Ecol. Monogr.* 16: 409–447.

————— 1947. Turbulence as an environmental determinant of relative growth in *Daphnia. Proc. Natl. Acad. Sci.* 33: 141–148.

————— 1957. Systematics of North American *Daphnia. Mem. Connecticut Acad. Arts, Sci.* 13: 1–180.

————— 1959. Cladocera *in* Edmondson, W.T. ed., *Ward and Whipple Freshwater Biology*, Wiley, New York, pp. 587–656.

————— 1965. Predation and relative helmet size in cyclomorphic *Daphnia. Proc. Natl. Acad. Sci.* 53: 119–126.

————— and Hutchinson, G.E. 1950. The rate of passive sinking of *Daphnia. Ibid.* 36: 272–277.

Cannon, H.G. 1935. Feeding mechanism of the Branchiopoda, with an appendix on the mouthparts of the Branchiopoda. *Philos. Trans. Roy. Soc. London* 222B: 267–352.

Chandler, A. 1954. Causes of variation in the haemoglobin content of *Daphnia* in nature. *Proc. Zool. Soc. London* 124: 625–630.

Dehn, M. von 1930. Untersuchungen über die Verdauung bei Daphnien. *Z. Vergl. Physiol.* 13: 334–358.

————— 1937. Experimentelle Untersuchungen über den Generationswechsel der Cladoceren. *Zool. Jahrb. Abt. Physiol.* 58: 241–272.

————— 1949. Untersuchungen über den Generationswechsel der Cladoceren, cytologische Untersuchungen bei *Moina. Chromosoma* 3: 167–194.

————— 1955. Die Geschlechtsbestimmung der Daphniden. Die Bedeutung der Fettstoffe, untersucht an *Moina. Zool. Jahrb. Abt. Allg. Zool.* 65: 334–356.

Eriksson, S. 1935. Studien über die Fangapparate der Branchiopoden. *Zool. Bidr. Uppsala* 15: 23–287.

Frey, D.G. 1955. Langsee: A history of meromixis. *Mem. Ist. Italiano Idrobiol.*, Suppl. 8: 141–164.

————— 1958. The late-glacial cladoceran fauna of a small lake. *Arch. Hydrobiol.* 54: 209–275.

————— 1959. Phylogenetic significance of the head pores of the Chydoridae (Cladocera). *Internat. Rev. Hydrobiol.* 44: 27–50.

————— 1960. The ecological significance of cladoceran remains in lake sediments. *Ecology* 41: 684–699.

————— 1964. Remains of animals in Quaternary lake and bog sediments and their interpretation. *Ergebn. Limnol.* 2: 1–114.

————— 1967. Phylogenetic relationships in the family Chydoridae. *Proc. Symp. Crustacea,* Ernakulum, 1: 29–37.

Fries, G. 1964. Die Einwirkung der Tagesperiodik und der Temperatur auf den Generationswechsel, die Weibchengrösse und die Eier von *Daphnia magna. Z. Morphol. Ökol.* 53: 475–516.

Fritsch, R.H. 1953. Die Lebensdauer von *Daphnia* sp. bei verschiedener Ernährung. *Z. Wiss. Zool.* 157: 35–56.

Fryer, G. 1963. The functional morphology and feeding mechanism of the chydorid cladoceran *Eurycercus lamellarus. Trans. Roy. Soc. Edinburgh* 65: 335–381.

————— 1968. Evolution and adaptive radiation in the Chydoridae. *Philos. Trans. Roy. Soc. London* (B) 244: 221–385.

Goulden, C. E. 1968. Systematics and evolution of the Moinidae. *Trans. Amer. Philos. Soc.* (N. S.) 58(6): 1–101.

Green, J. 1954. Size and reproduction in *Daphnia magna. Proc. Zool. Soc. London* 124: 535–545.

————— 1955. Haemoglobin in the fat cells of *Daphnia. Quart. J. Microscop. Sci.* 96: 173–176.

Harnisch, O. 1949. Vergleichende Beobachtungen zum Nahrungserwerb von *Daphnia* und *Ceriodaphnia. Zool. Jahrb. Abt. Syst.* 78: 173–192.

Heberdey, R.F. 1936. Der Farbensinn helladaptierter Daphnien. *Biol. Zentralbl.* 56: 207–216.

————— and Kupka, E. 1942. Das Helligkeitsunterscheidensvermögen von *Daphnia pulex. Z. Vergl. Physiol.* 29: 541–582.

Hoshi, T. 1959. Critical respiration from the dissociation of oxygen from haemoglobin of the daphnid *Simocephalus. Sci. Rep. Tohuku Univ.* (4)25: 239–245.

Hutchinson, G.E. 1967. *A Treatise on Limnology,* Wiley, New York, vol. 2.

Jacobs, J. 1961. Regulation mechanism of environmentally controlled allometry in cyclomorphic *Daphnia. Physiol. Zool.* 34: 202–216.

————— 1961. Cyclomorphosis in *Daphnia galeata mendotae. Arch. Hydrobiol.* 58: 7–71.

————— 1962. Light and turbulence as codeterminants of relative growth rates in cyclomorphic *Daphnia. Internat. Rev. Hydrobiol.* 47: 146–156.

————— 1965. Control of tissue growth in cyclomorphic *Daphnia. Naturwissenschaften* 52: 92–93.

————— 1965. Morphology and physiology of *Daphnia* for its survival in predator-prey experiments. *Ibid.* 52: 141.

————— 1967. Funktion und Evolution der Zyklomorphose bei *Daphnia. Internat. Arch. Hydrobiol.* 62: 467–541.

————— 1967. Jährliche Zyklen des Adaptivwertes und ökologische Einnischung bei Daphnien. *Verhandl. Deutschen Zool. Ges.* 30: 290–296.

Jander, R. 1966. Die Phylogenie von Orientierungsmechanismen der Arthropoden. *Verhandl. Deutschen Zool. Ges.* 29: 266–306.

Kaudewitz, F. 1950. Entwicklungsphysiologie von *Daphnia pulex. Roux Arch. Entwicklungsmech.* 144: 410–447.

Klotzsche, K. 1913. Kenntnis des feineren Baues der Cladoceren (*Daphnia magna*). *Jenaische Z. Naturwiss.* 50: 601–646.

Kuhnemund, E. 1929. Die Entwicklung der Scheitel-platte von *Polyphemus* von der Gastrula bis zu Differenzierung der aus ihr hervorgehenden Organe. *Zool. Jahrb. Abt. Anat.* 50: 385–432.

Lieder, U. 1950. *Bosmina coregoni thersites* in den Seen der Spree-Dalme-Havelgebieten. *Arch. Hydrobiol.* 44: 77–122.

———— 1951. Der Stand der Zyklomorphoseforschung. *Naturwissenschaften* 38: 39–44.

———— 1952. Über die kurzfristige Veränderung des Rassencharakters einer *Daphnia longispina* Population. *Schweizer Z. Hydrol.* 14: 358–365.

———— 1953. Bastarde zwischen einigen Formtypen des *Bosmina coregoni* Kreises. *Arch. Hydrobiol.* 47: 453–469.

————1955. Die Schwimmstellung planktischer Cladocera, insbesondere von *Bosmina thersites*. *Arch. Hydrobiol. Suppl.* 22: 422–425.

———— 1958. Ablaufsformen und Tempo der Evolution der polytypischen baltischen Planktoncladoceren. *Verhandl. Internat. Ver. Limnol.* 13: 789–798.

Lockhead, J.H. 1950. *Daphnia magna in* Brown, F.A. ed., *Selected Invertebrate Types,* Wiley, New York, pp. 399–406.

Löffler, H. 1964. Vogelzug und Crustaceenverbreitung. *Verhandl. Deutschen Zool. Ges.* 27:311–316.

McMahon, J.W. 1965. Physical factors influencing the feeding behaviour of *Daphnia magna. J. Zool.* 43: 603–611.

McNaught, D.C. 1966. Depth control by planktonic cladocerans in Lake Michigan. *Publ. Great Lakes Res. Div., Univ. of Michigan* no. 15.

Mordukhai-Boltovskoi, P.D. 1968. Taxonomy of Polyphemidae. *Crustaceana* 14: 197–209.

Mortimer, C.H. 1935. Generationswechsel der Cladoceren. *Naturwissenschaften* 23: 476–480.

———— 1936. Experimentelle und cytologische Untersuchungen über den Generationswechsel der Cladoceren. *Zool. Jahrb. Abt. Physiol.* 56: 323–388.

Ocioszynska-Bankierowa, J. 1933. Bau der Mandibeln bei *Daphnia magna. Ann. Mus. Zool. Polon.* 10: 33–40.

Pacaud, A. 1939. L'écologie des Cladoceres. *Bull. Biol. France Belgique Suppl.* 25: 1–260.

Ringelberg, J. 1964. The positively phototactic reaction of *Daphnia magna. Netherland J. Sea Res.* 3: 319–406.

Ryther, J.H. 1954. Inhibitory effects of phytoplankton upon the feeding of *Daphnia magna. Ecology* 35: 522–533.

Saalfeld, E. von 1936. Blutkreislauf bei *Leptodora hyalina. Z. Vergl. Physiol.* 24: 58–70.

Scheffer, D., Robert, P. and Médioni, J. 1958. Réactions oculo-motrices de la Daphnie en résponse à des lumières monochromatiques. *Compt. Rend. Soc. Biol. Paris* 152: 1000–1003.

Schwartzkopff, J. 1955. Vergleichende Untersuchungen der Herzfrequenz bei Krebsen. *Biol. Zentralbl.* 74: 480–497.

Siebeck, O. 1960. Untersuchungen über die Vertikalwanderung planktonischer Crustaceen. *Intern. Rev. Gesamt. Hydrobiol.* 45: 381–454.

———— 1960. Horizontalverteilung planktischer Crustaceen im lunzer Obersee. *Ibid.* 45: 125–131.

———— 1963. Experimente im Litoral zum Problem der "Uferflucht" planktonischer Crustaceen. *Verhandl. Deutschen Zool. Ges.* 27: 388–396.

———— 1964. Die "Uferflucht" planktischer Crustaceen eine Folge der Vertikalwanderung. *Arch. Hydrobiol.* 60: 410–427.

Smirnov, N.N. 1968. Functional morphology of limbs of Chydoridae. *Crustaceana* 14: 76–96.

Smith, F.E. and Baylor, E.R. 1953. Color responses in the Cladocera and their ecological significance. *Amer. Natur.* 87: 49–55.

Sterba, G. 1956. Die Lebensdauer von *Daphnia pulex* unter natürlichen Bedingungen. *Zool. Anz.* 157: 179–184.

———— 1956. Zytologische Untersuchungen an grosskernigen Fettzellen von *Daphnia pulex. Z. Zellforsch.* 44: 456–487.

_____ 1957. Die Riesenzellen der Daphnien-oberlippe. *Ibid.* 47: 198–213.

_____ 1957. Die neurosekretorisichen Zellgruppen einiger Cladoceren. *Zool. Jahrb. Abt. Anat.* 76: 303–310.

Storch, O. 1925. Cladocera *in* Schulze, P., *Biologie der Tiere Deutschlands,* Borntraeger, Berlin, 15: 1–102.

Sturm, F. 1936. Nierenphysiologie besonders der Cladoceran nach elektiver Vitalfärbung. *Z. Vergl. Physiol.* 23: 420–428.

Ubrig, H. 1952. Einfluss von Sauerstoff und Kohlendioxyd auf die taktischen Bewegungen einiger Wassertiere. *Z. Vergl. Physiol.* 34: 479–507.

Uhlmann, P. 1954. Zur Kenntnis der natürlichen Nahrung von *Daphnia magna* und *Daphnia pulex. Z. Fischerei* 3:449–478.

Vollmer, C. 1951. *Wasserflöhe.* Die Neue Brehm-Bücherei, Akad. Verlagsges. Geest & Portig, Leipzig.

Wesenberg-Lund, C. 1939. *Biologie der Süsswassertiere, Wirbellose Tiere.* Springer, Wien.

Wolken, J.J. and Gallik, G.J. 1965. The compound eye of a crustacean, *Leptodora kindtii. J. Cell. Biol.* 26: 968–973.

Wotzel, F. 1937. Zur Entwicklung des Sommereies von *Daphnia pulex. Zool. Jahrb. Abt. Anat.* 63: 455–470.

5.

Subclass Ostracoda

There are about 2,000 living species known, 12,000 species including those from Recent sediments. The largest species known is *Gigantocypris agassizi,* up to 2.3 cm long, from deep in the Pacific Ocean.

Most ostracods, called mussel shrimps or seed shrimps, are about 1 mm long. They have trunk and limbs completely enclosed within a bivalved carapace that resembles a clam shell. Ostracods have no more than seven pairs of appendages, including both pairs of antennae.

Ostracoda have an extensive fossil record extending back to Cambrian time. Two orders, Archaeocopida and Palaeocopida, are extinct (with two possible exceptions). About 70 families and 592 genera have no known living descendants.

Anatomy

The shape of the unsegmented trunk is unique among arthropods, much shortened and hardly if at all longer than the head (Figs. 5-3, 5-4). Its posterior end may bear a paired leaflike or rod-like furca, absent in some groups. The indistinct border between head and trunk is indicated only by the attachment of the valves, which are attached from the dorsum to the middle of the side. The adductors, which close the valves, are a pair of transverse muscles consisting of strands connected in the middle of the body by short tendons (Figs. 5-3, 5-4). The muscles leave discrete scars on the valves (Fig. 5-1).

The carapace originates embryologically from the head (Fig. 5-3) as folds of the integument. Though split longitudinally along the dorsum, the right and left valves are held together by a hinge ligament. The ligament opens the valves, acting as the antagonist to the adductors. In addition to the ligament the valves may be held together by interlocking teeth and sockets or ridges and grooves, the part of one valve fitting closely into or around that of the other. During post-embryonic development the interlocking devices as well as the shape of the valves may change. If the valves are closed, the trunk and appendages are completely protected from the environment (Fig. 5-1).

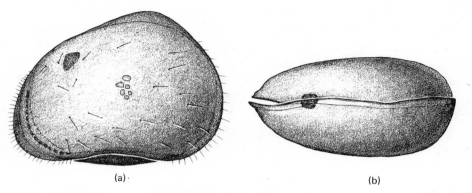

(a) ·   (b)

Fig. 5–1. *Physocypria pustulosa*; 0.6 mm long: (a) left, lateral view of female; (b) dorsal view with long hinge joint of the valves. Anterior (left) the nauplius eye shows through the valves; in the middle on (a), muscle scars. (After Pennak and after Kesling.)

The valves consist of an outer and an inner wall. Between the walls is a space within which blood circulates. The outer wall usually is thick and impregnated with calcium carbonate. The valves may be smooth (Fig. 5-1) or ornamented with humps, ridges, wings (Fig. 5-2). The inner wall, except for the border, is thin and unmineralized.

Within the flexible cuticle of the head and trunk there is a skeleton of cuticular ridges that stiffens the exoskeleton for the insertion of muscles and appendages. Also there is a tendonous endosternite.

The few appendages and the shortened trunk are unusual characteristics for a freeliving group of Crustacea. Cladocopa have only the five pairs of head appendages and the others have two pairs of thoracic limbs at most, never more than seven pairs of appendages. Corresponding to their small number, each pair is differently specialized for its function. At the same time the homologous limbs of different genera show many different adaptations to locomotion and feeding. These are excellent examples of the plasticity and adaptability of arthropod limbs (Figs. 5-3, 5-4). During postembryonic development there are changes in the form of appendages that accompany their functional changes.

The second antennae are the most important locomotor organs. The mandibles and first maxillae are variable in shape depending on the feeding habits.

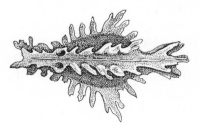

Fig. 5–2. *Cythereis jonesii*, dorsal view; 1 mm long. Anteriorly (left) two separated cups of the nauplius eyes show through the sculptured valves. The animal lives in the substrate. (After G. W. Müller.)

The second maxillae at their tip usually have strong setae, directed toward the mouth. They pass food forward to the mouth as indicated by the musculature, which provides only for forward and backward motion. Cytheridae lack these setae and the limbs resemble walking legs and are used for locomotion. The ends of these appendages form clasping organs in the males of many ostracods. These fifth appendages, which follow the first pair of maxillae, are here homologized with the second pair of maxillae of other crustaceans (rather than with the first thoracic limb, as thought by many ostracod specialists). Even in *Cyprideis,* in

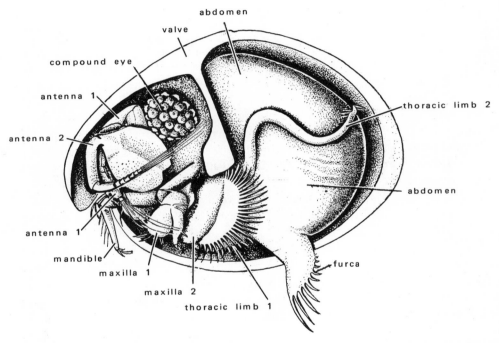

Fig. 5–3. *Cypridina levis,* a Myodocopina ostracod; left valve removed; 2.5 mm long. The epipodite of the second maxilla is a branchial plate. (After Cannon.)

which they resemble the walking legs, the fifth appendages appear as a primordium after the second molt. Furthermore, there is no additional ganglion between the one belonging to these appendages and that of the fourth pair of appendages. Only during late development do these limbs and their ganglia move posteriorly toward the 6th and 7th pairs of appendages, which belong to the thorax. The fact that in small species the primordia of the fifth appendages appear only after the third molt is not sufficient evidence to support the assumption that it is the 6th appendage, the first thoracic limbs, that usually appear at this stage in regular anamorphic development. Postponement of the appearance of appendages during postembryonal development is a common occurrence in Crustacea and can also be seen on other metameres in ostracods.

The first thoracic limbs are developed as walking legs in most ostracods, except in the Cladocopina which have lost the first as well as the second thoracic limbs. The second thoracic limbs serve in Darwinulidae and Cytheridae for walking while they have moved dorsally in Myodocopina and Cypridae and serve as grooming appendages, to clean the soft inner sides of the valves and the trunk (Fig. 5-3).

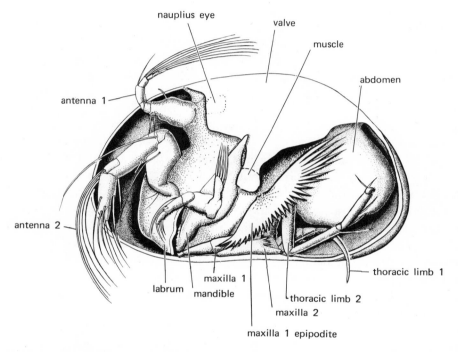

Fig. 5–4. *Cypridopsis vidua*, a Podocopina ostracod; left valve removed; 0.7 mm long. The epipodite of the first maxilla is a branchial plate. (After Cannon.)

Epidermal pigments or the orange gonads may show through the mineralized valves. Thus ostracods are white, gray to brown, black, rarely green, blue or red.

SENSE ORGANS AND NERVOUS SYSTEM. The most frequent sense organs found are sensory hairs on the limbs and the valves, the borders of which are particularly densely equipped. The mineralized areas are penetrated by pores, each filled by a cell whose seta continues to a sensory cell in the space between the walls.

The nauplius eyes, which usually consist of three cups each with 7 to 15 retinal cells, are close to the origin of the first antennae in the midline where they receive light if the shell opens (Figs. 5-1, 5-4). Paired compound eyes are found only in the Myodocopina (Fig. 5-3), being secondarily absent in others.

The nerve chain, corresponding with the few appendages, is very short.

DIGESTIVE SYSTEM.   The preoral cavity extends from the large vertical labrum (Fig. 5-4) to the metastomal area (labium). Within it work the vertical endites of the mandibles (Fig. 5-4). The adjacent sclerotized foregut bends and opens into the wide saclike midgut. At the anterior end of the midgut many species have a pair of ceca, often tubeshaped in the Cyprididae and usually extending on each side into the space within the valves. The anus is at the posterior end near the base of the furca.

EXCRETORY SYSTEM.   The antennal and maxillary glands have been studied in only very few species. Adults of *Doloria* and *Gigantocypris* have only the antennal glands left functional. Adult Podocopina have maxillary glands. The other glands opening on appendages are often, probably erroneously, considered to be nephridia-like excretory organs. For some of them the ectodermal origin has been demonstrated.

CIRCULATORY SYSTEM.   A heart is found only in the Myodocopina. It is short and has one pair of ostia, and it is located below the dorsal integument in the area of the valve attachment. Only in *Gigantocypris* have lateral arteries been found in addition to the anterior branched aorta.

REPRODUCTIVE ORGANS.   Except in Cladocopida the gonads are paired structures and of diverse relative size and shape. Cyprididae have them tubeshaped and they extend, as in some Cytheridae, into the space within the valves. Two oviducts open usually separately in front of the furca. The seminal receptacles open separately, as copulatory openings, in front of the furca. Often the duct of the seminal receptacles opens into the oviduct. The testes, diverse in shape, continue as vasa deferentia which connect to a complicated sclerotized penis.

The spermatozoa of some species are unusually long. An Australian *Platycypris* has the largest sperm known in animals, 1 cm long. *Pontocypris* has spermatozoa 8 times as long as the body; the body is 0.7 mm, the spermatozoa, 6 mm long. Human spermatozoa by comparison are only 0.06 mm long. The long tubelike testes of these species, their complicated efferent ducts, and the ejaculatory duct surrounded by thick bands of muscles and with radial tendons in the walls, may be functionally connected with the long spermatozoa.

Reproduction

Males differ from females in number and shape of setae on the appendages, in having a pair of endopodites transformed as grasping structures tipped by hooks, and in having the furca or valves differently shaped. The copulation position varies in different groups. It may be venter to venter, Cyprididae utilize a posterodorsal position, Entocytheridae use a dorsal position. *Notodromas* and some others observed climb on the posterodorsal part of the shells. The paired penes are introduced deep into the posterior opening between the valves, reach the copulatory openings, and mating is accomplished in several minutes. Female *Philomedes* become planktonic only for mating, gaining swimming setae on their molt to maturity. Swarms ascend 50 m to join planktonic males. After mating they descend and saw off their swimming setae with their second maxillae. The adult males have reduced mouthparts and presumably live for only a short period of time.

Freshwater inhabitants, especially among Cyprididae, have many species in which males have never been found and which reproduce only parthenogenetically; in others there are males only rarely, or in the southern portions of their range alone. In culture, *Herpetocypris reptans* has lived 30 years parthenogenetically without meiosis. In cultures of *Cyprinotus incongruens*, which is parthenogenetic in part of its range only, isolated females from a bisexual culture became hermaphroditic; starved hermaphroditic females gave rise to a bisexual generation.

Development

After leaving the oviduct, the eggs first reach the space between abdomen and valves. In most species they are soon deposited individually perhaps on plants or left in clumps. *Heterocypris* uses the second pair of antennae to attach the eggs to a substrate. The cement used comes from "shell glands" within the carapace that open near the second antennae. Podocopina species of all three freshwater families keep their eggs in the space enclosed by the valves up to hatching time. The movement of the thoracic limbs turns the eggs constantly and supplies oxygen. *Darwinula* extends the brood care to the third, *Cyprideis* to the second larval stage.

Cleavage is total, gastrulation is by invagination. There are no coelomic pouches. The first larva that hatches is an atypical but strong swimming nauplius having

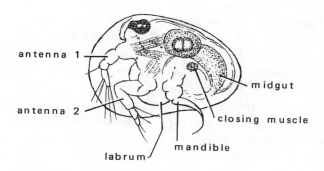

Fig. 5–5. First larval stage (nauplius) of *Cypris* in lateral view; 0.19 mm long. Valve is transparent; the mandible has a claw for walking. (After Claus from Korschelt and Heider.)

only uniramous cylindrical legs in addition to its carapace (Fig. 5-5). Usually after each molt the bud of a new appendage appears, which after the next molt it developed further.

In small species the appearance of the second maxillae may be postponed until after the third molt. *Heterocypris* postpones the primordia of the 6th appendages by one, *Darwinula* by two molts.

The number of larval stages varies; there are five in *Philomedes*, six in the Halocyprioidea, eight in many Podocopina. Adults do not molt again except for rare individuals of a few species. The exuviae of the extinct *Eridoconcha* and *Crytophyllus* are cemented to the new valves.

The change of function of appendages occurring after a molt is of interest. The nauplius can walk with the second antennae and the sickleshaped claws of the mandibles. In the second stage of *Cyprinotus* the mandibular endopodites are reduced into palps and a claw appears as the furca until the fourth stage. After the next molt there are claws on the long endopodites of the second maxillae. In the 6th stage these endopodites become small palps and claws develop on the first thoracic limbs, and at the same time the furca again becomes long and clawlike. The thoracic limbs generally attain their final shape in the instar before last.

The lifespan of freshwater ostracods depends on whether they overwinter as eggs or adults. Freshwater ostracods grow and mature faster than marine ones: some have been recorded to live 4 months, but others presumably live longer. Some marine inhabitants such as *Philomedes globosus* become 2.5 years old. The appearance of certain species may be seasonally limited, the animals passing the rest of the year within the eggs. Other species can be found all year, as long-lived individuals or as successive generations.

Habits

Myodocopina, Cladocopina, and Platycopina are found only in the ocean, while many Podocopina inhabit freshwater and several species of *Mesocypris* are terrestrial, in forest humus in New Zealand and South Africa. Only a few ostracods are pelagic, in the ocean *Gigantocypris* and the thin-shelled Halocyprioidea; in freshwater only a few tropical species. But *Cypridina* and some marine relatives may swim and rise to the surface. Some Cyprididae (*Notodromas monacha*, sometimes *Heterocypris incongruens* and *Cyprois marginata*) can be seen moving rapidly along, venter up, suspended on the surface film of the water, like the cladoceran *Scapholeberis*. Most ostracods live on plants or on the bottom, into which they may burrow 1–7 cm. They swim only short distances between algae.

There is a close correlation between the nature of the valves and habits in the many marine species. All marine swimmers have a convex curvature of the ventral edge of the shell (Fig. 5-3) and have thin valves. The coastal species that climb on plants are also thin-shelled (75% of Swedish species). Ostracods that dig in the substrate have thick-walled valves and (in the Swedish species examined) the valves are smooth; 73% of Swedish species of marine ostracods that live on the surface of the substrate have roughened or sculptured valves.

Freshwater ostracods are found in standing waters as well as on bottoms of brooks and streams; some species live in streams or pools periodically dry, temporary water in tree forks, etc. Some (*Candona*) survive in dry mud by closing their shells, and their eggs are resistant against drying and freezing for many years.

LOCOMOTION. Myodocopida swim well, venter down, by beating down forcefully with the second antennae extended between the valves. The basal article of the second antennae contains strong muscles. *Conchoecia* swims in jumps as does *Cyclops,* aided by the exopodites of the second antennae. Cyprididae, many of which inhabit freshwater, beat both pairs of antennae together, the first pair toward the posterior and up, the second pair posterior and down, forcing the body forward in a steady motion. Most push comes from the second antennae (Fig. 5-4). The second antennae of swimming species have long setae reaching beyond the claws. Cyprididae whose swimming setae reach only to the base of the terminal claw cannot rise in the water, but just glide above the bottom. Most members of the remaining families of Podocopina and Platycopina only walk. Podocopina usually have long slender first thoracic limbs as well as long second antennae with few clawlike setae. But within a genus the form of the limbs may differ, depending on habits. The first antennae in some species feel the way, in others they push obstacles aside or they remain immovable.

Cyprididae lift the body by pressing the sickle-shaped anteriorly directed claw of the first thoracic limbs on the substrate. At the same time they move the body forward, while the second antennae places its claws anteriorly into the ground to pull. The furca may take part by pushing.

Cytheridae are exclusively climbers and runners, and their locomotion is unlike that of other ostracods. The body is pulled forward by the second pair of antennae, assisted by three pairs of walking legs (the second maxillae, and first and second thoracic limbs), while the furca is reduced. The exopodites of the second antenna are reduced to long seta-like tubules from which issue silk secreted by a "shell" gland within the carapace. Dabbing the substrate with the silk the animal pulls along a dragline that enables it to climb on smooth surfaces and to climb back up a thread, spiderlike, in case of a fall.

With increasing size of the particles among which they live, the digging Cypridinidae, some Cyprididae and Cytheridae have a tendency to shorten articles of the antennae and walking limbs and at the same time to secrete a thicker exoskeleton (Fig. 5-3). The sand-inhabiting Cladocopina lack thoracic limbs completely.

The many Cypridinidae that have no limbs transformed into walking legs, pull themselves rapidly into the ground with their mandibular palps (*Cylindroleberis*), while the first antennae push sand grains up and posteriorly, providing needed space. As soon as the furca finds resistance, it aids by pushing. In *Philomedes* the second antennae also assist in motion. The Cypridinidae may also burrow.

FEEDING.   There is great diversity in feeding habits among different species. For the few for which the stomach contents have been examined, one can identify predators and carrion, plant and particle feeders.

The gut contents indicate that the following are predators: *Gigantocypris* feeds on copepods, small fish fry, and *Sagitta*. In the laboratory *Conchoecia* fed on dead copepods and malacostracan larvae. They handle food with the mandibular palps while keeping the valves open. Both maxillae and first walking legs push food to the mouth. Large particles are turned, exposing other sides to the mandibular chewing surfaces. The mandibles rotate around a point near the tip of the labrum, the endites rubbing against each other. The narrow esophagus permits only well chewed food to pass. Once 6 *Conchoecia* attacked a fairly large euphausiid killing it and feeding on it although otherwise only dead animals were used as food (Lockhead, 1968). *Cypridina castanea,* up to 6.6 mm long, feeds on minute mysids and heteropodid snails; *Cypridina norvegica,* up to 4 mm, feeds on minute polychaetes.

The carrion feeders observed are *Macrocypris,* which feeds on dead copepods and chews *Ulva,* and *Paradoxostoma,* which feeds on dead polychaetes, amphipods and plants. The mandibles of *Paradoxostoma* are modified (Fig. 5-6) to penetrate carrion and they suck with appressed mouthparts. Large groups of Cyprididae may flock to dead snails in freshwater.

Beech leaves that have fallen into the water are consumed, except for the skeleton, by large numbers of *Candona*. Algal filaments and desmids have been found in the stomach of *Heterocypris incongruens* and *Cypris pubera*. Diatoms, animal carrion and organic products of decay which accumulate on the bottom are eaten by running and burrowing Cytheridae, Darwinulidae, Candoninae, and *Cypridopsis.*

Philomedes and *Macrocypris minna* swallow mud, including such organic materials as foraminiferans, without selection, while other species are selective. In the midgut of some species of *Polycope, Cythereis, Loxoconcha* and *Cyprideis* a few mud particles with many diatoms were found. Gut contents of *Loxoconcha* and *Cytheresis* include also remains of animal carrion probably found in mud. Some ostracods chew also algae on the surface of barnacles. Filter-feeders are *Cylindroleberis, Cytherella,* and *Notodromas;* the first two filter stirred up detritus; *Notodromas* filters the scum on the water surface.

The mouthparts are so well hidden that their function has been difficult to observe. Only the structure of the mouthparts gives clues as to how they are used. Nothing is known about the predators or about those, for instance, that chew

Fig. 5–6. Mandible of *Paradoxostoma variabilis*; the endite forms a stylet. (After Sars.)

on rotting leaves. Many ostracods, especially those that remain on the substrate, collect particles out of the respiratory current that passes through the ventral part of the space between the valves. The particles can be increased by whirling up mud with the second antennae. The inner setae of the mouthparts presumably serves as a rake (*Cypridina*). *Cypridopsis vidua* has a large labrum that reaches to the edge of the valve and secretes mucus. Particles in the water (Fig. 5-4) are bound by strands of this mucus, which are then caught by setae of the mandibular palps. In moving back, the mandibular palps are combed by the first maxillae and captured particles are transported to the endites of the mandibles. The large setae of the ends of the second maxillae and the endites of the first thoracic limbs probably form a barrier behind the first maxillae.

The filter feeders, *Cylindroleberis* and *Cytherella,* which live on the substrate, do not move their filters, while *Notodromas* swings them rhythmically. In *Cylindroleberis,* the flow of the water is complicated and the filters highly differentiated, probably making the apparatus very efficient. At first the current produced by the fanlike exites of the second maxillae is forced into two well-defined channels. The paired first thoracic limbs have the form of two vertical plates, parallel to the median plane, with their ventral edges fitting into the edges of the valves below them. This arrangement produces a canal on each side, bordered below by the inside of the valve, above by the venter of the trunk, laterally by the long second maxillae with their exite fan and toward the midline by the first thoracic limbs. Between the first legs lies the furca, which prevents the flow of water into this space. In front of each channel is an immovable filter, the parallel setae, of which arise from the basal article of the first maxilla. These maxillae lie against the inner walls of the valves and move diagonally to the longitudinal axis of the animal, toward the anterior edge of the first leg (Fig. 5-7), thereby barricading the entrance to the water channel behind, and filtering all water. The accumulated filtrate is combed out by parallel setae of the anterior end of the second maxillae which move in a horizontal plane from the sides to the middle (Fig. 5-7). While the filter hardly moves, the first maxillae move freely, probably bringing the setae of the endites to the tip of the filter setae and taking the filtrate accumulated there to the mouth and the mandibles. The mandibles move regularly during feeding, stuffing the material deep into the foregut. No mucus secretions have been observed, in the preoral cavity.

Cytherella filters but uses different appendages. The mandibles have the shape of the filtering first maxillae of *Cylindroleberis* and function as filters. The mandi-

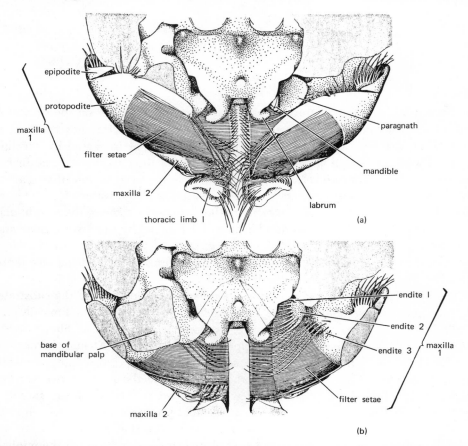

Fig. 5–7. Filters of *Cyclasterope hilgendorffi*; anterior view after removing valves; 2.6 mm wide. (a) The antennae and the mandibular palps have been removed. On the right side the mandible is cut off to expose the paragnath, on the left the mandibular palp has been cut off. (b) Anterior of same view as (a) after cutting transversely so as to cut off the anterior halves of the first maxillae, with their filter setae, the tips of the second maxillae and first trunk limbs. On the right side the paragnath has been removed to expose the underlying endites of the first maxillae. The faint outline on the labrum indicates the position of the scythe-shaped process of the mandible, the basis of which is shown on the right side of figure (a). The setae on the front inner edge of the first trunk limbs have been omitted. The furca (not shown) fits into the space between the first thoracic limbs. The light stippled areas are cut surfaces. (From Cannon.)

bles are curved longitudinally and cling laterally to the inner sides of the valves with their curved filter setae extending forward on each side to the ventrally projecting labrum. The labrum divides the ventral currents into two parts, each of which passes through a filter. Setae of the first maxillae comb the filtrate out, and inner setae of the endites as well as endopodites of the first maxillae transport the particles to the mouth. The passage through the mouth is made possible by dorsally bent endites of the mandibles and the first endites of the first maxillae.

Notodromas monacha filter the thin layer of scum on freshwater surfaces. The animal attaches itself, by the two planar gliding surfaces on each side of the ventral slit of the valves, to the surface film of the pond and glides along the surface dorsum down propelled by antennal beats (Fig. 5-8). The triangular exites of the first maxillae, by continuous fanning, move a current posteriorly through the ventral space between the valves. The mandibles remove bacteria from the current. The mandibular palps, which curve toward the venter and midline, bear a row of 10 parallel, feathered filter setae along the inner edges of the three distal articles (Fig. 5-8). These filter setae are bent posteriorly, forming on each side a comb that lies diagonally dorsolateral to the ventromedian surface. While taking in food the mandibular palps beat rhythmically 400 times per minute in a dorsomedian direction and at the same time posteriorly, the filtering surfaces extended toward the ventral current. At the end of each movement, the free ventral edges of the filter surfaces meet together below the labrum, thereby enclosing a space of triangular cross section, its ceiling formed by the large projecting labrum, its walls by the two filter surfaces. The filtered bacteria hang between the filter setae and are combed out by endite setae of the first maxillae. These endite setae press against the mandibular filter on each side, forcing themselves between the filter setae. As the mandibles move anterolaterally, the filters move too, and during this movement the maxillar endites, with their brushes of setae, sweep the filtrate posteriorly to the posterior border of the labrum. From here the food enters the preoral cavity within which the toothed mandibular endites work. The walls of the preoral cavity, the labrum and paragnaths also have teeth on an arched area which probably move dorsally transporting the food particles deeper toward the mandibles within the preoral cavity.

RESPIRATION. The large leaflike epipodites (exopodites according to some authors) the branchial plate, of the first or second maxillae create a continuous anterior–posterior current. The thin-walled carapace valves and the wall of the trunk transfer oxygen from the current to the hemolymph. Only in a few genera of the Cylindroleberididae are there gills in the form of seven paired, foliaceous folds of the trunk dorsum.

BIOLUMINESCENCE. A marine swimming species of *Cypridina* secretes in the labral gland a luminescent material which is sprayed out. The material becomes luminescent in the water and spreads as a bluish light. The presence of

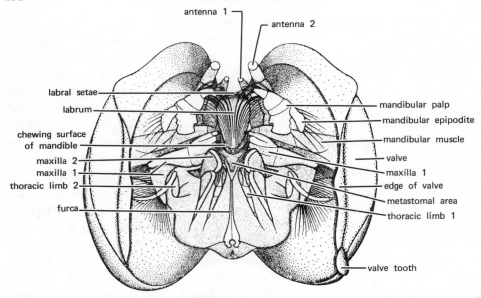

Fig. 5–8. *Notodromas monacha*, ventral view. Valves have been opened by force with a cover glass, spreading legs and body out unnaturally; first and second antennae cut off near their bases. (After Storch.)

luciferin and luciferase has been demonstrated but no symbiotic bacteria. Dried and pulverized Japanese *Cypridina* can be made to luminesce at any time by moistening the material and for a short time it produces a strong light that can be used for reading.

SYMBIOSIS.   *Entocythere* and relatives are symbionts of freshwater crayfish and freshwater isopods. Some have been found on the wood-boring marine isopod *Sphaeroma* and the amphipod *Chelura*. Some cytherids live also on the marine wood-boring isopod *Limnoria*. Live ostracods have been found within the gut of freshwater vertebrates, but they are probably accidental symbionts.

Ostracods are infested by a variety of parasites. In addition to ectoparasites such as ciliate protozoans, they may be hosts for nematodes, isopods, and larval copepods. Freshwater species also serve as intermediate hosts for certain cestodes and acanthocephalans. In *Candona* parasitized by the cestode *Hymenolepis* (the adult of which parasitizes the domestic black duck) the cysticerci attack the ovaries of the host, producing atrophy of the organs and marked hypertrophy of the carapace (R. V. Kesling, personal communication).

Classification

Ostracod taxonomy acquired practical importance when it was discovered that the material obtained in drilling oil wells is rich in ostracod fossils. These fossils permit the strata to be recognized and indicate the nature of brackish or freshwater deposits. It is one of many examples in which basic research, independent of practical aims, suddenly became of economic importance.

The classification of orders and suborders used here is that of the *Treatise on Invertebrate Paleontology*, Vol. Q.

ORDER MYODOCOPIDA

The valves have convex borders. The exopodites of the second antennae consist of several articles, and the second maxillae have a branchial plate. None of the appendages are walking limbs. There is usually, but not always, a pair of compound eyes and a heart with one pair of ostia (Fig. 5–3). The wide furcal rami are lamellate, with strong spines. All species are marine.

Suborder Myodocopina

The anterior end of the valve has a notch through which the antennae can be extended (Fig. 5–3). The basal article of the antennae, used for swimming, is very thick and pear shaped, its exopodite with nine, endopodite with one to three articles. The heart is always present. Lateral compound eyes usually present. The group is rich in species.

Cypridinoidea. Cypridina, 2–6 mm (Fig. 5–3), has some species that are luminescent. *Gigantocypris,* to 23 mm long, has soft, almost spherical red valves. The gape between the valves is short and, except for the anterior and a small posterior portion, very narrow. Both antennae and the mandibular palps have to extend through the anterior. Here the respiratory water enters. The animal floats, slowly swimming anteriorly at about 200–4000 m depth and preys on passing planktonic animals. One large *Sagitta* has been found, between the palps of a preserved specimen that had remains of two more *Sagitta* within its stomach. Another had a small fish in the stomach, and a third had nine copepods. The compound eyes are small, the naupliar eyes unusually large. The size of the limbs does not correspond to the large size of the body and their attachment region occupies only the anterior ventral quarter of the circular outline of the body. *Philomedes* is up to 3 mm long. *Cylindroleberis,* 1.8–8 mm long, and *Cyclasterope,* to 2.8 mm long, burrow below the surface and make a 1-cm tube of glued particles. The antennae beat posteriorly and push a belt of cement posteriorly. They are filter feeders, and presumably siphon off water from the surface through the tube. *Sarsiella,* like other genera mentioned, is cosmopolitan, reaching 0.8–2 mm long.

Halocyprioidea. The valves have an anterior notch. Unlike other Myodocopina the valves are only slightly mineralized. There is a heart; compound and naupliar eyes are absent. Hartmann (1966) raises this group to the suborder Halocypriformes. *Halocypris,** 0.8–8 mm long, has been collected with plankton nets. Another genus is *Conchoecia.*

Suborder Cladocopina

The valves lack a notch and are circular to oval. The second antennae have a multiarticulate exopodite and an endopodite with three articles. Thoracic limbs have been

* *Halocypris* is listed on the *Official List of Generic Names in Zoology.*

lost. The mandibular palps are weak, not leglike. Eyes and heart are absent. All live in interstices of the ocean floor. *Polycope,* 0.14–0.75 mm, is cosmopolitan in sand.

ORDER PODOCOPIDA

The valves have no anterior notch, their ventral border is straight or concave. The surface of the valves is sometimes sculptured. There are only vestiges of the exopodites of the second antennae. Swimming genera have the first article of the endopodites armed with terminal swimming setae. The second maxillae do not always have branchial plates; the branchial plates may appear on the first maxillae. The furca is absent, or when present is foot-like or flagellate. Compound eyes and heart are absent. Many species of Podocopida are marine, many freshwater, and a few terrestrial.

Suborder Platycopina

The valves have a straight ventral edge. The exopodites of the second antennae have two articles. The first thoracic limbs are leaflike in the female, chelate in the male. There are no second thoracic limbs. The furca is leaflike. The animals are marine and there is only one Recent genus, *Cytherella,* 0.5–1 mm long, worldwide in distribution.

Suborder Podocopina

The valves usually have a concave ventral edge. There is no exopodite on the second antennae or if present it is transformed. The first maxillae, and sometimes also the second, have branchial plates. The second thoracic legs are often present and are used as walking or grooming limbs. There are many species in marine and freshwater habitats.

Bairdiidae. The second antennae have a plate with setae in place of the exopodites. The second maxillae are leglike and have large branchial plates. The furca is rod-shaped. All are marine and walk. *Bairdia* and *Bythocypris* are cosmopolitan.

Darwinulidae. The valves are elongate, egg-shaped, narrower anteriorly. The exopodites of the second antennae are absent, only a seta being found on each one instead. Mandibles and second maxillae have branchial plates. There are two pairs of thoracic walking limbs. The furca is absent. Cosmopolitan *Darwinula stevensoni,* up to 0.75 mm long, is found on and in fresh water bottoms.

Cyprididae. Ventral border of valves is concave. The valves are only slightly mineralized. The exopodites of the second antennae are reduced to a scale with at most three setae. The first article of the endopodites is often armed with swimming setae. The second maxillae usually have branchial plates. The second pair of thoracic appendages is modified for grooming. The furca may be absent, shaped like a foot or filiform. Gonads usually lie in the blood sinus of the valves; the sperm is long and the ejaculatory duct has cylindrical muscles. The many species occur mostly in open water of lakes; some are marine, some terrestrial. *Pontocypris,* up to 0.85 mm long, and *Macrocypris,* up to 3 mm long, are marine. *Candona,** up to 1.6 mm long, and *Cyclocypris* inhabit freshwater. The anatomy of *Candona suburbana* has recently been studied in great detail (Kesling, 1965). *Notodromas monacha,* up to 1.2 mm long, is found in freshwater in Europe. *Cyprois,* up to 1.7 mm, *Cypris,* up to 2.6 mm, *Eucypris,* up to 2.5 mm, *Herpetocypris,** up to 2.6 mm, *Cypridopsis,* up to 0.8 mm long (Fig. 5–4), *Physocypria,* up to 0.7 mm (Fig. 5–1) and *Heterocypris,* up to 1.8 mm, are found in freshwater. *Platycypris* belongs here. *Mesocypris terrestris,* up to 0.9 mm, has been collected with a Berlese funnel from forest humus in South Africa, but seems to have few adaptations if any to the terrestrial habitat. The valves are covered by setae and close tightly; the second antennae are very strong with thick clawlike setae especially close to the tip. The furca is narrow with two large thick claws on the tip. *Mesocypris audax,* found in leaf mold of forests in New Zealand up to the subalpine region at 1000 m, closes in light if disturbed, then moves away by means of the 2nd antennae, 2nd pairs of limbs and furca. Short claws provide a good grasp, sufficient to permit climbing up the walls of glass dishes. The first antennae are used to investigate surroundings while the animal walks. When submerged in water, the animals survived for two weeks. Fecal material contains plant remains. The species is probably parthenogenic.

Cytheridae. The valves are mineralized and have sculptured surfaces (Fig. 5–2). The exopodites of the second antennae have been transformed into two- to three-articled spinnerets. There are no swimming setae. The second maxillae and both pairs of thoracic maxillae are transformed into walking limbs and lack branchial plates. The first maxillae have branchial plates. The furca is absent, usually replaced by one or two setae. The many species walk and are found in freshwater or marine habitats. *Cythere,* 1 mm long, is found in the North Atlantic. *Cythereis,* 1 mm long (Fig. 5–2), *Cytherura,* up to 0.6 mm long, *Loxoconcha,* 0.7 mm, are cosmopolitan, marine. *Microcythere minuta,* 0.23 mm long, in Heligoland, is the smallest ostracod species. *Paradoxostoma,* 0.7 mm, is cosmopolitan, marine. *Limnocythere,†* 0.9 mm, is cosmopolitan in freshwater. *Elpidium* is found in bromeliads in Brazil, *Entocythere* with many species are symbiotic on gills of crayfish and *Sphaeromicola* is found on isopods in caves in France. *Cyprideis* is found in brackish, rarely fresh water in Europe. *Xestoleberis* is cosmopolitan, marine.

References

Benson, R.H. *et al.* 1961. Ostracoda *in* Moore, R.E. ed. *Treatise on Invertebrate Paleontology.* vol. Q. Geol. Soc. Amer. Lawrence, Kansas.

Cannon, H.G. 1925. The segmental excretory organs of certain freshwater ostracods. *Phil. Trans. Roy. Soc. London* 214B: 1–27.

————— 1931. Anatomy of a marine ostracod, *Cypridina* (*Doloris*) *levis. Discovery Rep.* 2: 435–482.

————— 1933. Feeding mechanism of certain marine ostracods. *Trans. Roy. Soc. Edinburgh* 57: 739–764.

————— 1940. Anatomy of *Gigantocypris muelleri. Discovery Rep.* 19: 185–244.

————— and Manton, S.M. 1927. Segmental excretory organs of Crustacea. *J. Linnean Soc. London* 36: 439–456.

* The generic names *Candona* and *Herpetocypris* are listed on the *Official List of Generic Names in Zoology.*

† The genus *Limnocythere* is listed in the *Official List of Generic Names in Zoology.*

Chapman, A. 1961. Terrestrial ostracod of New Zealand *Mesocypris audax*. *Crustaceana* 2: 255–261.

Elofson, O. 1941. Marine Ostracoden Schwedens. *Zool. Bidr. Uppsala* 19: 215–534.

Fox, H.M. 1964. Larval stages of cyprids and on *Siphlocandona*. *Proc. Zool. Soc. London* 142: 165–176.

Franzl, W. 1940. Vitale Elektivfärbung und Funktion der "Schalendrüse" von *Cyprinotus incongruens*. *Z. Morphol. Ökol.* 37: 202–212.

Harding, J.P. 1953. A terrestrial ostracod, *Mesocypris terrestris*. *Ann. Natal Mus.* 12: 359–365.

Hart, C.W. 1962. Ostracods of the family Entocytheridae. *Proc. Acad. Natur. Sci. Philadelphia* 114: 121–147.

———— and D.G. 1967. Entocytherid ostracods of Australia. *Ibid.* 119: 1–51.

———— et al. 1967. Ostracod commensal on a wood-boring marine isopod from India. *Notulae Naturae*, Philadelphia. 409: 1–11.

Hartmann, G. 1955. Morphologie der Polycopiden. *Z. Wiss. Zool.* 158: 193–248.

———— 1961. Die Unterfamilie Cytherominae. *Zool. Anz.* 168: 66–79.

———— 1963. Zur Morphologie und Ökologie rezenter Ostracoden. *Fortschr. Geol. Rheinland Westfalen* 10: 67–80.

———— 1963. Phylogenie und Systematik der Ostracoden. *Z. Zool. Syst. Evolutionsforsch.* 1: 1–154.

———— 1963. Problem polyphyletischer Merkmalsentstehung bei Ostracoden. *Zool. Anz.* 171: 148–164.

———— 1966-1968. Ostracoda *in Bronns Klassen und Ordnungen des Tierreichs*. Akad. Verlagsges., Leipzig 5(1)2(4): 1–568.

Harvey, E.N. 1952. *Bioluminescence*. Academic Press, New York.

Hoff, C.C. 1942. The ostracods of Illinois. *Illinois Biol. Monogr.* 19: 1–196.

Hutchinson, G.E. 1967. *A Treatise on Limnology* vol. 2. Wiley, New York.

Iles, E.J. 1961. The appendages of the Halocyprididae. *Discovery Rep.* 31: 299–326.

Keilbach, R. 1952. Schalenasyemmetrien bei rezenten und fossilen Ostrakoden. *Zool. Anz.* 149: 147–157.

Kesling, R.V. 1951. The morphology of ostracod molt stages. *Illinois Biol. Monogr.* 21: 1–3.

———— 1961. Reproduction, ontogeny of Ostracoda *in* Moore, R.E. edit. *Treatise on Invertebrate Paleontology*. vol. Q: 19–20. Geol. Soc. Amer., Lawrence, Kansas.

———— et al. 1965. *Four Reports of Ostracod Investigations*. Authors, Ann Arbor.

Klie, W. 1926. Ostracoda *in* Schulze, *Biologie der Tiere Deutschlands*, Borntraeger, Berlin (22) 16: 1–56.

———— 1929. Ostracoda *in* Grimpe-Wagler, *Tierwelt der Nord-und Ostsee* 16: xbl–xb50.

Lochhead, J. H. 1968. The feeding and swimming of *Conchoecia*. *Biol. Bull.* 134: 456–464.

Moore, R.C. ed. 1961. *Treatise on Invertebrate Paleontology*. Q. Arthropoda 3, Crustacea, Ostracoda. Geol. Soc. Amer., Lawrence.

Morkhoven, F.P.C.M. van 1962-1963. *Post-Paleozoic Ostracoda*. vols. 1, 2. Elsevier, New York.

Müller, G.W. 1927. Ostracoda *in* Kükenthal, *Handbuch der Zoologie* 3: 399–434.

Pennak, R.W. 1953. *Freshwater Invertebrates of the United States*. Ronald, New York.

Scheerer-Ostermeyer, E. 1940. Entwicklungsgeschichte der Süsswasserostracoden. *Zool. Jahrb. Abt. Anat.* 66: 349–370.

Storch, O. 1933. Morphologie und Physiologie des Fangapparates eines Ostrakoden (*Notodromas monacha*). *Biol. Gen.* 9(1,2): 151–198; 355–394, 9(2,3): 299–330.

Tressler, W.L. 1959. Ostracoda *in* Edmondson, W.T. ed. *Ward and Whipple Freshwater Biology*, Wiley, New York, pp. 657–734.

Weygoldt, P. 1960. Embryologische Untersuchungen an Ostrakoden, die Entwicklung von *Cyprideis littoralis*. *Zool. Jahrb. Abt. Anat.* 78: 369–426.

6.

Subclass Mystacocarida

Three species of Mystacocarida are known, of which the largest is *Derocheilo-caris galvarina,* up to 0.5 mm long.

Mystacocarids are minute crustaceans with an elongate trunk consisting of 10 metameres. The first antennae are longer than the other appendages. The first pair of trunk limbs are maxillipeds; the following four pairs of appendages are very short, each consisting of only one article.

Anatomy

The elongate body shape of mystacocarids is probably an adaptation to their habitat, the spaces between sand grains (compare Fig. 6-1 with Fig. 7-11, a copepod from a similar habitat). A constriction divides the head into a short anterior portion bearing the first antennae, and an elongate posterior portion bearing the other cephalic appendages. The metamere of the maxillipeds is freely articulated, as are the following nine similar trunk metameres and the telson. On the posterior end of the head and on each metamere from 1-10 there is a pair of toothed furrows of unknown function. The first three pairs of appendages resemble those of metanauplii. The second antennae and the mandibles are distinctly biramous and are the important locomotory limbs. The first and second maxillae are uniramous; the maxillipeds sometimes have an exopodite. The maxillae and maxillipeds function as mouthparts. The caudal rami are large and clawlike.

The soft integument lacks pigment. Anteriorly there are usually four small, separate eyes, each with a lens. The tritocerebrum is far from the deutocerebrum and has a free commissure behind the foregut. The ganglia of the mouthparts are unusually large and are distinctly separated by connectives and commissures. The paired trunk ganglia 2 to 11 touch in the midline of each metamere, but each has two distinct connectives.

As in other minute crustaceans, the midgut lacks ceca. Nothing is known about the circulatory and excretory organs.

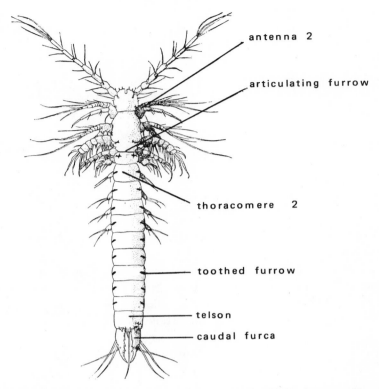

antenna 2

articulating furrow

thoracomere 2

toothed furrow

telson

caudal furca

Fig. 6–1. *Derocheilocaris remanei*, dorsal view; 0.35 mm long. (After Delamare and and Chappuis.)

The median ovary extends from the fourth or fifth somite posteriorly through one or two metameres and is continuous with a separate vitelline section which extends to the ninth metamere. The oviduct starts in metamere 5 or 6 and extends anteriorly to open on the fourth. Males resemble females except that the limb of the fifth metamere is modified for copulation. Their testis extends almost to the telson and has six to seven pairs of dorsal branches indicating an originally paired primordium. The median vas deferens opens ventrally on the fourth metamere.

Development

Mystacocarids have not yet been reared, and their postembryonal life is only known from the few stages that have been collected. Ten different stages have been collected of *Derocheilocaris remanei*, the smallest being a nauplius 0.15 mm long, which corresponds to the egg size and therefore is probably the youngest larva. The first stage of *D. typicus* is a metanauplius. Its body consists of an unsegmented anterior portion which bears first and second antennae and mandibles, three metameres, and the growth area with telson. The telson has a fully developed furca. It is of interest that the three anterior pairs of appendages resemble those of the adult, except that the coxae of the second antennae bear two long, bilobed endites, and the basis of the mandible bears a second shorter, simpler endite. These additional lobes (naupliar processes), found in all the early growth stages, are the major difference between larvae and juveniles or adults. The primordia of the first maxillae appear in the second stage and the buds of the second maxillae in the third, but both these pairs of appendages as well as more posterior ones attain their final shape only after several molts. The maxilliped first appears in the seventh stage, simultaneously with the buds of the fourth pair of thoracic limbs.

Relationships

The generalized morphology of the mystacocarid's cephalic appendages and the lack of fusion within the ladderlike nerve cord indicate a primitive condition. The close similarity of the adult to the metanauplius suggests that the adult morphology may have resulted from neotenic evolution. It is fairly certain on the basis of general tagmosis and morphology of the cephalic limbs that the Mystacocarida are most closely related to the Copepoda.

Habits

Derocheilocaris lives in the interstitial spaces in the sand of intertidal marine beaches. Only the Chilean species has been found in the sand of the sublittoral zone. *Derocheilocaris* has been found in eastern North America and western South America, the Atlantic coast of southern Europe, the western Mediterranean, the west coast of Africa, and the east coast of South Africa. Individual species have very wide distributions; all of the Eastern Hemisphere occurrences are of a single species, *D. remanei*.

On the French coast the thigmotactic *D. remanei* has been found in fine sands with a grain diameter of about 0.2 mm. By digging a small ditch in dry sand near the water line, one can collect the animals as water runs into the depression. The animals seem resistant to temperature changes between 10 and 25°C (but sensitive to temperatures above 27°C) and to variation in salinity between 10–40‰ Turbulence, low salinity and sand grain size are limiting factors. They live in capillary water among the sand grains and move forward rapidly by beating the second antennae, mandibles, and first maxillae. Also, the long setae of the second maxillae and the maxillipeds push the body among the sand grains, while the first antennae rapidly grope for the next space. The flexible, elongate trunk can bend double. It may telescope, then stretch out, using the furca as a prop, and thereby push forward. The short, stiff trunk limbs bend only at the base.

Food consists of detritus and microorganisms that are removed from sand grains by the long setae of the mouthparts. The particles are thrust under the large, posteriorly extending labrum into the mouth.

Classification

Derocheilocaridae. The first representative of this subclass was discovered in 1943 by R.W. Pennak and D.J. Zinn in an ecological study of beaches that had provided some unusual copepod species near Woods Hole, Massachusetts. It was named *Derocheilocaris typicus* and is found several centimeters below the surface in intertidal sand all along the eastern coast of the United States. In Chile, *D. galvarini* has been found intertidally and in coarse sand in water 25 m deep. The European and African species is *D. remanei.*

References

Armstrong, J.C. 1949. The systematic position of *Derocheilocaris* and the subclass Mystacocarida. *Amer. Mus. Novitates* 1413: 1–6.

Dahl, E. 1952. The Lund University Chile Expedition: Mystacocarida. *Lunds Univ. Arsskr.* (N.F.) 48(6): 1–41.

Delamare-Deboutteville, C. 1953. L'écologie et la répartition du mystacocaride *Derocheilocaris. Vie Milieu* 4: 321–380.

————— 1953. Revision des mystacocarides du genre *Derocheilocaris. Ibid.* 4: 459–469.

————— 1954. Le développement postembryonnaire des Mystacocarides. *Arch. Zool. Exp. Gén.* 91: 25–34.

————— 1960. Biologie des eaux souterraines littorales et continentales. *Vie Milieu* 9(Supplem.): 1–740.

————— and Chappuis, P.A. 1954. Morphologie des Mystacocarides. *Arch. Zool. Exp. Gén.* 91: 7–24.

Hessler, R.H. and Sanders, H.L. 1966. *Derocheilocaris typicus* revisited. *Crustaceana* 11: 141–155.

Jansson, B.-O. 1966. The ecology of *Derocheilocaris remanei. Vie Milieu* 17: 143–186.

Noodt, W. 1961. Crustaceous chilenos en aguas subterraneas. *Invest. Zool. Chilenas* 7: 97–99.

Pennak, R.W. and Zinn, D.J. 1943. Mystacocarida, a new order of Crustacea. *Smithsonian Misc. Coll.* 103(9): 1–11.

Zaffagnini, F. 1969. L' appareill reproducteur de *Derocheilocaris. Cah. Biol. Marine* 10: 103–107.

7.

Subclass Copepoda

There are 7,500 species of copepods known. One of the largest free-living species is the deep-water *Bathycalanus sverdrupi,* the female of which reaches a length of 17mm. The largest parasitic species is *Pennella balaenopterae,* on finback whales, with the female as long as 32 cm. One of the smallest is *Sphaeronellopsis monothrix,* a parasite of marine ostracods, in which the females may be as small as 0.23 mm and the males only 0.11 mm.

Copepods are small crustaceans that lack compound eyes and a carapace. The body is usually divided into an anterior appendage bearing part and a posterior part separated by a major articulation. The postgenital somites always lack appendages, except in one species of *Limnocletodes.* The first antennae are often longer than the other appendages. The first thoracic appendage is a uniramous maxilliped, the following four are biramous swimming legs, and the fifth is reduced and often uniramous. Females usually carry their eggs in one or two egg sacs. In their development copepods pass through a series of naupliar and copepodid stages. Parasitic copepods may lack many of these distinguishing features, but even though their bodies may be greatly transformed, they have developmental stages that are comparable to those of free-living species.

Free-living cyclopoid and harpacticoid copepods are known as fossils since the upper Tertiary, but parasitic forms are presumed to have existed earlier from the presence of cysts found on Jurassic echinoids.

Anatomy

Except for parasites and planktonic forms, most copepods are only about 0.5–2 mm long. In the discussion to follow, the morphology of free-living copepods is emphasized and the structure of parasites is discussed under symbiosis.

All together there are only 10 trunk metameres and the telson with the furca[*] consisting of two caudal rami. Of the six appendage-carrying thoracic somites one

---

[*] The term furca is also used for another structure in caligoid copepods, a bifurcate ventral projection between and posterior to the maxillipeds (Lewis, A.G. 1966. *Crustaceana* 10: 7–14).

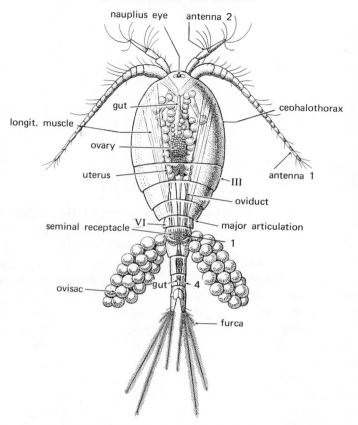

Fig. 7–1. *Macrocyclops albidus*, **dorsum of female; 2.5 mm long. Roman numerals denote thoracomeres, Arabic numerals pleomeres. (After Matthes.)**

or two are included within the cephalothorax, the others form the pereon, the four posterior ones the abdomen (pleon). The abdomen, except for a rudimentary pair of appendages in some species on the first pleomere lacks limbs. The gonopores are on the first abdominal somite, the anus dorsally on the border of the telson* (Figs. 7-1, 7-6).

The subdivisions of the copepod body are frequently complicated by fusions of somites and by differences in position of the major trunk articulation. In calanoid copepods this articulation is between the thorax and the abdomen, anterior to the somite bearing the gonopore. However, the cyclopoid and harpacticoid copepods have the major articulation anterior to the last thoracic somite while in caligoids it is one somite farther anterior (Fig. 7-2). Copepod specialists have terms to cover these variations. The body region anterior to the

* The telson bearing the furca may be a pleotelson containing the last pleomere in both Mystacocarida and Copepoda. At present there is no evidence one way or another.

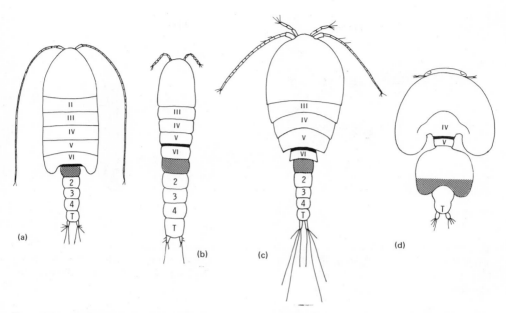

Fig. 7-2. Diagram of body regions of major copepod groups, represented by males with all pleomeres shown: (a) calanoid; (b) harpacticoid; (c) cyclopoid; (d) caligoid. Thick line indicates major articulation; stippled structure is genital somite; Roman numerals refer to thoracomeres, Arabic to pleomeres; T, telson. Region anterior to major articulation is called prosome; posterior, urosome. The external segmentation is often lacking. (After sketches made by J.R. Grindley.)

major articulation is the prosome, the region posterior to it is the urosome. The prosome consists of two regions. The first, the cephalosome, is the cephalothorax consisting of the head and one or two thoracomeres bearing maxillipeds and at times the first pair of swimming legs. The posterior part of the prosome, the metasome, includes most free thoracic somites and legs, and corresponds, in the calanoid copepods, to the pereon. However, in other copepods the last thoracic somites are included in the urosome (Fig. 7-2). In taxonomic works the somites of the metasome are numbered 1–5 (and often erroneously referred to as thoracic segments 1–5 whereas they actually are thoracomeres 2–6 or 3–6). The genital and first postgenital somites of the urosome are generally fused in female free-living copepods. However, various other fusions may be observed which greatly reduce the number of free somites.

The general shape of free-living copepods does not vary greatly. If it does, it is only in connection with different methods of movement. Harpacticoids show great variation in shape, including flattened and discoidal, elongated, fusiform, and other forms (Figs, 7-2, 7-11, 7-17). The bodies of parasites are often much

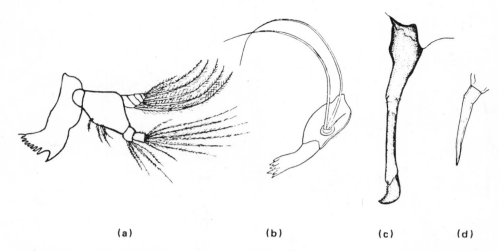

(a)   (b)   (c)   (d)

Fig. 7–3.   Copepod mandibles. (a) *Centropages typicus,* a filter feeder. (From Pesta after G.O. Sars.) (b) *Cyclops albidus,* a predator. (After Gurney, courtesy The Ray Soc.) (c) *Caligus randalli,* a parasite. (After Parker, Kabata, Margolis, and Dean.) (d) *Cardiodectes* sp., a parasite. (After Ho.)

transformed, becoming wormlike, sac-shaped, or developing long processes (Fig. 7-13).

The anterior projection of the head, called the rostrum (Fig. 7-4) may project forward as a short beak (as in many calanoids and harpacticoids) or is slung under the head as a posteroventrally directed lobe (as in most cyclopoids). Sometimes it is ornamented with setules or bears long filaments.

The first antennae are uniramous, often very long, and bear many sensory setae and esthetascs. Esthetascs are especially numerous in males. In many calanoid males the first antennae are modified on one side into a grasping structure. In most harpacticoids and cyclopoids they are geniculate on both sides for holding females. The second antennae are much shorter than the first. In calanoids and harpacticoids they are biramous, but in most cyclopoids uniramous, the exopodite usually being absent.

The mandibles in many species remain biramous (Fig. 7-3a), but the branches are short and resemble palps. The first maxillae lie close to the mandibles. Medially to these two pairs of appendages there may be a pair of paragnaths, which are small, often hairy lobes at either side of the metastomal area. The second maxillae represent the posteriormost head appendages. The maxillipeds, the first thoracopods, are usually the posteriormost appendages of the cephalo-thorax (cephalosome), since their somite is fused with the head. This is as in mystacocarids. The maxillipeds may be strongly prehensile as in some cyclopoids or may be reduced or absent.

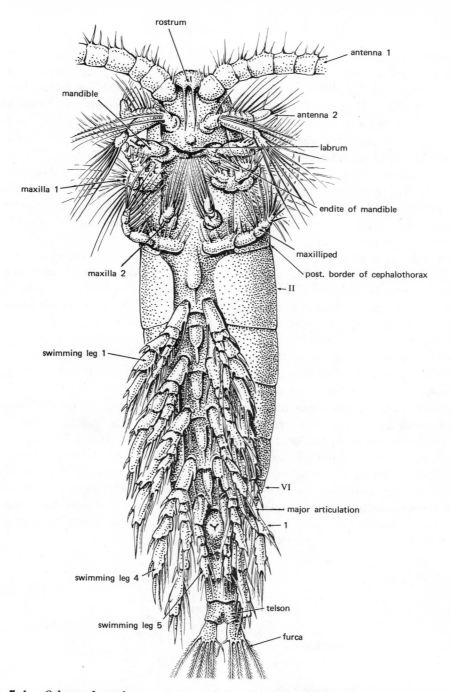

Fig. 7–4. *Calanus hyperboreus*; venter of female; 8 mm long. Swimming legs are in posterior position. This is a floater, with its major articulation on the anterior of the first pleomere. Parts of the first antennae and furcal rami are cut off. Roman numerals indicate the thoracomeres, Arabic numerals the pleomeres. The gonopore can be seen on the sternum of the first pleomere. (After Giesbrecht.)

The following four thoracic limbs, called swimming legs, are usually biramous, with the two-articled protopodite (coxa and basis) bearing an exopodite and endopodite, both usually with three articles. The two coxae are joined by an intercoxal plate (Figs. 7-4, 7-5), making the opposite legs a functional unit and forcing them to move together. The sixth pair of thoracopods (those on the last prosomal somite in calanoids or on the first urosomal somite in harpacticoids and cyclopoids) is often different in the two sexes. In calanoid females these legs may be reduced or absent, but in males they are transformed into a copulatory structure that grasps the spermatophores and transfers them to the female. In harpacticoids and cyclopoids the sixth thoracopod in both sexes is usually greatly reduced and may be absent in a few species. In adult parasites some or all legs may have been lost.

The coloration of copepods, frequently red or blue in calanoids, may be due to oil droplets from lipid reserves in the fat body or the oil sac. Pigments such as red zooerythrin or blue carotinalbumin occur in fresh-water copepods and probably also in the brilliantly colored free-living harpacticoids commonly found in tropical seas. In the marine cyclopoid *Sapphirina* there are interference colors on tiny platelets within hypodermal cells of the dorsum. Parasitic copepods may assume color based on the contents of the gut. Many copepods, however, are rather pale and transparent. The exoskeleton is thin and not mineralized. Generally it is smooth, but some species show minutely corrugated areas.

SENSE ORGANS AND NERVOUS SYSTEM.   The sensory organs include sensory setae, sensilla, esthetascs, and nauplius eyes. Esthetascs (also called aesthetes or esthetes by copepod specialists) are sense organs usually on the first antennae with a sclerotized articulated base, distally weakly sclerotized and with a rounded tip. Each nauplius eye consists of three ocelli, more developed than in other crustaceans compensating for lack of compound eyes. Rarely there may be more than three. Sometimes the lateral ocelli have moved apart and have large lenses near them (*Corycaeus, Sapphirina,* Fig. 1-16). There are no frontal organs and no brain centers for compound eyes.

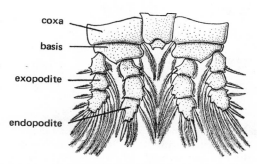

Fig. 7–5.   Swimming leg of female *Macrocyclops albidus*; 0.4 mm wide. (After Matthes.)

There is a supraesophageal ganglion, a pair of thick esophageal connectives, and a ventral nerve cord with ganglia (Fig. 7-6). The ventral nerve cord is only indistinctly segmented and extends into the last thoracic metamere. Parasites may have all ganglia so concentrated that they form a wide ring around the esophagus, from which leave lateral nerves.

Giant nerve fibers have been demonstrated in *Calanus*. The most important consists of a right and a left fiber that arise in the deutocerebrum and send off branches to the flexor muscles of the thoracic legs and to the longitudinal muscles

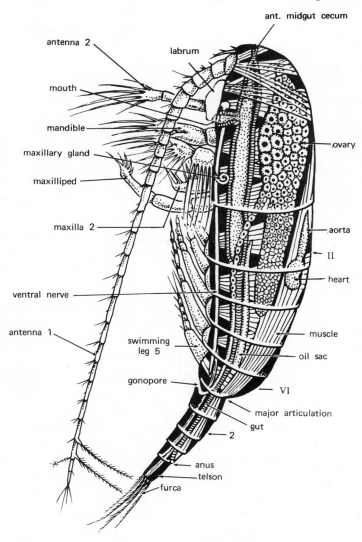

Fig. 7–6. *Calanus finmarchicus*, lateral view; 4 mm long. Roman numerals indicate thoracomeres, Arabic numerals pleomeres. (After Marshall and Orr slightly modified.)

of the pereon (metasome). These fibers probably facilitate rapid escape movements.

DIGESTIVE SYSTEM.   The mouth is inconspicuous. In siphonostomes the labrum and the metastomal area (behind the mouth) may be drawn out to form a siphon, sometimes as long as the body, along the sides of which are the styletlike mandibles.

The cuticularized, vertical foregut opens into a horizontal midgut that may have a median, anteriorly directed cecum (Fig. 7-6), sometimes with lateral pouches. The hindgut is very short, terminating in the anus on the dorsal side of the telson. In *Calanus* a conspicuous oil sac lies dorsal to the midgut, connected posteriorly to a cord of cells.

EXCRETORY SYSTEM.   A pair of maxillary glands (Fig. 7-6), opening on the second maxillae, performs the excretory functions in the adult. Antennal glands, located near the base of the second antennae, occur in naupliar stages.

CIRCULATORY SYSTEM.   Many copepods lack a heart. In harpacticoids and cyclopoids a circulatory system is absent. Strong pulsating movements of the gut provide for blood circulation in the hemocoel. Calanoids have a short heart (Fig. 7-6) with a pair of lateral ostia and a single median ventral ostium. An anterior aorta carries the hemolymph to the head. In the area of the five free thoracic somites there is a pericardial sinus separated as an upper tier of the body cavity by a membrane.

RESPIRATORY SYSTEM.   There are no special respiratory structures such as gills. Most copepods probably respire directly through the integument. In some genera (*Cyclops, Canthocamptus, Caligus,* and *Lepeophtheirus*) anal respiration seems to occur.

REPRODUCTIVE SYSTEM.   The dorsal gonads are usually unpaired, median structures in free-living copepods, but in parasites, which sometimes have large numbers of eggs, they are often paired. Two oviducts arise from the anterior end of the ovary. As the female matures, diverticula form where each oviduct loops posteriorly (Figs. 7-1, 7-6). The oviducts finally join into an ectodermal atrium and open on the genital somite, the first pleomere at the gonopore (in calanoids) or remain paired and open separately at two gonopores (in cyclopoids). Into the ectodermal atrium of the female calanoid open a pair of small seminal receptacles, via a fertilization duct. In cyclopoids ducts from a median seminal receptacle join each oviduct. In these copepods a median copulatory pore opens into the seminal receptacle, making three gonopores in all.

The vas deferens (single in calanoids but double in cyclopoids) arises from the anterior end of the testis. It turns posteriorly, becoming successively the seminal vesicle, the spermatophoric sac, and the ejaculatory duct, before opening on the genital segment. The duct is glandular and produces secretions which cement the spermatozoa together and form the spermatophore.

Reproduction

The two sexes usually differ in appearance, especially parasites, in which males may be dwarfs compared to females (Fig. 7-7). The first antennae, maxillipeds, and the sixth pair of thoracic limbs are often transformed into copulatory organs.

Sexual reproduction is usual in copepods, although a few cases of parthenogenesis have been reported in harpacticoids (*Elaphoidella* and *Epactophanes*). Although *Xenocoeloma* has long been considered hermaphroditic, recent observations indicate that there are males and females, as in the closely related *Aphanodomus*, also parasitic on polychaetes.

Males find females presumably by means of chemical senses. The male swims after a female and grasps her with the first (many calanoids) or rarely both pairs of antennae, maxillipeds, and then the sixth pair of thoracopods. The Diaptomidae and many other calanoids change the grasp and hold the female's abdomen with a hook on the sixth pair of thoracopods, letting go with the antennae. With his left forcepslike sixth thoracopod the male then attaches the spermatophore near the gonopore of the female. Copulation lasts about one-half hour. Sometimes several spermatophores may be attached to one female, indicating that the same female may be successively fertilized by several males.

In the cyclopoids the male embraces the fifth thoracic somite or the genital somite of the female with his two geniculate first antennae or with his maxillipeds (Fig. 7-8). During strong swimming movements spermatophores are attached at both gonopores. In the harpacticoids the male often grasps with his two prehensile first antennae the bases of setae on the caudal rami.

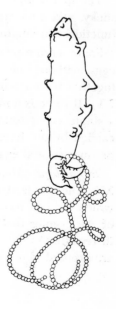

Fig. 7–7. Sexual dimorphism of *Akessonia occulta*; two small males attached to female; female 0.3 mm long. (After Bresciani and Lützen.)

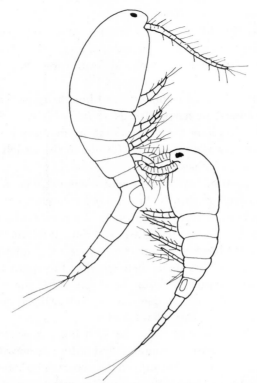

Fig. 7–8.   Mating pair of *Cyclops* sp.; right, male. (After Hill and Coker.)

The spermatophore of *Calanus,* an elongated sac more than 0.4 mm in length, is sticky, and after attachment the distal spermatozoa swell, pushing the proximal, functional spermatozoa into the seminal receptacles.

Some marine planktonic copepods (*Heterocope, Limnocalanus*) abandon their eggs, without holding them in sacs. In most copepods, groups of eggs are held together with a cement from glands of the distal oviduct, forming egg sacs (Fig. 7-1). If there is a median gonopore, or if the oviduct openings are close together, a single sac is formed, a pair if the gonopores are far apart. The egg sacs are carried by the female until the hatching of the nauplii. A fertilized female may produce several egg sacs.

In *Cyclops* (*Megacyclops*) *viridis,* the female has XY chromosomes, the male XX. The sex is determined at the time of egg laying and fertilization, during which the ova have a first division, a reduction division. At 18°C in about 44% of the zygotes the Y chromosome goes into the first polar body and is lost with it, resulting in 44% males. If the temperature is increased to 23°C and the eggs are exposed to ultraviolet light, in 87% of the oocytes the Y chromosome moves to the first polar body and is lost, resulting in 87% males ($2n = 10 + XX$). Thus in summer the number of males is greater than of females. Similarly, lack of food, pH 6, and pH 8 result in an increased percentage of males. In some other cyclopoids, the female, in addition to the autosomes, has only one X chromosome, the male two. In still other species no differences in structure between X or Y chromosomes can be seen with a light microscope.

Development

Eggs are laid several hours (*Eucyclops*) to several days (*Diaptomus*) after copulation. Each egg sac of *Cyclops viridis* contains about 50 eggs, but the female can produce 12–13 pairs of egg sacs, one after another. The egg sacs of parasites are often especially large, sometimes longer than the body. In some parasites the eggs are flattened and stacked like coins (Fig. 7-22). In *Lernaeocera branchialis*, the coiled up egg sacs, 15–20 cm long, may be five times the body length (Fig. 7-13d).

Development begins with total cleavage. Some parasites (lernaeopodoids) may have superficial cleavage. Free-living species usually pass through six naupliar stages and five copepodid stages before molting to the adult. The newly hatched nauplius (Fig. 7-9) lacks external segmentation. In succeeding nauplii the first maxillae, second maxillae, maxillipeds, and rudiments of the following two leg pairs gradually appear (Fig. 7-10a). A marked change in body form occurs at the molt to the first copepodid (Fig. 7-10b), with the larva assuming a shape more like that of the adult. At this instar segmentation becomes evident, with next limbs, thoracopods 4, being usually added. Additional metameres and appendages appear in subsequent copepodids. Secondary sexual characters and

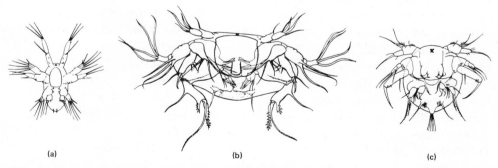

(a) (b) (c)

Fig. 7–9. Copepod nauplii. (a) Planktonic filter feeding first nauplius of calanoid, *Labidocera trispinosa*; 0.13 mm long. (After Johnson.) (b) One-week-old browsing nauplius of harpacticoid, *Stenhelia palustris*; 0.9 mm long. (After Bresciani.) (c) Browsing nauplius of harpacticoid, *Laophonte brevirostris*. (After Gurney, courtesy The Ray Soc.)

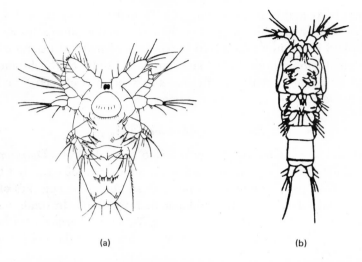

(a)   (b)

Fig. 7-10.   Copepod larvae. (a) Last naupliar stage of *Cyclops hyalinus*, venter. (After Gurney, courtesy of The Ray Soc.) (b) First copepodid stage of *Epactophanes richardi*, venter. (After Borutzkii.)

sexual dimorphism appear usually in the third copepodid. With the final molt the adult characters are fully expressed. Although in calanoids the full number of 12 instars (6 nauplii, 5 copepodids, and adult) is present, in many cyclopoids and most harpacticoids there are only five nauplii or rarely less. In parasites the number of developmental stages is often reduced, and some may have no free naupliar stages but hatch as a copepodid (*Salmincola*).

Planktonic nauplii of the Calanoida (laterally compressed, Fig. 7-9a), and Cyclopoida (often pear-shaped, Fig. 2-22) have appendages with long usually feathered setae and feed on plankton. The flattened nauplii of the Harpacticoida (Fig. 7-9b), which often creep or browse on the substrate have no feathered setae and are provided with prehensile setae and hooks at the basis of the second antennae and mandibles. With these attachments they take individual particles and move them below the labrum to the mouth. Detailed observations have not been made in most species. Along with the transformation to the first copepodid the feeding behavior changes, since the appendages now work as in the adult. Nauplii, with a large amount of yolk, do not feed, but subsist on yolk reserves.

Depending on the season, *Cyclops viridis* may live as long as 6–9 months (including the larval development of about 3 weeks). *Diaptomus vulgaris* may reach an age of 10–13 months. The adult female of *Lernaeocera branchialis,* a parasite on the gills of cod, probably stays on the final host for 1 year. The time elapsed from egg to egg varies with the species and environmental conditions. Thus, for *Calanus finmarchicus* this period is two summer months in the English Channel, but a year or more in East Greenland.

Habits

Unlike cladocerans, most copepods are marine, although many species inhabit freshwater. A few harpacticoids live out of water in moist moss, in moisture at the bases of sessile leaves, and in humus. Many calanoid and cyclopoid genera are found in marine plankton or in lakes and ponds. The depth at which they swim depends not only on the species and sex but on the temperature of the water, the season, the hour of the day and amount of light, and physiological condition (presence of eggs, etc.). In the northeastern Atlantic, between latitude 30° and 60°, most calanoids live in the upper 50 m (22 individuals/m³). In depths below 2,000 m fewer than 1/m³ were caught. In depths of 0–1,000 m there were 129 species; between 1,000–2,000 m, 71 species; 3,000–4,000 m, 25 species; and below 5,000 m, only 16 species.

Many species make daily vertical migrations which are undoubtedly controlled by light intensity, but may be influenced by other physical factors and endogenous rhythms of activity. Numerous studies have been made of the vertical migration of copepods by observations in the field based on sampling at different levels and by experiments both in the laboratory and in various experimental tubes lowered into the sea (Hardy, 1956; Hutchinson, 1967).

In general copepods rise toward the surface in the late afternoon as a result of swimming toward the light source of decreasing intensity. This upward migration is continued into the night oriented by gravity. During the hours of darkness position may be maintained by alternately swimming and sinking, or sinking may result from inhibition of activity. There may be a temporary rise at dawn due to positive movements toward early morning light of low intensity; however, with increasing light active, directed swimming results in downward migration. At the depth occupied during the day the migratory movement is replaced by a "hop and sink" behavior by which the copepods maintain their position until the upward migration of the next evening. Vertical migrations range from a few meters to 150 m or more.

Vertical migrations are carried out not only by copepods but by almost all zooplankton organisms. This behavior pattern appears to have evolved independently in all these groups, suggesting that it is of profound significance in their lives. Several suggestions regarding the significance of these diurnal vertical migrations appear to be of limited validity. It has been suggested that the downward migrations during the day are to avoid predators hunting by sight. However, many of the predators carry out parallel migrations and many plankton organisms including some copepods reveal their positions in the depths by luminescence. The hypothesis that zooplankton avoid dense phytoplankton which might produce exotoxins through photosynthesis during daylight has not been supported by experimental studies. It has even been suggested that vertical migration results from using light intensity to maintain optimum depth, but this appears unlikely when unvarying factors such as pressure might be more effectively employed. That each species stays in, and moves with, a zone of optimum light intensity is not supported by correlation of positions with photometer readings. That vertical migrations are a form of social display allowing copepods to sense in some way their population density and modify their reproductive rate appears highly unlikely.

The most probable explanation for this behavior is the difference in velocity and temperature of surface and deeper water masses. In most seas, estuaries, rivers, and even some lakes, surface currents are stronger. Therefore, by going down daily into the depths copepods and other zooplankton reduce their rate of transport into unfavorable regions and also ensure that they come up to feed on fresh patches of phytoplankton each night. It has also been suggested that such movements between water masses may prevent the isolation of plankton patches and facilitate gene flow through the population.

Finally, it has been established that vertical migration may serve to increase fecundity. It can be shown that copepods and other zooplankton reach their greatest size at low temperatures, although their growth is slower. As fecundity is an exponential function of adult size, it may be increased by living in cold deep water. By means of vertical migration copepods can offset the effects of this and obtain an energy bonus by feeding near the surface where the water is usually warmer.

Both this increase in fecundity and the distributional advantages of moving between water masses at different depths would appear to be of sufficient selective advantage to Copepoda and other plankton organisms to explain the widespread distribution of this behavior.

The marine planktonic copepods are often represented in one area by many species (for example, 104 species in a tow from 750 m to the surface made by the *Siboga* in the Banda Sea, west of New Guinea). However, in lakes the species of planktonic copepods number perhaps only two or three. Yet, as in the sea, freshwater copepods may at times occur in enormous numbers of individuals, up to 1,000/liter.

Freshwater free-swimming copepods (especially *Cyclops*) inhabit lakes, ponds, and slower flowing streams, either in open water or among plants. Bottom-dwelling copepods, especially harpacticoids, are found both in the ocean and in fresh water. Many harpacticoids and some cyclopoids live interstitially in spaces between sand grains of marine and lake beaches. These are often cylindrical in shape as an adaptation to the habitat (Fig. 7-11).

Some species can be found all year, others only during the warm or the cold season. In the freshwater calanoid *Epischura massachusettsensis* the single annual generation lasted from early April to the end of June, with no free-swimming individuals found in the pond during the rest of the year.

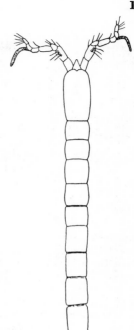

Fig. 7–11. *Stenocaris arenicola,* a sand dwelling harpacticoid; 0.6 mm long. (After Wilson.)

Cyclomorphosis is of little importance in copepods, in comparison to cladocerans. However, in many species variations occur in the size of individuals developing at different times of the year. Cyclomorphosis may also be indicated by particular features such as enlargement of a membranous projection in the second antennae in *Diaptomus* species.

Environmental factors may affect the size and morphology of copepods. Adults of *Cyclops* reared experimentally at low temperatures (7.7–8°C) were as much as 50% longer than those reared at higher temperatures (28–30°C).

Resistant stages are produced by some freshwater species if conditions become adverse, if mosses or puddles dry up, if the water temperature becomes too warm (for cold-water harpacticoids) or too cold (for calanoids and cyclopoids), or if the H_2S content increases. Some *Diaptomus,* near the end of the reproductive period, produce resistant eggs with a double chitinous shell (Fig. 7-12). These lie on the bottom and remain viable even if the water dries up. Copepodids or adults of cyclopoids and harpacticoids can crawl into mud and cover themselves with a secretion, in which condition they survive cold, heat, or drying-up (Fig. 7-12). Dry soil taken from a former pool and covered with water soon yields copepodid and adult cyclopoids and nauplii of *Diaptomus.* Probably these resistant stages in dried mud are important in the dispersal of copepods.

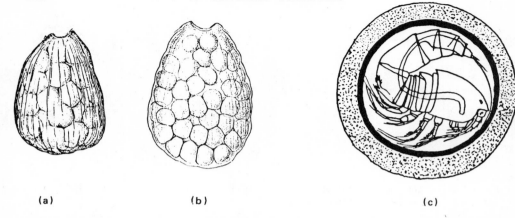

(a)   (b)   (c)

Fig. 7–12.   Egg sacs and cysts: (a) egg sac of *Diaptomus castor*, with resting or resistant eggs 0.7 mm long; (b) egg sac of *D. castor* with nonresistant eggs 0.9 mm long. (Both after Røen.) (c) Cyst of *Canthocamptus microstaphylinus* showing outer and inner layers. (From Spandl, after Lauterborn and Wolf.)

SYMBIOSIS.   Many copepods are internal or external parasites, or commensals, of fishes, whales, and invertebrates (sponges, coelenterates, mollusks, polychaetes, crustaceans, echinoderms, and ascidians). In freshwater there are only fish parasites, some of them of considerable economic importance. In the sea there are many more kinds of parasitic copepods, ranging from ectoparasites which may have only a loose or temporary association with the host (especially those associated with invertebrates) to those which are partly embedded in the host or actually live inside the body and have a true parasitic relationship (fishes and invertebrates).

Copepod parasites, such as *Ergasilus* (Fig. 7-13a), may still resemble free-living relatives. Other genera show all possible stages in reduction of body segmentation, leading to wormlike or saclike bodies (Fig. 7-13). The adaptations of copepods to a parasitic way of life involve morphologically both simplifications and specializations. Simplifications include the loss of body segmentation and the reduction or loss of mouthparts and legs. Specializations include the development of sucking discs and claws for attachment, modifications of the oral area (as the siphon of siphonostomes or the greatly enlarged second maxillae of lernaeopodoids), and deformations of the body (as in the long processes seen on the anterior region of *Lernaea*). Specializations sometimes occur in the larval stages, as in the chalimus (copepodid) stage of caligoids and lernaeopodoids, where the larva develops a threadlike filament on the front of the head for attachment to the host (Fig. 7-14).

Copepods may be parasitic as larvae but free-living as adults, as in monstrilloids, or may be free-living as larvae but parasitic as adults, as in many cyclopoids, notodelphyoids, caligoids, and lernaeopodoids. In some parasites the free-living

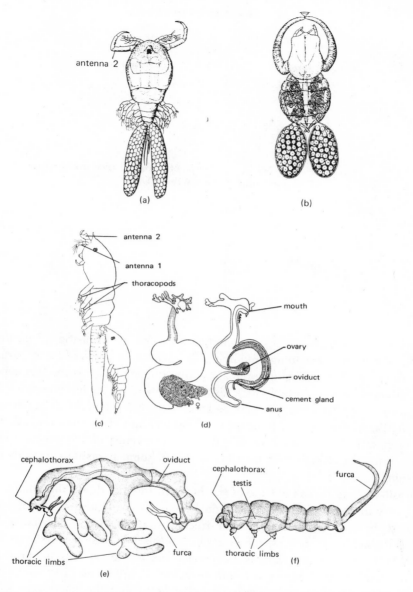

Fig. 7-13. Progressive changes in parasitic copepods. (a) *Ergasilus sieboldi*; the second antennae have transformed into holdfasts; trunk 2 mm long. (After Schaeperclaus.) (b) *Achtheres percarum*, female with second maxillae united into an attachment bulla; 4 mm long. (After Scott from Wagler.) (c–d) *Lernaeocera branchialis*; (c) mating of young adults as yet little transformed, female 3 mm long. (After Claus from Giesbrecht.) (d) Egg-carrying transformed female; left with eggs, right longitudinal section; up to 40 mm long. (After Wilson from Pesta.) (e) *Cucumaricola notabilis*, a holothurian parasite; left: mature female; right: mature male; trunk of female 35 mm long, male 4.5 mm long. (After Paterson, courtesy of Cambridge Univ. Press), copyright from N.F. Paterson 1958, External features and life cycle of *Cucumaricola notabilis*. *Parasitology* **48**: 272, 274.)

Fig. 7–14.   Chalimus of *Caligus rapax* attached to a fish scale; 3 mm long. (After Wilson.)

larval stage is very short. In some only the females are parasitic (*Ergasilus*, Fig. 7-13a), the males remaining free-living. In *Achtheres ambloplitis,* a parasite on the gills of rock bass, all nauplii remain within the egg and it is the first copepodid that hatches. It must attach to a fish usually within 36 hr, otherwise it dies. Sometimes more than one host may be involved in the life cycle, as in *Lernaeocera branchialis,* where the larvae occur on flounders and the adults on cod. Infestation of new hosts is made possible by the occurrence of free-swimming larval stages, transient though they may be. Males of some parasitic copepods may be neotenic.

Parasites living on the outside of the host (in the mouth, on the gills, or on the skin) vary in their degree of attachment. Some, like *Caligus,* may scuttle over the fish host, or may leave it and occur free in plankton. Other fish copepods are deeply embedded in tissues, for example, *Collipravus,* which penetrates through tissues of the throat and embeds its anterior end in the aortic bulb. True internal parasites include such forms as the monstrilloid *Cymbasoma* (Fig. 7-21), with its larval stages in the blood vessels of polychaetes and the cyclopoid *Mytilicola* (Fig. 7-18) in the intestine of mussels and oysters.

The developmental stages and life cycles of many parasitic copepods remain unknown. Indications are, however, that these stages and cycles are similar within closely related groups.

Not only are many copepods themselves parasitic, but many serve as intermediate hosts for helminth parasites of vertebrates, including man. Procercoids of *Dibothriocephalus latus,* the broad fish tapeworm of man, develop in *Cyclops* or *Diaptomus,* and plerocercoids form in plankton-feeding fishes which in turn are eaten by larger game fishes. The plerocercoids remain viable in fish muscles and infestation of man occurs when insufficiently cooked fish is eaten. *Lintoniella* cestodes of sharks have their plerocercoids in copepods. Copepods eat the coracidium of the cestode *Haplobothrium;* the coracidium then transforms into a procercoid, develops into a plerocercoid in a copepod-eating fish, and becomes adult in a ganoid fish. In the cat nematode *Gnathostoma* the first intermediate host is a copepod, the second a fish, frog, snake, or bird, and the final host a cat or other mammal, including man. In the Guinea worm, *Dracunculus, Cyclops* eats the first stage larva. The final host, usually man, becomes infested by swallowing the *Cyclops* in drinking water.

Copepods themselves may be affected by external parasites and diseases. Some microorganisms may multiply within the copepod until they cause its body to appear completely discolored. Various organisms may be attached to copepods including protozoans such as *Zoothamnium* and suctorians, diatoms such as *Licmorphora*, dinoflagellates such as *Blastodinium*, and some stages of certain isopods.

Monogenetic trematodes (Udonellidae) live on *Caligus.* Hydroids may be found on *Lernaeocera,* barnacles often on *Pennella.*

POPULATIONS. Some marine plankters live together in swarms during developmental and reproductive periods, and are important as food for many fishes. The center of a *Calanus finmarchicus* swarm, observed 66 days in succession, moved from the latitude of Edinburgh, Scotland, 148 km south toward the Dogger Bank in the North Sea. Ameba-like, the swarm changed its shape. At first it was 100 km wide and 20–40 km long. By following the growth of various larval stages within the swarm it was possible to identify the group. Collections made from 3.5 m above the bottom to the surface gave the following densities. Below 1 m² of surface between 24 March and 11 April there were more than 500 females of *Calanus;* on 14 April, more than 10,000 *Calanus* of various stages; on 24 April, more than 20,000; on 10 May, more than 100,000 of various stages. While on 25 March 1 m³ averaged 9.5 females and 0.75 larval stages, the number of larvae increased to 1,367 on 10 May, but only 11 male and 7 female adults in 1 m³.

BIOLUMINESCENCE. Luminescent organs are found only in marine calanoids (*Leuckartia, Metridia, Oncaea, Pleuromamma,* etc.). The structures consist of groups of secretory cells of greenish color on various parts of the body, even on the legs. The luminous secretion is expelled when the animal is disturbed, and perhaps serves to distract predators. The secretion is a mixture of two substances that luminesce from 2 to 37 sec, even after the copepod has departed. Two cells, 0.1 mm in length, open on each pore. Their secretions are different, the chemical compositions not known.

LOCOMOTION. The calanoids are floaters, with a long stiff prosome and a short abdomen (Fig. 7-2a, 7-4). Their long first antennae, at least half-body length, sometimes longer than the body, are held stiffly straight out from the body. The body often hangs diagonally, venter up. Friction, increased by the many setae, is so great that the copepod sinks only slowly, jerking itself up with the appendages at intervals.

Many tropical and Mediterranean species show setae on their appendages and caudal rami elaborately modified, for example, the feathered setae on the caudal rami of *Calocalanus* (Fig. 7-15). These setae may noticeably increase resistance. In some marine forms such as *Calanus,* the specific weight is reduced by oil droplets from a large oil sac.

When *Diaptomus* has sunk 2–4 cm, the legs make one paddling movement. A fraction of a second before this movement the first antennae relax and fold to the sides of the body. At the same time the body is propelled 2–4 cm up, and the first antennae stiffen again. Some species can move rapidly through the water by a steady movement. *Calanus finmarchicus* at times can make oblique or hori-zontal leaps of several feet. Many calanoids can progress by gentle swimming movements. For example *Diaptomus* progresses horizontally with the venter toward the surface. This motion is produced by the second antennae, mandibles, and first maxillae, while the legs are held anteriorly, immobile, at an angle of 45° (Fig. 7-6). The active head appendages do not beat metachronally but vibrate rapidly. Observers using stroboscopic lights and a motion picture camera have recorded that the second antennae make 1,200-2,700 movements per minute. At the same time each protopodite rotates at a right angle around its longitudinal axis, while the exopodite and endopodite are spread out. The movment can be imitated by extending the lower arm and hand in one direction and then with thumb extended turning the hand barely 90°. In the water this causes currents on each side which carry the copepod forward. *Diaptomus* swims in this manner horizontally, *Calanus* usually vertically.

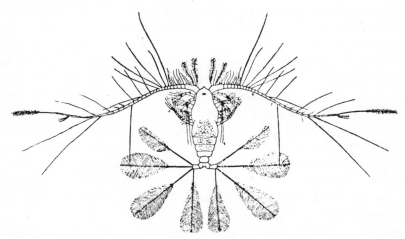

Fig. 7–15.   *Calocalanus pavo,* a planktonic calanoid with well-developed furcal rami; body 1.2 mm long. (After Giesbrecht.)

The swimmers, represented by the Cyclopidae, have the urosome anteriorly elongated by narrowing of the 6th thoracomere and an anterior shift of the major articulation to the borders between the 5th and 6th thoracomere (Figs. 7-1, 7-2). Consequently the center of gravity is shifted to the middle of the body and the species swim in all possible positions, venter up or down, directly down, up, obliquely, or horizontally. The most important locomotor organs are the legs, but the locomotion has not been studied in detail. The first antennae are used as a rudder by bending them to one side; they are almost always less than half the body length (Fig. 7-1). The individual species differ in their movements.

Cyclops sinks as soon as the motion of the limbs stops. It then stiffly extends the first antennae, while its body slowly rotates around the antennal basis, as if on a joint, until the anterior end faces down. It appears that the frictional resistance of the setae on the caudal rami is greater than that of the first antennae.

The swimming movement of the thoracic legs is alike in floaters and swimmers. The legs, folded anteriorly, are spread out laterally with their setal fans forming a wide swimming surface and then are beaten backward in a rapid metachronal rhythm. The last pair begins, and the next to the last joins immediately. Then all the limbs are appressed against each other, and all move forward together. Because they all move forward together, exposing a smaller surface, there is no change in the direction of movement started by their backward stroke. Although the angle of excursion of the legs is about 140°, forward and backward strokes together last about 1/60 sec. At this time the body jerks anteriorly or up. The first antennae first relax and hang at the sides. At each jerk the abdomen bends and with the caudal rami acts as a rudder. The first antennae may also serve as a rudder and depending on positions, the *Cyclops* can move forward, laterally, up or down, with the venter up or down.

The harpacticoids living in mud or sand and among sphagnum or aquatic plants are more wormlike. The body in these species shows no sharp division into a wide anterior prosome and a slender posterior urosome, but gradually narrows from head to telson, except in sand inhabitants, which may be cylindrical (Fig. 7-11). This worm shape facilitates the crawling motion accomplished by successive movements to left and right by lateral bending of the body, while the legs move the body forward at the same time. In swimming also the bending of the body facilitates locomotion, although the bending is mostly dorsoventral. The swimming motion of the narrow legs is not very efficient in these substrate-dwelling animals.

It has been found that some calanoid copepods such as *Pseudodiaptomus* may also burrow in bottom mud and detritus although they are generally regarded as planktonic.

In highly modified parasitic copepods the swollen unsegmented bodies, with legs reduced or absent, are often incapable of locomotion, but may show contractions and twisting movements. Many of these animals are unable to swim if removed from their hosts.

SENSES.   The senses have been little studied. *Calanus* keeps the typical naupliar longitudinal position due to the direction of light. *Diaptomus* separates two colors, a short- and a long-wave absorption. Many nauplii are positively photokinetic. Some copepodids of parasite copepods are known to be positively photokinetic in the first copepodid but negatively so in the second. *Centropages, Acartia,* and *Calanus* swim upward when placed in water of greater depth. They orient themselves in their vertical migration according to differences in pressure of the surrounding water, but pressure receptors have not been found. Geotaxis makes it possible for many cyclopoids to swim more or less horizontally with the body in a diagonal or horizontal position. That the body position is in this case not stabilized by a light sense can be demonstrated in aquaria by moving the light source from above to the side. (In such an experiment vertical swimmers immediately change direction, indicating that vertical swimming is light oriented.)

FEEDING.   The feeding habits of only a few copepods have been studied in detail.

Only calanoids are known to filter floating material, but not all members of the group are filter feeders. The water currents that arise on each side of the body when the head appendages rotate in swimming are used to draw in food (Fig. 7-16). In many cases, locomotion and obtaining food are one operation; only once has a copepod (*Eurytemora*) been observed to filter feed while sitting on

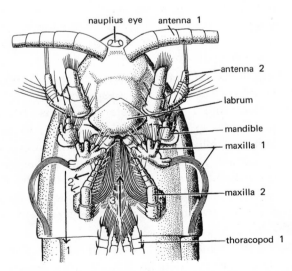

Fig. 7–16.   Anterior of *Diaptomus,* venter; 1.2 mm long. The second maxillae have a filter comb toward the inside of the proximal article. Above the filter comb is the maxilliped. Arrows indicate the most important water currents. 1, main current, part of the whirl; 2, the filtered water drawn toward main curent; 3, food current drawn by suction toward the filter channel. (Slightly modified after Storch.)

algae. Although the filtering may differ in details, the filter is always formed by the second maxillae. From the maxillary protopodites a row of setae, one above the other, leads toward the labrum. The setae of the left and right sides form two converging walls that close the filter channel (Fig. 7-16). The currents moving laterally past the filter comb draw water out from the filter chamber from between the setae. The filter comb along its mesal wall removes floating material. The pressure of the water running into the vacuum of the filter channel carries the filtrate further to the mandibles, aided by the endites of the first maxillae. The long setae on the exites beating transversely to the body axis and drive water in the filter channel. The maxillipeds in many genera take part in the motion.

The taking up of food, however, is not an automatic action produced by swimming. Calanoids swim for some time before taking up floating material. *Diaptomus gracilis* in a pond full of planktonic algae may have only bottom-dwelling desmids in the gut. Evidently they select. The rate of pumping of *Calanus finmarchicus* has been estimated by various methods to be 40 ml per summer day.

The gut contents of 3,000 *Calanus* examined consisted of 30–40% green or brown particles, mainly digested detritus and flagellates without cell walls. There were also skeletal parts of diatoms, radiolarians, shells of siliceous flagellates, coccolithophorids (Chrysomonadina), and tintinnids (spirotrichous ciliates). Fragments of small crustaceans indicated that *Calanus* can take larger prey.

Many cyclopoids and some calanoids (*Acartia, Anomalocera,* female *Euchaeta,* and *Candacia*) are predators. Some calanoids (*Calanus, Temora,* and *Eurytemora*) are predators in addition to being filter feeders.

Cyclops viridis, C. albidus, and *C. fuscus* swim just above the bottom in vertical flat arcs. In sinking they hit the bottom or vegetation with their two pairs of antennae, often finding by touch small oligochaetes or chironomid larvae. They do not react to prey a few millimeters away. The prey, perhaps a chironomid, may be as long as the predator. It is handled by the spiny first maxillae, while the second maxillae and maxillipeds help to hold it. If it escapes it is attacked again, 15 times within 10 min of observation. The first maxillae then move the prey between the open mandibles, of which the chewing surfaces below the mouth are parallel to the long axis of the animal. The mandibles squeeze the prey, and by moving anteriorly and up they tear off a piece and pass it to the mouth. These rapid movements, carried out during 2–4 sec, are followed by a one minute pause before the larva is handled again, after being pushed forward by the first maxillae. The copepod does not chew, but each time tears a piece from the larva and pushes it into the very extensible pharynx. The pharyngeal muscles produce strong suction. A 2-mm long chironomid may be torn apart and swallowed within 9 minutes, but a 3-mm larva takes about half an hour to be torn apart. A soft 4-mm *Nais* disappears much faster, within 3.5 min. If the prey is twice as big as the predator, only the viscera may be torn out and fed on, while the integument and head are abandoned. Cyclopidae have been seen to catch other copepods, and daily consume five first or second instar larvae of *Culex* or *Anopheles*. They also have been observed to chew fish carrion. They will at times follow very small swimming prey. *Cyclops strenuus* may swallow coracidia and cercariae whole.

The feeding behavior of some cyclopoids is intermediate between that of predators and parasites. Some Oncaeidae and Corycaeidae hold the prey with their maxillipeds, open the integument, and take in body fluids.

Plant feeders, such as *Eucyclops macruroides* and *E. macrurus,* grasp floating clumps of microscopic green algae, for example, *Scenedesmus* 20 μm in diameter), *Micractinium* (4–8 μm in diameter), and push them into the mouth, using the same methods as predators, the mandibles tearing off pieces. The cells may be found undamaged in the gut. Algal filaments 20 μm in diameter are handled as are chironomid larvae, and are swallowed in about 3 min if 0.3 mm long. Thicker ones, 50 μm in diameter, are pressed out or are cut by the first maxillae and then sucked out. Many plant feeders swim along plants or on the bottom, feeding on one-celled green algae or diatoms.

Details of how harpacticoids feed are mostly unknown. *Diarthrodes cystoecus* chews on marine algae, preferring red algae over brown and generally Rhodymeniales over other Rhodophyceae. When the adult feeds the oral cone is appressed to the alga. The first maxillae tear the cells of the alga, each stroke of the appendage conveying torn off food material to the mouth. Only when food has accumulated in the preoral space do the mandibles push it on into the esophagus. The second maxillae and maxillipeds do not participate in actual food handling. The feeding mechanism in the nauplius of *D. cystoecus* is very different from that in the adult.

Parasites use their sharp, often styletlike mandibles to penetrate the host, and then suck on the host tissue (as in some cyclopoids including siphonostomes, in *Caligus,* etc.). Others (*Ergasilus*) secrete enzymes that externally digest the gill epithelium of the host. *Lernaeocera* pushes its anterior region into the host tissue taking up blood. The exact nature of the food and feeding mechanism in most parasitic copepods is unknown.

Classification

The parasitic groups cannot all be satisfactorily derived from the free-living ones. The adoption of a parasitic way of life can be seen to varying degrees in all orders except the Calanoida.

ORDER CALANOIDA

Members of this free-living order are recognized by having the major articulation between the thorax and abdomen. Thus the urosome corresponds to the abdomen and the gonopore is the first urosome somite. The abdomen is much narrower than the prosome (Figs. 7–2a, 7–4, 7–6). The first antennae, with 22–25 articles each (26 in

Ridgewayia), are at least half the length of the body and often longer than the body. The right first antenna of the male (left first antenna in some genera) is often transformed into a grasping structure (neither first antenna is modified in *Platycopia*). The biramous second antennae and the mandibles with their biramous palps are locomotory structures. A heart is present. The vas deferens is developed only on one side. Some females deposit eggs free in the water, others carry them in an egg sac.

Calanoids are usually pelagic, in the ocean, brackish water, and freshwater. *Calanus finmarchicus* (Fig. 7–6), up to 5.4 mm long, is widespread in the North Atlantic Ocean to lat 79°N, and has been found in the Mediterranean, Black Sea, South Atlantic Ocean, Indian Ocean, and Pacific Ocean. In the eastern North Atlantic great red swarms of these copepods may appear at the surface in summer. They are one of the most important foods for herring, and in the Barents Sea during certain months this species provides 65% of the herring diet. The Sei whale also feeds largely on this copepod.

Some other marine planktonic genera are *Centropages, Pseudocalanus, Calocalanus, Heterocope, Acartia, Euchaeta, Temora, Candacia, Anomalocera,* luminescent genera such as *Metridia, Pleuromamma,* and *Leuckartia,* and deep-water genera like *Bathycalanus* to 2,500 m depth. *Limnocalanus* and *Eurytemora* are euryhaline found in the sea and also in brackish and fresh water. Fresh-water genera include *Diaptomus,** *Osphranticum,* and *Epischura. Diaptomus,* up to 5 mm in length, is a large genus, with more than 50 species known from the United States, and many others from Europe, Africa, Asia, Australia, and South America. *Labidocera* is another genus.

ORDER HARPACTICOIDA

Most harpacticoids are free-living and of small size, less than 2 mm long. Usually the body is slender and often cylindrical, with thorax and abdomen tapering gradually (Figs. 7–2, 7–11, 7–17), though some species have flattened or compressed bodies. The somite of leg 5, the last thoracic somite, is joined with the genital somite, and the major articulation is between the somite of leg 4, the 5th thoracomere, and that of leg 5, the 6th thoracomere (Fig. 7–2). The first antennae, with rarely more than eight articles, are usually shorter than half the body length; in males both first antennae are transformed into a grasping structure. The second antennae are biramous, but their exopodite is often very small. The mandible has a palp. The maxillipeds may be prehensile, sometimes rudimentary. The first leg may be a grasping organ, with terminal clawlike spines. Legs 2–4 are swimming appendages. Leg 5 usually shows an inner expansion on the basal part, but is often rudimentary. The heart is absent (except in *Misophria*). The female carries one, sometimes two, ventrally attached egg sacs.

Relatively few species occur in plankton. The many species can swim short distances, but usually they crawl on the substrate, in sand, or among algae, eelgrass, or sphagnum. *Ectinosoma, Tisbe,* and *Miracia* occur in marine plankton. Some littoral marine genera are *Harpacticus, Thalestris, Nitocra, Stenhelia,* and *Mesochra* (Fig. 7–17a), all with more or less elongated bodies; *Alteutha* and *Porcellidium* (Fig. 7–17b), both with dorsoventrally flattened and isopod-like bodies; *Tegastes,* laterally flattened

---

* *Diaptomus* is on the *Official List of Generic Names in Zoology.*

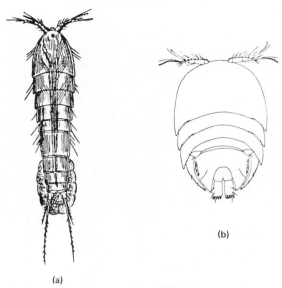

(b)

(a)

Fig. 7–17. Harpacticoid copepods. (a) *Mesochra lilljeborgi*, female with eggs; 0.5 mm long. (From Borutzkii after Sars.) (b) *Porcellidium echinophilum*, female; 0.6 mm long. (After Humes and Gelerman, courtesy of E. J. Brill Ltd.)

and amphipod-like; and *Leptomesochra, Evansula, Cylindropsyllus*, and *Stenocaris* (Fig. 7–11), all sand-dwellers with greatly elongated, cylindrical bodies. *Amphiascus* often occurs in brackish water. *Canthocamptus* is widespread in freshwater. *Bryocamptus* inhabits moist moss, springs, brooks, ponds, swamps, and caves.

Cancrincola and *Antillesia* inhabit the gill chambers of land crabs. *Balaenophilus* lives on the baleen of whales, *Cholidya* on octopuses, *Neoscutellidium* on Antarctic fishes, and a species of *Sacodiscus* on lobsters. *Sunaristes* inhabits the shells occupied by hermit crabs. Other harpacticoid genera are: *Diarthrodes, Limnocletodes, Laophante* and *Epactophanes*.

ORDER CYCLOPOIDA

The cyclopoids may be either free-living or parasitic. The major articulation is between the last two thoracomeres, between somites of swimming leg 4 and that of swimming leg 5 (Figs. 7–1, 7–2c). The last thoracomere is not firmly joined to the genital somite. The prosome, the trunk anterior to the major articulation is much wider than the urosome (the last thoracomere and abdomen). The first antennae, often modified for grasping in males, are moderately long, usually shorter than the prosome, and often have more than eight articles (usually not more than 17, but as many as 26 in *Cyclopicina*), sometimes fewer. The second antennae usually lack an exopodite, though a rudimentary one is present in many siphonostomes. The mandibles have a palp in some, the palp is reduced to a few setae in others, and in many the palp is absent. The basal part of leg 5 lacks an inner expansion. There is no heart. The testes are paired.

The females carry two dorsally or dorsolaterally attached egg sacs (a single egg sac in *Corycaeus*).

Most cyclopoids are free-living, in the ocean, and in brackish and fresh water, but many are parasitic. The order is often divided into three suborders. The position of some genera, such as *Akessonia*, is uncertain.

Suborder Gnathostoma

The Gnathostoma have both first antennae geniculate in the male. The second antennae lack an exopodite and terminal claws. The mandibles and first maxillae have masticatory areas. The maxillipeds are not subchelate, and does not show sexual dimorphism. Legs 1–4 are well-developed, with usually 3-articled branches. Most species are free-living, and range from about 0.6–5 mm in length. Among the marine genera are *Sapphirina*, *Cyclopina*, and *Oithona*. *Halicyclops* lives in brackish water. *Cyclops* (Fig. 7–1), *Mesocyclops*, *Macrocyclops*, *Eucyclops*, and *Paracyclops* inhabit freshwater.

Suborder Poecilostoma

The Poecilostoma have nonprehensile first antennae. The second antennae lack exopodites and are often prehensile. The mandibles lack masticatory areas, each may be produced into a slender process. In a few parasites mandibles are absent. The first maxillae are reduced. The maxillipeds are prehensile in the male, often reduced or even absent in the female. Legs 1–4 are generally present and well developed, with the branches 3-, 2-, or 1-articled, the reduction usually in the endopodites. Many species are free-living, but many also are parasitic on invertebrates and fishes. *Oncaea*, with luminescent organs, and *Corycaeus*, with corneal lenses, occur in marine plankton.

The female of *Ergasilus* (Fig. 7–13a) is parasitic on the gills of marine and freshwater fishes, to which it attaches by apical claws on the second antennae; the male is free-living. The egg sacs are as long as the body and contain many small eggs. *Ergasilus sieboldi*, up to 1.7 mm long, is free-living in slowly running water during postembryonal development (five naupliar stages and five copepodid stages) and until mating (at 2–2.5 months). During this period the animals have blue pigment. After mating, the females attach to the gills, rarely the skin, of freshwater fishes or herring. Rheotaxis leads them to the respiratory water of fish gills. They climb among the gills, holding on with their second antennae, and digest and swallow the epithelium. The damaged gills are susceptible to mold infection, or in healing adhere, reducing their surface. *Ergasilus* is the most important fish parasite in Central Europe. As many as 3,000 specimens have been taken from a tench (*Tinca tinca*) 25 cm long. All slowly moving species of fishes are parasitized. With 100–1,000 parasites a fish loses weight.

Mytilicola intestinalis (Fig. 7–18), with males to 3.5 mm, females to 8 mm long, lives as reddish adults in the gut of mussels (*Mytilus*) in Europe. It also occurs in oysters and other bivalves. The weakly segmented body is greatly elongated, with the prosome and urosome nearly the same width. The second antennae are prehensile, and the mouthparts (mandibles absent) and legs reduced. The copepods move in a worm-like fashion. The first three larval stages (first nauplius, second nauplius, and first copepodid) are free-living. The remaining four copepodids and adults are parasitic in

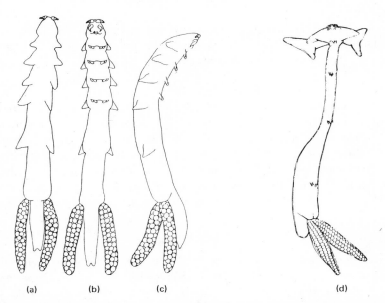

Fig. 7–18. **Cyclopoid parasites of the suborder Poecilostoma. (a–c)** *Mytilicola porrectus*, **female 3 mm long: (a) dorsal, (b) ventral, (c) lateral. (After Humes, courtesy of J. Parasitol.) (d)** *Lernaea cyprinacea*, **female 7 mm long. (After Yamaguti.)**

mussels. In parasitized mussels protein digestion is speeded up, the O_2 requirement is increased, and at the same time the capacity for filtration and absorption of food is lowered, but whether the parasite is the sole cause of these physiological changes is unknown. Production of edible mussels in Europe has been greatly reduced in areas where they are heavily parasitized by the copepod. Other species of *Mytilicola* occur in America and Japan, but are of less importance.

Lernaea (Fig. 7–18d) is found on freshwater fishes in all continents of the world. The adult female, up to 22.5 mm long in *L. cyprinacea,* has an elongated straight body. The head is a small knob, near which there are one or two pairs of short soft horns, embedded along with part of the slender neck in the flesh of the fish. Legs 1–4 are small, biramous, and located far apart. The adult male retains the more usual cyclopoid form. There are three free-living naupliar stages, followed by the first copepodid which settles on the skin or gills of the fish, attaching with the aid of the second antennae and maxillipeds. Second, third, fourth, and fifth copepodid stages follow. The next molt produces males and females which for a time are free-living (these females of *L. cyprinacea* up to 1.1 mm, males 0.72 mm). After copulation the males die and the females begin penetration of the fish. The subsequent elongation of the body of the female is mainly in the four leg-bearing somites.

Other poecilostomes include symbionts like *Hemicyclops* loosely associated with crustaceans and polychaetes; *Lichomolgus* associated with a variety of hosts such as octocorals, madreporarians, pelecypods, nudibranchs, echinoderms, and ascidians; *Octopicola* on cephalopods; *Stellicola* on starfishes; and *Scambicornus* (Fig. 7–19a) on holothurians. Examples of more strictly parasitic genera are *Sphaeronellopsis* in marine ostracods, *Bomolochus* on the gills of marine and brackish water fishes, *Chondracanthus* in the mouth or on the gills of marine fishes, and *Tucca* on the skin of marine fishes. *Meomicola* is another poecilostome genus.

Suborder Siphonostoma

The Siphonostoma usually have geniculate first antennae in the male. The second antennae often have a very small exopodite. The labrum and metastomal area are usually produced to form a cone (in some cases a very long siphon, as in Fig. 7–19b).

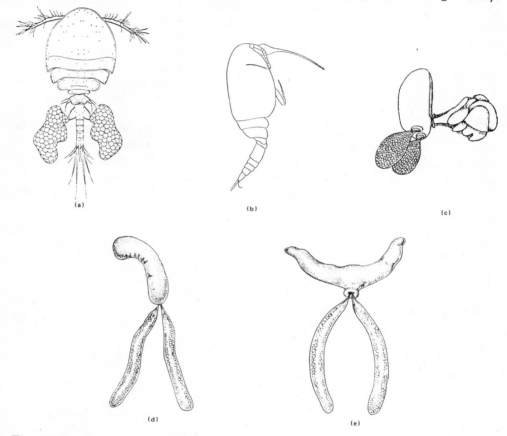

(a)   (b)   (c)

(d)   (e)

Fig. 7–19. **Cyclopoid parasites of the suborder Siphonostoma. (a)** *Scambicornus lobulatus,* **female, dorsum; 1.5 mm long. (After Humes.) (b)** *Myzopontius australis,* **female, lateral view showing long siphon; 0.9 mm long. (After Nicholls.) (c)** *Herpyllobius arcticus,* **female with anterior part of body buried in an annelid; 1.5 mm long. (After Brian.) (d)** *Xenocoeloma brumpti,* **female; 3.5 mm long. (After Bresciani and Lützen.) (e)** *Aphanodomus terebellae,* **female; 5 mm long. (After Bresciani and Lützen.)**

The mandibles are stylets, sometimes with a palp. The second maxillae and maxillipeds are subchelate and prehensile. Legs 1–4 are well-developed, usually with 3-articled branches. These copepods are commonly associated with marine invertebrates, but are often collected free. *Asterocheres* lives commonly on sponges, madreporarians, and echinoderms; *Micropontius* occurs on cake urchins; and *Cancerilla* on ophiuroids.

Possible members of the Cyclopoida are very highly modified genera such as *Herpyllobius* (Fig. 7–19c), *Xenocoeloma* (Fig. 7–19d), and *Aphanodomus* (Fig. 7–19e), all parasitizing polychaetes.

ORDER NOTODELPHYOIDA

This order is artificial, but for practical reasons it is used here. Probably the various genera should be divided between the Gnathostoma and the Poecilostoma, and the order abandoned. It may be characterized as follows: The body is cyclopoid, although often modified, sometimes exceedingly swollen, or elongate and wormlike. A brood pouch may be present in the dorsal part of the thorax (Fig. 7–20). The first antennae are either clasping organs or unmodified. The second antennae, mandibles, and legs 1–4 are either unmodified or reduced. In some, the maxillipeds are prehensile and the caudal rami specialized. Males are often, though not always, free-swimming, but females stay with the host.

Notodelphyoids are marine, associated chiefly with ascidians, where they inhabit the pharynx. *Notodelphys affinis*, living in *Ascidia*, has five free naupliar stages, followed by five copepodids and the adult. The first copepodid is free-living, and the second is presumed to be the infective stage since it and the remaining stages occur in the ascidians. Other examples are *Notopterophorus*, *Demoixys* (Fig. 7–20b), and *Scolecimorpha*.

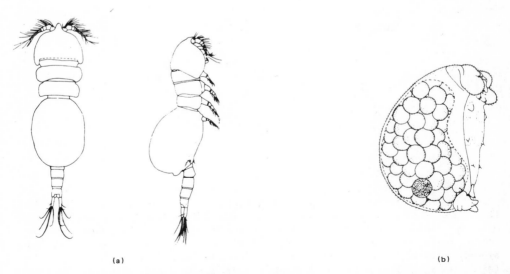

(a)   (b)

Fig. 7–20.   Notodelphyoid parasites: (a) *Notodelphys reducta*, female, 2.3 mm long; (b) *Demoixys chattoni*, female, 1.7 mm long. (After Illg and Dudley.)

ORDER MONSTRILLOIDA

The second antennae and mouthparts are absent in adults. The first antennae and legs 1–4 are well-developed (except in *Thespesiopsyllus*, where the small fourth pair is uniramous). There is no functional gut. The major articulation occurs in the Thespesiopsyllidae between thoracomeres 4 and 5 (the somite of leg 3 and that of leg 4), but in the Monstrillidae between thoracomeres 5 and 6 (the somite of leg 4 and that of leg 5). The eggs are carried in two egg sacs.

The adults occur in marine plankton. The immature stages, as far as known, are parasitic in polychaete annelids, gastropods, or ophiuroids. The life history of *Thespesiopsyllus* is not known in detail, but nauplii have been observed in the stomach of the ophiuroid *Ophiopholis*.

In the Monstrillidae there is a free naupliar stage with first antennae, second antennae, and mandibles, but without a mouth or gut. After entering the host the nauplius dedifferentiates, losing its appendages, and in the process of molting develops characteristic long processes. The postnaupliar stages probably represent much modified nauplii and copepodids. The adult breaks through the skin of the host to emerge free in the water.

Cymbasoma rigidum (=*Haemocera danae*), female 3 mm long, is elongated, with the urosome slightly narrowed (Fig. 7–21d). The female has two postgenital somites, the male three. The rarely collected adults evidently feed on stored food, since there is no gut. The eggs hatch, releasing a free-swimming nauplius (Fig. 7–21a). This breaks through the skin of a polychaete into the body cavity and then into the dorsal blood vessel. Here it undergoes a loss of the ocelli and the three pairs of appendages, develops two taillike processes (Fig. 7–21b), and begins to form a cuticle. Probably several molts occur, during which the larva, bathed in blood, absorbs sufficient food to last at least until the adults reproduce. The two processes become very long, and within the cuticular envelope the body of the adult develops (Fig. 7–21c). The adults escape by breaking through the blood vessel and skin of the worm to begin their brief pelagic reproductive period.

Monstrilla helgolandica, female 1.4 mm long, develops in the body cavity of the prosobranch snail *Odostomia scalaria*, where the larva forms three processes, one of them bifid.

ORDER CALIGOIDA

These are parasitic on marine and freshwater fishes. The major articulation is between the somites of leg 3 and leg 4, the 4th and 5th thoracomeres. In some parasitic females the major articulation has been lost. The body is flattened, depressed, with the prosome either wider or narrower than the urosome (Figs. 7–2d, 7–22). The first antennae are reduced to one or two articles. The second antennae, second maxillae, and maxillipeds are prehensile. The mouth region is suctorial, with very small mandibles. Legs 1–4 differ in form. There are two egg sacs, often with a single row of flattened eggs.

Both sexes are parasitic on fishes (including sharks and rays), aquatic mammals, and rarely invertebrates (e.g., cephalopods). Males and in some cases females may swim freely in the plankton. *Lepeophtheirus dissimulatus*, female (Fig. 7–22a) 2.6 mm, male 1.1 mm long, is a parasite of marine fishes. The prosome is broad, flattened,

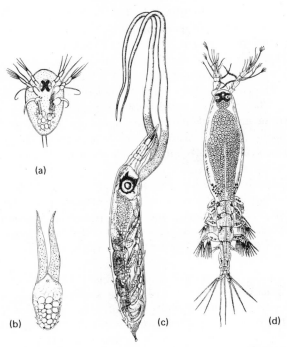

(a)

(b)   (c)   (d)

Fig. 7–21. Monstrilloid parasite, *Cymbasoma rigidum*: (a) nauplius; (b) later stage with two processes; (c) copepodid still inside annelid; (d) freeswimming adult female, 3 mm long. (From Baer, after Malaquin, courtesy of Univ. of Illinois Press.)

and shieldlike, wider than the genital somite. The first and second nauplii are planktonic. The next stage, the copepodid, is free-living at first, but soon attaches to the upper surface of the mouth of the fish by means of the second antennae and by a frontal filament which it secretes. The first four chalimus stages which follow remain attached there, but the fifth and sixth break free and migrate into the gill cavity or onto the skin. Copulation occurs just after the last molt. The adults move freely over the surface of the host.

Caligus rapax, female up to 7 mm, male up to 4 mm in length, has developmental stages quite similar to those of *Lepeophtheirus*. The bases of the first antennae are fused to the anterior section of the cephalothorax, forming the frontal plates, on each side of which there is a suckerlike frontal crescent-shaped structure. It feeds on blood obtained from punctures in the skin or gills of marine fishes such as Gadidae or Salmonidae. The well-fed adults may leave the host and live in the plankton, later to attack another fish. Migrating fishes may carry the copepods into fresh water.

Pandarus is a common parasite of sharks. *Cecrops* is a large marine caligoid, female up to 30 mm, male up to 17 mm in length, found on the gills of the sunfish. *Lernaeocera** branchialis*, called the gill maggot, is a large, S-shaped, red parasite, female

---

* The name *Lernaeocera* is on the *Official List of Generic Names in Zoology*.

(Fig. 7–13c, d) up to 3.5 cm if straightened out, living on the gills of cod *(Gadus)*. All thoracic legs are close together behind the head. The body is red because of a hemoglobinlike respiratory pigment in the body cavity. In the life cycle there are two free-swimming stages, the single nauplius and the single copepodid. The next stage, the first chalimus attaches to the gills of a flounder *(Pleuronectes)* by means of the secretion of a filament from the anterior part of the head. The second, third, and fourth chalimus stages remain fixed there by the filament. Just after the next molt to free-living adult males and females copulation occurs. The fertilized female is not a good swimmer and probably depends on currents to carry her out of the gill chamber of the flounder and up to the water occupied by the cod. She there attaches to the gill filaments and metamorphoses gradually to the swollen twisted condition during allometric growth. One or two such debilitating parasites reduce the host's weight by 30%. The male (Fig. 7–13c) remains free and eventually dies. Although the females on the cod gills superficially do not look much like copepods, their larval development and type of mouthparts show them to be caligoids.

Collipravus embeds its anterior end deep in the throat area of marine fishes, reaching even the aortic bulb. *Penella* (Fig. 7–22b) lives on marine teleosts and mammals. It is often found on flying fishes and swordfishes, where its head and neck are buried in the flesh. *Penella balaenopterae*, up to 32 cm long, is embedded 5–7 mm into whale blubber. The developmental stages may be parasitic on cephalopods. Another genus is *Cardiodectes.*

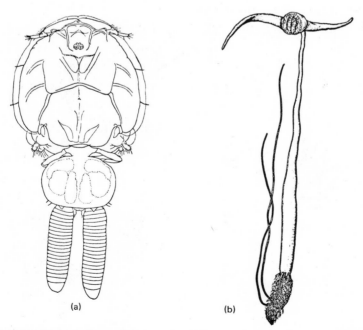

(a)   (b)

Fig. 7–22. Caligoid parasites. (a) *Lepeophtheirus dissimulatus,* **female, dorsum; 2.5 mm long. (After Shiino.) (b)** *Penella instructa,* **female. (After Yamaguti.)**

7. SUBCLASS COPEPODA

ORDER LERNAEOPODOIDA

Members of this order are parasites of marine and fresh-water fishes. The bodies of both sexes often lack external segmentation and there is no major articulation in the trunk. Sexual dimorphism is often pronounced, with a dwarf male attached to the female in some. The first antennae are minute. The second antennae have a very small exopodite. The second maxillae in the female are modified for grasping, often these two appendages are united at their tips in an attachment bulla (Fig. 7–13b). (In *Naobranchia* the second maxillae are flattened membranous appendages which are not united in a bulla, but instead wrap around the gill filament). The relative position of the second maxillae and maxillipeds becomes reversed during development, so that in the adult the maxillipeds are located anterior to the second maxillae. Legs 1–4 are often lacking, but the first two pairs may be present in some males (*Achtheres*). There are two egg sacs, with the eggs not in a line. In development there are no free naupliar stages. The stage which hatches from the egg is a free-swimming copepodid. This contains a preformed frontal filament inside the head. The copepodid molts to a chalimus which attaches to the fish gills by means of the filament now extruded. At the next molt young males and females appear, both having lost the filament and attaching by means of their maxillipeds. These males develop directly to sexual maturity. The females molt once more before reaching maturity, at which time they become attached by the second maxillae and bulla.

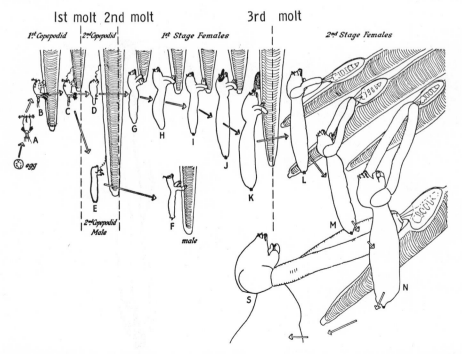

Fig. 7–23. Lernaeopodoid parasite. Diagram of life cycle of *Salmincola salmonea*, drawn to scale with arrows indicating succeeding stages; there are 3, possibly 4, molts between D and G; N, 4 mm long. (After Friend.)

Salmincola salmonea (Fig. 7–23) is a common parasite of the Atlantic salmon. The female, up to 8 mm long, is found attached to the gills by two very large second maxillae joined at their tips in the attachment bulla. The male, about 1.4 mm long, is attached to the gills by its maxillipeds. The egg sacs, which may be 11 mm long, with a pair containing about 900 eggs, are produced while the fish is in fresh water. The eggs hatch, releasing the copepodid which attaches to the gills of a salmon by means of its second maxillae and maxillipeds. This occurs at the time when the salmon first return from the sea to the spawning areas in the upper reaches of rivers. The frontal filament, at first inside the head, is extruded, and is planted in a hole rasped by the head appendages. The next stage, the chalimus (second copepodid) holds onto the filament with its second maxillae. At the second molt the young males are produced (which quickly become mature) and also the young first stage females. Both of these are attached by their maxillipeds, but are free to move about. It is at this stage that copulation occurs, following which the male dies. The fertilized females now extrude a head bulla which has been secreted by frontal glands; this is pressed against the gill to which it adheres. The female molts to the second stage and a small plug appears at the end of each elongated second maxilla. These two plugs are forced into sockets in the single head bulla, and the parasite thus becomes permanently attached. Eggs may be produced before the host salmon returns to the sea. During the marine journey the female *Salmincola* grow, but reproduction is inhibited until the next trip to freshwater, where infestation of young copepod-free salmon occurs. Other species of *Salmincola* occur on trout in North America. *Achtheres* (Fig. 7–13b) is a parasite in the mouth or on the gills of several fresh-water fishes. *Lernaeopoda* often occurs on sharks and dogfishes.

References

Baer, J.G. 1951. *Ecology of Animal Parasites.* Univ. Illinois Press, Urbana.

Bainbridge, R. 1953. Interrelationships of zooplankton and phytoplankton. *J. Marine Biol. Ass.* 32: 385–447.

Banse, K. 1959. Die Vertikalverteilung planktischer Copepoden in der Kieler Bucht. *Ber. Deutschen Wiss. Komm. Meeresforsch.* (N.F.) 15: 357–390.

Berner, Å. 1962. Feeding and respiration in the copepod *Temora longicornis. J. Marine Biol. Ass.* 42: 625–640.

Borutskii, E.V. 1952. [Freshwater Harpacticoida.] Fauna SSSR 3(4): 1–424.

Brehm, V. 1927. Copepoden and Branchiura *in* Kükenthal, W. and Krumbach, T. *Handbuch der Zoologie,* De Gruyter, Berlin 3(4): 435–496.

Bresciani, J. 1961. Larval development of *Stenhelia* (Copepoda Harpacticoida). *Vidensk. Medd. Dansk Naturh. Foren.* 123: 237–247.

Cannon, H.G. 1929. Feeding mechanism of the copepods *Calanus* and *Diaptomus. J. Exp. Zool.* 6: 131–144.

Carlisle, D.B. and Pitman, W.J. 1961. Diapause, neurosecretion and hormones in Copepoda. *Nature* 190: 827–828.

Caspers, H. 1939. Vorkommen und Metamorphose von *Mytilicola intestinalis* in der Nordsee. *Zool. Anz.* 126: 161–171.

Chappuis, P.A. and Delamare Deboutteville, C. 1958. La fauna interstitielle littorale du Lac Érie. *Vie et Milieu* 8: 366–376.

Clarke, G.L. *et al.* 1962. Comparative studies of luminescence in copepods. *J. Marine Biol. Ass.* 42: 541–564.

Cole, H.A. and Savage, R.E. 1951. The effect of the parasitic copepod *Mytilicola intestinalis* upon the condition of mussels. *Parasitology* 41: 156–161.

Corner, E.D.S. 1961. The feeding of the marine copepod *Calanus helgolandicus*. *J. Marine Biol. Ass.* 41: 5–16.

Costanzo, G. 1960. Sullo suiluppo di *Mytilicola intestinalis*. *Arch. Zool. Italiano* 44: 151–163.

Cowey, C.B. and Corner, E.D.S. 1963. The nutrition and metabolism of *Calanus*. *J. Marine Biol. Ass.* 43: 485–495.

Cushing, D.D. and Tungate, D.S. 1963. The identification of a *Calanus* patch. *Ibid.* 43: 327–338.

Digby, P.S.B. 1950. The biology of the small planktonic copepods of Plymouth. *Ibid.* 29: 393–438.

Deevey, G.B. 1960. Effects of temperature and food on seasonal variations in length of marine copepods. *Bull. Bingham Oceanogr. Collect.* 17: 54–86.

Dudley, P.L. 1966. *Development and Systematics of Some Pacific Marine Symbiotic Copepods. The Biology of the Notodelphyidae.* Univ. Washington Press, Seattle.

Dussart, B. 1967. *Les Copépodes des Eaux Continentales; Calanoides et Harpacticoides.* Boubée, Paris.

Einsle, V. 1967. Die äusseren Bedingungen der Diapause planktisch lebender Cyclopsarten. *Arch. Hydrobiol.* 63: 387–403.

Elgmork, K. 1962. A bottom resting stage in the planktonic freshwater copepod *Cyclops*. *Oikos* 13: 306–310.

Fahrenbach, W.H. 1962. The biology of a harpacticoid copepod. *Cellule* 62: 301–376.

_____ 1964. The fine structure of a nauplius eye. *Z. Zellforsch.* 62: 182–197.

Ferris, G.F. and Henry, L.M. 1949. The nervous system and a problem of homology in Copepoda. *Microentomology* 14: 113–119.

Friend, G.F. 1941. Life-history and ecology of the salmon gill-maggot *Salmincola salmonea*. *Trans. Roy. Soc. Edinburgh* 60: 503–541.

Fryer, G. 1957. The feeding mechanism of some freshwater cyclopoid copepods. *Proc. Zool. Soc. London* 129: 1–25.

_____ 1957. The food of some freshwater cyclopoid copepods. *J. Anim. Ecol.* 26: 263–286.

Fuchs, K. 1914. Die Keimblätterentwicklung von *Cyclops viridis*. *Zool. Jahrb. Abt. Anat.* 38: 103–156.

Gauld, D.T. 1953. Diurnal variations in the grazing of planktonic copepods. *J. Marine Biol. Ass.* 31: 461–474.

_____ 1957. A peritrophic membrane in calanoid copepods. *Nature* 179: 325–326.

_____ 1957. Copulation in calanoid copepods. *Ibid.* 180: 510.

Grabda, J. 1963. Life cycle and morphogenesis of *Lernaea cyprinacea*. *Acta Parasit. Polonica* 11: 169–198.

Graininger, J.N.R. 1951. The biology of the copepod *Mytilicola intestinalis*. *Parasitology* 41: 135–142.

Grice, G.D. and Hulsemann, K. 1965. Abundance, vertical distribution and taxonomy of calanoid copepods. *Proc. Zool. Soc. London* 146: 213–262.

Gurney, R. 1931-1933. *British Freshwater Copepoda.* 3 vol., Ray Soc., London.

_____ 1948. Morphology of the Copepoda. *Ann. Mag. Nat. Hist.* (11)14: 711–714.

Halisch, W. 1939. Anatomie und Biologie von *Ergasilus minor*. *Z. Parasitenk.* 11: 284–330.

Hardy, A.C. 1956. *The Open Sea.* Collins, London.

_____ and Bainbridge, R. 1954. Experimental observations on the vertical migrations of plankton animals. *J. Marine Biol. Ass.* 33: 409–448.

Hintz, H.W. 1951. The role of certain arthropods in reducing mosquito populations. *Ohio J. Sci.* 51: 277–279.

Humes, A.G. 1951. Harpacticoid copepod from the gill chambers of a marsh crab. *Proc. U.S. Nat. Mus.* 90: 379–386.

————— 1955. The postembryonic stages of a freshwater calanoid copepod. *J. Morphol.* 96: 441–472.

Hutchinson, G.E. 1967. *A Treatise on Limnology,* vol. 2. Wiley, New York.

Illg, P. 1958. North American copepods of the family Notodelphyidae. *Proc. U.S. Nat. Mus.* 107: 463–649.

Jacobs, M. 1925. Entwicklungs-physiologische Untersuchungen am Copepodenei. *Z. Wiss. Zool.* 124: 487–541.

Kiefer, F. 1929. Crustacea Copepoda in *Das Tierreich.* De Gruyter, Berlin. 53: 1–102.

————— 1960. *Ruderfusskrebse (Copepoden).* Kosmos Verl., Stuttgart.

Lang, K.A. 1946. The mouthparts of the Copepoda. *Ark. Zool.* 38(A5): 1–24.

————— 1948. *Monographie der Harpacticiden.* Stockholm.

Lochhead, J.H. 1950. Cyclops *in* Brown, F.E. ed. *Selected Invertebrate Types.* Wiley, New York. pp. 406–413.

Lowe, E. 1936. Anatomy of a marine copepod. *Trans. Roy. Soc. Edinburgh* 58: 561–603.

Lowndes, A.G. 1935. The swimming and feeding of certain calanoid copepods. *Proc. Zool. Soc. London* 1935: 687–715.

Marshall, S.M. and Orr, A.P. 1955. *The Biology of a Marine Copepod Calanus.* Oliver & Boyd, Edinburgh.

————— 1955. Food uptake, assimilation and excretion in *Calanus. J. Marine Biol. Ass.* 34: 495–529.

————— 1956. Feeding and digestion in the young stages of *Calanus. Ibid.* 35: 587–603.

————— 1957. The life history of the copepod *Calanus finmarchicus* in different latitudes. *Année Biol.* (3)33(1–2): 43–47.

————— 1957. Seasonal changes in respiration of copepods. *Ibid.* (3)33(5–6): 221–226.

————— 1960. Observations on vertical migration especially in female *Calanus. J. Marine Biol. Ass.* 39: 135–147.

————— *et al.* 1961. The phosphorous cycle: excretion, egg production, autolysis of *Calanus. J. Marine Biol. Ass.* 41: 463–488.

Metzler, S. 1957. Die Beeinflussbarkeit des Geschlechtsverhältnisses von *Cyclops viridis* durch Aussenfaktoren. *Zool. Jahrb. Physiol.* 67: 81–110.

Monod, T. and Dollfus, R.P. 1932. Copépodes parasites de mollusques. *Ann. Abt. Parasit.* 10: 129–204.

Noble, E.R. and Noble, G.A. 1964. *Parasitology.* Lea and Febiger, Philadelphia.

Noodt, W. 1957. Ökologie der Harpacticoidea des Eulitorals der deutschen Meeresküste. *Z. Morphol. Ökol.* 46: 149–242.

Oorde-de-Lint, G.M. van and Schuurmans-Stekhoven, J.H.S. 1936. Copepoda parasitica *in* Grimpe, G. and Wagler, E., *Tierwelt der Nord-und Ostsee* (Lief. 31) 10c: 73–198.

Pennak, R.W. 1953. *Fresh-water Invertebrates of the United States.* Ronald, New York.

Pesta, O. 1927. Copepoda non parasitica *in* Grimpe, G. and Wagler, E., *Tierwelt der Nord-und Ostsee* (Lief. 8) 10c: 1–72.

Raymont, J.E.G. and Gauld, D.T. 1953. The respiration of some planktonic copepods. *J. Marine Biol. Ass.* 31: 447–460.

Rose, M. and Vaissière, R. 1951. Le système excréto-glandulaire des Sapphirines. *Arch. Zool. Expér. Génér.* 87: 134–138.

Rüsch, M.E. 1960. Geschlechtsbestimmungsmechanismus bei Copepoden. *Chromosoma* 11: 419–432.

Sandercook, G. 1967. Selected mechanisms for the coexistence of *Diaptomus* spp. *Limnol. Oceanogr.* 12: 97–112.

Schaperclaus, W. 1954. *Fischkrankheiten,* 2 ed. Berlin.

Schröder, R. 1962. Vertikalverteilung des Zooplanktons in Abhängigkeit von den Strahlungsverhältnissen in Seen. *Arch. Hydrobiol. Suppl.* 25: 414–429.

Scott, T. and Scott, A. 1913. *The British Parasitic Copepoda,* 2 vol. Ray Soc., London.

Siebeck, O. 1960. Die Bedeutung von Alter und Geschlecht für die Horizontalverteilung planktische Crustaceen. *Int. Rev. Ges. Hydrobiol.* 45: 125–131.

————— 1960. Die Vertikalwanderung planktonischer Crustaceen unter Berücksichtigung der Strahlungsverhältnisse. *Ibid.* 45: 381–454.

Smyly, W.J.P. 1957. The life history of the harpacticoid copepod, *Canthocamptus. Ann. Mag. Natur. Hist.* (12)10: 509–512.

Spandl, H. 1926. Copepoda *in* Schulze, P., *Biologie der Tiere Deutschlands,* Borntraeger, Berlin 19: 1–82.

Stekhoven, J. H., Jr. and Punt, A. 1937. Zur Morphologie und Physiologie der *Lernaeocera branchialis. Z. Parasitenk.* 9: 648–668.

Storch, O. 1928. Des Nahrungserwerb zweier Copepondennauplius *Diaptomus gracilis* und *Cyclops strenuus. Zool. Jahrb. Abt. Allg. Zool.* 45: 385–436.

————— 1929. Die Schwimmbewegung der Copepoden auf Grund von Mikro-zeitlupenaufnahmen. *Verhandl. Deutschen Zool. Ges.* 33: 118–129.

————— and Pfisterer, O. 1925. Der Fangapparat von *Diaptomus. Z. Vergl. Physiol.* 3: 330–376.

Umminger, B. 1968. Polarotaxis in copepods. *Biol. Bull.* 135: 239–261.

Vaissière, R. 1961. Morphologie et histologie comparées des yeux des crustacés copepodes. *Arch. Zool. Exp. Gén.* 100: 1–125.

Walter, E. 1922. Die Lebensdauer der freilebenden Süsswasser-Cyclopiden. *Zool. Jahrb. Abt. Syst.* 44: 375–420.

Wickstead, J.H. 1959. A predatory copepod. *J. Anim. Ecol.* 28: 69–72.

————— 1962. Food and feeding in pelagic copepods. *Proc. Zool. Soc. London* 139: 545–555.

Wierzbicka, M. and Kedzierski, S. 1964. The dormancy of some species of Cyclopoida. *Polsk. Arch. Hydrobiol.* 12: 47–80.

Wilson, C.B. 1932. The copepods of the Woods Hole region. *Bull. U.S. Nat. Mus.* 158: 1–635.

Wilson, M.S. 1949. Branchiura and parasitic Copepoda *in* Edmondson, W.T. ed., *Ward and Whipple Freshwater Biology,* Wiley, New York. pp. 862–868.

————— and Yeatman, H.C. 1949. Free-living Copepoda. *Ibid.* pp. 735–861.

Yamaguti, S. 1963. *Parasitic Copepoda and Branchiura of Fishes.* Wiley, Interscience, New York.

8.

Subclass Branchiura

There are about 120 species of Branchiura described, of which the Japanese *Argulus scutiformis,* up to 3 cm long, is the largest.

Branchiura, fish lice, are small crustaceans with a wide flat carapace and an unsegmented abdomen. There are compound eyes and four pairs of biramous thoracic limbs.

Anatomy

The animal is dorsoventrally flattened (Fig. 8-2) and its body consists of two parts. The anterior part is modified for parasitic life, the posterior part for locomotion. At the end of the trunk there is an unsegmented finlike abdomen with a posterior median notch. At the base of the notch is the anus and behind it a minute furca (Fig. 8-1). The anterior part is a cephalothorax containing the head somites to which the first thoracic somite is more or less fused. A flat fold, the carapace, extends anteriorly and to the sides; posteriorly it differs considerably from species to species. The lateral extensions can be moved up and down slightly, winglike. The pereon (the somites of the thorax not fused to the head) has three limb-bearing metameres (Fig. 8-3) and is partly or completely covered by the carapace dorsally. The hypodermis contains mainly individual gland cells and glands and produces a delicate exoskeleton.

The head appendages, adapted to parasitic life, differ considerably from those of other crustaceans. Both pairs of antennae if present are very short and may bear hooks in *Argulus.* Here both are attachment organs and have only scattered innervated sense organs (Fig. 8-4). *Chonopeltis* lacks first antennae and the second antennae have no hooks. Neither antenna pairs have hooks in *Dipteropeltis.* Behind the small antennae the much larger two pairs of maxillae can be seen. They are uniramous without endites and have lost their function as mouthparts. The first maxillae of *Dolops* have a long clasping hook at the tip. The base of the first maxilla in *Argulus* has changed into an attachment organ, a deep tubelike sucker; young animals still have vestiges of the distal articles (Fig. 8-5). The

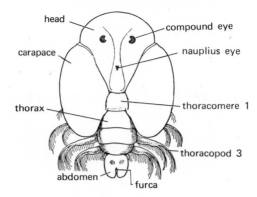

Fig. 8–1.   *Argulus laticauda*, dorsal view; 6 mm long. The tergite of the first thoracomere is fused to the head. (After Snodgrass.)

mandibles can be found only by careful examination. They are hidden between labrum and labium and form a pair of transverse toothed hooks, the concave surfaces and tips of which face anteriorly. In *Dolops* they lie directly behind the mouth between the papillate raised lips. However, in *Argulus* distal from the mouth there is a posteriorly directed proboscis on the tip of which, under the borders of the lips, the mandibles are hidden (Fig. 8-4). The mandibles can extend out of the lip folds. The proboscis lies flat posteriorly and can be extended ventrally. In *Argulus,* but not in other genera there is in the midline of the body an anteriorly directed hollow spine, the so called sting or poison spine, in some species reaching to the anterior border of the head (Fig. 8-4). The distal tip of the spine bears the openings of large cephalothoracic glands and can be withdrawn into the base. Since the spine is innervated by the labral ganglion (sympathetic nerve) it can be concluded that the spine is derived from the labral primordium in the embryo.

The morphological significance of the head appendages was not recognized for a long time. It can be studied by comparing *Argulus* with larval instars and with the less specialized *Dolops.* The metanauplius, which is an embryonic stage and does not leave the egg, has all the five pairs of anterior appendage primordia typical in form and location of the rest of the Crustacea. There is no evidence of the large poison spine of *Argulus,* suggesting that this spine is a secondary acquisition, and not the vestige of an appendage. The newly hatched stage has the total complement of appendages, in form and position as in other crustacean subclasses (Fig. 8-7). The mandible has a large endopodite, a palp, which has an obvious relationship to the hookshaped endite. This endite, unlike those of other

Fig. 8–2.   *Argulus foliaceus,* seen from in front; 4.7 mm wide. (After Herter.)

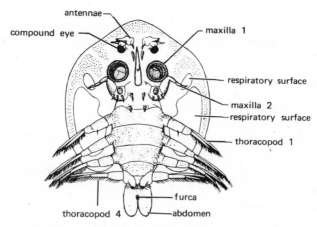

Fig. 8–3. *Argulus foliaceus*, ventral view; 7 mm long. (After Wagler.)

crustaceans, is movable. At the same time, it can be seen that the main part between endopodite and endite will contribute to the lateral wall of the future proboscis. Also on the anterior of the metastomal area (labium) one can see two small swellings, out of which arise the labial spines of the adult, two lobes each with a hollow spine. These spines have been interpreted as maxillary vestiges by some authors. However, if they were maxillary vestiges, one would expect them to be large in a larval stage. As this is not the case they are probably of secondary origin and the two large head appendages of the adult correspond to the first and second

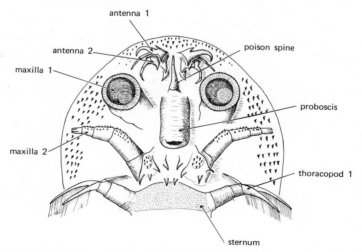

Fig. 8–4. *Argulus foliaceus*, venter of cephalothorax; 3.5 mm long. There are numerous posteriorly directed spines; the sternum belongs to thoracomere 1. (After Martin.)

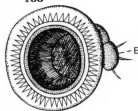

<-E Fig. 8-5. *Argulus foliaceus*, maxilla of a young male showing vestigial distal articles (E); 0.2 mm in diameter. (After Claus.)

maxillae, despite their lack of endites. Further evidence is that the maxillary glands open on the second maxillae.

In marked contrast to the modified cephalic appendages, the four pairs of thoracopods (the first one belonging to the cephalothorax) are biramous swimming limbs, having their surfaces increased by setae (Fig. 8-3). The thoracic limbs are modified for clasping in males. The abdomen lacks appendages.

SENSORY ORGANS AND NERVOUS SYSTEM.   Sensory hairs of diverse form are scattered over the body. The three naupliar eyes situated above the brain and touching each other are similar in structure to those of ostracods. Each of the paired compound eyes in *Argulus* has 30–75 ommatidia. In *Argulus* the cornea lies in a blood sinus under the integument within which the compound eyes are kept in trembling motion by muscles, an adaptation that can be compared with the ectodermal eyesac of *Triops*.

The nervous system consists of a supraesophageal brain and a mass of ganglia which barely extend to the first legs. The mass has constrictions marking a sub-esophageal ganglion and five additional ganglia. The subesophageal ganglion innervates the mandibles and the first maxillae; the other ganglia innervate the following appendages, the last innervating also the abdomen.

DIGESTIVE SYSTEM.   The ectodermal foregut continues as a wide midgut. Into the anterior of the midgut open two ducts which come from the branched ceca. The ceca lie within the carapace and grow with it (Fig. 8-6), permitting the intake of large quantities of food at one time, an adaptation found in other temporary parasites. The anus opens at the base of the notch on the abdomen.

EXCRETORY AND CIRCULATORY SYSTEMS.   One pair of large maxillary glands, which open at the base of the second maxillae, constitute the excretory system.

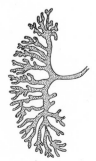

Fig. 8-6. *Argulus foliaceus*, right gut cecum of a female; 4.2 mm long. (After Claus.)

The heart lies in the last thoracic somite and through one pair of ostia takes in blood that comes from the platelike abdomen. The abdomen makes pumping movements of importance to circulation. The blood flows through a long cephalic aorta toward the head; the aorta opens at the end but has no branches. *Dolops* has hemoglobin in the blood.

REPRODUCTIVE SYSTEM. The large median ovary in the thorax produces numerous eggs. The ovary opens through paired oviducts, only one of which is functional at a time. The oviducts open into an ectodermal pouch behind the last swimming limbs. Each of the two seminal receptacles present has its own opening on the tip of a minute sclerotized spine. These spines are located near the opening of the oviduct.

The testes lie in the abdomen. Their efferent ducts extend anteriorly to fuse in the thorax into a large seminal vesicle, from the anterior of which a pair of sperm ducts originate, pass posteriorly, and fuse into a median duct before opening between the last limb coxae.

Reproduction

During the mating season the male *Argulus* swims actively. He settles on the back of a female while she is feeding on a fish, and surrounds her last limbs with his. After grasping the female from above, the male twists its posterior end beneath the female, first on one side, then the other, presumably filling each seminal receptacle in turn. The males of some species have a simple genital papilla but it is not known how it fits over the spine. *Dolops ranarum*, unlike *Argulus*, employs spermatophores which form as material flows out from the male sperm duct. The single spermatophore is impaled on the two spines bearing the opening of the seminal receptacles of the female permitting the sperm to pass into the seminal receptacles. As each egg is laid it is pricked and injected with sperm.

After 3 weeks, the time probably depending on the temperature, the female of *Argulus foliaceus* leaves the host and seeks aquatic plants or stones on which, with a secretion of the oviduct, she attaches a double row of thick shelled eggs. This is a departure from the brood care practiced by other crustaceans, most of which carry their eggs in brood chambers or on appendages. About 20 to 480 eggs are laid at a time; altogether 800 to 1200 are deposited. In an aquarium the female parasitizes fishes between clutches.

Development

Naupliar and metanaupliar stages of *Argulus foliaceus* are passed within the egg, in 3 weeks at 19°C, in 5 weeks at 14°C. The first free stage has a carapace and immediately attaches itself to a fish. Eight more juvenile instars follow. All nine larval stages are passed in 5 weeks at 14°C, in 4 weeks at 16°C.

The first free stages of *Argulus* differ from the adults (Fig. 8-7) mainly by their small size (0.6 mm), the shortness of the carapace (which does not cover the

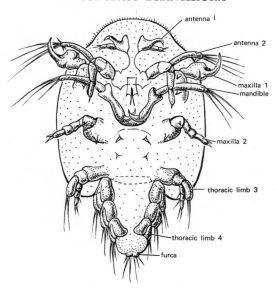

antenna 1

antenna 2

maxilla 1
mandible

maxilla 2

thoracic limb 3

thoracic limb 4
furca

Fig. 8–7. *Argulus foliaceus*, second larval stage, ventral view; 0.7 mm long. The hooked endites of the mandible are covered by the transparent labrum whose convex posterior border is distinct. Anteriorly is the poison spine. (After Stammer.)

three pereomeres), the shortness of the abdomen (which does not extend beyond the furca), the rudimentary condition of the three most posterior thoracic limbs, the large branches of the second antennae and mandibular palps, and the absence of a sucker on the first maxillae (Fig. 8-7).

There are two types of newly hatched individuals in Branchiura even in the genus *Argulus*. In one the larvae swim jerkily with the second antennae, the mandibles, and the first limbs. The other swims with thoracopods 1–4. The first becomes like the second after one molt. Some species of *Argulus* and *Dolops ranarum* remain in the egg until all thoracic limbs are functional.

The larvae swimming with antennae attach themselves to the host with the distal hooks of the first maxillae and first antennae. After about 8 days they molt and after the second molt all four pairs of swimming limbs are functional, and the second antennae and mandibular palps are shortened. In the fifth stage the sucker begins to form, but becomes functional only after the seventh molt. In the meanwhile the animal grows to 2 mm. After two more molts at intervals of 3 to 4 days, *Argulus* becomes sexually mature and the carapace and abdomen achieve their final form. The adults continue to molt.

The earliest stages of *Chonopeltis brevis* differ from the adult mainly in having smaller size and in lacking suckers. Instead of suckers they use segmented appendages for attachment. Although later instars have increasing numbers of setae on the limbs, neither larvae nor adults can swim. All stages are parasitic. The adult changes hosts (bottom-dwelling fishes) by an unknown method.

Relationships

The presence of a carapace and compound eyes are characters that distinguish Branchiura from Copepoda. The differentiation of the limbs distinguishes Branchiura from the Branchiopoda and Ostracoda. The similarity with *Caligus* is probably due to convergent evolution.

Habits

Branchiura are temporary fish parasites that feed on blood or mucus. Many are active swimmers, found in the ocean as well as in freshwater. *Chonopeltis* and some species of *Dipteropeltis* cannot swim, judging from their anatomy. *Argulus foliaceus* swims rapidly, gliding smoothly through the water propelled by the thoracic limbs. Usually the dorsum is turned toward the light, a reflex that depends on the compound eyes. But *Argulus foliaceus* can swim up and down at any time. To rest they attach to objects, even glass, with their suckers. On glass, they can glide forward by pushing the suckers forward. Finding of the host by *Argulus foliaceus* does not depend on chemical stimuli or passing shadows. But an *Argulus* touched by a fish shoots in a curved path through the water and then settles, often reaching the fish that disturbed it. The animal remains attached depending on tactile and chemical stimuli. Receptors of such stimuli probably are sensory hairs, of which various kinds have been described without knowledge of the functions.

If clean filter paper moved through water touches an *Argulus,* which is not attached to a host, the animal will shoot off but immediately it attaches to the paper for a second before swimming off again. If the experiment is repeated with filter paper rubbed over the scales of a fish, the fish louse will move about on it for a while.

A sense of taste has not been demonstrated. After attaching to the host, rheotaxis is of importance. *Argulus* turns soon until its longitudinal axis is parallel to the host and then crawls toward the head of the fish. It settles where there is little friction with the water, usually behind gill operculi, behind fins, or in other thin-skinned places. When swimming in the water the animal moves against the current. The seat of the rheotactic sense may be hairs that have thin filamentous tips. These cover the head and dorsum of the carapace and are innervated by a long nerve from the tritocerebrum. Although fish lice are known to be temperature-sensitive the sense organs are not known.

Hosts are diverse fish species. It is certain that during the course of their life they not only change hosts, but also host species. It is puzzling how the non-swimmers change hosts. Some *Argulus* species have been found on tadpoles. *Argulus arcassonensis* has even been found on dead cuttlefish (*Sepia*). *Argulus* attaches to the host mainly with the large sucker of the first maxillae, but may increase adhesion by appressing itself tightly. If the host swims rapidly, the parasite is pressed backward toward the substrate, forcing the many thorns (Fig.

8-4) to hook and anchor. In *Argulus* and *Dolops* the antennae help attach parasite to host.

Larger carp, tench and pike do not react to *Argulus foliaceus,* but carp 4.5 cm long and tadpoles shoot through the water when this fish louse attaches. It is assumed that the sting of the poison spine and the metastomal spines is irritating.

The parasite remains on the skin, cutting through with its mandible, and drills into blood vessels. In a relatively short time the digestive ceca are filled with blood.

Several fish lice on a tadpole or small fish may be fatal to the host. But carp 4 cm long and pickerel 10 cm long do not show obvious damage from the presence of 20 *Argulus* for a few days. However, heavy infestations may occur, and 420 *A. foliaceus* have been counted on the skin of a 1 kg pickerel (*Esox* sp.). Especially fish in weirs or fish ponds may be heavily attacked. A tench 28 cm long had 4250 parasites. Besides the damage caused by the poison, the wounds provide entrances for fungi. An infestation may make the host so restless that it feeds little. Trout may succumb to the parasites. Central European species of fish lice transmit a fish disease, *Pseudomonas punctata.*

"Flagella," structures, of dubious homology, occur on some of the legs of some species of *Argulus* and *Dolops* and are used to wipe the ventral surface of the carapace.

Ceca and gut absorb the food, and a well-fed *Argulus foliaceus* feeds again after 2 weeks at 3°C, or 1 week at 14°C. The attached *Argulus* beat with the swimming limbs, constantly supplying fresh respiratory water to their venter. If *Argulus* is placed in a 1% silver nitrate solution, the Ag^- ions are immediately reduced and darken some areas of the carapace (Fig. 8-3) that are presumed to be of respiratory function. The anterior respiratory areas of the carapace are smaller than the posterior ones (Fig. 8-3). Each area is framed by a sclerotized ridge, but the cuticle of the area itself is very thin, its hypodermis differentiated as are amphipod gills. Below it is a large blood sinus. The failure of other areas to reduce the salt indicates that they are of less importance.

Both abdominal plates have musculature that pumps the blood from the venter of the trunk posteriorly and then anteriorly and up, toward the heart. These "tails" thus function mainly as accessory hearts, less in respiration.

Classification

Argulidae. All Branchiura are placed within the single family. *Argulus* has over 100 species, common in freshwater and oceans around the world. The European *Argulus foliaceus,* the carp louse, attacks many species in freshwater including tadpoles and newts; *A. coregoni* also of Europe has a shorter carapace, up to 13 mm long, and is mainly a parasite of trout. There are about 15 species of *Argulus* in North America. The African *Chonopeltis* lacks a poison spine but has suckers (Fig. 8–8). The largely South American *Dolops* lacks both the poison spine and suckers and its preoral cavity consists only of labrum and metastomal area. Other *Dolops* species are known from Africa and Tasmania. *Dipteropeltis* (Fig. 8–9) is South American.

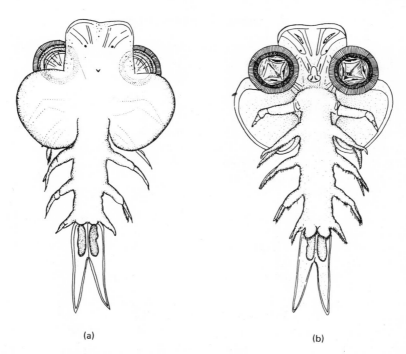

(a)   (b)

Fig. 8–8.   *Chonopeltis inermis*: (a) dorsal view; (b) ventral view; 11 mm long. (From Yamaguti after Fryer, courtesy of Wiley-Interscience.)

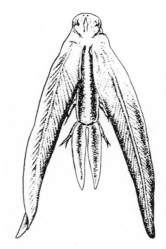

Fig. 8–9.   *Dipteropeltis hirundo*, dorsal view; 14 mm long. (From Yamaguti, courtesy of Wiley-Interscience.)

References

Argilas, A. 1937. La présence d'*Argulus arcassonensis* sur *Sepia filliouxi*. *Proc. Verb. Soc. Linn. Bordeaux* 88: 145–147.

Bower-Shore, C. 1940. The common fish louse, *Argulus foliaceus*. *Parasitology* 32: 361–371.

Debaisieux, P. 1953. Histologie et histogenèse chez *Argulus foliaceus*. *Cellule* 55: 243–290.

Fryer, G. 1958. Occurrence of spermatophores in the genus *Dolops*. *Nature* 181: 1011–1012.

————— 1960. The spermatophores of *Dolops*, their structure, formation and transfer. *Quart. J. Microscop. Sci.* 101: 407–432.

————— 1961. Larval development in *Chonopeltis*. *Proc. Zool. Soc. London* 137: 61–69.

Herter, K. 1927. Die Reizphysiologie des Karpfenlaus (*Argulus foliaceus*). *Z. Vergl. Physiol.* 5: 283–370.

————— 1928. Die Komplexaugen des Karpfenlaus (*Argulus foliaceus*). *Zool. Jahrb. Abt. Physiol.* 45: 159–176.

Jirovec, O. and Wenig, K. 1934. Atmung von *Argulus foliaceus*. *Z. Vergl. Physiol.* 20: 450–453.

Kollatsch, D. 1959. Die Biologie und Ökologie des Karpfenlaus, *Argulus foliaceus*. *Zool. Beitr.* (N.F.) 5(1): 1–36.

Madsen, N. 1964. The anatomy of *Argulus* integument, central nervous system, sense organs, praeoral spine, and digestive organs. *Lunds Univ. Årskr.* (N.F. Avd. 2) 59: 3–32.

Martin, M.F. 1932. Morphology and classification of *Argulus*. *Proc. Zool. Soc. London* 1932: 771–806.

Meehean, O.L. 1940. *Argulus* in the collections of the United States National Museum. *Proc. U.S. Natl. Mus.* 88: 459–522.

Pennak, R.W. 1953. *Fresh-water Invertebrates of the United States*. Ronald, New York.

Ringuelet, R. 1943. Los Argúlidos Argentinos con el catálogo de las especies neotropicales. *Rev. Mus. La Plata* (N.S.) 3: 43–99.

————— 1948. Argúlidos del Museo de la Plata. *Ibid.* (N.S. Zool.) 5: 281–296.

Romanowsky, A. 1956. Die Identität der Arten *Argulus japonicus* und *Argulus pellucidus*. *Zool. Anz.* 157: 264–265.

Stammer, J. 1959. Morphologie, Biologie und Bekämpfung der Karpfenläuse. *Z. Parasitenk.* 19: 135–208.

Tokioka, T. 1936. Larval development and metamorphosis of *Argulus japonicus*. *Mem. Coll. Sci. Kyoto* (B) 12: 93–114.

Wagler, E. 1935. Die deutschen Karpfenläuse. *Zool. Anz.* 110: 1–10.

Wilson, C.B. 1944. Parasitic copepods in the United States National Museum. *Proc. U.S. Natl. Mus.* 94: 529–582.

Wilson, M.S. 1949. Branchiura and parasitic Copepoda *in* Edmondson, W.T. ed., *Ward and Whipple Freshwater Biology*, Wiley, New York, pp. 862–868.

Yamaguti, S. 1963. *Parasitic Copepoda and Branchiura of Fishes*, Wiley, Interscience, New York.

Zacwilichowska, K. 1948. The nervous system of the carp louse *Argulus foliaceus*. *Bull. Int. Acad. Cracovie* B II: 117–128.

9.

Subclass Cirripedia

About 900 species of Cirripedia are known. The largest, *Lepas anatifera* may reach 80 cm in length including stalk.

Cirripedia are sedentary or symbiotic crustaceans. While all species are marine, some Thoracica spend much of their lives in air or nearly freshwater. The body is primitively composed of head, thorax, and abdomen contained within a bivalved carapace. But almost all Cirripedia are as adults much modified.

Some Ascothoracica retained the primitive body divisions, the thorax consisting of six, the abdomen of five somites. The first abdominal somite bears the male intromittant organ, the last the furca. This 6–5 plan is shared with the Copepoda and from it can be derived the Branchiura, Mystacocarida, and possibly the Ostracoda, the group constituting the Maxillopoda of Dahl (1963). This plan became modified in the evolution of the subclass. The head, in all orders, is reduced and the abdomen lost except in the Ascothoracica and the cyprid larval stage. The caudal appendages found in most orders are considered homologous with the furca of the Ascothoracica and other crustaceans. The carapace (mantle), enclosing the body, is bivalved in Ascothoracica and cyprid larvae (Figs. 9-1, 9-15). In Thoracica and Acrothoracica it forms a sac. In the Thoracica the sac bears calcareous plates forming an outer, permanent wall or shell (Figs. 9-4, 9-5). All cirripeds attach initially by the first antennae, and later, in permanently attached forms, by the preoral region of the head. The second antennae are present as swimming and feeding appendages only in nauplii. The biramous thoracic appendages are swimming limbs in Ascothoracica and in cyprid larvae (Figs. 9-1, 9-14a, 9-22b), but they are cirriform in adult Acrothoracica and Thoracica, and are used in setose feeding (Figs. 9-4, 9-20). In the Rhizocephala only larval stages have appendages; the adult consists of a food absorbing interna, a network of tubes penetrating the host, and an externa, a sac containing the reproductive parts (Figs. 9-22, 9-23).

Development usually includes six naupliar stages and a cyprid larva before metamorphosis into an adult. Except in the Ascothoracica the nauplii have the anteriolateral margins of a dorsal shield produced into horns (Fig. 9-22). Such horns are not found in the nauplii of other crustaceans. In the last stage the lateral margins of the dorsal shield bend ventrally and, following the next molt, the shield is replaced by the bivalved carapace of the cyprid larva. The cyprid resembles an ostracod, even with similar compound eyes (Fig. 9-14a). But the cyprid larva differs from ostracods in lacking 2nd antennae, in the presence of six short pairs of biramous swimming thoracopods, and in having the valves united anterioventrally as well as dorsally, and in lacking a distinct dorsal hinge.

The subclass is divided into four orders: (1) Ascothoracica, anatomically generalized or primitive forms that are semi- or wholly parasitic on coelenterates and echinoderms; (2) Thoracica, or barnacles, usually provided with mineralized shells; (3) Acrothoracica, minute forms without calcareous plates found burrowing in calcareous substrates; and (4) Rhizocephala, wholly parasitic, mostly on decapod crustaceans, and recognized as Cirripedia only by their larval stages.

Cirripeds have a well-documented fossil record beginning in the Silurian. The radiation and diversity we see today began to occur in the Upper Cretaceous and was essentially completed by the end of the Miocene.

Darwin's monograph on the Cirripedia, published before *The Origin of Species*, remains the basic work for the group.

ORDER ASCOTHORACICA

There are nearly 30 species known, the largest having a mantle up to 8.5 cm wide.

Ascothoracica are parasitic crustaceans, most of them minute. They lack compound eyes in all stages and nauplii lack frontolateral horns. The carapace, consisting of two valves, covers the entire body. The first antennae are subchelate the second absent. The mouthparts are modified for piercing and sucking, or in endoparasitic forms, they are rudimentary. The thoracic limbs are biramous swimming appendages, or they may be vestigial or absent.

Anatomy

The Ascothoracida have an elongated body with large abdomen of five somites bearing a furca. The body is contained in a bivalved, uncalcified carapace; the valves are connected along a short seam (Fig. 9-1). The total number of thoracomeres is 6 giving a total of 11 postcephalic somites, but both parts are variously reduced in females of most genera.

The internal anatomy of *Ascothorax* has been carefully studied and the following description is based mainly on this genus. The body has a head, six thoracomeres and five pleomeres, the last of which bears a furca. The valves are attached to the head and can be closed tightly by a ventral adductor muscle. Digestive ceca and gonads lie within the valves and the valves surround a large cavity

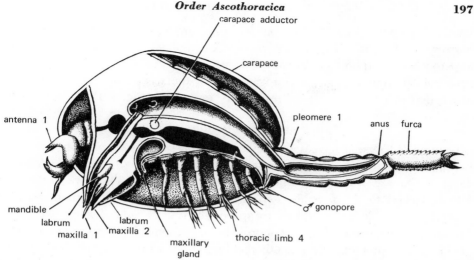

Fig. 9–1. Male of *Ascothorax ophioctenis*, viewed from left side; 3.2 mm long. The internal anatomy shows the digestive system and the nervous system (black). (After Wagin.)

within which the eggs develop to the ascothoracid stage (equivalent to the cyprid).

The first antennae are long and thick, the distal articles forming a strong pincer (Fig. 9-1). The second antennae are present in the naupliar stages which are frequently passed in the egg. All 3 pairs of mouthparts are styletlike structures enclosed in the conical labrum modified for piercing and sucking. The foregut has strong dilating muscles and opens into a straight midgut that has branched ceca. In the female of *Ascothorax* the second to fifth thoracic limbs bear an endopodite and exopodite. The first and sixth, however, are uniramous, as are the limbs of the male.

The ventral ganglia are concentrated and fused within the thorax (Fig. 9-1). A pair of maxillary glands are excretory organs. Blood circulates within spaces; there is no heart nor respiratory organs.

Most ascothoracicans are dioecious. In males lobes of the testes, in the thorax and valves, empty into vasa efferentia which open into a vas deferens running posteriorly to the penis in the first abdominal somite (Fig. 9-1). In females, the ovary is median, above the gut and its lobes reach into the expanded valves. The two oviducts open on two papillae just posterior to the basal article of the first thoracic limb.

Ascothoracicans that live as endoparasites in the body cavities of echinoderms or below the coenosarc of antipatharians, gorgonians, and zooantharians, have less distinct body segmentation than *Ascothorax*. Both mouthparts and thoraco-pods tend to be reduced or absent. In female endoparasites, the carapace is much larger than the trunk, resulting from the expansion of the gut and ovarian lobes (Fig. 9-2). Also its free borders may grow together and leave only a narrow slit into the cavity, hiding the trunk completely. Such coelomic parasites apparently take liquid food for the gut is incomplete and there is no anus. In some coelom-inhabiting genera (Dendrogasteridae), the valves may be lobed for the absorption of nutrients and for respiratory exchange (Fig. 9-3).

Reproduction

Generalized genera have separate sexes, while specialized parasitic forms are hermaphroditic, some perhaps protandric. Males of *Ascothorax ophioctenis* are ⅓ to ⅛ the size of females, which are 3–4 mm in length. They live in the same host, attached to the females. Fertilization is external, in the brood chamber. Nauplii with a large shieldlike carapace filled with yolk develop within the brood chamber, where they metamorphose into a cyprid-like, so-called ascothoracid stage, 0.4 mm long. The ascothoracid stage has a bivalved carapace, like that of adult males. The second antennae are lacking, and the mouthparts and biramous thoracic swimming limbs are formed as in adults. This stage leaves the brood chamber of the female and searches for a new host. In some genera (e.g., *Laura*) larvae leave the female as nauplii.

Relationship

The Ascothoracica are Cirripedia, as indicated by the location of the female genital apertures at the base of the first thoracopods, the prehensile first antennae, loss of second antennae and, the general resemblance of the ascothoracid stage to cyprid larvae. Differences are the persistence of the abdomen and bivalved carapace seen only in the cyprid of other cirripeds, the lack of cement glands for permanent attachment of the adult and the absence of setose feeding. The Ascothoracica are more primitive than other Cirripedia, except for being semipredaceous carnivors and in having made specializations for parasitism rather than for setose feeding.

Fig. 9–2. Female of *Ascothorax ophioctenis* viewed from below; 3.5 mm wide. The ventral surface of the body can be seen through the aperture. The arrows indicate the direction in which the mantle of the Dendrogasteridae grows and branches. (After Wagin.)

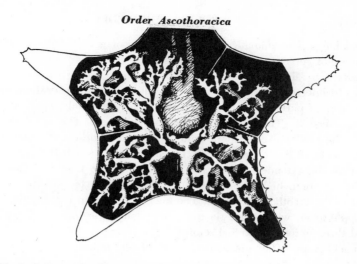

Fig. 9–3. *Myriocladus arbusculus* in the starfish *Hippasteria californica*. Branching lobes of the bilateral carapace of the parasite can be observed extending into four of the five arms of the host. Distance from center to tip of arm, 7 cm. (After Fisher.)

Habits

Synagoga mira swims freely, attaching temporarily and feeding on corals. All others have become permanently fixed to their hosts. Usually a male is found with each female. *Ascothorax ophioctenis* lives in the bursa at the base between two arms of the serpent star *Ophiocten sericeum*. It is immovably attached, extending the bursa of the host so much that the presence of the parasite can be recognized from the outside. One or both antennal chelae are extended and attached to the bursal wall. The slit between the carapace valves toward the bursal opening, permits the limbs to circulate a current of respiratory water. The labrum is appressed to the bursal wall followed by penetration of the mouthparts in feeding. The gutterlike first maxillae form a tube through which cell debris and fluids, perhaps coelomic fluid, are drawn from the host. The parasitized serpent star does not develop gonads. The habits of some other species are discussed below.

Classification

Synagogidae. The trunk is typical of the Maxillopoda; the thorax supports six pairs of biramous limbs and abdomen is distinctly 5-segmented. The two valves of the carapace are connected to the head region. *Synagoga,* up to 4 mm, is an ectoparasite of *Antipathes.* Adults of *S. mira* are free swimming. *Ascothorax ophioctenis* (Figs. 9-1, 9-2) with a mantle to 3.5 mm long, is an endoparasite in the bursal cavities of serpent stars.

Lauridae. There are only three to six free uniramous, unsegmented thoracic limbs, four pleomeres. In *Laura,* there are five pairs of uniramous thoracic appendages. The saclike, dorsoventrally expanded carapace is much more voluminous than the trunk. Except for a small slit, the margins are fused. The trunk is up to 12 mm, the mantle up to 40 mm long. The animal has the appearance of a flattened cigarette and lives in the coenosarc of gorgonians and antipatharians.

Dendrogasteridae. The external segmentation is more or less evident, abdomen and thorax either distinct or fused. The abdomen may show three or four somites. The furca is reduced to partially fused rudiments, or absent. Only rarely do vestiges of thoracic limbs remain. The mouthparts also are reduced. The sac-shaped mantle is fused around its border except for a minute slit. In *Dendrogaster* and *Myriocladus,* endoparasitic on starfish and brittle stars, the slit through which the first antennae and mouth cone can be extended is opposite the mouth. The lateral walls of the valves consist of hollow, branched lobes (Fig. 9-3). The gut is incomplete and the larvae are released in the ascothoracid stage. The males live within the mantle cavity of the female. *Ulophysema,* with a mantle to 22 mm long, has four to five pairs of limb vestiges and three to five abdominal somites. The young stay in the brood pouch until the ascothoracid stage. They are endoparasites of sea urchins.

ORDER THORACICA, BARNACLES

The Thoracica include the goose, acorn and wart barnacles. Approximately 650 species are known. *Lepas anatifera* with a peduncle as long as 80 cm and *Balanus aquila* with a basal diameter of about 12 cm are among the largest, but most are 1–2 cm and some are only about 1 mm high.

Thoracica have six pairs of serially arranged thoracopods (cirri), arranged evenly along the thorax in generalized forms. The carapace (mantle) is usually covered by permanent calcareous plates.

All Thoracica attach permanently to inanimate objects, algae, or other animals, by the preoral region. The first antennae function only briefly as attachment organs when the cyprid larva is undergoing metamorphosis into the adult stage (Figs. 9-8, 9-13, 9-14). The exterior of the carapace (mantle) supports permanent calcareous plates, but in a few species these may be secondarily reduced or lost (Figs. 9-4, 9-5, 9-16). The six pairs of thoracic limbs (cirri) are more or less evenly

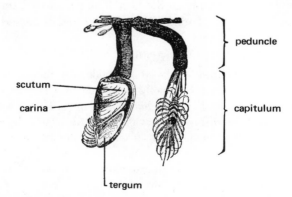

Fig. 9–4.   Goose barnacle, *Lepas anatifera*; left, lateral view; right, ventral view showing the cirri; 8 cm long. (After Hertwig.)

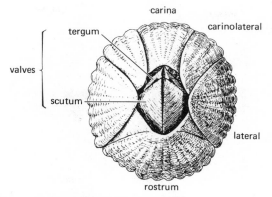

Fig. 9–5. Acorn barnacle, *Balanus balanoides*, seen from above. Tergum and scutum on each side form an operculum, closing the mantle cavity. When they are rotated laterally, they form the aperture through which the cirri can be extended; 1 cm, basal diameter. (After Lameere.)

distributed on the thorax (Figs. 9-6, 9-13), a characteristic that distinguishes them from the Acrothoracica. Most species are hermaphrodites, but some hermaphrodites are accompanied by dwarf males, termed complementals. When males occur in dioecious species they are also much reduced in size and structure and frequently do not feed; they are termed dwarfs.

Anatomy

The preoral region, the area anterior to the mouthparts, becomes an attachment organ. The development of this region determines the general appearance of the barnacle; if elongated, the resultant body shape is that of the stalked (pedunculate) goose barnacles, Lepadomorpha (Figs. 9-4, 9-6, 9-10a, 9-16); if flattened into a broad attachment disc (basis), the form is that of sessile barnacles: Verrucomorpha and Balanomorpha (Figs. 9-5, 9-17). Thus in goose barnacles the preoral stalk is proportionately large compared to the capitulum, and may surpass it in size. The capitulum consists of the postoral head region and trunk and investing carapace (mantle). The flattened attachment disc of the Verrucomorpha and Balanomorpha may be either membranous or calcareous. In these the preoral region is not visible externally, but is hidden beneath the plates covering the mantle.

The pedunculate and sessile barnacles, so different in appearance, are closely related. The difference in form relates to differences in their adaptations to environmental conditions, as variations under special conditions illustrate. When crowded many sessile barnacles elongate several times their usual length by allometric growth of the wall plates. The preoral region under these conditions becomes equivalent to the stalk (Figs. 9-8, 9-9), containing the ovaries and elevating the feeding portion as in pedunculate barnacles. An extreme case is

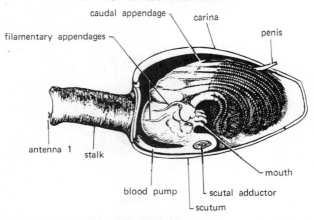

Fig. 9–6.   Goose barnacle, *Lepas*; plates of left side removed; 8 cm long. (After Darwin.)

seen in *Xenobalanus,* which occurs on small cetaceans. Here the rudiments of the shell are embedded in the skin of the host, the feeding appendages and associated organs being suspended by a long fleshy stalk. On the other hand, the peduncles of some lepadomorph genera, such as *Lithotrya* burrowing in limestone and *Megalasma* living on objects in the deep-sea, hardly elongate at all (Figs. 9-7, 9-16c) and to this extent resemble barnacles lacking a stalk. These resemblances in form are secondary, but they suggest the trends that lead to the evolution of one group from the other. Detailed studies have shown the direction has been from pedunculate to types without stalk and that stalkless types such as the Balanomorpha and Verrucomorpha evolved independently from different pedunculate ancestors.

Each short first antenna of the cyprid has a disclike swelling for temporary attachment and a cement gland for permanent attachment (Fig. 9-14). During metamorphosis the preoral portion of the head becomes separated by a constriction from the oral and postoral portion of the body (Figs. 9-14a, b). The mantle is attached to the preoral portion, below the constriction, and encloses the body as in Conchostraca, Ostracoda and Leptostraca. The ventral opening forms the aperture through which the limbs can be extended (Figs. 9-4, 9-13).

The cyprid valves consist of thick noncalcareous cuticle. The calcareous plates found in nearly all Thoracica form between this cuticle and the underlying hypodermis, during metamorphosis, and, in general, act as protective armor for animals unable to escape predators (Figs. 9-15, 9-16). Formation of calcareous elements is not limited to this region, however. Some Lepadomorpha, such as *Lithotrya,* have a region at the base of the stalk which secretes calcareous material and the stalk of all scalpellids is covered to varying degrees by calcareous scales (Figs. 9-7, 9-10a, 9-16a). In molting, only the integument of the trunk limbs and inside lining of the mantle is shed; the plates and scales are not. In

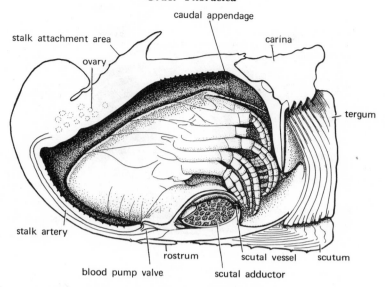

Fig. 9–7. The coral boring pedunculate barnacle, *Lithotrya valentiana*, plates of left side removed; 2 cm long. (Redrawn after Cannon.)

Thoracica, the plates enlarge by deposition of minerals and cuticular material around their periphery. Thickness also increases with age.

During metamorphosis in Lepadomorpha and Verrucomorpha, five noncalcareous primordial plates appear on the mantle; a median carina covering the dorsal edge as a keel, the terga posteriorly and the scuta anteriorly as paired plates on either side of the aperture (Fig. 9-14c). The ventral edges of the scuta will form the greater border of the (occludent) margin of the aperture. Calcification follows, around and beneath the chitinous primordia. In the Lepadidae and Poecilasmatidae this arrangement of five plates is retained, but in the Scalpellidae the number is increased by the appearance of the anterioventral rostrum, and numerous laterals around the margin separating the capitulum from the peduncle (Figs. 9-10, 9-16a). However, these additional plates are not preceded by primordial plates, and in the Balanomorpha all plates appear directly.

The Balanomorpha can easily be derived from the Lepadomorpha through a form such as *Pollicipes* (Fig. 9-10). If the stalk were shortened and the lateral plates were reduced in number and the principal ones increased in size, the animal would resemble the balanomorph genus *Catophragmus* (Fig. 9-11). Further evolution of the Balanomorpha includes reduction in the number of lateral plates (Fig. 9-11).

In the Balanomorpha the scutum and tergum on each side of the body are articulate and are supported by an arthrodial membrane to form the operculum, the details of which are generally characteristic for each species. The apertural margin of the opercular valves is homologous with the ventral border of the

9. SUBCLASS CIRRIPEDIA

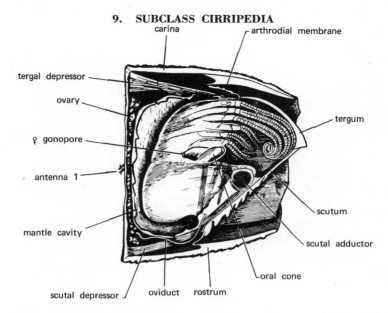

Fig. 9–8. *Balanus tintinnabulum*, plates and valves of left side removed; 5 cm diameter. (After Darwin.)

mantle aperture (Figs. 9-5, 9-8). The flexible noncalcareous membrane permits the valves to act as a unit, the operculum, independent of the wall plates.

The aperture in most Thoracica is closed by a transverse adductor muscle between the scuta (Figs. 9-8, 9-13). Depressor muscles (Fig. 9-8), acting on the basal margins of the valves in Balanomorpha, are the largest single muscle units known and are used extensively in neuromuscular studies. Extension of the thorax and cirri is accomplished by muscles and blood pressure.

The oral cone projects from behind the scutal adductor muscle. It comprises a wide labrum followed by three pairs of mouthparts supplied by serially arranged nerves from the subesophageal ganglion. The mandible consists of opposed, toothed coxal endites each supporting a one-articled palp. The bladelike first

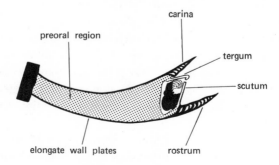

Fig. 9–9. A much elongated barnacle, *Balanus balanoides*, from a crowded colony, plates and valves of the left side removed. (After Gutmann.)

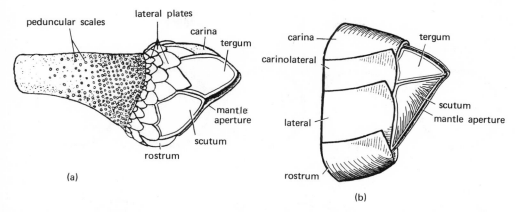

Fig. 9–10. Comparison between Lepadomorpha and Balanomorpha. (a) *Pollicipes polymerus*, 5 cm total length. (After Nussbaum.) (b) *Balanus*. (After Broch.) Both viewed from left side. The terga and scuta guarding the aperture as shown with a double border to facilitate comparison.

maxillae have a spiny chewing edge. The soft, setose second maxillae tend to fuse basally and close the back of the mouth region (Fig. 9-12). Each supports openings of maxillary glands. The ventral surface of the thorax bears six pairs of biramous limbs (cirri), each ramus composed of many short articles supporting long setae used for straining food from surrounding water (Figs. 9-4, 9-13). However, one to three pairs may be modified to serve as maxillipeds. Blood pressure alone extends the cirri.

The abdomen has three somites in cyprid larvae, but all are lost in adults, so that the anus comes to reside at the posterior end of the thorax in the adult. As the result of bending, mouth and anus approach the same level, an adaptation often seen in sessile animals (e.g., Entoprocta, Bryozoa) (Figs. 9-6, 9-7, 9-8).

NERVOUS SYSTEM AND SENSORY ORGANS. The reduced supraesophageal ganglion lies anterior to the pharynx and long connectives extend from it to large subesophageal ganglia (Fig. 9-13). In *Lepas*, the ventral nerve cord is ladderlike, ganglia 2, 3, and 4 being separated by connectives. A single ganglionic mass contains nerves of the 5th and 6th limbs. In *Balanus*, however, the postoral ganglia are completely fused. Compound eyes, found in the cyprids of all orders except Ascothoracica, are lost in the adults. The nauplius eye in front of the brain in nauplii and cyprids is retained in some adults. It is frequently divided bilaterally into units that are among the simplest known photoreceptors used in neurological research.

DIGESTIVE AND EXCRETORY SYSTEMS. The cuticularized muscular foregut continues into a U-shaped midgut having a pair of ceca at its anterior end (Fig. 9-13). The ectodermal hindgut is also cuticularized. Maxillary glands, opening on the second maxillae, function as excretory organs.

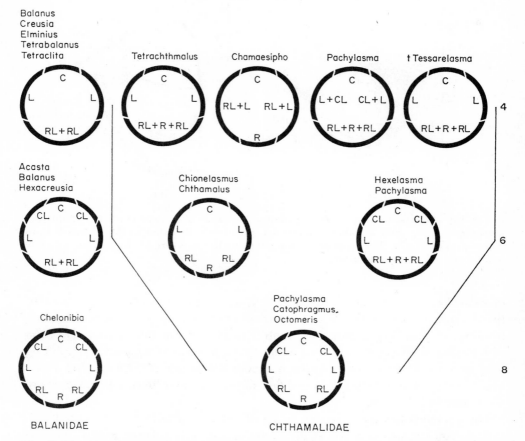

Fig. 9–11. Progressive fusion and reduction of wall plates as shown in different genera of Balanomorpha. The maximum of 8 plates includes the rostrum (R), two rostrolaterals (RL), two laterals (L), two carinolaterals (CL), and the carina (C). The minimum of 4 plates includes compound rostrum, laterals and carina. (From Newman after Darwin, courtesy the author and by permission of the Zool. Soc. London.)

CIRCULATORY AND RESPIRATORY SYSTEM. The circulatory system lacks true arteries but in larger Lepadomorpha the main sinus is covered by an anucleate fibrous membrane. There is no true heart, its place being taken by a large vessel-like structure, the blood pump, lying between the scutal adductor muscles and esophagus (Fig. 9-6). The principal vessel from this pump runs anteriorly to the stalk and contains a valve that prevents backflow (Fig. 9-7). Circulation through lacunar spaces results mainly from movements of the body musculature. When the mantle aperture is closed, metabolism is very low. Oxygen is taken up by the thin inner wall of the mantle, the surface of which in the Balanomorpha is increased by many accordionlike folds near the base of the opercular valves. Probably the fingerlike extensions known as filamentary appendages on the trunk wall of *Lepas* also have a respiratory function (Fig. 9-6).

Reproduction

Most Thoracica are hermaphroditic, a condition that is relatively rare among Crustacea but common in sessile and parasitic animals and those adapted to other specialized habitats. A bisexual condition found in some primitive species of *Scalpellum* and *Ibla*, dwarf males occurring with females, is probably primitive. That hermaphroditism is generally secondary is further attested to by the appearance of males in some hermaphroditic species. These are called complemental to distinguish them from dwarf males occurring with true females. Males are minute (0.4–2 mm long) and are always attached to individuals acting as the opposite sex. In some species the males have reduced plates, but in others they are saclike, the trunk not set off from the stalk. The limbs of the males are usually reduced, and in extreme cases the male is a minute hirsute sac, and except for a pair of antennae and gonads, both the external and internal features are completely lost. It has recently been discovered that some balanids (*Conopea* and a species of *Solidobalanus*) have complemental males, but none is known to have separate sexes.

The male gonads of hermaphrodites lie among other organs in the thorax, and open into a long, terminal, extendible penis (Fig. 9-13). The ovaries, strangely, are preoral. The oviducts pass into the trunk and open into the mantle cavity at the base of the first thoracic limb, the most anterior position is any crustacean (Fig. 9-8). In *Balanus* colonies it has been observed that a large number of individuals act as males and simultaneously extend their penes beyond the mantle aperture and into neighbors acting as females. Certain weak organic acids are known to elicit this behavior.

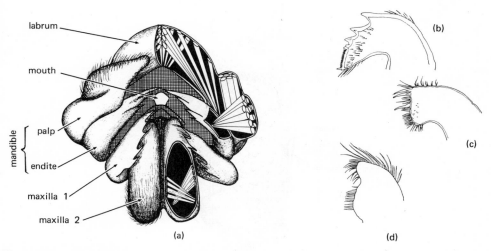

Fig. 9–12. **Oral region and mouthparts of (a)** *Lepas fascicularis.* **(From Petriconi, courtesy of the author.) (b-d)** *Tetrachthamalus obliteratus:* **(b) right mandible, (c) right first maxilla, (d) right second maxilla. (From Newman, courtesy of author and by permission of the Zool. Soc. London.)**

Development

The eggs are fertilized in the mantle cavity where they remain until the nauplii hatch. Freshly hatched nauplii may be either released immediately or retained until favorable conditions prevail. In some species the naupliar stages are passed through in the egg, and the larvae hatch as cyprids.

In colonies of *Balanus balanoides* of northern oceans, individuals have been observed to expel their nauplii at the same time. Well-nourished individuals release a substance into the mantle cavity which stimulates the embryos to become active. Thus the nauplii are released into the plankton when food is abundant. Synchronous release also increases the probability of dense settling, insuring adequate numbers in close proximity for reproduction. The water may resemble a thick brown soup when full of cyprids ready to settle and attach.

The cyprid does not feed, but swims jerkily by synchronous beats of the limbs as it seeks a place of attachment. Upon locating an appropriate substrate, it walks about slowly on the first antennae, as on stilts, alternately attaching them. It may

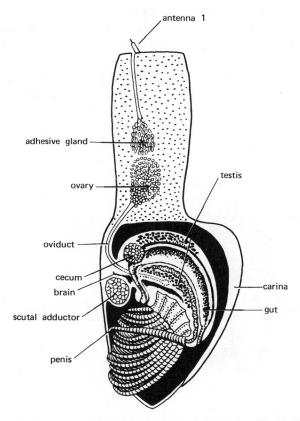

Fig. 9–13.   Goose barnacle in sagittal section; mantle cavity black; 8 cm long. (After Broch.)

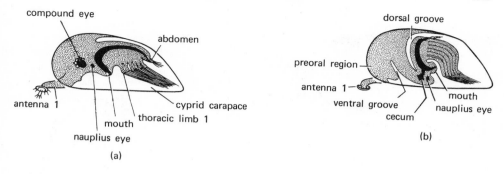

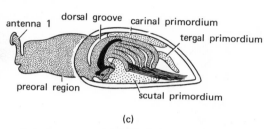

Fig. 9–14. **Diagram of metamorphosis of the goose barnacle, *Lepas*, viewed from left. Carapace of bivalved cyprid shell transparent in illustration. (a) Free swimming cyprid, 1.5 mm long; (b) attached cyprid, pupal stage undergoing metamorphosis. Part of the animal has already rotated. (c) Later stage; cyprid has not yet shed carapace, but by allometric growth the preoral region has become the stalk and the uncalcified primordial plates have appeared on the hypodermis. Thoracic swimming limbs are not yet cirriform, but the abdomen and furca have been lost. (After Korschelt.)**

attach temporarily several times by means of the suckerlike processes on the 1st antennae before secreting a cement for the final permanent attachment. Now motionless and without feeding, it transforms within its cuticle, as a so-called pupal stage (Figs. 9-14, 9-15). The adult form develops with the molting of the cyprid shell. The larvae of most *Balanus* spend 7–14 days in the plankton before seeking out and permanently attaching to the substratum.

The bent body shape of the pedunculate barnacles results from two deep, transverse, folds of the body of the cyprid larva (Fig. 9-14). A dorsal groove forms a cleft between the preoral and postoral portions that permits folding by 90° from the original longitudinal axis of the larva. At the same time, the anterior ventral groove constricts the preoral region, assisting in the rotation. Finally the preoral portion anterior to the ventral cleft enlarges and extends out from the mantle to form the peduncle. At the same time, the new mantle cuticle under the cyprid shell forms the five noncalcareous primordial plates (Fig. 9-14c). The compound eyes are lost and the biramous swimming limbs are transformed into feeding appendages, the cirri.

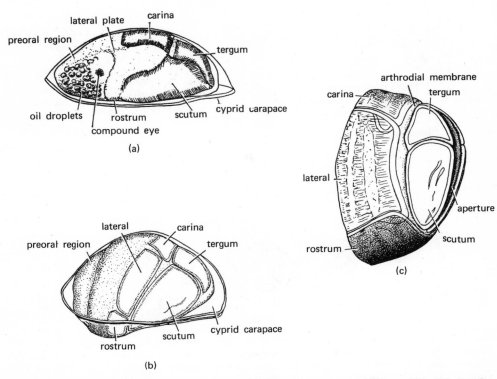

Fig. 9–15. **Metamorphosis of the acorn barnacle** *Balanus balanoides,* **viewed from the left side. (a) Attached cyprid or pupal stage beginning to transform; under the bivalved cyprid carapace, 5 calcareous primary plates can be seen; the shell is 1.2 mm long. (b) The "pupa" still within the cyprid carapace takes a spherical shape, two pairs of lateral plates having appeared. (c) The preoral region has flattened to a basal disc, completely enclosed by the wall plates including compound rostrum; 1.2 mm diameter. (After Runnström.)**

The metamorphosis of *Balanus* species is better known and more complicated. Just after attachment, histolysis of the muscles, the maxillary glands and compound eyes of the cyprid larva can be observed. The fragments are taken up by phagocytes, clumps of which can be seen in the swimming larva and are found above the gut and in the blood spaces of each mantle extension. About 8 hr after attachment, *B. amphitrite* loses its cyprid shell and cuticle from the appendages from which all tissues have withdrawn. For the next 4–6 hr the animal is a pear-shaped sac, with its stalk cemented to the substrate. Soon, in the middle of the upper side, a cone-shaped opercular dome rises, and around its margins skeletal plates will form. In *B. balanoides,* two pairs of plates, the laterals, begin to appear after the carina, terga, and scuta (Fig. 9-15b). Under the old cyprid shell the calcified plates arrange themselves in two rings: the lower whorl includes the two laterals, the median carina and the newly added rostrolaterals, the upper whorl the terga and scuta, form the operculum. Then the two rostro-laterals fuse to become the compound rostrum. Thus the wall includes the unpaired rostrum, unpaired carina and the laterals, and together these soon cover completely the preoral disc, attached to the substrate (Fig. 9-15). There are no noncalcareous plate primordia in the Balanomorpha.

The whole metamorphosis takes place in 24 hr in *Balanus improvisus*. Later, as in *Balanus balanoides,* each lateral plate splits off a carinolateral, so that the final number of wall plates is 6. In *Balanus improvisus* the cirri develop on the 4th day after attachment of the cyprid and feeding begins on the next day. The mortality during this period is high. If within a week 6000 attach per 100 cm², as few as 300 may remain after a month, each having a diameter of 3 mm.

Balanus amphitrite molts for the first time 3–5 days after metamorphosis from the cyprid larva and every 2–3 days thereafter, averaging about 19 molts in the first 47 days. *Balanus balanoides* (Fig. 9-5) occurs on both sides of the Atlantic, settling on a wide variety of solid objects, including pilings, rocks, shells, and crabs, from the low tide line to areas that are splashed only occasionally. This species is capable of withstanding a considerable amount of drying. On rocky coasts they form a wide band stretching for miles. Gravid individuals each contain 6000–13,000 nauplii in the mantle cavity. On the Island of Sylt in the North Sea, animals grow to a diameter of 3–4 mm within 26 days after attaching, 6–7 mm after 58 days. At this locality all young molt synchronously, making the water almost opaque with floating exoskeletons. At Port Erin, Isle of Man, Great Britain, *B. balanoides,* in the high intertidal, reaches an age of about 3 years, but individuals at median tide level become twice as old or older. *Balanus improvisus* is found subtidally in marine situations, but they occur intertidally in estuaries where other species are absent; in Holland and Norway they survive only 1 year.

Habits

Barnacles can move about freely only as larvae; all adults are attached. The balanids use other organisms, rocks, pilings, or boats as substrates. Colonies on rocks form bands for miles in the tidal zone with densities as great as 1000–2500 individuals per 100 cm². They also attach to mussels, on the carapaces of crabs, and on some corals.

Barnacles of the genus *Chelonibia* attach to sea turtles, *Coronula* and *Xenobal-anus* to the skins of whales. The Lepadomorpha are also found on inanimate substrates, floating logs, ships, or stones. *Lepas fascicularis* constructs a float of bubbles made from secretions of the cement glands. Clumps of secretion are formed, within which gas bubbles appear. Some species attach to fish, crabs, seasnakes, even to jellyfish, sponges, antipatharians, and gorgonians. The symbiosis of the barnacles and their hosts has been little studied. In corals only dead individuals are grown over by host tissue. Teeth on the articles of the cirri help maintain an opening to the exterior, but other mechanical and metabolic processes are involved.

Obligate commensalism is common, the whale and turtle barnacles having already been mentioned. Among *Balanus,* species of *Membranobalanus* occur in sponges, *Conopea* on alcyonarians, certain *Austrobalanus, Armatobalanus,* and *Hexacreusia* with corals. *Creusia* and *Pyrgoma* are found exclusively in corals and one species, *P. monticulariae,* not only exploits coral for support and protection, but feeds on its tissues as well, thus being the only wholly parasitic balanomorph known.

In very dense colonies, young balanids may not be able to form their normal conical form. Usually the wall plates then tend to grow more rapidly at the base and become elongate (Fig. 9-9). Such groups often form a dome in which the central animals are much longer than those around the periphery and the individuals bend outward from the center like a bunch of flowers. Table 9-1 gives an idea of how dense the clumps can become.

An individual may produce up to 20,000 offspring in one reproductive period, generally in spring in northern species. Of these perhaps five may survive to attach. The cyprid generally prefer a substratum in currents.

The rate of growth depends on the species, the climate, and the availability of food. It takes 16 days for *Balanus amphitrite* to reach 8.8 mm in diameter in Madras, India; in the Adriatic it takes 5 months to reach 10 mm. Turbulence and waves are of importance as the currents bring plankton, and in quiet places fewer *Balanus balanoides* settle than in areas of strong waves.

Dense settling of large swarms of cyprid larvae occurs in areas in which other living or dead barnacles occur. The shell of barnacles contains proteins collectively called arthropodin, because of similarities to proteins found in the exoskeleton of other arthropods. Arthropodin is attractive to larvae and assures settlement in situations that were or are favorable for the species and insures that individuals will be close enough for cross-fertilization to take place.

Table 9–1.   Densities of barnacles in number of individuals per m² (after Moore)

Height above mean tide	*Balanus balanoides*			*Chthamalus stellatus*			Both species, total
	Adults	Very young	Total	Adults	Very young	Total	
+3.4 m	0	0	0	0	0	0	0
+2.7 m	0	0	0	15,200	9,200	24,400	24,400
+1.8 m	0	400	400	54,000	38,000	92,000	92,400
+0.8 m	4,000	12,400	16,400	55,600	35,200	90,800	107,200
−0.2 m	40,400	20,400	60,800	400	4,800	5,200	66,000
−2.1 m	0	0	0	0	0	0	0

Barnacles are very resistant to abiotic factors. The pressure of the surf, which can move rocks of 15 tons, does not remove a living *Balanus perforatus* from the rock. *Balanus balanoides* in the tide zones of Arctic oceans can survive months of freezing during the winter, as well as daily drying for 6–9 hr between high tides during the summer. Such severe conditions are also survived by eggs and nauplii in the mantle cavity. In experiments on the Swedish coast, larvae survived in the mantle cavity though the hermaphrodite was kept out of water 24 days. Extreme conditions are survived by *Chthamalus stellatus* of the sun-baked Sicilian coast, where some colonies are covered by water only one-sixth of the time, and otherwise are reached only by spray. Survival during periods of exposure to strong sun and rain (freshwater) is accomplished by closing the aperture to a very narrow slit. The walls of the mantle cavity act as respiratory organs and after the water it contains has been exhausted, atmospheric respiration may be resorted to. Curiously, while many intertidal balanids can maintain their internal osmotic concentration well above dilute seawater for several days, truly estuarine barnacles conform very closely to the tonicity of the external environment. This is unusual for a crustacean but it is characteristic of many brackish-water and freshwater mollusks.

Barnacles are of economic importance when they foul ships and clog sea water systems. *Balanus amphitrite* is now found in virtually all warm-water regions of the world and is believed to have reached most non-Indo-Pacific localities by ships. In recent years the south Australian and New Zealand *Elminius modestus* was thus inadvertently introduced to the European coasts, to the near exclusion of a native form, *B. improvisus* in some localities. Barnacles settle on the ship's underside and reduce the speed by as much as 35% by increasing friction. In a survey of 250 ships in the Atlantic, 70% were covered with cirripeds. Of *Balanus amphitrite* there may be as many as 150–300 individuals per 100 cm², of *Lepas,* 400–450 in the same area. The attachment of balanids occurs mostly in harbors, but *Lepas* fouls ships at sea. Growth is rapid in warm waters. On a boat traveling between Roscoff, France and Spain, *Lepas anatifera* was recorded as growing 0.87 mm daily, attaining a capitular length of 3 cm and stalk length of 7 cm (contracted) within 113 days. Between July and September the daily growth was more than 1 mm. Balanids on boats in the tropics reach a length of 4 mm 1 month after attaching, and 12 mm after 3 months. Most feed to a large extent on diatoms, other small invertebrates and their larvae. In the course of travels, boats get completely covered by various *Balanus* species, algae, hydroids, or oysters. If after unloading the boat rises, the animals above the waterline die. But in the process the protective paints are usually destroyed so that the metal is exposed to corrosion. In freshwater harbors the marine growth is killed and cleaning and repainting is necessary.

SENSES. The cirri and often the stalk and mantle in Lepadomorpha are covered by sensory bristles or setae, stimulation of which causes withdrawal of the body and closing of the aperture. It has also been observed in *Scalpellum* that cirri will go through the feeding motions only if the setae on the concave, anterior side are stimulated. All species tested could differentiate between large and small particles, and in most only small ones were transported to the mouth; large particles being discarded. Rheotaxis has been observed in *Balanus* and *Pollicipes,* and the cirri are extended toward wave produced and other currents. Rejection of unpalatable materials indicates a chemical sense, and in aquaria some inorganic salts or urea added to the water cause the cirri to stop their movements. Also a generalized light sense is present. *Balanus* may close if a shadow passes over it during the day and this happens even if in a dark chamber a small beam of light is turned off that previously illuminated only a skeletal plate, far from the nauplius eye.

FEEDING.   As is shown by the gut contents, the particles taken in, algae, bacteria, and detritus, are 2 μm to 1 mm in diameter. These are collected by filtering. *Balanus* opens its aperture by rotating the opercular valves laterally and extends the fourth to sixth pairs of cirri. The cirri are spread and curve outward as a funnel shaped net, the base of which nearly surrounds the mouth. Their setae form the mesh of a sieve, with openings about 33 μm wide. As the cirri together beat toward the mouth, the water within the cone is forced out and at the same time large particles are rejected. The first three pairs of cirri, also arranged around the oral cone with their setae form a net with a mesh of about 1 μm that collects the material from the posterior cirri and transfers it to the mouthparts. The cirri continue their rhythmic casting and folding for hours. In currents some barnacles simply keep their cirri extended, filtering water, folding them only at intervals. In completely quiet water many species will not extend their cirri to feed at all.

The activity of a species depends on its size, the temperature, the tides, the period of acclimation, and the geographical race. Species occurring next to each other usually show differences. At Port Erin (Isle of Man, Great Britain) the southern *Chthamalus stellatus* is active at temperatures from 6 to 36°C. The beat at 30°C reaches a maximum of 60 movements per minute. In contrast the northern *Balanus balanoides,* active from 2 to 31°C, reaches a maximum of 33–37 beats/min at 17°C.

The Lepadomorpha also filter by rhythmical extensions of the cirri but they do not have as rapid a beat as most balanids. In the guts of *Lepas* and *Scalpellum* mostly nauplii and copepods have been found. *Pollicipes* living on wave washed shores hold the cirri immobile in the backwash and planktonic animals touching the cirri are captured and carried to the mouth. If a little piece of meat is held near the cirri of *Lepas, Scalpellum,* or *Verruca,* a cirrus will coil around it like a tendril and press it against the mouthparts. The functioning of the mouthcone in feeding can be observed in *Lepas fasciculoris* (Petriconi, 1969). One can see how the two 2nd maxillae (Fig. 9-12) move to the side and toward the middle, widening and narrowing the food groove between them. At the same time mucus secreted from glands entangles small food particles. The spine bordered endites of the first maxillae pass these particles to the mandibular endites, which in turn break up large particles and push the material into the mouth. The palps prevent food from escaping over the edge of the labrum. In the dogfish parasite *Anelasma,* a relative of *Lepas,* the cirri are degenerate rudiments and the species feeds by nutritive processes of the peduncle invading the host.

RESPIRATION. Feeding movements pump water into the mantle cavity over the eggs and the inner lining of the mantle, the respiratory tissue. The so-called branchia of certain balanomorphs, and filamentary appendages found in some lepadomorphs are thought to be respiratory in function, but they may also be involved in the incubation of the eggs and embryos. The cirri themselves apparently serve as respiratory organs during periods of high activity.

Classification

The order Thoracica contains three Recent suborders.

SUBORDER LEPADOMORPHA

The 350 Recent species known are arranged in seven families and 25 genera. The body is composed of a stalklike peduncle and a capitulum containing the labrum, postoral portion of head, and thorax (Figs. 9–4, 9–6, 9–10a, 9–12). Four of the most common families are characterized here.

Scalpellidae. The capitulum has more than five plates and the peduncle is more or less covered by plates or scales (Figs. 9–7, 9–10a, 9–16a). *Scalpellum* and allied genera are generally found in moderately deep to very deep water on stones, clams, crabs, on the stem of the sponge *Hyalonema,* and coelenterate colonies. Some have a stalk as much as 9 cm long and a mantle 6 cm high. *Pollicipes* (=*Mitella*) forms dense colonies along rocky shores (Fig. 9–10a), and the closely related *Lithotrya,* bores stalk first into limestone, coral, and mollusk shells along tropical shores (Fig. 9–7). Intertidal genera are usually hermaphroditic. In deep water hermaphrodites frequently are accompanied by complemental males, or sexes are separate.

Iblidae. This family has only four plates arming the capitulum and the stalk, covered by long cuticular processes, is not demarcated from it; most species have caudal appendages. Members of the only genus, *Ibla,* with species up to 2 cm in length, attach to rocky substrate or other lepadomorphs along tropical shores and in relatively deep water (Fig. 9–16d). Species are either hermaphrodites accompanied by complemental males, or have separate sexes.

Lepadidae. Goose barnacles are generally oceanic rather than benthic. The capitulum has five plates, some of which may be reduced or missing. The stalk is naked, lacking setae or scales. *Lepas*° has many species occurring on a variety of floating objects such as pumice, floating snails and seaweed (*Macrocystis* and *Sargassum*). *Lepas anatifera* (Fig. 9–4), usually several cm long but up to 75 cm, is cosmopolitan. *Lepas fascicularis* initially attaches to *Fucus, Janthina, Vellela,* or small pieces of floating wood, objects that would eventually sink due to the weight of the growing barnacles,

° The name *Lepas* is on the *Official List of Generic Names in Zoology.*

and a few to hundreds of individuals eventually form a common bubble-float of their own. *Conchoderma*, found attached to floating objects and occasionally on pilings or buoys, has the plates variously reduced or lost. *Conchoderma virgatum*, up to 5 cm in total length, attaches to ships, cables, wood, turtles, seasnakes, swimming crabs, and on the large parasitic copepod *Penella* attached to fish. *Conchoderma auritum*, with a stalk to 13 cm long, is usually found on the sessile barnacle *Coronula* attached to the skin of whales. *Alepas*, with only Y-shaped scuta arming the capitulum, attaches to scyphomedusae. *Anelasma*, with its plates completely lost, attaches around the base of the dorsal spine of dogfish (*Squalus*, *Etmopterus*). It is wholly parasitic, sending rootlike processes into the tissues of the host. All lepadids are believed to be hermaphroditic and none have articulated caudal appendages.

Poecilasmatidae. The capitulum usually supports five plates and the peduncle lacks calcareous scales. They differ from the lepadids in having a pair of uniarticulate caudal appendages and in being benthonic, or associated with benthic organisms such as crustaceans and sea urchins, rather than pelagic. A closely related family (Oxynaspididae) is found on Antipatharia. Distribution is tropical to warm temperate, in shallow and deep water. Those in exposed situations have completely armored capitula, but those that receive protection from their hosts have undergone considerable reduc-

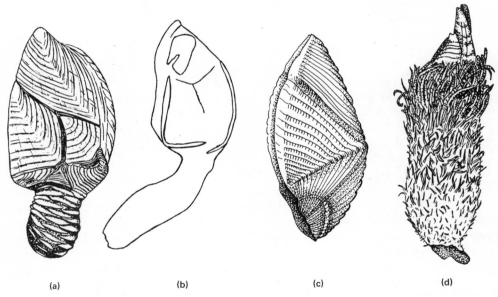

(a)   (b)   (c)   (d)

Fig. 9–16. **Pedunculate barnacles viewed from the right side: (a)** *Arcoscalpellum regium* **from the deep sea; note lower whorl of 3 latera, upper latus interposed below tergum and between scutum and carina and calcareous plates on peduncle; 13 mm long. (After Pilsbry.) (b)** *Octolasmis californiana* **from the gill chamber of the spiny lobster; note reduction of plates and naked peduncle; 8 mm long. (After Newman.) (c)** *Megalasma striatum* **from the deep sea, note how lower margins of scuta and the carina have developed downward to enclose most of the naked peduncle; 10 mm long. (After Hoek.) (d)** *Ibla cumingi*, **female, intertidal, from the tropical Indo-Pacific; note paired terga and scuta surrounding aperture and nonmineralized spines covering spindle-shaped capitulum and peduncle; 22 mm long. (After Newman.)**

tion. *Poecilasma* is found on the carapace of the deep-sea crab *Geryon* in the Atlantic and Indian oceans. Related forms live on the mouthparts of spiny lobsters along tropical shores and have recently been shown to develop modified appendages and feed on the food being manipulated by their hosts. A species of *Octolasmis* occurs in the gill chambers of *Geryon*; on the pleopods of the giant deep-sea isopod *Bathynomus* and other species occur in the gill chambers of crabs and lobsters along the shore (Fig. 9–16b). *Megalasma* lives on a variety of objects in relatively deep water (Fig. 9–16c). Like the lepadids, all are hermaphroditic.

SUBORDER VERRUCOMORPHA

The 54 living species known are all included in a single genus and family. Most species have caudal appendages. In being sessile they resemble the balanomorphs, except that the wall is not bilaterally symmetrical (Fig. 9–17). The scutum and tergum of one side are articulated with the rostrum and carina forming an asymmetrical box-like wall of four plates, the free or movable scutum and tergum forming the lid. All are hermaphrodites and most species are found in the deep sea. *Verruca stroemia*, one of the few intertidal species, has a diameter up to 1 cm and is found on the inside of empty oyster shells and on shells of *Buccinum*.

SUBORDER BALANOMORPHA

There are about 250 living species in two families and 25 genera (Fig. 9–5, 9–8, 9–10b). Fundamentally there are eight plates forming the wall: the carina; paired carinolaterals, laterals and rostrolaterals; and the rostrum. The plates become variously fused or lost (Fig. 9–11), and in *Pyrgoma* the wall is made of a single plate. Both pairs of scuta and terga form the movable operculum. Members of each pair may become fused (*Pyrgoma* and *Creusia*), reduced (*Chelonibia* and *Coronula*) or lost (*Xenobalanus*). All are hermaphroditic, but a few balanids have complemental males.

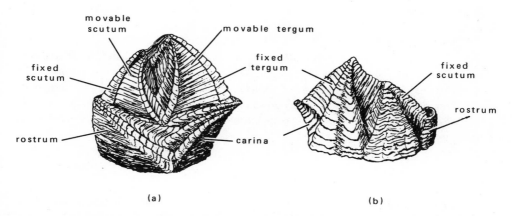

Fig. 9–17. The wart barnacle, *Verruca gibbosa;* illustrating the bilateral asymmetry of the shell; (a) viewed from right side; (b) viewed from left side. From the deep sea; 6.5 mm diameter. (After Nilsson-Cantell.)

Chthamalidae. The third pair of cirri resembles more the fourth than the second. A few species have caudal appendages, the labrum is never notched, the wall plates are solid. *Catophragmus*, with a diameter of 3.5 cm, has an eight plated shell (Fig. 9–11) with several rows of small imbricate plates around the base, derived from the peduncular scales of scalpellid ancestors. *Octomeris* and *Pachylasma* also have eight plates, but the imbricate plates have been completely lost. In general, the rostrum is simple and is overlapped by adjacent plates, but in *Tetrachthamalus* the rostrolaterals have fused with the rostrum and thus this composite plate overlaps adjacent plates (Fig. 9–11). In *Chthamalus* the carinolaterals are missing, so that the wall is of but six plates. Most genera and species are intertidal but *Pachylasma* is found in the deep sea. The most common is *Chthamalus* with a diameter up to 1 cm, living on *Mytilus*, pilings and rocks to the level of median spring tides where individuals get wet only once every other week.

Balanidae. The third pair of cirri resemble more the second than the fourth (Fig. 9–8). None have caudal appendages. Only *Chelonibia* has eight plates (Fig. 9–11). However, three of these are fused so that superficially the genus appears to have but six plates as does *Balanus*. The fusion has been between the rostrolaterals and the rostrum, forming what is known as the compound rostrum. The sutures can be seen both on the exterior and interior of the plate but are best demonstrated by grinding a cross section of the shell. The compounding of the rostrum results in the plate overlapping rather than being overlapped and this generally identifies members of this family; except in *Chelonibia*, the fusion sutures have completely disappeared. In *Chelonibia*, as in most balanids, the labrum is deeply notched.

The genus *Balanus* (Fig. 9–5) has been divided into 10 subgenera, based primarily on the structure of the wall. In primitive forms (*Austrobalanus, Solidobalanus,* etc.), the wall plates are solid; in more advanced form (*Balanus, Armatobalanus, Megabalanus,* etc.), they are permeated by longitudinal canals. Certain species of *Conopea* and *Solidobalanus* have recently been reported to have small complemental males attached to the rostral ends of the scuta and adjacent rostral surfaces of the wall.

Further reduction in the number of plates from the six in *Balanus* has occurred in different lines (Fig. 9–11). *Tetrabalanus*, occurring in bays along the west coast of South America, has four plates but is otherwise indistinguishable from *Balanus* (*Balanus*). The intertidal forms, *Tetraclita* and *Elminius* on the other hand, also have four plates making up the wall but otherwise they are very different from *Balanus*. Both are represented by intertidal species; *Tetraclita* being mainly tropical-warm temperate and having a wall permeated by many rows of irregularly arranged longitudinal canals; *Elminius* occurring in the southern hemisphere and having a solid wall permeated by longitudinal slips of cuticular material.

ORDER ACROTHORACICA

The approximately 30 species known are included in three families and eight genera. The largest, *Trypetesa lampas*, may reach a length of 2 cm, and it is also the most specialized.

Acrothoracica are small cirripeds which bore into virtually any calcareous substratum including snail shells, the plates of chitons and the calcareous skeletons of corals. They themselves do not form plates on their carapace and the cuticle is not mineralized. The preoral region forms an attachment disc rather than a stalk. The first pair of cirri are attached to the sides of the mouth field, the remaining five or fewer pairs being separated by some distance and attached to the end of the thorax (Figs. 9-18, 9-20). Dioecious; females accompanied by very reduced, dwarf males.

Anatomy

Trypetesa is the least typical acrothoracican known, but it occurs in gastropod shells inhabited by hermit crabs on both shores of the Pacific and the Atlantic in the northern hemisphere and is therefore among the better studied. The preoral region becomes enlarged some time after metamorphosis of the cyprid larva into the adult (Fig. 9-18) to accommodate the ovary. At the same time a dorsal extension of the mantle cavity pushes into it. In the course of growth the preoral region widens into a flat disc that may be irregularly lobed, and the ventral wall forms a thick nonmineralized plate, the attachment disc, which is cemented to the snail shell. When molting, the nonmineralized layers of the disc are not shed. The connection between the preoral and postoral regions and the thorax is similar to that in goose barnacles. The thorax is similarly bent but the segmentation is generally less distinct. The mouthparts consist of the usual three pairs of appendages, the mandibles and two pairs of maxillae, plus the much reduced first pair of cirri. The remaining cirri collect the food and form a group of three reduced, uniramous pairs at the end of the thorax. The gut of *Trypetesa* is blind and its ceca reach into various parts of the body even the attachment disc and cirri. Other genera have a complete gut, as many as five pairs of normal terminal cirri and frequently caudal appendages, and they all occupy less specialized habitats (Fig. 9-20).

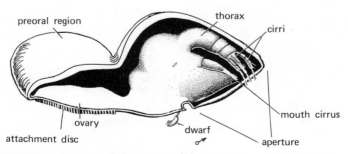

Fig. 9–18. Adult female of *Trypetesa lampas*, viewed from left side, left mantle removed; 8 mm long. Note dwarf male. (After Berndt and Darwin.)

Reproduction and Development

The Acrothoracica are all bisexual. The male attaches as a cyprid to the exterior of a female and there undergoes metamorphosis into a dwarf male. The males of *Trypetesa*, 0.4–1.2 mm long, are bottle-shaped (Fig. 9-19), and the first antennae are the only appendages present. Except for muscles, a ganglion, eye spot, and testes that open into an extensible penis, there are no internal organs. The method of fertilization is unknown. Eggs are telolecithal and cleavage is total, very unequal, spiral, and determinant. There are four naupliar stages and a typical cyprid with six pairs of thoracic limbs and compound eyes.

Habits

Most Acrothoracica bore into corals. *Trypetesa* lives in snail shells inhabited by hermit crabs, inside the aperture and usually on the columella. It burrows with nonmineralized spines of the mantle and chemical secretions (carbonic anhydrase). *Trypetesa* burrows, dorsal side first, into the substratum. The mantle aperture thus comes to fill the opening of the burrow. The attachment disc, its surface parallel to the shell surface, commences growth and enlarges inside the hole; it often can be seen from outside through the translucent shell. *Trypetesa* takes in water for both respiration and filter feeding by expanding the mantle cavity and bending the body away from the mantle slit as often as 60 times a minute. Contraction of the mantle cavity, the rocking of the trunk against the mantle and the beat of the cirri drives the water out again. The cirri only once in a while extend out of the aperture to remove foreign bodies. Currents forced in and out of the mantle cavity draw food bearing water over the posterior cirri. The mouth cirri are drawn against them and sweep particles to the mouthparts.

Hemoglobin has been identified in the blood of *Cryptophialus*.

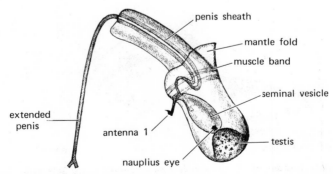

Fig. 9–19.   Dwarf male of *Trypetesa lampas* in lateral view; 1 mm long. (After Hertwig.)

Classification

Suborder Apygophora

These are Acrothoracica with an incomplete gut.

Trypetesidae. Representatives occur in northern Pacific and Atlantic Oceans. *Trypetesa lampas* (Fig. 9–18) with females 7–15 mm long and males 1.2 mm, bore into the columnella of dead *Buccinum, Natica, Fusus, Tegula,* etc. shells inhabited by hermit crabs.

Suborder Pygophora

These are Acrothoracica with a complete gut.

Lithoglyptidae. This family lacks a gastric mill. There are six genera of which the following are most abundant and have the widest distribution: *Lithoglyptes* in corals in the Indo-Pacific Ocean and the Caribbean; *Kochlorine* in comparable substrata and having a comparable distribution but also known from West Africa. *Berndtia,* is found in corals from Japan (Fig. 9–20).

Cryptophialidae. This family has a gastric mill. *Cryptophialus* lives in the shells of other barnacles, gastropods, and in the plates of chitons in the Indo-Pacific and the southern hemisphere.

ORDER RHIZOCEPHALA

There are about 200 species of rhizocephalans known, the largest being *Briarosaccus callasus,* 10 cm long, 5.5 cm wide, and 3 cm thick.

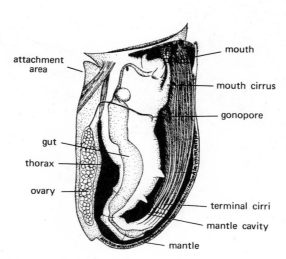

Fig. 9–20. Acrothoracican, *Berndtia purpurea,* female, with mantle of right side removed; 0.3 mm long. This is one of the least specialized of the Pygophora, having 6 pairs of biramous cirri. Compare with *Trypetesa* (Fig. 9–8), which has only 4 uniramous cirri. Note arrangement of terminal cirri at end of thorax, as compared to with a thoracican cirriped. (After Utinomi.)

Rhizocephala are recognized as Crustacea, specifically Cirripedia, by their larval stages. Two suborders are recognized, the Kentrogonida and Akentrogonida. Kentrogonids generally hatch as nauplii and become females or hermaphrodites endoparasitic on decapod crustaceans (Figs. 9-21 to 9-23). The kentrogon stage invades the host and develops into an internal network of tubes. The brood sac subsequently erupts outside of the host. The brood sac has a mantle cavity, ganglia, and large ovaries, but no appendages. The males, at the end of the cyprid stage, migrate into a special pocket within the brood pouch of the female, transform into spermatozoa and thus cease to exist as individuals. Akentrogonid females or hermaphrodites are ectoparasitic on cirripeds, isopods and decapods. They generally hatch as cyprids and do not form a kentrogon stage. The account to follow will be concerned primarily with the Kentrogonida.

The life history of *Sacculina carcini* (Fig. 9-21), a parasite of crabs, is well known. Nauplii hatch out of the eggs in the brood pouch of the female and during the first 8 days free in the water, pass through further nauplius stages

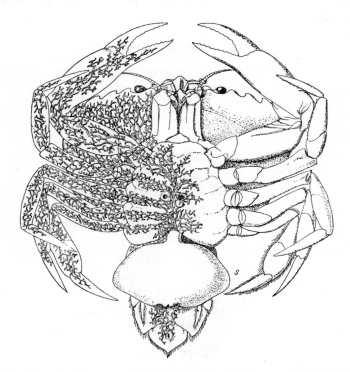

Fig. 9–21. Green crab, *Carcinus maenas,* **infected with** *Sacculina carcini,* **viewed from below. The right side of the crab is shown as if transparent, illustrating the extensive interna or root system of** *Sacculina;* **the externa or reproductive external part(s) seen attached between the abdomen and the thorax of the crab is 1.8 cm wide. (After Boas.)**

that feed entirely on their stored yolk. Unlike the larvae of Thoracica, they lack a gut and median tail spine (Fig. 9-22a). After a fourth or fifth molt, a cyprid larva, also without gut, forms a bivalved shell and prehensile first antennae without cement gland attachment disc and compound eyes. The female cyprid swims for 4–5 days and finally, at night, attaches by its first antennae to a young crab about 1.2 cm in length, at the base of a seta of the anterior body or appendages (Fig. 9-22b, c). Then the transformation of the free-living cyprid into a kentrogon commences (Fig. 22c-f). At first the hypodermis withdraws from the cuticle and contracts to a small sac, losing all appendages. The sac contains only undifferentiated cells, already present in the nauplii between the bases of the antennae. The epidermis of the sac produces a new cuticle and the old cyprid cuticle with all limbs, thorax, mantle, and nauplius eyes, are shed. The kentrogen consists of a hypodermal sac with still undifferentiated cells and the first antennae. In the base of the antennae a hollow spine (kentron), forms, and is then pushed forward into the base of a seta of a crab. Subsequently, the entire body content of undifferentiated cells migrates through the passage into the thorax of the crab where the youngest stages of *Sacculina* are found as a group of cells with small processes attached to the midgut, behind the gastric mill (Fig. 9-22g). The processes first grow in length, branch and surround the gut, and later extend into the muscles of the trunk and legs, the nervous system, and the gonads. They remain on the surface of organs and do not penetrate inside, presumably taking up nutrients by diffusion from the blood. They may also attack the organs with enzymes, but few details are known. Little too is known about the structure of the tubes, which consist of tissues under a thin cuticle and contain a lacuna filled with mesenchyme. The anastomosing roots, called the interna, live within *Carcinus* for 7–8 months. During this time a tube grows posteriorly, ending in a sac. When the sac has a diameter of 2 mm, the tissues between it and the host's hypodermis and cuticle disintegrate. The sac then grows through the hole to the outside, as the externa, and within 4 days it reaches 4 mm in diameter and is ready to receive a male cyprid should one swim to it. Six weeks later it has become a 12-mm wide knob under the base of the crab's abdomen and if fertilized, will begin to produce nauplii. Since infection, 9 months have passed.

The external sac remains connected with its stalk. It consists of a mantle surrounding the mantle cavity within which there originates a membrane that has a knoblike trunk on its lower border, containing a ganglion with some nerves, a much branched ovary and paired male cell receptacles (formerly called the testes). The male presses its whole cell content into the receptacle. The cell content differentiates into sperm. There are no sense organs, appendages, gut or excretory organs. Three-year-old individuals of *Sacculina* die. The host survives the parasite. However, 6% of the infected male *Carcinus maenas* do not form testes, and in many the form of the abdomen becomes feminized. Males of

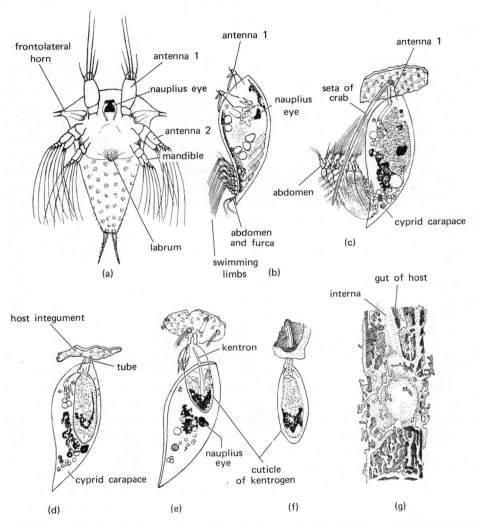

Fig. 9–22. The development of *Sacculina carcini*: (a) Ventral view of nauplius; 0.23 mm long. The 2nd antennae and mandible lack endites, the labrum is vestigial, and there is no gut (rhizocephalan nauplii do not feed). (b) Cyprid larva, viewed from left side; 0.2 mm long. (c) Cyprid attached at base of a seta of the crab, undergoing internal reorganization and shedding the thorax, its appendages and the abdomen. (d) The kentrogon being assembled from among the remaining cells. (e) Kentron appearing within the kentrogon; cyprid shell and remaining debris being shed. (f) Kentron thrust through host cuticle forming canal through which cellular content of kentrogon will migrate into host. (g) Initial establishment of interna around the host's gut. (After Delage from Gieslbrecht.)

Pachygrapsus marmoratus are also feminized by the parasite, and the male gonopods do not appear.

Reproduction is similar in the Japanese *Peltogasterella*, parasitic on *Pagurus* (Fig. 9-23). Populations of this parasite include some females that exclusively produce small eggs which, in their first maturation division contain 15 bivalent chromosomes and one univalent chromosome. Certain other individuals always produce large eggs without univalent chromosomes. Small larvae hatch from the small eggs. They attack hermit crabs and develop into females with male cell receptacles. The large nauplii that hatch from large eggs change into a large cyprid larvae, the larval males. A male cyprid attaches to the mantle opening of a female parasite sac, its body cells shrink into a cluster of cells which then moves through the lumen of the first antenna into the mantle cavity of the female, and enters a male-cell receptacle. Within the receptacle the cells differentiate into spermatozoa that, fertilize the eggs in the mantle cavity when the female has become sexually mature.

Classification

Suborder Kentrogonida

Kentrogonida have a kentrogon stage; the development generally includes naupliar stages. These are parasitic on decapod crustaceans.

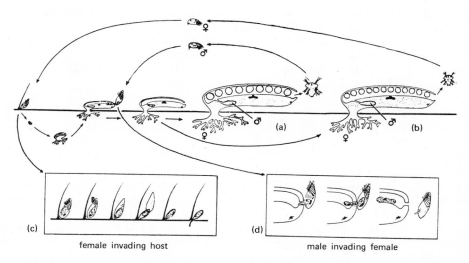

Fig. 9–23. Life cycle of *Peltogaster gracilis*: (a) female producing large eggs, nauplii and male cyprid larvae; (b) male producing small eggs, nauplii and female cyprid larvae; (c) female kentrogon invading host crab; (d) male cyprid cells entering mantle cavity of female and migrating toward male-cell receptacles. (After Newman, Zullo and Withers; after Yanagimachi.)

The suborder includes five families, each characterized by anatomical details usually observed after sectioning. The largest, the Peltogastridae, contains 13 genera divisible into gregarious and nongregarious species parasitic on pagurids and galatheids. The Sacculinidae contains six genera found parasitic on anomurans (Hippoidea and Galatheoidea) and brachyurans (Gymnopleura, Dromiacea, Brachygnatha). Lernaeodiscidae, with five genera, occurs on galatheids and thalassinoideans. Clistosaccidae and Sylonidae are represented by monotypic genera; the former on pagurids, the latter on Caridea.

Suborder Akentrogonida

Akentrogonida have no kentrogon stage; the penetration is superficial or with moderate to extensive nutritive processes; there are no naupliar stages during development. Females lack male cell receptacles. This group includes *Thompsonia* parasitic on Caridea, Galatheoidea, Paguroidea, Thalassinoidea, and Brachygnatha; *Duplorbis* is found on isopods; *Mycetomorpha* on carideans; and *Chthamalophilus* and *Boschmaella* (=*Microgaster*) on balanomorph cirripeds.

References

Cirripedia

Dahl, E. 1963. Evolutionary lines among Recent Crustacea *in* Whittington, H.B. and Rolfe, W.D.I., *Phylogeny and Evolution of Crustacea.* Mus. Comp. Zool., Cambridge, Mass.

Darwin, C. 1851, 1854. *A Monograph on the sub-class Cirripedia.* Ray Soc. London. (Reprinted 1964.)

Krüger, P. 1940. Cirripedia *in Bronns Klassen und Ordnungen des Tierreichs.* Akad. Verl., Geest & Portig, Leipzig 5(1)3(3): 1–560.

Newman, W.A., Zullo, V.A. and Withers, T.H. 1969. Cirripedia, *in* Moore, R.C. ed. *Treatise on Invertebrate Paleontology.* Part R. Univ. of Kansas, Lawrence.

Tarasov, N.I. and Zevina, G.B. 1957. Usonogie raki [Cirripedia Thoracica] morei SSSR. Fauna SSSR, Zool. Inst. Akad. Nauk, SSSR (N.S.) 69, 6(1): 1–268.

Ascothoracica

Brattström, H. 1947. Ecology of the ascothoracid *Ulophysema öresundense. Lunds. Univ. Årsskr.* N.F. (2) 43: 4–75.

Krüger, P. 1940. Ascothoracida *in Bronns Klassen und Ordnungen des Tierreichs.* Akad. Verl., Geest & Portig, Leipzig 5(1)3(4): 1–46.

Okada, Y.K. 1939. Les Cirripedes Ascothoraciques. *Trav. Sta. Zool. Wimereux* 13: 489–514.

Wagin, V.L. 1946. *Ascothorax ophioctenis* and the position of Ascothoracida in the system of the Entomostraca. *Acta. Zool.* 27: 155–267.

——————— 1964. [On *Parascothorax synagogoides,* parasitic on *Ophiura quadrispina* and on geographical distribution of Ascothoracica] *Akad. Nauk Okeanol. Trud. Inst. Okean. SSSR,* 69: 271–284.

Thoracica

Barnes, H. 1957. Spring diatom increase and "swarming" of the common barnacle *Balanus balanoides. Année Biol.* (3) 33: 67–85.

——————— 1958. The growth rate of *Verruca stroemia. J. Marine Biol. Ass.* 37: 427–433.

——————— and Barnes, M. 1954. Biology of *Balanus balanus. Oikos* 5: 63–76.

———— 1956. The egg-mass in *Balanus balanoides. Arch. Soc. Zool. Bot. Fennica Vanamo* 11: 11–16.

———— 1957. Resistance to desiccation in intertidal barnacles. *Science* 126: 358.

———— and Costlow, Jr., J.D. 1961. The larval stages of *Balanus balanus. J. Marine Biol. Ass.* 41: 59–68.

———— and Reese, E.S. 1959. Feeding in the pedunculate cirripede *Pollicipes polymerus. Proc. Zool. Soc. London* 132: 569–585.

Batham, E.J. 1946. *Pollicipes spinosus,* embryonic and larval development. *Trans. Proc. Roy. Soc. New Zealand* 75: 405–418.

Bernard, F.J. and Lane, C.E. 1962. Early settlement and metamorphosis of the barnacle *Balanus amphitrite. J. Morphol.* 110: 19–39.

Bocquet-Védrine, J. 1965. Tégument et de la mue chez le Cirripède Operculé *Elminius modestus. Arch. Zool. Exp.* 105: 30–76.

Bott, R. 1953. Die Seepocken der deutschen Nordseeküste. *Natur Volk* 83: 93–100.

Bowers, R.L. 1968. Observations on the orientation and feeding behavior of barnacles associated with lobsters. *J. Exp. Marine Biol. Ecol.* 2: 105–112.

Cannon, H.G. 1947. Anatomy of the pedunculate barnacle *Lithotyra. Phil. Trans. Roy. Soc. London* 233B: 89–136.

Connell, J.H. 1961. Interspecific competition of the barnacle *Chthamalus stellatus. Ecology* 42: 710–723.

Costlow, J.D., Jr. 1956. Shell development in *Balanus improvisus. J. Morphol.* 99: 359–415.

———— and Bookhout, C.G. 1956. Molting and shell growth in *Balanus amphitrite. Biol. Bull.* 110: 107–116.

———— 1957. Larval development of *Balanus eburneus* in the laboratory. *Ibid.* 112: 313–324.

———— 1958. Larval development of *Balanus amphitrite* reared in the laboratory. *Ibid.* 114: 284–295.

Crisp, D.J. 1955. The behaviour of barnacle cyprids in relation to water movement. *J. Exp. Biol.* 32: 569–590.

———— 1956. A substance promoting "hatching" and liberation of young in cirripedes. *Nature* 178: 263.

———— and Meadows, P.S. 1963. Adsorbed layers: the stimulus to settlement in barnacles. *Proc. Roy. Soc. London* 158B: 364–387.

———— and Patel, B. 1967. The influence of the contour of the substratum on the shapes of barnacles. *Symp. Crustacea, Ernakulum* 2: 612–626.

———— and Southward, A.J. 1961. Different types of cirral activity of barnacles. *Phil. Trans. Roy. Soc. London* 243B: 271–308.

Daniel, A. 1957. Illumination and its effect on the settlement of barnacle cyprids. *Proc. Zool. Soc. London* 129: 305–313.

Evans, F. 1958. Growth and maturity of the barnacles *Lepas hillii* and *Lepas anatifera. Nature* 182: 1245–1246.

Gutmann, W.F. 1960. Funktionelle Morphologie von *Balanus balanoides. Abhandl. Senckenbergischen Naturf. Ges.* 500: 1–43.

———— 1961. Die Siedlungsweise der Seepocke *Balanus balanus. Natur Volk* 91: 171–178.

———— 1962. Formproblem der Seepocken—Schale. *Natur Mus.* 92: 193–200.

Henry, D.P. and Mclaughlin, P.A. 1967. The subgenus *Solidobalanus. Crustaceana* 12: 43–58.

Hentschell, E. 1925. Das Werden und Vergehen des Bewuchses an Schiffen. *Mitt. Zool. Staatsinst. Zool. Mus. Hamburg* 41: 1–51.

Hickling, C.F. 1963. Deep-sea shark *Etmopterus spinax* and its cirripede parasite *Anelasma squalicola. J. Linnean Soc. London* (Zool.) 45: 17–24.

Howard, G.K. and Scott, H.C. 1959. Predaceous feeding in two common gooseneck barnacles. *Science* 129: 717–718.

Hoyle, G. and Smyth, T. 1963. Giant muscle fibers in the barnacle *Balanus nubilus*. *Comp. Biochem. Physiol.* 10: 291–314.

Johnstone, J. and Frost, W.E. 1927. *Anelasma squalicola*, its general morphology. *Proc. Trans. Liverpool Biol. Soc.* 41: 29–91.

Lochhead, J.H. 1936. Feeding mechanism of the nauplius of *Balanus perforatus*. *J. Linnean Soc. London* (Zool.) 39: 429–442.

Newman, W.A. et al. 1967. Recent concepts of growth in Balanomorpha. *Crustaceana* 12: 167–178.

Norris, E. and Crisp, D.J. 1953. Distribution and planktonic stages of the cirripede *Balanus perforatus*. *Proc. Zool. Soc. London* 123: 393–409.

Patel, B. 1959. The influence of temperature on the reproduction of moulting *Lepas anatifera*. *J. Marine Biol. Ass.* 38: 589–597.

Petriconi, V. 1969. Vergleichend anatomische Untersuchungen an Rankenfüsslern. *Zool. Jahrb. Abt. Anat.* 86: 67–83.

Pilsbry, H.A. 1907. The barnacles (Cirripedia) contained in the collections of the U.S. National Museum. *Bull. U.S. Nat. Mus.* 60: 1–122.

_____ 1916. The sessile barnacles (Cirripedia) contained in the collections of the U.S. National Museum; including a monograph of the American species. *Ibid.* 93: 1–366.

Runnström, S. 1925. Biologie und Entwicklung von *Balanus balanoides*. *Bergens Mus. Aarbok Naturvidenskap.* 5: 1–40.

Rzepishevsky, I.K. 1962. Mass liberation of the nauplii of the common barnacle, *Balanus balanoides*. *Int. Rev. Ges. Hydrobiol.* 47: 471–479.

Southward, A.J. 1955. The relation of cirral and other activities to temperature and tide level. *J. Marine Biol. Ass.* 34: 403–433.

_____ 1955. Feeding of barnacles. *Nature* 175: 1124–1125.

_____ 1957. Observations on the influence of temperature and age on cirral activity. *J. Marine Biol. Ass.* 36: 323–334.

Walley, L.J. 1964. Histolysis and phagocytosis in the metamorphosis of *Balanus balanoides*. *Nature* 201: 314–315.

_____ 1967. Cirral glands in cirripedes. *Crustaceana* 11: 151–158.

Acrothoracica

Kühnert, L. 1934. Entwicklungsgeschichte von *Alcippe lampas*. *Z. Morphol. Ökol.* 29: 45–78.

Tomlinson, J.T. 1955. Morphology of an acrothoracican barnacle *Trypetesa lateralis*. *J. Morphol.* 96: 97–114.

_____ 1969. The Burrowing Barnacles (Cirripedia: Acrothoracica). *Bull. U.S. Nat. Mus.*, No. 296.

Turquier, Y. 1968. Le mécanisme de perforation du substrat par *Trypetesa nassarioides*. *Arch. Zool. Exp. Gén.* 109: 113–122.

Utinomi, H. 1957. Biology and external morphology of the female of *Berndtia purpurea*. *Publ. Seto Marine Biol. Lab.* 6: 1–25.

_____ 1960. Internal anatomy of the female of *Berndtia purpurea*. *Ibid.* 8: 223–279.

Rhizocephala

Boschma, H. 1953. The Rhizocephala of the Pacific. *Zool. Med.* 32: 185–201.

Bocquet-Védrine, J. 1961. Monographie de *Chthamalophilus delagei* rhizocephale parasite de *Chthamalus stellatus* Cah. *Biol. Marine* 2: 455–593.

———— 1964. Embryologie précoce de *Sacculina carcini*. *Zool. Med.* 39: 1–11.

Foxon, G.E.H. 1940. Life history of *Sacculina carcini*. *J. Marine Biol. Ass.* 24: 253–264.

Ichikawa, A. and Yanagimachi, R. 1958. The nature of the testes in *Peltogasterella*. *Annot. Zool. Japan* 31: 82–96.

———— 1960. The reproductive function of the larval cypris males of *Peltogaster* and *Sacculina*. *Ibid.* 33: 42–56.

Reinhard, E.G. 1942. The entoparasitic development of *Peltogaster paguri*. *J. Morphol.* 70: 69–79.

———— 1942. The reproduction role of the complemental males of *Peltogaster*. *Ibid.* 70: 389–402.

Yanagimachi, R. 1961. The mode of sex-determination in *Peltogasterella*. *Biol. Bull.* 120: 272–283.

———— 1961. The life cycle of *Peltogasterella*. *Crustaceana* 2: 183–186.

10.

Subclass Malacostraca and Superorder Phyllocarida

SUBCLASS MALACOSTRACA

There are about 19,100 species of living malacostracans known. The largest malacostracans are species of decapods (see Chapter 13).

Malacostraca have eight thoracic somites and six (rarely seven) abdominal somites. In only very few species does the abdomen bear a furca; in most genera there are abdominal appendages, pleopods. The last pair of pleopods is almost always differentiated into uropods. The endopodites of the thoracopods are always jointed and resemble the walking legs of arachnids and insects.

Malacostraca for the most part have a complex morphology and many groups surpass the other subclasses in the capabilities of their central nervous system and sense organs. Also their average size is much larger than that of the other crustaceans. Many of them, in addition, have an unusually heavily armored exoskeleton.

Anatomy

The constancy of metameres is characteristic. Besides eight thoracomeres, there are almost always six pleomeres. Only the Phyllocarida and the lophogastrids (Mysidacea) retain the seventh abdominal somite; in all others this segment is absent although traces of it have been recognised in some Peracarida and Eucarida. All Malacostracans have the gonopores in the same somite, the eighth thoracomere in the male, the sixth in the female.

The compound eyes, usually stalked, are connected with three optic centers, which have a medulla interna in addition to the lamina ganglionaris and the medulla of other crustaceans. Also, in most families, these optic centers lie in paired lateral tubular projections of the protocerebrum that extend into the eyestalks (Fig. 1-18).

The gut in the Malacostraca is distinguished by having the end of the ectodermal foregut differentiated into a chewing and filtering apparatus. This so-called stomach directs fine particles into the branched ceca (midgut gland, hepatopancreas) while large particles are directed into the gut, where they cannot plug a lumen. This function is comparable to that of the digestive tract of some bivalves and gastropods. Generally the breaking up of food particles is accomplished by enzymes from the midgut gland in the stomach. The enzyme saturated food particles are squeezed between longitudinal folds derived from folds of the esophagus. In many crustaceans these folds resemble those of arachnids (Fig. 10-1). Filtering is accomplished as the enzyme-treated, chewed mash falls through setae into the longitudinal grooves and moves posteriorly. The stomach is simplest in *Anaspides*. In *Anaspides* the lateral and longitudinal folds continue uninterrupted from the esophagus to the entrance of the midgut and the stomach is one uniform chamber. In all other malacostracans (indistinctly in phyllocarids), the stomach and its folds are divided into two successive groups: the anterior, called the cardiac chamber, is used for chewing; the posterior, the pyloric chamber, separates fine from coarse particles (Fig. 10-1). The different malacostracan orders display different structures and specializations.

Figure 10-1 gives the generalized morphology of the stomach. On its floor is the ventral fold, T-shaped in cross section. Posteriorly it loses its attachment from the floor and forms a cone-shaped valve between the two stomach chambers. At each side of the ventral fold a lateral fold extends deep into the lumen, producing a deep ventral groove on each side. Transverse parallel setae fence these grooves off from the stomach lumen. The grooves guide enzymes from the more posterior midgut gland toward the cardiac chamber and also take up the fine, partly digested mash. Between cardiac and pyloric chambers the stomach lumen is constricted, not only by the folds but also by the cone-shaped valve and a similar valve from the roof. Coarse particles, shells, and fishbones are thereby prevented from entering the pyloric chamber. This sorting is of great importance in crustaceans that take in coarse food. For instance in crayfish only a narrow Y-shaped slit connects cardiac and pyloric chambers (Figs. 10-2, 10-3b). Also in the pyloric stomach food is broken down by the lateral folds. There is more filtering as the chewed up food in the lateral grooves has to pass a second filter before reaching the midgut gland. This second filter is formed by combs covering the T-shaped grooves of the ventral fold (Figs. 10-1, 10-2). In the passage between pyloric stomach and midgut, the stomach contents are guided into two separate levels, one above for coarse, undigested material, and one below for the filtrate. This separation is made possible by the funnel-shaped pyloric walls which continue into the midgut to form the upper tier (Figs. 10-1, 10-2). The funnel is formed by extensions of the ventral, lateral and dorsal folds, which separate from the walls and extend posteriorly as free lamellae, guiding the material into the midgut. The lower tier, the filtrate channel, leads into a

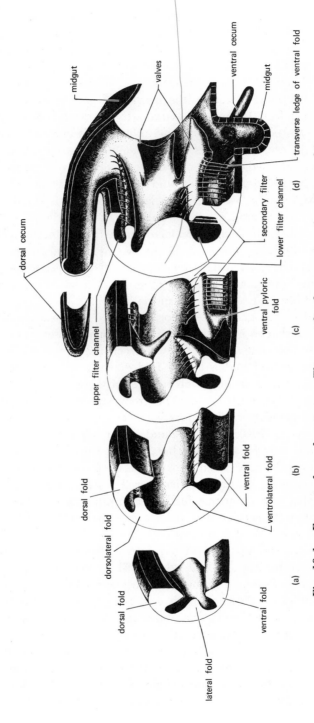

Fig. 10–1. Foregut of a malacostracan. The organ has been cut into 4 sections and its left wall removed. The closely spaced filter setae are widely spaced in the diagram. (a) Esophagus with four wall folds; (b) cardiac chamber, the lateral folds divide in two; (c) area connecting cardiac and pyloric regions; (d) pyloric chamber and its connection to the midgut. The valves are projections of the ventrolateral fold and ventral pyloric fold. (After Siewing and Zimmer.)

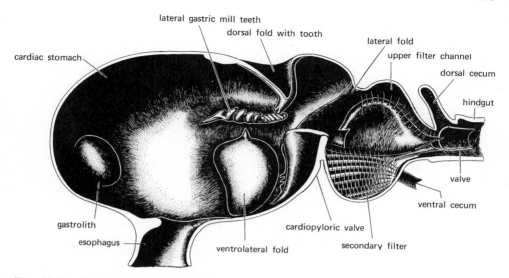

lateral gastric mill teeth
dorsal fold with tooth
lateral fold
upper filter channel
dorsal cecum
hindgut
cardiac stomach
valve
ventral cecum
gastrolith
esophagus
ventrolateral fold
cardiopyloric valve
secondary filter

Fig. 10–2. **Stomach of crayfish** *Astacus*: **left wall removed; 2.5 cm long. For clarity the pyloric chamber is shown slightly enlarged, the grooves of the secondary filters are enlarged, and their number and the number of the setae are reduced. In the cardiac stomach the swellings of the sclerotized wall are omitted: folds are found only in the posterior third. The lateral fold is at the entrance to the pyloric stomach: the valve (right) is the continuation of the pyloric stomach. (Illustrated by Dr. E. Popp in part after Jordan.)**

ventral pouch of the stomach, within which lie the entrances to the two midgut glands (Figs. 10-1, 10-2). Because the combs of the second filter on both sides separate from the ventral fold and turn laterally, each groove is closed at its posterior end toward the midgut gland (Figs. 10-1, 10-5). In the dorsal part of the pyloric chamber the setae along the paired dorsolateral folds fence off the median channel and guide the finest material into the midgut ceca.

Of the many specializations in decapod stomachs we shall select the one of the crayfish (*Astacus*) to serve as a detailed example. The cardiac chamber is large and spherical, to accommodate the large quantities of coarse food taken in (Figs. 10-2, 13-9). It contains paired, wide ventrolateral folds (Fig. 10-3a) which squeeze the food. Above are two dorsolateral folds bearing strong teeth along their edges (Figs. 10-2, 10-3a). Between these two a dorsal fold from the roof hangs into the lumen, and from its posterior tip extends a hooked calcareous tooth (Figs. 10-2, 10-4). The complicated stomach musculature can not only press the walls of the stomach together and extend them but can also push the middle tooth posteriorly and against two rows of lateral teeth. The whole structure, called the gastric mill, can readily be examined in preserved animals; in life its function can be observed only with difficulty. From the anatomy it can be seen that the ventrolateral grooves guide the digestive juices from the midgut

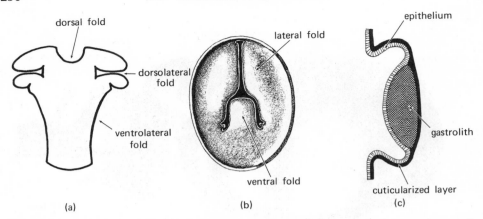

Fig. 10–3. Stomach of the crayfish *Astacus*. (a) Cross section through posterior section of cardiac stomach; the dorsolateral folds bear teeth. (b) Cross section showing slit into the pyloric stomach in anterior view. (c) Cross section through a gastrolith within cardiac wall. (a and b after Huxley and Jordan.)

toward the food particles in the cardiac stomach between the teeth. Enzymes attack the compressed shreds of meat which are ground between the teeth. Large shells and fishbones are probably ejected through the mouth. The Indian crab *Parathelphusa hydrodromus,* however, can grind mollusk shells into powder. By marking the food with a dye it can be seen that food remains about 4 hr in the cardiac chamber of this crab.

The pyloric stomach is separated by the large cardiopyloric valves (Figs. 10-2, 10-3b), consisting of paired thick lateral folds, between which the partly digested food is pressed. This partly digested material then enters the two paired filters, one above the other. The upper filter consists of a long row of setae along the upper margin of each flat lateral fold, separating the main pyloric lumen from dorsal longitudinal channels that lead to the midgut (Figs. 10-2, 10-5). The ventral paired filters are more complicated. Its lumen extends as shown in Fig. 10-1 on each side between the high longitudinal ventral fold and the lateral folds of the pyloric stomach (Fig. 10-2). Transverse setae form a grate that prevents large particles from entering. In *Astacus* there is also a secondary filter before the entrance to the midgut. To deal with the large quantities of food taken in there are numerous parallel filter grooves within the lateral walls of the ventral fold (Figs. 10-2, 10-5). The back wall of each such groove is formed by the lateral walls of the ventral fold; the side walls are formed by ridges that have parallel vertical setae along their edges (Fig. 10-5). Posteriorly these ridges separate from the wall and converge to form a basket of setae in front of the cecal openings, permitting only the finest particles to pass into the glands. Coarse particles are held back and reach the midgut through a funnel valve. In the lobster the material passes into the midgut. In *Astacus,* which has a very short midgut, it passes almost immediately into the long ectodermal hindgut.

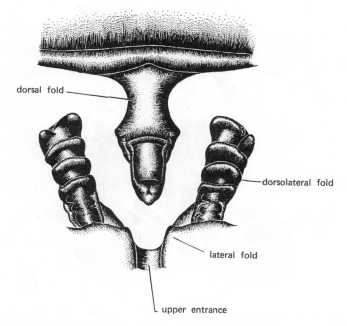

dorsal fold

dorsolateral fold

lateral fold

upper entrance

Fig. 10–4. Dorsal wall of the posterior portion of the cardiac stomach with gastric mill seen from the lumen in the lobster *Homarus gammarus*, the dorsal and dorsolateral folds have teeth, the upper entrance leads to the pyloric stomach; dorsal fold 15 mm long. (Illustrated by Dr. E. Popp.)

The filtrate, consisting of minute particles, enters the duct system of the midgut glands and is pushed into all branches by contraction of a fine muscle network in the outer walls of the ceca. In the walls of these blind ducts absorption takes place, also storage of glycogen, fat and in some species calcium. Many enzymes are produced periodically by the cecal cells.

Relationships

The 13 Recent orders of the Malacostraca are easily grouped into 6 super-orders. Closest to ancestral forms are the few species of the many-segmented groups. The most primitive superorder is the Phyllocarida. Even in the adult these have a seventh abdominal somite as well as a furca, a simple chewing stomach, and well-segmented nervous and circulatory systems (Fig. 10-7). The nonphyllocarid malacostracans are often grouped as the Eumalacostraca. Fossil phyllocarids are the oldest known Malacostraca, ranging from at least the Ordovician to Triassic periods. It seems likely that the Eumalacostraca were derived from this archaeostracan stock. Hoplocarida, though with segmentation similar to that of the Phyllocarida, are already specialized from the ancestral form: their seventh pleomere and furca have been lost and the uropods are typically malacostracan lateral extensions, part of a tail fan. Internally the stomach

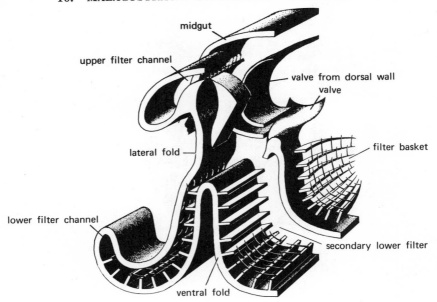

Fig. 10–5. Dorsal and ventral pyloric filters of the crayfish *Astacus*, shown as two diagrams. In the anterior (left) diagram only the upper half of the right wall is shown, and in the posterior (right) diagram the left and part of the right walls are removed and the right side of the secondary filters are shown; for clarity the numbers of filtering grooves and setae are reduced. The central tube carries unfiltered material directly to the hindgut. Posteriorly the ridges of the filter basket are not attached to the wall; the valve is a projection of the ventral fold dividing the pyloric chamber into two parts which assures the passage of the unfiltered fraction into the hindgut. (After Jordan.)

is separated into cardiac and pyloric chambers. The few species of Syncarida exhibit primitive traits in the segmented nervous system and the long dorsal vessel (Fig. 12-1). Otherwise, by modification of the two anterior male pleopods into male copulatory appendages, they reveal their relationship to the successful Eucarida. The Eucarida have a much shorter heart, a single descending artery, and compact gonads, unlike those of the other three superorders (Fig. 1-18). In these characters they seem more specialized. The most primitive order included (Euphausiacea), was once placed in the Peracarida (Mysidacea), indicating the close relationship of the two successful superorders Eucarida and Peracarida. The two orders are otherwise easily separated as all Peracarida have a brood pouch consisting of medially directed epipodites. The small superorder Pancarida also practices brood care, but the carapace forms the brood chamber, belying the suggested affinity to the Peracarida.

SUPERORDER PHYLLOCARIDA

The Leptostraca is the only Recent order included in the Phyllocarida. It comprises 10 known species, which range in size from less 1 cm long (*Paranebalia* and *Nebalia typhlops*) to 4 cm long (*Nebaliopsis typica*).

Phyllocarida are malacostracans with a large carapace that completely encloses the anterior body including appendages. Although attached only at the head, the two valves of the carapace reach to the abdomen. The eight thoracic limbs are leaflike turgor appendages. The 7th abdominal segment continues as a telson having a long furca. There are no uropods.

Anatomy

The two valves of the carapace are fused dorsomedially but have no hinge. They stand to each other at an acute angle but a pair of transverse adductor muscles connected in the middle by a tendon can close the valves tightly. The anterior end of the carapace dorsum continues as a movable articulated rostral plate. Several muscles controlling movement of the anterior of the head, the eyestalks and first antennae can indirectly move the rostrum.

Both the first and second antennae are very long and protrude from the carapace (Fig. 10-6). The mouthparts are like those of other malacostracans. The eight leaflike limbs of the thorax are much alike. Except in the deviant *Nebaliopsis* the flat endopodites are long, narrow, and bear many setae (Fig. 10-8). The platelike exopodites are directed posteriorly and, with the epipodites (absent in *Nebaliella*), close the lateral space between successive limbs (Fig. 10-6). The anterior six pleomeres have paired appendages. The first four pairs are large, biramous, setose, swimming limbs. The very short posterior pairs are uniramous with two articles at most (Fig. 10-6). The integument is thin and flexible.

SENSE ORGANS AND NERVOUS SYSTEM. Esthetascs are found mainly in the first antennae, a smaller number on the second. Sensory hairs are more abundant.

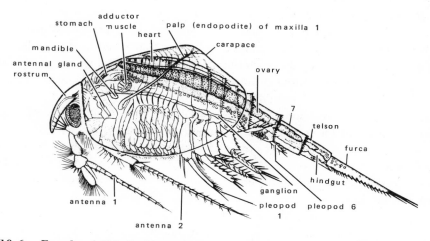

Fig. 10–6. Female of *Nebalia bipes*, lateral view; 7, seventh pleomere; 8 mm long excluding furca. (After Claus.)

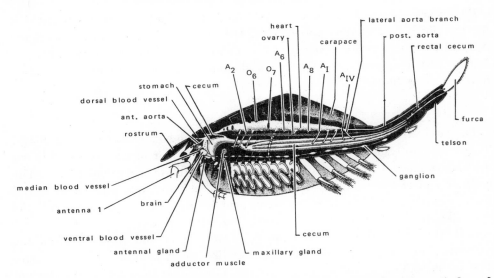

Fig. 10-7. Anatomy of *Nebalia*, longitudinal section; 6 mm long. O, ostium; A, lateral artery; Arabic numerals refer to thoracic arteries, Roman, abdominal. (After Siewing and Gruner.)

The compound eyes are on the end of long movable stalks. They are absent or reduced in *Nebaliella* and *Nebalia typhlops*. There are 17 somatic ganglia; in the head and thorax they are closely spaced, and the connectives between ganglia are visible only in the abdomen. As the embryo has 18 neuromeres it is assumed that the last ganglion of the adult developed by the fusion of the two most posterior embryonic ganglia.

DIGESTIVE SYSTEM.   The ectodermal foregut comprises an esophagus and stomach. The cardiac stomach is well developed in *Nebalia*. Behind it two pairs of short, anteriorly directed dorsal and ventral ceca orginate from the midgut. Two ventral openings here lead into a pair of short ducts each of which extends backwards as three very long midgut gland tubes that reach to the hindgut. The midgut reaches almost to the posterior end of the body where there is a short dorsal cecum before it continues into the short cuticularized hindgut. The anus opens between the furcal rami.

The bathypelagic *Nebaliopsis* has a completely different method of feeding and the ectodermal foregut is not differentiated and there is no cardiac stomach, but a strong circular and dilator musculature is present. In place of the tube-like midgut evaginations, there is only a single large ventral midgut cecum which may fill most of the thorax and abdomen. The gut is very narrow, widening only just before the hindgut.

EXCRETORY SYSTEM.   Nephridia are present in the somites of the second antennae and second maxillae. As in the Mysidacea, strips of gland cells occur on the basal and other articles of the thoracic limbs. They overlie blood sinuses and are possibly nephrocytes.

CIRCULATORY SYSTEM.   The circulatory system appears very primitive. The heart, surrounded by the pericardial sinus, extends from the area of the mandibles to the fourth pleomere (Figs. 10-6, 10-7). All told there are seven pairs of ostia, three in the head, the others in the thorax in the same metameres in which they first appeared in the embryo. The most posterior and largest is in the sixth thoracomere and opens laterally, as do the first three, while the others open on the dorsum of the heart. There is an anterior aorta at the head end, a posterior aorta at the hind end, and 12 pairs of lateral arteries. Each lateral artery probably belongs to the somite whose appendages it supplies; however, during development their origin in the heart shifts one somite posteriorly (Fig. 10-7). The 12 lateral arteries supply the internal organs, the extrinsic muscles of the thoracopods and four anterior pairs of pleopods. The appendages themselves are penetrated only by lacunae of the trunk's ventral sinus.

RESPIRATORY SYSTEM.   The epipodites and exopodites of the thoracic limbs and the large internal surface of the carapace are respiratory surfaces. All are underlain by a much-branched sinus system. The very long posteriorly directed palp of the first maxillae is used to clean the respiratory surfaces of the carapace (Fig. 10-6).

REPRODUCTIVE SYSTEM.   The testes and ovaries are long, paired, unsegmented tubes lying dorsally on each side of the gut and extend from the chewing stomach deep into the abdomen (Fig. 10-7). Each short vas deferens opens on a cone-shaped genital papilla on the coxa of the last, the eighth thoracic limb. The oviducts open at the base on each side of the sixth thoracopod. The eggs are emptied into a brood pouch formed by the long setae of the endopodites of the thoracopods of the mature females, so brood carying females barely move these appendages and do not feed.

Development

Nebalia bipes develops directly, and after hatching the young remain in the brood pouch with incomplete limbs and carapace. During development here they molt twice and leave the mother when about 1.5 mm in total length. In this way young are protected by the mother throughout two stages and part of a third, and independent life is started only after the yolk is used up. The juveniles resemble adults, except that they are minute, their carapaces relatively shorter, their pleopods incompletely formed, their antennae with fewer articles and their most anterior cardiac ostia missing.

The larvae of *Nebaliopsis*, distinguished by a large spherical yolk-filled stomach that causes their sides to bulge, have been found in plankton collections. In this genus the eggs may not be carried in a brood pouch.

Relationships

The regular segmentation of the body, the trunk musculature, nervous system, circulatory system, presence of a furca, and absence of uropods indicate that the

Leptostraca are primitive malacostracans. The survival of the seventh abdominal somite is another primitive feature; however, it is found in other Malacostraca and is not alone proof that this is an intermediate group. The leaflike limbs also cannot be considered primitive, as they are probably a secondary adaptation for feeding, quite different from those of Branchiopoda.

Habits

All Leptostraca are marine and with the exception of the bathypelagic *Nebaliopsis*, found at depths down to 6000 m, are bottom-dwellers. The well-studied *Nebalia bipes* lives in shallow coastal areas where rotting material accumulates. It tolerates pollution. Other species and genera are found at depths of 10 to about 400 m and occasionally much deeper, down to 2400 m.

LOCOMOTION.   Young animals and females rest in the same place for hours, their thoracic limbs beating rhythmically, moving respiratory currents of water through the carapace. The mature males may swim long distances propelled by the four anterior pairs of pleopods. Each of these pleopods is hooked to its partner so that each pair functions as a unit when swimming. Males differ from females by their slenderness, stronger abdominal muscles, larger eyes and more (about 1000) esthetascs on the first antennae, characters permitting dispersal to find sexual partners.

Both males and females can bend the body dorsally so that rostrum and furca touch. When burrowing in mud, the rostrum is bent down, closing the anterior of the carapace partly or completely and preventing possible clogging of the respiratory organs. *Nebaliopsis*, which has been collected alive only twice, probably swims permanently.

FEEDING.   *Nebalia bipes* mainly filters whirled up detritus but may use the mandibular palps to take up larger detritus particles, dead animals, even other *Nebalia* from the surface of the mud or from decaying vegetation, shells and stones. *Nebaliella* lives in mud and its gut and the spaces between mouthparts are filled by particles, including mud particles, suggesting that it is an indiscriminate mud feeder. Additional evidence is that the thoracic filter channel* and the filters of the stomach are not as well developed as those of *Nebalia*. *Nebaliopsis*, a deep sea animal, seems to feed a very different way. The huge cecum in one individual was filled with an orange-red formless mass, resembling yolk. It is thus assumed that it feeds on floating eggs, readily available during breeding periods of some animals. The large gut cecum permits gorging when food is abundant for survival during periods when food is scarce.

Nebalia bipes disturbs the surface of the mud and by metachronal movements of the thoracic limbs sucks the particles into a filter channel between the left and

---

* To avoid confusion with the filtering mechanism of other crustaceans, the H.G. Cannon term "filter chamber" of *Nebalia* is here called "filter channel."

right limbs. The current from the anterior first meets the rostrum, which can be lowered somewhat, dividing the current. The dorsal current passes over the animal's back, and the ventral current enters the filter channel below the lower edges of the carapace. The lower the rostrum, the narrower the entrance to the filter channels, preventing clogging with masses of mud. *Nebaliella* uses the eyestalks for current control.

The method of current production is the same as in the Anostraca (and indeed as in all Branchiopoda and many Malacostraca). On each side of the median filter channel *Nebalia* has a number of interlimb spaces, the anterior and posterior walls of which are formed by the vertically hanging protopodites and bases of the endopodites (Figs. 10-8, 10-9). The filter channel is closed in dorsally by the sternites and laterally by the proximal parts of the trunk limb endopodites. Ventrally the channel is closed by the narrow endopodites which curve sharply backward and form a mat with their long setae (Fig. 10-8). Posteriorly the channel is sealed by the eighth thoracic limbs, linked together by feather setae. The exopodites and epipodites close the spaces between successive limbs and overlap posteriorly, thus acting as valvular flaps permitting the movement of water out of, but not into, the filter channel (Fig. 10-6). A filter wall is formed by the interlocking of one row of setae on each limb with a second row of setae on the limb in front, to form a grid across the inner entrance to the interlimb space (Fig. 10-10).

The functioning of the filter channel depends on the metachronal beat of the limbs and has two phases, a suction phase and a compression phase. The suction phase starts when a limb begins its backstroke and enlarges the interlimb space immediately in front. Water cannot enter laterally because the suction closes the

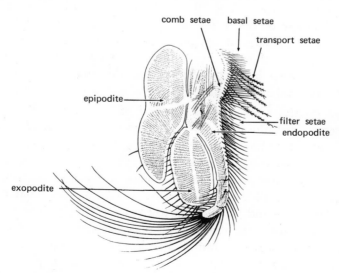

Fig. 10–8.   First thoracic limb of *Nebalia bipes*, 1.4 mm long. (After Claus.)

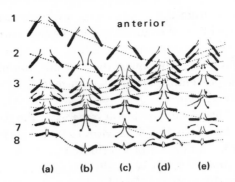

Fig. 10–9.   Diagram showing the positions of the thoracic limbs of *Nebalia* in four successive movements, in horizontal section; exopodites omitted; arrows indicate water currents. (After Cannon.) (a) Limbs 1 and 2 move forward, the interlimb spaces behind them therefore expand and water is sucked in anteriorly. (b) Limb 8 backstrokes rapidly and water is sucked posteriorly through the chamber; both limbs 8 close the filter chamber posteriorly. At the same time limbs 1 and 2 move backward and reduce the anterior interlimb spaces. In (d) this movement has gone so far that limbs 7 are in backstroke; the interlimb space immediately in front widens and sucks back water from the anterior; the last interlimb space is compressed and water is pumped out laterally. (e) Limbs 6 begin their backstroke while 1 and 2 continue their forestroke.

valves formed by the epipodites and exopodites. It can therefore only enter the filter channel anteriorly where it is controlled by the rostral plate. The water sucked from the filter channel is filtered by the setae, leaving the filtrate on the walls of the channel. The second phase commences when a limb starts its forward stroke; the interlimb space in front of it is reduced and water is pumped out through the lateral "valves" (Fig. 10-9). This happens only in the posterior half of the channel (between limbs 3–8 on Fig. 10-9), since the relatively stronger and more rapid backstrokes of the posterior limbs during phases one and two suck water straight to the rear of the channel before it can be pumped out laterally by the anterior limbs 1–3. Thus while the limbs act as suction pumps, only the posterior ones act as exhalant pumps.

The filtrate on the setal grids is combed off by posteriorly directed, stiff, short comb setae along the posterior edge of the endopodites (Fig. 10-10). The particles fall into the filter channel onto long, strong, anteromesally directed, feathered transport setae of the protopodites. These setae, in forward stroke of their limb, transport the particles to the setae of the next anterior limb (Fig. 10-11), brushing the articles off in backstroke, so that they are finally combed out by the second maxillae and are brought to the mouth by the first maxillae. Only fine particles reach the filter setae; large ones are caught immediately by the transport setae and carried anteriorly. Very large particles cannot enter the filter channel at all. They are held at the anterior by setae of the first limbs and are picked up by the mouthparts directly.

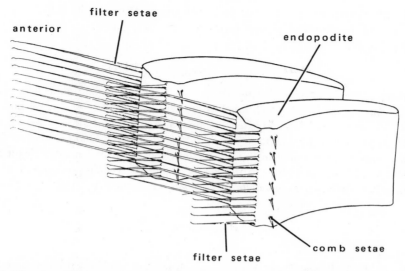

Fig. 10–10. A portion of the right filter wall as viewed from the filter channel. The filter setae hook into one another. (After Cannon.)

The limb structure makes it probable that the genera *Nebaliella* and *Paranebalia* feed similarly.

Classification

Three orders are included. Of these, the orders Archaeostraca and Hymenostraca comprise about 50 genera of exclusively fossil phyllocarids. Less than half of these are known from more than one tagma, and in only a few can the

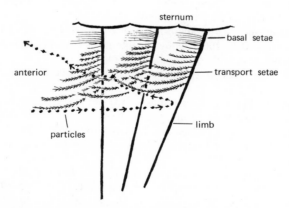

Fig. 10–11. Diagrammatic longitudinal section through the middle of the filter channel of *Nebalia* showing the path of a food particle carried forward and upward by the transport setae toward the midventral line. Only the basal and transport setae are shown; the axes of three thoracic limbs are indicated as straight lines. (After Cannon.)

tagmosis be demonstrated to be malacostracan. The fossils differ from the Leptostraca in having the telson produced dorsally between the furcal rami. Some of these fossils are large and reach a length of 75 cm.

ORDER LEPTOSTRACA

Nebaliidae. Partly cleaned lobster shells can be used as bait to collect *Nebalia bipes,* 6–11 mm, (Fig. 10–6) which inhabits sea weeds in shallow waters on both sides of the Atlantic and elsewhere, and often is common to a depth of 200 m. At times they are found on the bait in lobster pots. The blind *N. typhlops* is found in deeper waters, *Epinebalia pugettensis* is found in the intertidal region of rocky shores in Puget Sound and south to California as well as in the Sea of Okhotsk. *Nebaliopsis typica,* up to 40 mm long, is bathypelagic. *Paranebalia* and *Nebaliella* are other genera.

References

Malacostraca

Degkwitz, E. 1957. Proteolytische Verdauungsfermente bei verschiedenen Crustaceenarten. *Veröff. Inst. Meeresforsch. Bremerhaven* 5: 1–13.

Haffer, K. 1965. Der Kaumagen der Mysidacea im Vergleich zu den verschiedener Peracarida und Eucarida. *Helgoländer Wiss. Meeresunters.* 12: 156–206.

Hanström, B. 1947. The brain, the sense organs, and the incretory organs of the head in the Malacostraca. *Lunds Univ. Arsskrift.* (2) 43: 1–44.

Siewing, R. 1956. Morphologie der Malacostraca. *Zool. Jahrb. Abt. Anat.* 75: 39–176.

Phyllocarida

Cannon, H.G. 1927. The feeding mechanism of *Nebalia. Trans. Roy. Soc. Edinburgh* 55: 355–370.

——————— 1931. Nebaliacea. *Discovery Rep.* 3: 199–222.

——————— 1946. *Nebaliopsis typica. Ibid.* 23: 213–222.

——————— 1960. Leptostraca *in Bronns Klassen und Ordnungen des Tierreichs,* Akad. Verl., Geest & Portig, Leipzig, 5 (1) 4 (1): 1–81.

Duveau, J. 1957. Les glandes tégumentaires de *Nebalia. Arch. Biol.* 68: 45–64.

Hessler, R.R. and Sanders, H.L. 1965. Bathyal Leptostraca from the continental slope of the northeastern United States. *Crustaceana* 9: 71–74.

Manton, S.M. 1934. The embryology of the crustacean *Nebalia. Phil. Trans. Roy. Soc. London* (B) 223: 163–238.

Rolfe, W.D.I. (in press) Phyllocarida *in* Moore, R.C. ed. *Treatise on Invertebrate Paleontology,* Pt. R., Geol. Soc. Amer., Univ. Kansas Press, Lawrence.

Rowett, H.G.Q. 1943. The gut of Nebaliacea. *Discovery Rep.* 23:1–17.

——————— 1946. The feeding mechanisms of *Calma glaucoides* and *Nebaliopsis typica. J. Marine Biol. Ass.* 26: 352–357.

Siewing, R. 1959. Morphologie des Kaumagens von *Nebalia bipes. Zool. Anz.* 162: 325–339.

Thiele, J. 1927. Leptostraca *in* Grimpe, G. and Wagler, E. *Die Tierwelt der Nord - und Ostsee.* Akad. Verl., Geest & Portig, Leipzig, 10 (7): Xg1-Xg8.

——————— 1927. Leptostraca *in* Kükenthal, W. and Krumbach, T. *Handbuch der Zoologie,* de Gruyter, Berlin 3 (5): 567–592.

11.

Superorder Hoplocarida

The Hoplocarida includes the order Stomatopoda, with more than 250 species, the largest of which is *Harpiosquilla raphidea*, up to 33.5 cm long.

The Stomatopoda, called mantis shrimps, are elongate, fairly large Malacostraca, with a short carapace that does not cover the posterior thoracic tergites, and a well-developed abdomen. The raptorial second maxillipeds are especially long and often armed with a spined subchela. The third to fifth maxillipeds also are subchelate, but are much weaker. The nervous and circulatory systems are metameric.

Stomatopoda are known as fossils since the Jurassic; a separate order of Hoplocarida, the Palaeostomatopoda, comprises two Carboniferous genera characterised by having the subchelae 2–5 subequal in size.

Anatomy

Adult mantis shrimps have a noticeably short carapace that covers only the head region, hardly extends laterally, and does not form a respiratory chamber (Fig. 11-1). The carapace is longer in the swimming larvae, reaching posteriorly to the anterior edge of the eighth thoracomere, while in adults it roofs over at most the soft anterior four thoracic tergites (Figs. 11-1, 11-4). The anterior border of the carapace has a median projecting movable rostrum (Fig. 11-2) overhanging the eyestalks and, in larvae, fused to the carapace.

The very short anterior five thoracomeres are appressed to each other and barely movable against one another although they have soft tergites. Each has a pair of subchelate limbs, the maxillipeds. These are followed by three movable somites with pereopods, walking limbs. The very flexible abdomen consists of six large somites, and there is a large telson, often sculptured with thorns, humps, or ribs.

The animals are well adapted to inhabit burrows. They enter their burrows head first and turn inside by bending the trunk sharply, then protruding the anterior end from the burrow. This habit would be impossible if the long, wide carapace of the larva were retained.

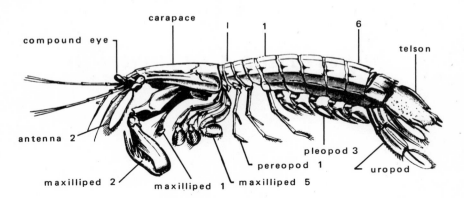

Fig. 11-1. *Squilla mantis*, lateral view; 18 cm long. Roman numerals indicate pereomeres, Arabic pleomeres. (After Claus, Grobben.)

The first antenna has three flagella. The second antenna has the second article
of the exopodite modified into a long, wide scale resembling those of many
decapods (Figs. 11-1, 11-2). The mandibles have a vertical shaft which rotates
forward and back on dorsal and ventral articulations. In the direction of the
shaft there is a toothed incisor process that meets its partner behind the labrum.
From the base of each mandible a long, anteriorly directed, horizontal toothed
molar process pushes under the long labrum so that its tip reaches the mouth at
the base of the labrum, extending into the foregut. The two pairs of maxillae are
unspecialized.

The first five thoracopods do not function in locomotion. They lack an exopodite,
but usually have an epipodite. The first is specialized as a grooming limb with
setae. The other four pairs are subchelate, on each the distal, clawed article
being folded against the previous one as a jackknife. The first of these raptorial
limbs is very large and the chelae may have strong spines (Fig. 11-1). The three
most posterior thoracopods, 6–8, the walking limbs, have a three-articled protopodite bearing an exopodite and endopodite (Fig. 11-1).

The five pairs of leaflike pleopods are swimmerets (Fig. 11-3). The short stalk
of each bears tubular gills attached to the exopodite (Fig. 11-3b). The endopodites of each pair hook together by a mesally-spined appendage (appendix
interna) so that a pair of swimming limbs forms a functional unit. The males have
the first pair of pleopods modified into copulatory organs. The appendages of the
sixth pleomere are the uropods: leaflike, strongly sclerotized plates which,
together with the telson, form the tail fan. The uropod exopodite has two
articles and, the distal article, like the one-articled endopodite, usually is fringed
by feathered setae (Figs. 11-1, 11-4).

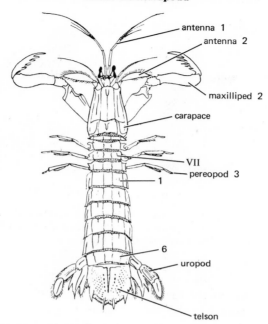

Fig. 11–2. *Oratosquilla oratoria*, **dorsal view; to 20 cm long. Roman numeral indicates thoracomere, Arabic pleomeres. (After de Haan, from Balss.)**

Some mantis shrimps such as *Squilla mantis* are gray; most are brightly colored: olive green with red setae, dark brown with patches or bands of orange-yellow or dark blue. Some limbs may contrast with the body in color.

SENSE ORGANS. The chemical sense organs are the esthetascs of the first antennal flagellae. In six thoracic and six abdominal somites there are muscle receptors that resemble those of the Paguridae. A median naupliar eye between the eyestalks consists of three ocelli. The very large compound eyes have been studied in detail only in *Squilla mantis*. Cylindrical, they lie with the longitudinal axis perpendicular to the eyestalk (Fig. 11-6). Ommatidia fill three-fourths of the cross section of the cylinder, they are absent from the area of attachment to the stalk. There are no statocysts.

NERVOUS SYSTEM. The supraesophageal ganglion has unusually long connectives to the ventral chain of ganglia (Fig. 11-4). These fuse in the metameres of the mouthparts and five maxillipeds into a long subesophageal ganglion with lateral constrictions. However, the ganglion pairs pushed together in the midline of the pereon and pleon form a long chain with its nine apparently median ganglia separated by long connectives (Fig. 11-4).

DIGESTIVE SYSTEM. The digestive system has been studied only in *Squilla*. The cardiac stomach projects anteriorly beyond the mouth, almost to the tip of

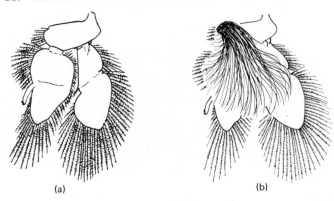

(a)   (b)

Fig. 11-3.   Right first pleopod of female *Meiosquilla desmarestii*; the small appendix
interna (on the left) on the endopodite hooks the pleopods together; (a) gill cut off;
(b) with gill, 7 mm long. (After Giesbrecht and Calman.)

the head (Fig. 11-4). Its anterioventral part may have originated from fusion
with the esophagus. Anteriorly the gut branches on each side into long midgut
gland tubes that lie parallel to the gut. Each lateral glandular tube has a lateral
cecum in each somite of the four posterior thoracic somites and abdomen, altogether 10 on each side, and fingerlike tubules reach into the telson in a fanlike
distribution. The gut is extremely narrow, and the anus opens on the telson.

EXCRETORY SYSTEM.   There are maxillary glands. Nephrocytes lie in the
sinuses (as in mysids) which extend from the gill-bearing pleopods to the
pericardium. Antennal glands may be present in the larvae.

CIRCULATORY SYSTEM.   The well-developed circulatory system is metamerically arranged, not just as an adaptation to large body size but presumably of
primitive character. The heart of *Squilla* extends from the region of the second
maxillae to the end of the fifth pleomere (Fig. 11-4). There are at least 13 pairs
of ostia, 6 pairs in the thorax, 7 in the abdomen.

The embryo has eight pairs of thoracic ostia, of which the second and third are
lost while the first becomes much larger. The seventh pair of ostia of the abdomen
belongs to a seventh pleomere which has been lost in stomatopods but is
present in the Phyllocarida and some other malacostracans.

In addition to an anterior and a posterior aorta there are 15 pairs of lateral
arteries, the most anterior of which supplies the mouthparts and belongs to the
head, while the 14 others belong to the trunk metameres. The lateral arteries go
directly into the limbs after sending branches into muscles and other structures.

The metameric arrangement of the lateral arteries has been shifted in places.
The second pair of lateral arteries forks and supplies the first and second thoracic
limbs. The third pair supplies the third limbs (Fig. 11-4). The ninth pair of
lateral arteries enters not the first pleopods as might be expected but the musculature of the last thoracic and first abdominal metameres. The 10th lateral arteries
supply the first pleopods. The significance of these shifts is not known.

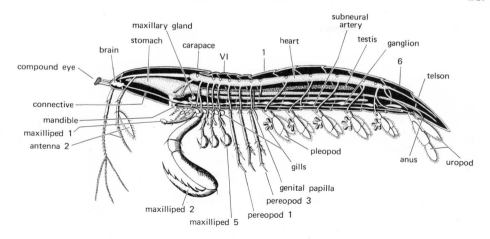

Fig. 11–4. Anatomy of male *Squilla*. Roman numerals indicate thoracomeres, Arabic, pleomeres. (After Siewing.)

Secondary branches of the lateral arteries lead to a large subneural artery. The subneural artery starts below the mandibular ganglion and ends in the sixth pleomere (Figs. 11-4, 1-21a).

Returning to the heart, the blood passes through a large longitudinal lacuna between gut and ventral nerve cord. The lacuna guides the hemolymph into the pleopod gills. From there the hemolymph flows to the pericardial sinus through metamerically arranged, dorsally directed sinuses bordered by membranes.

REPRODUCTIVE SYSTEM. The wide, paired ovaries of *Squilla* lie next to each other in the middle under the pericardial sinus and above the midgut glands, and extend through most of the trunk. In the telson area they are fused into a median tube. Anteriorly the ovaries open into the corners of a pouch which in turn opens through a median slit of the sixth thoracic sternite.

The two thin, coiled testes extend from the third pleomere deep into the telson, where they fuse into a tube. In the third pleomere a pair of coiled vasa deferentia arise that open on the coxae of the eighth thoracopods at the tip of a long genital papilla. The coiled efferent duct of a paired accessory gland also opens on the papilla. The gland itself lies in the thorax below the heart. The first two pairs of pleopods of the male are transformed into a copulatory organ, very different from that of decapods.

Reproduction

Mating has been observed in *Gonodactylus glyptocercus*. The female lies on her back, doubled up, head and abdomen raised. The male crouches over the female, across her body, with his head and abdomen down. The ventral regions between the end of the carapace and the second pleomere of the two animals are tightly appressed. Spermatophores are believed to be transferred during this time to the seminal receptacles.

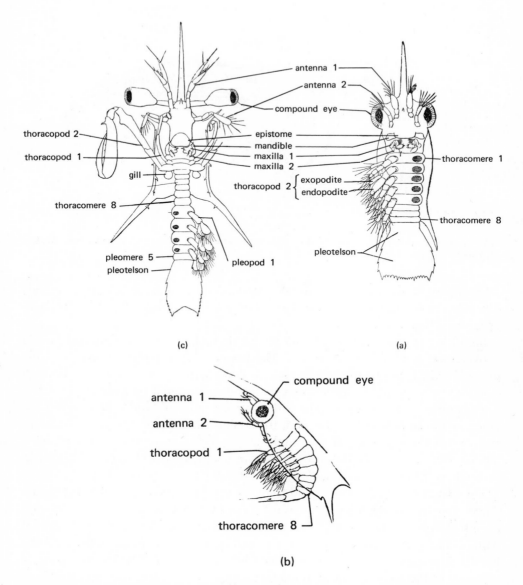

(c)

(a)

(b)

Fig. 11–5. Stomatopod larvae: (a) first pelagic antizoea of *Lysiosquilla*, 2 mm long. Erichthus type: the abdomen between the 2nd lateral and the posterior spine of the telson has only one additional spine; (b) second pelagic antizoea of *Lysiosquilla* in lateral view; carapace and rostrum 2–4 mm long; (c) first pelagic stage of pseudozoea of *Squilla*, 4.7 mm. Alima type: the abdomen between the 2nd lateral spine and the posterior border has four more spines. (a, c, after Giesbrecht from Korschelt and Heider; b, after Giesbrecht.)

In oviposition *Squilla mantis* doubles up its trunk, the abdominal venter below the thorax, the shrimp supported by the maxillipeds and outstretched uropods. The eggs emerge from the median gonopore, during the course of 4 hr, in the form of soft mass held together by secretions from cement glands of the three posterior thoracomeres. The egg mass is picked up by the three posterior pairs of small chelate maxillipeds and formed into an elliptical mass, 10–14 cm in diameter, containing about 50,000 eggs in three layers. The egg mass is held by the mother for about 10-11 weeks and often turned so that fresh water is provided from all sides. During this brood care the female does not feed. Species of *Gonodactylus* have a much smaller egg mass with diameters of only 8–14 mm.

Development

Transparent larvae hatch from the eggs which in regular or irregular anamorphic development grow into a synzoea. The pelagic antizoea larva (Fig. 11-5a) is found in *Lysiosquilla* and *Coronida*. Those of *Lysiosquilla* transform in the course of five molts when 2.2 mm long, stepwise by formation of abdominal somites and pleopods, reduction of the exopodites of thoracopods, formation of eyestalks into synzoeae about 8 mm long.

The pseudozoea larvae (Fig. 11-5c) are found in most genera. The abdomen is well developed at an early stage. In *Squilla* the pseudozoea has two bottom-dwelling stages, 2–2.5 mm long, then to become pelagic at about 100 m depth, where it passes through 4 stages to reach 8 mm in length before molting into a synzoea after the thoracic limbs 3–5 have been formed.

The synzoea is also pelagic and has four stages in *Squilla*. By transformation of the chelate appendages of limbs 3–5, walking limbs 6–8, gills, and reduction of the carapace, they become adult stomatopods.

The pelagic period of *Lysiosquilla* lasts 2–3 months in the Gulf of Naples. *Squilla mantis* seems to develop faster: each of its first two prepelagic stages lasts 2–3 days, the second pelagic stage lasts only 4–5 days. Because the pseudozoea in its first stage has a large raptorial limb, as does the antizoea in its fourth stage, we can assume the larvae are predators. The first three stages of the antizoea hold their prey with the wide, distal, spined ends of the first maxillae. The larvae at times appear in huge numbers.

Relationships

Certain primitive characters, the persistence of the nauplius eye in the adult, the metameric trunk musculature, the long heart with its numerous ostia and lateral arteries, and the extended ovary suggest that stomatopods resemble possible ancestors of Malacostraca. The peculiar cephalothorax is a specialization not present even in the larvae, and the five anterior chelate limbs are also specializations.

Habits

Stomatopoda are marine coastal bottom-dwellers. Rarely they are found as deep as 1300 m. A few tropical species may enter brackish water; a Formosan species possibly ascends rivers. The habits of only very few species are known. Inhabitants of cliffs and coral reefs stay in cavities or fissures. *Gonodactylus oerstedii* turns its head toward the exit waiting for passing prey. *Protosquilla guerini,* if disturbed, closes its cavity with its large, domed telson, which is fused to the sixth abdominal somite and armed, porcupinelike, with spines. Species living in sand make their own burrows. This below the low tide line *Coronis excavatrix* burrows 1 m deep into the bottom, head end first, tearing at the sand with the raptorial limbs. It turns by pushing its head along the venter of the abdomen and as it emerges, flattens the tube with its back. The sand brought up is thrown to the side before turning back into the tube to remove more sand. Only rarely does the animal leave its burrow. It usually lies in wait for prey at the entrance. Some *Squilla* species make U- or Y-shaped burrows. *Squilla mantis* in aquaria dig burrows as deep as 15 cm, with an opening at each end, one of them used as a lookout. Grooves of the maxillipeds and ridges and rows of tubercles on the trunk exoskeleton permit the circulation of respiratory water. As it is often captured with dragnets, it apparently does not retreat deeply. Some stomatopods live as symbionts in worm tubes; *Acanthosquilla vicina,* 2.7 cm long, lives with *Balanoglossus.*

PROTECTION.   In addition to the protection of their burrows stomatopods can defend themselves effectively with their raptorial legs, telson and uropods. Large individuals, with their jackknife claws, can wound fingers.

LOCOMOTION.   The adults, like the larvae, are excellent swimmers, rapidly moving any direction with their pleopods. The large exopodite of the second antennae and also the uropods are used as rudders. The three pairs of pereopods are pulled in, the dorsum turned toward the water surface. In forestroke the exopodites of a pleopod come to lie directly in front of their endopodites. They spread out in backstroke to produce a large paddling surface.

By strong bending followed by stretching, mantis shrimp can jerk backward or forward.

Mantis shrimp can also walk on the substrate with their sixth to eighth thoracopods. In this locomotion *Squilla* may support itself on the carpus of the raptorial limb as on an elbow.

COORDINATION.   Some information about the functioning of the central nervous system has been obtained by severing individual connectives. If both circumesophageal connectives are cut, the shrimp loses its inhibitions and becomes more active than control animals, e.g., cleaning itself endlessly. It can walk and respiratory movements continue, but the ability to swim is lost.

If the connectives are cut between the subesophageal ganglion and the first ganglion of the pereon, the respiratory movements of the pleopods slow to two-thirds, the animal falls on its back and cannot turn. The pereopods will move only if the posterior body half is stimulated. Probably there is a loss of excitation behind the cut. Sensory stimulation results, it is assumed, by excitation of the brain or subesophageal ganglion, which leads in turn to excitation of the centers in the ventral chain.

SENSES. The senses are well known only in *Squilla mantis*. *Squilla* always swims with its back up, although the center of gravity is such that dead or paralyzed animals fall on their backs. The equilibrium of the living animal is due to active movements not dependent on statoliths or light reflexes, but mainly to stimulation by currents. If the pleopods are tied down or removed, the animals sink in any position, without attempting to turn over. The eyestalk, however, will make compensatory movements.

A sense of vibration is present, in part in the long antennal setae. Chemical stimuli are received mainly by the first antennae, and these touch the visually located prey before it is attacked. A mantis shrimp will follow with the eyestalk a clay sphere moved by a glass rod. As the object approaches an antenna of the shrimp examines it, and thereafter the object is ignored, in contrast to a piece of meat. Hungry mantis shrimp whose first antennae have been amputated, refuse food for several weeks or may never feed again.

Despite living in semi-darkness at depths of 60–100 m depth with 0–100 lux light (usually diffuse light of 20–30 lux from above), *Squilla* depends heavily on its eyes. The eyes are used to find prey at these intensities, but *Squilla* rests in brighter light, although sometimes they respond at 2500 and more lux. The ommatidia are very narrow and close, the ommatidial angle very small, only 12′ (Fig. 11-6). Thus the picture of an object 10 cm away falls on as many visual elements as in the eye of a man 1.2–1.5 m away from the object. One can assume pictorial vision, at least for prey capture, as usually shrimps and other crustaceans 3–5 cm long are taken. *Squilla* stares at an individual resting shrimp and finally attacks, probably on seeing movement of the prey's thin appendages. Probably vision of movement is of importance to *Squilla*, which follows its prey with its

Fig. 11–6. Longitudinal section through compound eye of *Squilla mantis*; 6.5 mm high. Only the crystalline cone and cornea are shown; S, optic axis of an ommatidium; T, tangent of the cornea of the ommatidium. (After Demoll.)

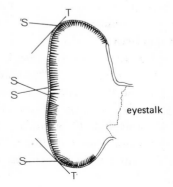

eyestalk

eyes, turning the eyestalks or even the whole anterior body. Even small objects at fair distances are noticed.

At a distance of 1.5 m, in semidarkness on the bottom, discs 2.5 cm in diameter will cause only weak responses. In darkness there is response to a moving beam from a flashlight with an opening 0.5 cm in diameter, moved to and fro across 50 cm at a distance of 2–3 m. The shrimp tend to turn or move toward moving light in semidarkness or darkness. Perhaps in nature prey appears as light spots on a dark background.

A judgment of distance is indicated in the reactions. Within twice the range of the first antennae, objects less than 5 cm away are watched and examined. In response to larger moving objects *Squilla* flees. The structure and position of the eyes make it probable that the distance is judged by binocular stereoscopic vision. In the normal position of the eyes, vertical to the substrate, the stalks diverge by 110°, and many objects in front of the animal must be seen simultaneously by both. Perhaps due to this, living prey is attacked with excellent aim as far distant as 10 cm.

The substrate on which *Squilla mantis* lives is not chosen visually, and there is no color change to match the background. Only amputation of the eyestalks combined with removal of endocrine glands causes the dark pigments to contract and the animal to pale.

FEEDING.   Stomatopods are predators. The favorite prey of *Squilla* is crustaceans; *Lysiosquilla* also takes fish and mollusks, and *Gonodactylus* is said to prefer actinians. Catching of prey has been seen in only a few species. *Coronis excavatrix* waits at the mouth of its burrow and grasps small crustaceans or fish with its raptorial limbs without leaving its burrow, but may stalk prey for 15–20 cm. *Squilla* also ambushes prey at dusk; only rarely it stalks its prey, slowly, with the antennae flat on the ground. In captivity *Squilla* takes shrimp 3–5 cm long. Following it with the eyes, it gives a touch with the antennae before shooting toward the prey by a beat of thoracic limbs and pleopods and beating the raptorial legs from below, the spines penetrating the shrimp. The jumps may be 8–10 cm long. Sometimes the prey is missed on the first try. The prey is taken by the three pairs of small maxillipeds, aided by the raptorial limbs, and pressed against the mouth region.

RESPIRATION.   The gills are attached to the pleopods. Small disc-shaped epipodites usually on the first five pairs of thoracic limbs may also function in respiration (Fig. 11-4). The epipodites are relatively unimportant (in a 15-cm long *Squilla* they have only 1 cm² surface altogether), while the numerous pleopod tubules have greater importance. A 15-cm *Squilla*, on each of its 10 pleopod gills has 30 secondary and 25-30 tertiary branches, altogether about 10,000 gill branches 1 cm long, 0.02 mm diameter, with a total surface of about 600 cm². Among malacostracans otherwise only isopods have gills on abdominal limbs.

The pleopods are moved continuously for long periods at about 25–40 beats/min; the 5th starts the backward movement, a little later the 4th begins. But individuals that are at rest may make a respiratory pause of half an hour.

ECONOMIC USES. *Squilla mantis* is fished and eaten in all Mediterranean countries. In Japan the widespread *Oratosquilla oratoria* is used as food, and in the tropical Pacific *Lysiosquilla maculata* is considered a delicacy.

Classification

ORDER STOMATOPODA

The order contains five families, of which one, the Sculdidae, is exclusively fossil.

Bathysquillidae. The telson has a median carina and all four pairs of marginal teeth of the telson have movable tips. This family contains one genus, *Bathysquilla*, and two species, one Indo-West Pacific, one Atlantic. Both species are large and occur in deep water, to 1300 m.

Gonodactylidae. The telson has a median carina and has no more than two denticles between the marginal teeth. Thirteen genera and approximately 95 species are assigned to this family; they are usually found in shallow, tropical waters. *Gonodactylus*, from 1–10 cm in length, is a characteristic inhabitant of coral reefs and rocky inshore habitats; *G. oerstedii* is the most abdundant West Indian species. Members of the genus apparently have evolved a highly complex series of behavioral patterns which may be specific. *Hemisquilla ensigera californiensis*, up to 20 cm long, occurs off southern California. It is translucent blue and orange in color. An inhabitant of bottom mud, it constructs a burrow 5 cm wide and 1.2 cm broad at a 45° angle. It is occasionally caught by fishermen. *Pseudosquilla lessoni* of the southern California coast lives deep in the substrate of intertidal rocks. *Protosquilla* belongs here.

Lysiosquillidae. The telson lacks a median carina and the propodi of the last three maxillipeds are broad and are beaded or ribbed ventrally. Approximately 55 species in nine genera are now recognized in the Lysiosquillidae. Most are soft-bodied, loosely articulated forms living in burrows in soft mud or sand; a wide range of sizes is found in adults, from 2 cm or less in *Nannosquilla* to 27.5 or more in *Lysiosquilla*.* *L. scabriacauda* is a common western Atlantic species, ranging from Brazil to the northern Gulf of Mexico. *Coronida*, 1.2–4 cm, occurs in habitats more typical of gonodactylids, coral reef areas. Some and perhaps all species of *Acanthosquilla* are commensal with balanoglossids. *Coronis* belongs in this family.

Squillidae. The telson has a median carina and has more than four denticles between the large marginal teeth. This family contains over 100 species in 14 genera; most species are 2–20 cm long, but members of *Harpiosquilla* are as large as 33.5 cm. *Squilla* mantis, to 20 cm long, is found on mud and detritus bottoms in the Mediterranean and adjacent Atlantic; its biology and anatomy have been investigated in more detail than any other stomatopod. *Oratosquilla oratoria*, the most common squillid in Japanese waters, is fished commercially there; it may get as large as 20 cm.

---

* The name *Lysiosquilla* has been placed on the *Official List of Generic Names in Zoology.*
* *Squilla* has been placed on the *Official List of Generic Names in Zoology.*

Squilla empusa, 20 cm or less, is the most common species in the western Atlantic; it lives on sand or mud in shallow waters and is found from Massachusetts to northern South America; it is often taken in nets by commercial fishermen. *Meiosquilla* is found in the Atlantic and eastern Pacific.

References

Alexandrowicz, J.S. 1951. Muscle receptor organs in the abdomen. *Quart. J. Microscop. Sci.* 92: 163–199.

Balss, H. 1938. Stomatopoda. *Bronns Klassen und Ordnungen des Tierreichs,* Akad. Verl., Geest & Portig, Leipzig 5 (1) 4 (2): 1–173.

Brown, F.A. 1948. Color changes in the stomatopod crustacean, *Chloridella empusa. Anat. Rec.* 101: 732.

Chaudonneret, J. 1957. Système nerveux des derniers segments thoraciques de la squille. *Ann. Sci. Natur. Zool.* (11) 19: 225–232.

Gurney, R. 1946. Notes on stomatopod larvae. *Proc. Zool. Soc. London* 116: 133–175.

Hanström, B. 1947. The brain, the sense organs, and the incretory organs of the head in the Crustacea. *Lunds Univ. Årsskr.* (N.F.) 43: 1–44.

Knowles, F.G.W. 1954. Neurosecretion in the tritocerebral complex of crustaceans. *Pubbl. Staz. Zool. Napoli* 24: 74–78.

Manning, R.B. 1963. Preliminary revision of *Pseudosquilla* and *Lysiosquilla. Bull. Marine Sci. Gulf Caribbean* 13: 308–328.

————— 1963. Embryology of *Gonodactylus. Ibid.* 13: 422–432.

————— 1968. The family Squillidae. *Ibid.* 18: 105–142.

————— 1968. Stomatopod Crustacea from the western Atlantic. *Stud. Trop. Oceanogr., Univ. Miami.*

————— and Provenzano, A.J. 1963. Early larval stages of *Gonodactylus oerstedii. Bull. Marine Sci. Gulf, Caribbean* 13: 467–487.

Pilgrim, R.L.C. 1964. Stretch receptor organs in *Squilla mantis. J. Exp. Biol.* 41: 793–804.

————— 1964. The anatomy of *Squilla mantis. Pubbl. Staz. Zool. Napoli* 34: 9–42.

Sandemann, D.C. 1964. Distribution between Oculomotor and Optic Nerves in *Carcinus. Nature* 201: 302–303.

Schaller, F. 1953. Verhalten und sinnesphysiologische Beobachtungen an *Squilla mantis. Z. Tierpsychol.* 10: 1–12.

Serène, R. 1954. Observations biologiques sur les stomatopodes. *Ann. Inst. Océanogr. Monaco* 29: 1–94.

Siewing, R. 1956. Morphologie der Malacostraca. *Zool. Jahrb. Abt. Anat.* 75: 39–176.

12.

Superorder Syncarida

There are about 60 species of Syncarida known. The largest is *Anaspides tasmaniae*, 5 cm long.

Syncarida are primitive Malacostraca. They lack a carapace; the first thoracomere is fused to the head or free. The remaining, almost cylindrical somites are unspecialized and similar to each other. Thoracic limbs 2–6, at least, are biramous, each with two platelike epipodites, of which one or both may be lacking in the smallest species.

Syncarida had been known to paleontologists for 40 years from freshwater deposits of the Carboniferous and the Permian, before the first Recent representative was discovered in Tasmania. Later it was found that the minute Bathynellacea belong to this group. Corresponding to their small size these dwarf crustaceans, living in subterranean water between sand grains have a secondarily much simplified body structure. Each of the three orders is considered separately. A fourth order, the fossil Palaeocaridacea (Carboniferous and Permian), differs from the Anaspidacea in having all eight thoracic somites free; some fossil genera have the telson and uropods styliform. Perhaps Anaspidacea give us an idea how primitive malacostracans might have looked.

ORDER ANASPIDACEA

The order contains five known species of which *Anaspides tasmaniae*, 5 cm long, is the largest.

Anaspidacea are Syncarida whose first thoracomere is more or less fused with the head to form a cephalothorax. There are six free pleomeres. The first antennae have statocysts. The first two pairs of pleopods of the male form a copulatory organ.

Anatomy

The head of *Anaspides* is fused firmly to the first thoracomere although separated by a transverse groove. Even the longitudinal muscles are missing between head and the first thoracic somite. The coxa of the first thoracic limbs, besides two epipodites, also has one to two endites. The thin exopodites of the thoracopods 2–6 have many short articles and are densely setose. The endopodites usually have five to six articles. The third to fifth pleopods, besides a strongly setose, long, many-articled exopodite, have a small endopodite with at most two articles (Fig. 12-1). In *Koonunga* and *Micraspides* these have disappeared (Fig. 12-2). As in decapods the first two pairs of pleopods have been transformed into gonopods in the male; in *Anaspides* females they have short endopodites. The flattened uropods are fringed with setae.

NERVOUS SYSTEM AND SENSE ORGANS.   The internal anatomy has only been studied in *Anaspides*. The supraesophageal ganglion is mostly like that of other malacostracans. However there is no subesophageal ganglion (Fig. 12-1). All postantennal ganglia, even those of mouthparts, are separated by connectives. There are statocysts without statoliths in the basal article of the first antennae. Also there are compound eyes, which are sometimes on stalks.

DIGESTIVE AND EXCRETORY SYSTEMS.   The foregut has a simple cardiac and pyloric stomach. Many thin, tube-shaped ceca arise from the anterior end of the midgut. The maxillary gland, with its long excretory tube, is the excretory organ.

CIRCULATORY SYSTEM.   The primitive heart, which reaches from the first thoracic to the 4th abdominal somite, has only a single ostium opening in the third thoracomere (Fig. 12-1). In addition to an anterior and posterior aorta, there are seven pairs of lateral arteries, each of which gives off a visceral branch

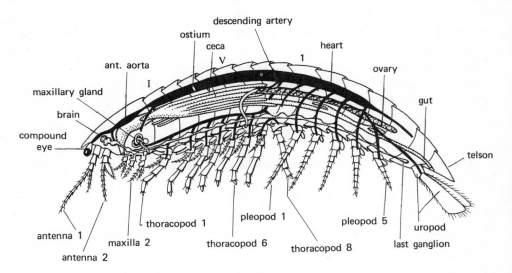

Fig. 12–1.   Anatomy of female *Anaspides*. The thoracic limbs have a small exopodite and 2 epipodites; the pleopods have a short one-articled endopodite. Roman numerals refer to thoracomeres, Arabic to pleomeres. (After Siewing.)

near its origin. The most anterior pair originates next to the anterior aorta; the next pair arises at some distance, in the eighth thoracic somite. Its lateral branch is well developed on one side only, forming the descending artery (Fig. 1-21) that continues just above the nerve chain as a supraneural artery with branches supplying the nervous system to the mouthparts and thoracopods (Fig. 1-22b). The third lateral arteries do not connect to limbs while the other four pairs supply all of the abdominal limbs.

RESPIRATORY SYSTEM.   Respiration takes place through the plate-shaped epipodites and also probably through other thin areas of the integument.

REPRODUCTIVE SYSTEM.   The testes extend as paired thin coiled tubules from the end of the fifth thoracomere to the telson. Anteriorly the tubules continue as vasa deferentia which open at the base of the last thoracopods. The ovaries are simple paired tubes that extend from the sixth thoracomere to the last pleomere. The anterior oviducts open on the mesal side of the sixth thoracic limbs. The eighth thoracic sternite forms a seminal receptacle.

Reproduction

The female of *Anaspides,* unlike most other Crustacea, glues her eggs, each about 1 mm in diameter, to pieces of wood or other solid objects. The eggs undergo direct development and the young are 2.7 mm long on hatching. They resemble the adults but still lack eyestalks, rostrum, and endopodites of all pleopods. As other juvenile characters they have a vestigial nauplius eye and a deep notch in the telson suggesting a furca. During embryonic development there are indications, in the form of a posterior ganglion, of a seventh pleomere. This posterior ganglion later fuses with the sixth ganglion of the abdomen.

Relationships

The uniform trunk somites, the almost diagrammatic shape of the thoracic limbs (Fig. 12-2), the metameric nervous system and long heart indicate that the Anaspidacea are very primitive Malacostraca. Another primitive characteristic is the presence in the embryos of a vestigial seventh abdominal somite. Their relict distribution, Tasmania and neighboring Australia, indicates that they are very old survivors of the past. The absence of a carapace may be a secondary character of the superorder. The closest relatives appear to be the Eucarida.

Habits

Three of the Recent species are found in Tasmania, the fourth, *Koonunga,* in Victoria, South Australia, all in freshwater nearly free of predators. *Anaspides* and *Paranaspides* live in cold mountain streams and lakes rich in vegetation in Tasmania, usually at about 1200 m elevations. *Micraspides* live in spaces among sphagnum moss, while *Koonunga* (Fig. 12-2) can be found in small ephemeral turbid waters with a mud bottom. *Koonunga* can burrow with the first thoracopods.

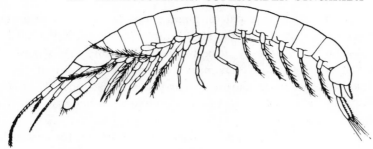

Fig. 12–2.   Anaspidacean, *Koonunga*, lateral view; 8 mm long. (From Noodt, courtesy of W. Noodt.)

LOCOMOTION.   *Paranaspides* rarely leaves plants, while *Anaspides* frequently moves along the bottom partly swimming, partly walking. In both genera the exopodites of the thoracopods are in constant motion. In *Paranaspides* they beat 250–300 times per minute, in *Anaspides* 100 times per minute, producing a posterior current on both sides of the animal that supplies the epipodites with fresh water and probably also pushes the animals forward. In swimming, however, the laterally extended, feathered pleopods are most important. They paddle in metachronous rhythm, the most posterior starting the movement. By suddenly bending the trunk and beating the tail fan, the animal can shoot backward a considerable distance, even out of the water. Members of both genera swim only short stretches, always with the back toward the water surface. Usually they walk or climb on the uneven bottom with the second to eighth thoracopods aided by the pleopods. The first pair of limbs does not touch the ground but makes swinging or vibrating movements.

FEEDING.   *Paranaspides* is microphagous (feeding on small particles). With both sets of setae of the preischium of the first thoracopod it brushes up the particles from the sides of the animal, diatoms, protozoans, etc., from stalks or leaves toward the mouth. Near the mouth the particles are taken over by distal endites of the second maxillae and brought to the mouth. The maxillae have a filtering function, as do those of *Hemimysis,* drawing in water and filtering it. Particles of iron saccharate put in the water are soon found among mouthparts and in the gut. Unlike *Hemimysis* the constantly moving thoracic exopodites do not contribute to the suction which is produced only by the maxillae.

Anaspides guides detritus and algae from leaves and stones to the mouth using mainly the endites of the first limbs. Maxillary suction and filtering are effective only near the mouth as can be demonstrated with iron saccharate. *Anaspides* can grasp worms, or even tadpoles found while moving about or digging in the mud with limbs 2–5. Only sensory hairs on the limbs, not the eyes, help to find food. The field of vision does not include the area just in front and below the animal. Some Paleozoic species (Palaeocaridacea) have one or more endopodites transformed into strong raptorial limbs.

Classification

Anaspididae. The compound eyes are stalked (Fig. 12–1). The second antennae have a scalelike exopodite. The exopodites of thoracopods 2 to 6 have several articles, 1 and 7 are unsegmented. The 8th thoracopod is uniramous. The pleopods have short, cone-shaped endopodites. In Tasmania are found *Anaspides tasmaniae,* to 5 cm long, and *Paranaspides lacustris,* to 5 cm long.

Koonungidae. The compound eyes are not stalked (Fig. 12–2). The second antennae lack an exopodite. The 7th thoracopod lacks an exopodite. The pleopods lack endopodites, except the male gonopods. *Koonunga cursor,* to 0.8 cm long, occurs near Melbourne, Australia; its first thoracopod is modified for digging (Fig. 12–2). *Micraspides calmani,* to 0.8 cm long, is found in small ponds and streams in Tasmania.

ORDER STYGOCARIDACEA

Four species are known to belong to this order. The largest is *Parastygocaris andina,* up to 4.2 mm long.

Stygocaridacea are minute, blind Syncarida, which differ from the Anaspidacea by the following characteristics: the head is completely fused to the first thoracomere; the first thoracopod has become a maxilliped; the pleopods are absent except for uropods and male gonopods (Fig. 12-3); there are dentate setae on the endite of the mandible. The telson, consisting of two short setose stumps on each side of the anus, could be interpreted as a vestigial furca.

The elongate, wormlike, flexible stygocarids move rapidly in the ground water between sand grains of southern South America south of lat 30° and New Zealand. Removed from the sand, they can neither swim nor walk. The youngest animals found had all limbs of the adult, indicating direct development. Probably these animals, first described in 1963 by W. Noodt, have habits similar to those of the Bathynellacea.

Classification

Stygocaridae. There are two genera, *Stygocaris* of Chile and *Parastygocaris* of central Argentina (Fig. 12–3). Representatives of the order have recently been found in New Zealand groundwaters.

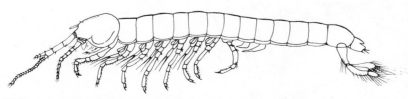

Fig. 12–3. Stygocaridacean, *Parastygocaris,* lateral view; 4 mm long. (From Noodt, courtesy of W. Noodt.)

ORDER BATHYNELLACEA

There are more than 60 species of Bathynellacea known. The largest is *Bathy-nella magna,* up to 5.4 mm long.

Bathynellacea lacks eyes. The head and thorax are not fused. The sixth pleomere is fused to the telson, which bears a furca. In addition to uropods there may be one or two pairs of pleopods.

Anatomy

The Bathynellacea are the smallest malacostracans. Their body is adapted to life in the spaces between sand grains, being dwarfed, worm-shaped, lacking eyes, and having reduced pleopods, which in most Malacostraca are well developed for swimming. The absence of a swimming larva can also be considered an adaptation to interstitial life (Figs. 1-1, 12-4).

The last thoracopods are very small and in the male probably function as gonopods. In some genera they are present only in males. In some species also thoracopod 7 and sometimes 6 are lost. The second to fifth pleopods (exceptionally 3rd–5th) are lacking, but there are uropods (Figs. 1-1, 12-4). In some species of *Bathynella* each of the first seven thoracopods has two epipodites, while in other genera there is only one epipodite and only on some limbs, perhaps an adaptation for a small body with a proportionately large surface.

There are no eyes nor statocysts, and the brain has no optic center. But the nervous system, as in other dwarf forms, is very large. The number of cells apparently is never decreased below a certain number (as also in Pauropoda and other small arthropods). The ventral cord has a swelling in each somite, but reaches only to the 2nd, 3rd, or 4th pleomere because the posterior ganglia are pushed together anteriorly.

The gut is a straight tube, without cardiac or pyloric or ceca, an adaptation to small body size and to the food. There is a dorsal blood vessel through most of the body. Its muscular heart varies in length among species even within populations. In *Bathynella* it may be limited to the 4th thoracic metamere, and thus be

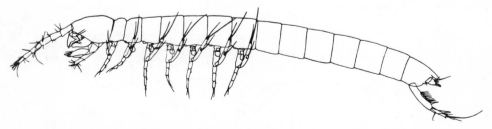

Fig. 12–4. Bathynellacean, *Parabathynella neotropica,* lateral view; 1.2 mm long. (From Noodt, courtesy of W. Noodt.)

almost spherical, or it may extend from the second to fourth metamere. Ostia have not been found so far. There are no arteries except for the anterior and posterior aorta, another adaptation to small size. The maxillary gland has a long excretory canal, looped to the fourth thoracic metamere. Its sacculus and opening are in the somite of the second maxilla. Each metamere has one pair of nephrocytes. The gonads are simple paired sacs in the abdomen. The male gonopore opens on the basis of the eighth thoracic limb, the female on the sternite of the sixth thoracic limb.

Reproduction

As in other minute invertebrates, the eggs are relatively large, 0.11 mm in diameter, and are individually deposited. After 2 months at temperatures of 11°C (optimum of adults 8–10°C) the young *Bathynella natans* hatches as a larva, 0.45 mm long, which does not move much. It has antennae, mouthparts, the first thoracopods, and only 10 somites. The helpless first stage lives off yolk and develops anamorphically. After the molt to the second stage there are 11 metameres, the first two thoracopods and limb buds the third and fourth thoracic and first abdominal somites. Now the larva can move and feed. After further molts different individuals have a different number of new limbs and limb buds. At 9°C it takes 9 months to pass the six juvenile stages and obtain all somites and appendages. After further molts, one in the male, two in the female, the larvae mature. But mature individuals continue to molt, males probably four, females five times. In the laboratory they live to become 2 years old at temperatures of 12–14°C.

Habits

All are freshwater animals. *Bathynella magna* and *B. baicalensis* live in Lake Baikal at depths of 20–1440 m. All others live between sand grains of ground water. They can be collected by letting water from a well run for 30–45 min through a plankton filter. They are found in the ground water and in sand where the ground water comes in contact with surface water, e.g., springs, lakes, and streams. Habitats are the sandy beaches of Lake Tanganyika and at a distance of only 1 m from the water level in lagoons of Madagascar and slightly brackish water at the mouth of the Amazon.

While most seem to have a low temperature optimum, *Thermobathynella adami* lives in water 55°C. The unusual habitats seem to be a refuge to this old group of Malacostraca. None have as yet been collected from Australia and North America although one might expect them there as in the ground water of all continents.

LOCOMOTION. *Bathynella* moves rapidly between the water-logged sand grains. By bending and stretching the body it can make some emergency swimming movements. Weak currents can be resisted only by holding on.

FEEDING. Bathynellacea feed on detritus but can overpower protozoans with the sharply-toothed mandibles. The sandy substrate near the coast and river shores may have a high content of organic detritus and bacteria. Ground water currents also may provide detritus, bacteria, fungi on which *Bathynella* can live.

Classification

At present the classification is fluid as new discoveries are being made. There are three families.

Bathynellidae. There are paragnaths and the second antennae have exopodites. One or 2 pairs of pleopods are found in *Bathynella*; *B. natans* (Fig. 1–1) 1.5–2 mm long, is very variable, races probably evolved as a result of isolation. They are found in many wells in Europe. Other species are known from Lake Baikal, Japan, southern Argentina, southern Chile and southern New Zealand.

Parabathynellidae. There are no paragnaths nor exopodites on the 2nd antennae and the animals mostly lack pleopods (Fig. 12–4). *Parabathynella* is known from southern and southwestern Asia, Europe, Japan, Africa, Madagascar, New Zealand, and South America. Madagascan species are found near the coast. *Thermobathynella adami* is found in waters of Lake Upemba, Katanga, Congo at 55°C.

Leptobathynellidae. These animals have paragnaths and have sometimes vestigial exopodites on the five articled second antennae. There are no pleopods. *Leptobathynella* has five known species in South America, some known from several localities far apart. *Leptobathynella amyxi*, to 0.55–0.7 mm long, is the smallest malacostracan, found in sand near the slightly brackish mouth of the Amazon, 1 m from the water level. Two species of *Brasilibathynella* are known from the Brazilian coastal region.

References

Syncarida

Brooks, H.K. 1962. On the fossil Anaspidacea, with a revision of the classification of the Syncarida. *Crustaceana* 4: 229–242.

Delamare Deboutteville, C. 1960. *Biologie des eaux souterraines.* Hermann, Paris.

Noodt, W. 1965. Natürliches System und Biogeographie der Syncarida. *Gewässer und Abwässer* 37–38: 77–186.

Schminke, H.K. and Noodt, W. 1968. Discovery of Bathynellacea, Stygocaridacea and other interstitial Crustacea in New Zealand. *Naturwissenschaften* 55: 184–185.

Siewing, R. 1959. Syncarida *in Bronns Klassen und Ordnungen des Tierreichs.* Akad. Verl., Geest & Portig, Leipzig. 5(1) 4(2): 1–121.

Anaspidacea

Cannon, H.G. and Manton, S.M. 1929. The feeding mechanism of the syncarid Crustacea. *Trans. Roy. Soc. Edinburgh* 56: 175–189.

Gordon, I. 1961. The mandible of *Paranaspides lacustris. Crustaceana* 2: 213–221.

Hickman, V.V. 1937. Embryology of the syncarid *Anaspides tasmaniae. Pap. Proc. Roy. Soc. Tasmania* 1936: 1–36.

Manton, S.M. 1931. Photograph of a living *Anaspides tasmaniae. Proc. Zool. Soc. London* 37: 1079.

——— 1931. On the maxillary glands of the Syncarida. *J. Linnean Soc. London* 37: 467–472.

——— 1964. Mandibular mechanisms and the evolution of arthropods. *Philos. Trans. Roy. Soc. London* (B) 247: 1–183.

Nicholls, G.E. 1931. *Micraspides calmani* from the west coast of Tasmania. *J. Linnean Soc. London* 37: 473–488.

Siewing, R. 1954. Verwandtschaftsbeziehungen der Anaspidaceen. *Verhandl. Deutschen Zool. Ges.* 18: 240–252.

——— 1956. Morphologie der Malacostraca. *Zool. Jahrb. Abt. Anat.* 75: 39–176.

Williams, W.D. 1965. Zoological notes on Tasmanian Syncarida. *Int. Rev. Ges. Hydrobiol.* 50: 95–126.

Zerbib, C. 1967. La glande androgène chez un crustacé Syncaride, *Anaspides. Compt. Rend. Acad. Sci. Paris* (D) 265: 415–418.

Stygocaridacea
Gordon, I. 1964. The mandible of Stygocaridae and some other Eumalacostraca. *Crustaceana* 7: 150–157.

Noodt, W. 1963. Anaspidacea in der südlichen Neotropis. *Verhandl. Deutschen Zool. Ges.* 26: 568–578.

Bathynellacea
Jakobi, H. 1954. Biologie, Entwicklungsgeschichte und Systematik von *Bathynella natans. Zool. Jahrb. Abt. Syst.* 83: 1–62.

Noodt, W. 1967. Biogeographie der Bathynellacea. Symp. on Crustacea, Ernakulum, Marine Biol. Ass. India 1: 411–417.

13.

Superorder Eucarida

There are about 8600 known species in the two orders of Eucarida.

The large carapace is almost always firmly fused to the dorsal side of all thoracomeres. The compound eyes are always stalked. The peduncle of the second antenna usually consists of two articles. The mandibles of the adult lack a lacinia mobilis (a movable process near the mandibular teeth). The midgut gland (hepatopancreas) consists of many tubules. The heart is short. There are usually several larval stages during development.

The order Euphausiacea resembles in some respects the Mysidacea and appears to be the older eucarid group although euphausiids are not known fossil. The Decapoda are more diverse, both in morphology and in habits, and are known as fossils since the Permo-Trias.

ORDER EUPHAUSIACEA

There are 90 species of euphausiaceans known. *Thysanopoda cornuta,* 8 cm long, is the largest.

Euphausiacea are shrimplike. The carapace is fused to all thoracomeres but is so short laterally that the gills are exposed (Fig. 13-1). No thoracic limbs are specialized into maxillipeds.

The group is distinguished from the decapods by lack of maxillipeds, from the mysids by the complete fusion of the carapace to all thoracic tergites, and the exposed gills.

Anatomy

The body is laterally flattened, and the cephalothorax is only half or one-third the length of the abdomen. The gills are exposed, not in branchial chambers (Fig. 13-1). An anterior carapace extension, the rostrum, forms a roof over the head. The first five pleomeres are always free and movable and have well-developed epimera (lateral extensions) of the tergities. The elongate telson bears a pair of subapical processes.

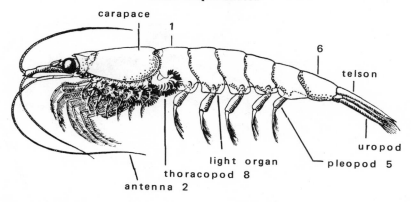

Fig. 13-1. *Meganyctiphanes norvegica*, female; 3.5 cm long. The five pairs of pleopods are stronger than the thoracic appendages; Arabic numerals are pleomeres. (After Holt and Tattersall from Zimmer.)

The long, soft thoracic limbs have endopodites with five articles and an exopodite with two articles; the distal one is a flat swimmeret with a setal fringe (Fig. 13-4). The coxa bears a branched tubular gill, the epipodite, and also at least one endite on the first limbs (Fig. 13-4). The seventh and eighth pairs of limbs are always reduced in size (Figs. 13-1, 13-2). In many genera the 2nd or sometimes the third or both thoracopods are raptorial by having strong basal articles and elongate endopodites, specialized armature or are chelate (Fig. 13-2). The five pairs of pleopods are narrow and flattened anterioposteriorly. Their protopodite consists of two articles, their swimming branches are one-articled and are fringed by setae. A retinaculum can hook both members of a pair together. The male has the first two pleopods modified as copulatory organs. The dorso-ventrally flattened uropods have a one-articled peduncle and one-articled branches. Together with the elongate pointed telson, they form a narrow tail fan.

The flexible, soft cuticle is only slightly if at all mineralized. Only *Bentheuphausia* has an opaque red body; usually the animals are transparent, although with larger or smaller patches of red, purple or brown. The coloration is derived in part from chromatophores and from pigments within the cuticle.

SENSE ORGANS. There are sensory hairs, esthetascs, and compound eyes. Esthetascs are groups of thin hairs underlain by sensory cells on antennae. The adults at times still have a nauplius eye under the rostrum, between the eyestalks. The eyestalks have two articles. The compound eyes, except in *Bentheuphausia*, are very large, their long diameter at times one-sixth to one-eighth body length. In some predators (e.g., *Nematoscelis* and *Stylocheiron*) the compound eyes are divided by a constriction into a part directed toward the sides and a part directed dorsally (Vol. 2, Fig. 16-23, p. 320).

NERVOUS SYSTEM. The esophageal connectives are long and are followed by a chain of postoral ganglia, which are separated from each other in members of many genera. However, the three pairs of mouthpart ganglia are fused.

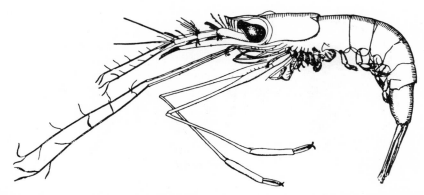

Fig. 13–2. *Stylocheiron suhmi*, a deep-sea predator; 2 cm long. The endopodite of the 2nd antenna and the 3rd thoracic limb are extremely long. (After Chun.)

DIGESTIVE SYSTEM.   The esophagus opens into a large cardiac stomach different in predators (*Nematoscelis, Stylocheiron*) and small plankton feeders. Between the cardiac stomach and the midgut opens one pair of dorsal, anteriorly directed cecal tubes and also the midgut gland. The midgut gland consists of paired ventral, much-branched tubules which fill much of the cephalothorax.

EXCRETORY AND CIRCULATORY SYSTEMS.   The basal article of the second antenna contains the antennal gland. The heart is in the posterior part of the cephalothorax and has two pairs of ostia. Toward the head is the anterior aorta, which forms a cor frontale, a muscular structure which aids in pumping the blood. Also there are four pairs of lateral arteries. The third pair is unequal; on one side a thin artery which may be absent, on the other side is a well-developed descending artery having branches and forking into the subneural artery supplying the nervous system and the cephalothoracic appendages. The pleopods and uropods are supplied from metameric branches of the paired or sometimes median posterior aorta.

REPRODUCTIVE SYSTEM.   The horseshoe-shaped testis has lateral diverticula. Its two branches push between the midgut gland and continue as winding vasa deferentia. The ovary is in the same location as the testis and is also horseshoe shaped. But both lateral branches end blindly and each oviduct originates at the level of the sixth limb from the lateral ovarian walls. The winding oviduct goes to the venter, opening on the coxa of the sixth thoracic limbs or on a separate coxal plate. The gonopores are hidden by a fold which, from the posterior border of the sixth thoracic sternite, is anteriorly directed, covering the coxa of these metameres.

Reproduction

Mating has never been observed. Members of some genera drop the eggs individually (e.g., *Thysanoessa, Euphausia,* and *Meganyctiphanes*). *Nyctiphanes* has two pear-shaped ovisacs attached to the ischium of the sixth thoracic limb. *Nematoscelis* and *Stylocheiron* have a single ovisac. About 60 eggs have been found in each ovisac of *Nyctiphanes*, 35–60 in *Nematoscelis*, only 18 in *Stylocheiron*. Some species are believed to move during the spawning period to certain areas in the North Atlantic, forming huge swarms. Finback whales (*Balaenoptera*) during these periods feed on *Thysanoessa inermis* and *Meganyctiphanes norvegica*.

Development

The free eggs have the same specific weight as the water and float in water wherever they have been released. The post embryonal development starts with two nauplius stages and a single metanauplius; all three stages live from stored yolk and do not feed. The following three stages, 1.3–4 mm long, are called calyptopis and feed on minute plankton. Their name refers to their eyes being hidden by the carapace. Only during the next, furcilia stages, do the second to eighth thoracic limbs and the pleopods form. There are about eight furcilia stages followed by an undetermined number of 7–15 mm long cyrtopia stages which swim with pleopods; the second antennae, previously a swimming organ, now has its adult shape. Eggs kept within ovisacs have an abbreviated development, one or perhaps several nauplius stages being eliminated.

Habits

Euphausiacea are entirely marine. They survive when temporarily swept into brackish water but cannot reproduce there. Most species are inhabitants of the open ocean, very few living along coasts (e.g., *Meganyctiphanes*, *Nyctiphanes*, and *Pseudeuphausia*). All are permanent swimmers and never rest on the bottom. Some species appear in enormous swarms, and are fed on by fish and whales. Thus they are of some economic importance. The stomach contents of herring in the Barents Sea are 28% *Euphausia*, and at times blue whales and finback whales in Antarctic waters may feed exclusively on *Euphausia*. A 26 m long blue whale was found to have about 5 million *E. superba* in its stomach.

Observations of *Euphausia superba* swarms have been made continuously for 15 years in the Weddell Sea and near South Georgia Island. During daytime older animals (from the fourth furcilia stage) can be seen just below the surface; younger animals descend quite regularly to 200 m with increasing light. The swarms spread over an area of 45 m²; and extend 100–500 m in one direction. The outline of the swarm may change ameba-like, clusters splitting off and rejoining the main group. The upper 1–2 m are densest. Each individual occupies about 16 ml water, so that individuals swim at distances of only 1.6 cm apart. Thus a swarm 54 × 36 m and 1 m thick is estimated to contain about 96 million individuals. Whales feed only in the dense upper layers. Groups of swarms may be scattered over an area of 380 km². While neighboring swarms 400–500 m apart may contain different age groups, an individual swarm contains animals all of about the same age. There is reason to believe that individuals of one swarm stay together throughout their postembryonal life of about 2 years.

Nauplii and metanauplii are found only at depths of 700–1500 m in warm waters. After molting at a depth of about 750 m to become first calyptopis larvae, the animals slowly rise during a period of about 1 month to the colder surface water. Here they remain for the rest of their lives. The swarms of South Georgia Island drift there in currents from the Weddell Sea in the south at an average speed of 16 km a day.

As in all planktonic animals it is difficult to get an accurate picture of the vertical distribution of individual species. This might be obtained from numerous collections made at various depths. However, such collections are not available, and furthermore the preferred depth of a species probably varies in different seasons and also in different oceans. A deep haul may yield animals that are not ordinarily found at such depths. *Euphausia krohnii,* with a depth preference of 200 m, has occasionally been collected at 2200 m, and under a surface swarm of *E. superba* some animals may swim as deep as 500, 700, even 1000 m.

While horizontal migrations are completely dependent on currents, many species ascend at night, as do many other planktonic animals. But not all individuals participate in vertical migrations. Whether such deep sea species as *Bentheuphausia* ever ascend is not known.

BIOLUMINESCENCE.   Light organs are found in both sexes of all species (except *Bentheuphausia*). Usually numbering 10, their locations are one pair in the eyestalk, one in each coxae of the second and seventh thoracopods and median ones on the anterior four abdominal sternites (Fig. 13-1). All organs have the same structure; only the one in the eyestalk lacks a lens. In the back is a concave reflector consisting of concentric layered lamellae which continues toward the hypodermis as a lamellar ring (Fig. 13-3). The light rays go through a biconvex lens toward the hypodermis. A rod-shaped mass made up of many lamellae produces light, the secretions coming from cells in the space between reflector

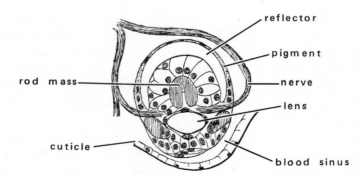

Fig. 13–3.   Section through the light organ of *Nematoscelis.* (After Chun.)

and lamellar ring. On the outside the reflector is covered by red pigment and a blood sinus. The cause of the light has not been explained. That it originates in the rod mass is assured by the rod's continued glow, even when excised. The blue-green light is produced spontaneously. Swimming *Meganyctiphanes norvegica* glow once in a while for 2–3 sec. The luminescent organs can be turned around their axis by muscles. *Meganyctiphanes* has an annual and daily rhythm of luminescence depending on its swarming and mating period.

LOCOMOTION. Pleopods are the swimming organs. Rapid retreat is effected by a beat of the tail fan. The exopodites of the thoracopods have the sole function of creating the respiratory current. While *Euphausia* in an aquarium swim rapidly in various directions without resting, predacious *Stylocheiron* and *Nematoscelis* hover vertically for long periods head up. Probably all species have to swim continuously to avoid sinking.

FEEDING. There are microplankton filter feeders and predators. Some species filter feed and can also feed as predators. *Euphausia superba* feeds almost exclusively in unicellular organisms, and in its stomach only prey less than 0.04 mm diameter has been found, mainly diatoms, eaten by both young (1–4 cm) and adults (4–6 cm long). Other Euphausiacea take a mixed diet: *Meganyctiphanes norvegica,* depending on its age and season, feeds on dinoflagellates, diatoms, and copepods, besides organic and inorganic particles. The algae contain carotene which, within the body, is transformed into vitamin A, as probably is the astaxanthin of the copepods consumed. Thus *Meganyctiphanes,* like other Euphausiaca, is an important vitamin source for fish (e.g., herring). The stomach content of *Nyctiphanes couchii* consists of *Sagitta* parts, copepods, other small crustaceans, many diatoms and dinoflagellates.

Nematoscelis, Stylocheiron (Fig. 13-2) must be predators, judging by the structure of their limbs and cardiac stomach.

The filtering apparatus has been studied only in living *Euphausia superba*. It is made up of the six anterior well-developed thoracopods, whose parallel basal articles are held diagonally anterior (Fig. 13-5). The coxa, ischium, and merus bear rows of long, parallel, filter seta (Fig. 13-4). Each seta touches the mesal surface of the anterior thoracic limb and rests here on a row of short comb setae (Fig. 13-4). In this way the six limbs form the wall of a filter funnel on each side of the body. The distal funnel wall is formed by the posteriorly bent three distal articles of the limbs whose setae are directed caudally and lie against the mesal wall of the next posterior limb, each seta lying between two short comb setae (Figs. 13-4,13-5).

The filter setae are feathered in two rows, forming a wide mesh distally, a small mesh proximally (Fig. 13-5). The distance between setal branches is 35 μm on the propodus, 7 μm on the ischium.

The net is closed above by the ventral surface of the body. Its anterior wall is formed by medially directed setae of the ischium and merus of the first thoracic

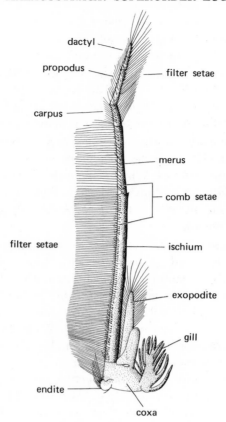

Fig. 13–4. Thoracic limb of *Euphausia superba*, lateral view; 24 mm long. (After Barkley.)

limbs which close together, forming a fence. The posterior wall is made up of a comb of long setae of the ischium and merus of the sixth limb.

It is not known how the current is produced, but it is assumed that when the shrimp moves forward the animal produces a continuous current of water through the filter. But it is also possible that the exopodites move water through the filter setae. Beating movements of the exopodites in an anterioposterior direction have been observed in *Meganyctiphanes*. The transport of the filtrate to the mouth has not been described. It is known that in backstroke of limbs the short comb setae, comb the filter setae of the neighboring posterior limbs and also straighten them out should they get entangled. In dead specimens large clumps of particles are attached to the comb-setae of the basal limb articles, and also the ventral groove between the coxae is filled with a row of particles which pile up anteriorly near the maxillae. This suggests that the particles move to the base of the limbs and from there are transported to the mouth.

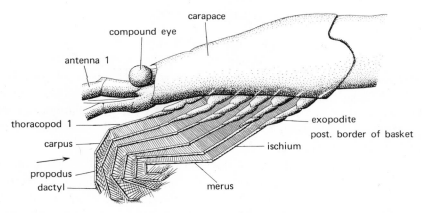

Fig. 13–5. Filter basket of *Euphausia superba*; cephalothorax from the left side with gills and the last 2 pairs of thoracic limbs removed; carapace 22 mm long. The filter setae of the ischium and merus of the first thoracopod are directed toward the midline (perpendicular to the plane of the page) and unlike those of the other legs can not be seen. The posterior border of the filter basket is formed by long comb setae of the ischium and merus of the 6th thoracopod, also facing the midline and perpendicular to the plane of the page. The arrow indicates the direction of the filter current that enters through the wide mesh of the 3 distal articles of the first thoracopods. Large particles are prevented from entering the funnel, but can escape laterally from the distal basket wall formed by the carpus, propodus and dactylus. (After Barkley.)

Morphological studies of *Nyctiphanes* make it likely that here too, as in some peracarids, there is a filtering apparatus on the second maxilla toward which the current carries the particles in the ventral groove. The filtrate on the setae of the second maxillae is combed out by the comb of spines on the basal part of the first thoracic limbs and is taken to the first maxillae.

RESPIRATION. Epipodial gills, small and unbranched on the first limbs, increase in size and branching on the following seven metameres (Fig. 13-1). The movement of the exopodites of the same legs circulates the water.

Classification

Bentheuphausiidae. The animals lack luminescent organs and have only weakly-developed eyes. *Bentheuphausia amblyops* may reach 48 mm in length.

Euphausiidae. *Thysanopoda* is 19–81 mm long. *Meganyctiphanes norvegica* (Fig. 13-1), to 44 mm long, lives on the slope of the continental shelf and spawns from May until July, forming huge swarms around the edge of the Gulf Stream at the latitude of Norway. *Nyctiphanes* reaches 14–17 mm in length. *Euphausia krohnii* is 26 mm long; *E. superba* 50–60 mm is found in the Antarctic Seas. These are strong swimmers, resisting some currents. The swarms at the surface may live for 10 months at 0°C, some in winter at −2°C. Finback whales take also only 1-cm long furcilia early in the feeding period; later they take only individuals 2 cm long. *Pseudeuphausia* is 8–12 mm, *Thysanoessa inermis* averages 20 mm long. *Nematoscelis* is 14.5–28 mm, *Stylocheiron* 6.7–32 mm (Fig. 13–2) is found in depths of 100 m.

ORDER DECAPODA

There are 8500 known species of decapods. The largest are the spiny lobster *Jasus verreauxi,* up to 60 cm long, and the Australian crayfish, *Euastacus serratus,* up to 50 cm long. The American lobster, *Homarus americanus,* is known to attain a weight of 20 kg (44.5 lb). Although the body length of the largest specimen is not known, a 16 kg (35 lb) lobster measured 60.3 cm. The giant spider crab of Japan, *Macrocheira kaempferi,* has a body width of a little more than 45 cm but its outstretched claws may span 365 cm.

Decapoda are eucarids that have the carapace fused to all thoracic metameres dorsally and extending on each side to the legs, completely enclosing the gills within branchial chambers. The first three thoracic limbs are maxillipeds, the others walking legs. Usually the 4th thoracopod (1st pereopod) is a cheliped, and is used to handle prey and in defense (Fig. 13-6).

Anatomy

The extensive body segmentation and presence of appendages on all metameres gives many decapods a primitive appearance. But the nervous, sensory, circulatory, and digestive systems are complex, as is necessary for animals of large body size. The head metameres and those of the entire thorax are almost always fused and immovable. The sternum of the last thoracomere may sometimes be an exception (in Astacidae, Parastacidae, Anomura, and Dromiacea); it may be separated by a connective tissue membrane from the anterior metameres, making a very flexible joint between thorax and abdomen. Also in some species the dorsum of the last one or two thoracomeres is free, not fused to the carapace (Figs. 13-18, 13-24, some Dromiacea, *Porcellana, Typton,* and Paguridae).

Decapods appear in "long-tailed" and "short-tailed" body types. Between are some Anomura.

1. The "long-tailed" decapods, once united as Macrura (Natantia, Palinura, Astacura), have a well-segmented abdomen and a tail fan. The tail fan is used for locomotion and consists of a telson and uropods (Fig. 13-6). The carapace is approximately Ω-shaped in cross section (Figs. 1-4, 13-6, 13-8). The sternites as a rule are narrow but those of the abdomen, at least in Palinura are Astacura, are wide (Fig. 13-6).

2. The "short-tailed" decapods, the crabs, have a flat abdomen which is carried folded against the ventral surface of the cephalothorax (Figs. 9-21, 13-7). The center of gravity has thus shifted to the area of the pereopods, of advantage for walking. Some crabs can run with surprising speed, especially amphibious and terrestrial ones. This change in body form has evolved independently several times: in Brachyura (Fig. 13-32), Coenobitoidea, Paguroidea, Lomisidae, Lithodidae (Fig. 13-24c), and Porcellanidae (Fig. 13-25c).

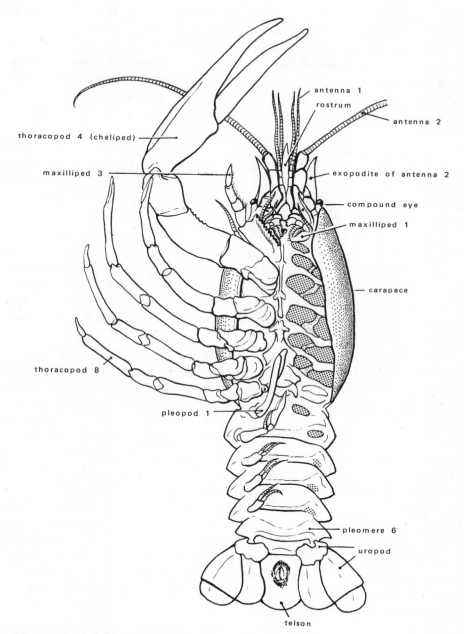

labels on figure:
antenna 1
rostrum
antenna 2
thoracopod 4 (cheliped)
exopodite of antenna 2
maxilliped 3
compound eye
maxilliped 1
carapace
thoracopod 8
pleopod 1
pleomere 6
uropod
telson

Fig. 13–6. Crayfish, *Astacus astacus*; ventral view of a male; about 10 cm long. On the animal's left, thoracic limbs 2–8 and the pleopods have been removed, their insertions stippled. The insertions of the 3rd–7th thoracic limbs are bridged by part of the internal skeleton. (After Hatschek and Cori.)

The Brachyura form the largest group, containing half the known decapod species. The cephalothorax has become wide (Fig. 13-32), and in the middle of the wide sternum there is a shallow depression into which the folded abdomen fits (Fig. 13-7). The abdomen has lost its locomotory function, and the extent of loss of abdominal appendages varies. The large lateral extensions of the carapace are V-shaped in cross section and frame the sternum and mouthparts. Usually the mouth area is completely covered by the basal articles of the third maxillipeds (Fig. 13-7).

3. The Anomura are intermediate between the "long-tailed" and "short-tailed" decapods. In most species of Galatheoidea and Hippoidea there is an abdomen with a tail fan (Figs. 13-25, 13-26), but it is always bent ventrally. Some Coenobitoidea and Paguroidea have a soft, sac-shaped abdomen that contains most of the viscera (Figs. 13-23, 13-24).

APPENDAGES.   The specializations of the very diverse limbs are discussed in the systematic section. The mandibles often bear a palp (Fig. 1-11). The second maxillae have a scaphognathite (a bailer) (Fig. 13-11). The anterior three thoracopods are always maxillipeds and their basal articles usually are flat in crabs (Fig. 13-7). They usually have exopodites, but exopodites are only rarely found on the pereopods, only in some adult Natantia and on the first pereopod of Dromiacea.

In most decapods the first pair of pereopods usually is not used for locomotion but bears a large chela (chelipeds) for handling prey and defense (Fig. 13-32). The size of this cheliped and its shape are adapted to the animal's habits. Often left and right chelae are slightly different. In lobsters, for instance, one chela has many small teeth on the cutting surface, the other has a few large tubercles, more useful for cracking mollusks open. This cracking appendage is generally larger. The difference in size may be great, as in the snapping pincer of some *Alpheus* (Fig. 13-18c), the digging cheliped of *Callianassa* (Fig. 13-22b), and *Upogebia,* the displaying cheliped of the fiddler crabs, *Uca* (Fig. 13-34d), or in those hermit crabs in which one cheliped is used as an operculum for the shell. If the large cheliped is removed, during the course of several molts, crabs of many genera regenerate a new small cheliped while the formerly small cheliped becomes the large one. In such regeneration the age of the crab may be of importance. If both chelipeds are simultaneously autotomized, there is no reversal in size in some species of snapping shrimps, *Alpheus.* If the large cheliped is autotomized at least 40 hr after the small one, there is reversal of size and shape. In fiddler crabs (*Uca*) right or left "handedness" is believed to be genetic, as removal (at an early asymmetrical stage) of the large claw is followed by regeneration of another large claw in its place (Vernberg and Costlow, 1967).

The biramous pleopods have few articles and are strong only in some Natantia and some burrowing Anomura. Usually they are thin, especially in the Palinura and Astacura. The first two pairs of pleopods of the male are usually modified to

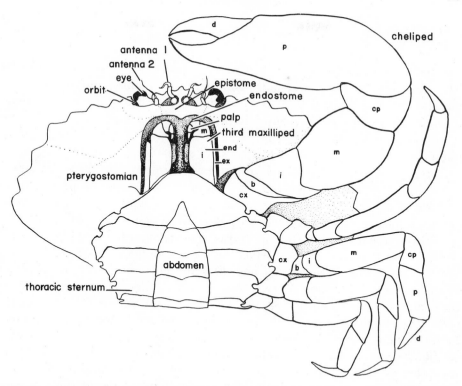

Fig. 13–7. A brachyuran crab in ventral view, left pereopods removed. Pterygostomian area of carapace forms the outer wall of the branchial chamber. Abbreviations: **b,** basis; **cp,** carpus; **cx,** coxa; **d,** dactyl; **end,** endopodite; **ex,** exopodite; **i,** ischium; **m,** merus; **p,** propodus. (Modified from Williams.)

form gonopods. The uropods with the telson form a tail fan. In hermit crabs the uropods have become a holdfast organ, and in most Lomisidae, Lithodidae, and Brachyura they have been lost or become rudimentary (Figs. 13-6, 13-7, 13-23c, 13-24a).

EXOSKELETON. The cuticle consists of three different chitinous layers and a very thin nonchitinous epicuticle lacking wax. The two outer chitinous layers, except in many Natantia, are heavily mineralized with calcareous deposits. Of course the lining of the branchial chamber, gills, joint membranes, and ectodermal gut (except the gastric mill) remain soft.

The chitinous layers contain noncollagenous proteins. Because the outer layer contains melanin it is called the pigment layer. The middle layer is thickest, under it lies a layer covering the hypodermis. Pores or gland canals traverse the cuticle, as do plasma cords that continue into setae. The calcareous deposits start in the innermost layer of the epicuticle and traverse the pigment layer to the base of the thick layer (the calcified layer) from outside in. The innermost layer is not mineralized. In the species examined there is about seven times as much $CaCO_3$

as $Ca_3(PO_4)_2$. In the exoskeleton of crayfish species studied there is 47% $CaCO_3$, 6.7% phosphates, 34% chitin, and 6% organic substances and some salts.

COLORS AND COLOR CHANGE.   Various pigments in the pigment layer determine the coloration and markings. In addition there is a blue pigment, an astaxanthin protein complex, that turns red on being denatured, as in boiled lobsters. In species with a thin cuticle (many Natantia, some Brachyura, e.g., *Uca*) the color is due to chromatophores of the subhypodermal connective tissues. These chromatophores are star-shaped or branched cells, all at one level or slightly overlapping. The pigments may be clumped in the center or may be dispersed into the branches, forming large colored spots. The pigment movements are due to light, primary color changes resulting from direct action of light on the chromatophores and secondary color changes operating through the eyes by action of hormones. In early experiments with shrimps it was demonstrated that pigment movement within chromatophores stopped when the blood supply was blocked. Cutting the ventral nerve cord, however, did not affect the pigment movement in posterior chromatophores. All parts of the central nervous system secrete neurohormones that act on pigments. However, the most important neurosecretory organs are cells in the protocerebrum and optic ganglia that send most of their secretions into the sinus gland, and also the neurosecretory cells of the tritocerebrum, secretions from which move to the circumesophageal connectives. There tend to be differences among species of different suborders but usually the hormones of these two organs are antagonistic. The chemical structures of these chromatophorotropins is not known, but they can be separated by electrophoresis and paper chromatography. To study their functions, the components can then be injected into decapods whose eyestalks have been removed (see also Chapter 1). After extirpation of the eyestalk and the hormone storing sinus gland, the dark pigments of Natantia expand while in Brachyura the black and white pigments concentrate.

For information on the nature of color adaptation, the pigment function in the amphibious ghost crab *Ocypode macrocera* has been studied (Rao, 1968). The three pigments, black, white, and red in the adult crab exhibit primary and secondary responses to light. In the zoea only the black chromotophores have secondary responses, in the megalopa also the red ones. Extracts of eggs, zoeae, and megalopa eyestalks and different sized adults not only show the presence of these hormones, but also show that their titers change in different stages. Extracts of eggs and zoeae cause dispersion of black and red pigment and concentration of red in eyestalkless *Ocypode*, although functional red chromatophores are not present in zoeae. The white chromatophores of zoeae have only primary responses to light as also in the megalopa. But in the megalopa there is a white pigment dispersing hormone. In the adult there are dispersing and concentrating hormones for all three pigments. *Ocypode macrocera* can thus adapt to color and shade of the background. If individuals, 20–25 mm in carapace width, adapted to a white dish are transferred to a black dish during daytime, within an hour the black pigment disperses and white concentrates. *Ocypode* transferred from black to white concentrate black and red and disperse the white.

The dispersing activity of alcoholic extracts of glands for all three kinds of chromatophores is weak in small crabs, 5 mm in carapace width, increasing rapidly in specimens 10–25 mm. The activity of black and white pigment dispersion decreases later with increase in size of the donor but the activity of red pigment dispersion increases. The dispersion activity is related to the behavior: small *O. macrocera,* measuring 5 mm in carapace width, adapt their coloration to the surroundings, their only defense when pursued. Crabs, 10–15 mm in size, ran swiftly to burrows. The largest, 30–40 mm, behave likewise or take a threatening attitude, chelae raised. The smallest respond more quickly to background change than larger ones. If larger than 30 mm they fail to concentrate their red when transferred to a white background and remain dark red.

Similarly many other decapods can adapt to the color, some shrimps even to mottling of the background, e.g., *Hippolyte varians,* adapts to green, red, and brown algae. (See also Classification, below.) That factors other than substrate color are important can be shown by placing a number of dark-adapted *Crangon* with a light-adapted one: the light one will become dark despite the white background.

In addition to their environmental response, the chromatophores of many decapods also have a daily rhythm. In *Palaemonetes, Palaemon* and *Hippolyte* the pigments concentrate at night and the animals become pale and translucent. Brachyura (*Uca* sp. and *Carcinus maenas*) that have a rhythm are darker by day than at night. In *Ocypode macrocera, Carcinus maenas,* and other crabs the adaptation to the background is not interrupted by the daily rhythm. But, in the brachyurans *Uca pugilator, Sesarma reticulatum,* and *Callinectes sapidus,* the rhythm response is greater than the response to the background.

INTERNAL SKELETON. The Reptantia have a strong internal skeleton on the ventral side of the cephalothorax providing attachments for the limb muscles. This endoskeleton is formed by four longitudinal rows of mineralized metameric apodemes (infoldings of the integument) of the ventral side, which by transverse and longitudinal projections combine into a complicated scaffolding (Fig. 13-8). In the Scyllaroidea and Astacura the ventral nerve cord is enclosed within an endoskeletal tube.

SENSE ORGANS AND NERVOUS SYSTEM. The decapod sense organs resemble those found in other orders. Many Brachyura lack nauplius eyes. The compound eyes often have areas, barely visible from outside, that lack crystalline cones or rhabdoms.

The subesophageal ganglion contains at least the ganglia of the three pairs of mouthparts and the maxillipeds (Fig. 1-17a). The arrangement of ganglia of the following metameres differs in different groups. Astacura have a chain of 11 ganglia whose connectives are paired or fused into a median cord (Fig. 1-17a).

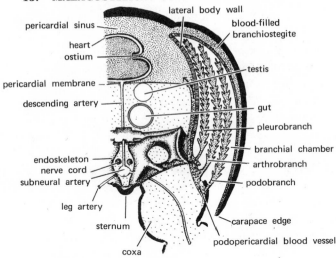

Fig. 13–8. **Cross section through the cephalothorax of** *Astacus* **in the area of the first walking limb; epipodite lamellae have been omitted; the podopericardial blood vessel carries oxygenated blood. (After various authors.)**

Caridea and *Palinurus* have all cephalothoracic ganglia joined with the subesophageal ganglion. In most Anomura the first abdominal ganglia, in Brachyura all of them, move anteriorly and fuse with the posterior of the subesophageal ganglion to form a compact center for all postantennal metameres (Fig. 1-17b). Only in the Dromiidae and Raninidae are all abdominal ganglia in the cephalothorax, though they are not fused. In some Anomura the medulla interna and optic centers are on the sides of the protocerebrum rather than in the eyestalks (*Emerita, Petrolisthes, Porcellana, Callianassa* and in some deep sea crabs). This may be due to a general reduction of the eyes.

DIGESTIVE SYSTEM.   The mouth, between labrum and paragnaths (Fig. 1-10), opens into the ectodermal foregut whose posterior part is a large cardiac and pyloric stomach (see Chpt. 10). Corresponding to the size of the decapod there are many branched midgut ceca (midgut gland, hepatopancreas) which take up the filtered food particles (Figs. 13-9, 13-14). Undigested large particles reach the mid- and hindgut. While the crayfish *Astacus* has a very short endodermal midgut limited to the place where the ceca open, others, such as lobsters, have a long one.

EXCRETORY SYSTEM.   Excretion and osmoregulation take place for the most part in the antennal glands (green glands) opening into the basal article of the second antenna. Close to the opening is the sacculus, as in many other arthropod nephridia. *Crangon* and *Alpheus* have a short excretory canal. Most other decapods, before the widened duct to the outside, have a complicated excretory canal that surrounds the sacculus and may consist of two different parts. The wide end is of ectodermal origin and in Astacura is of medium size; in other

Decapoda it may be equipped with numerous diverticula. In Caridea the diverticula are long and especially so in the Paguridae where they reach into the abdomen, and in Brachyura they spread over most of the cephalothorax. While a lobster produces urine amounting to 1 ml/hr per 500 g body weight, the 700 g crab *Maja*, 1 hr after voiding, already had 5 ml urine. Most N excretion, however, occurs in the form of NH_3 through the gills.

RESPIRATORY SYSTEM.   There are usually many gills, with a large surface area, indicating that the gills are the main respiratory organ and that the branchial chamber walls of the carapace are of less importance than in other crustaceans.

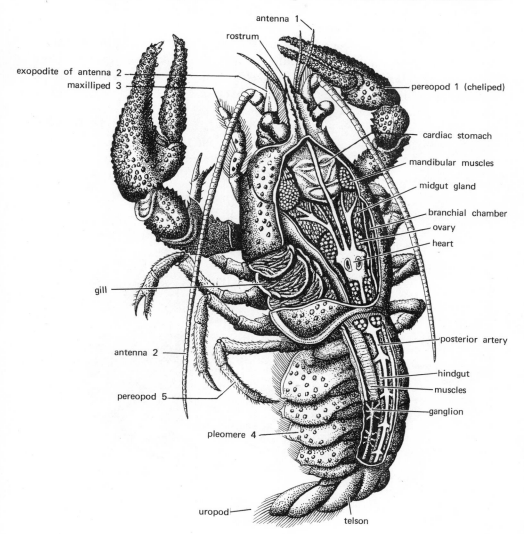

Fig. 13–9.   Crayfish *Astacus astacus* in dorsal view, carapace and tergites cut open; 10 cm long. (After Pfurtscheller.)

(Some Decapoda, e.g., *Lucifer*, have no gills at all.) The gills are the epidodites of the thoracopods. There are three types according to the place of implantation in the adult. In adult decapods there is a gill not only on the coxa, the podobranch, but also one or two on the joint membrane between trunk and coxa, the arthro-branchs and one above it on the integument of the trunk, the pleurobranch (Fig. 13-8). Beside these there may also be a lamella (Figs. 13-9, 13-10). That the gills and lamellae are all epipodites was shown by Claus (1886). By studying penaeid larvae, he found that the primordia arise as three adjacent buds on the coxa before differentiating into mature gills and the lamellae.

The pleurobranch gills and perhaps also the arthrobranch gills may be considered preepipodites of the proximal leg article, the precoxa, which is fused to the body wall. The number of gills on the different limbs is quite variable in different species so that a gill formula can help the taxonomist separate species.

As an adaptation to the heavy exoskeleton, the gills have evolved a large surface. The gill shaft contains the blood vessels and has many branches through which the blood flows in a complicated manner.

The gills may be of three different structures: dendrobranchs, trichobranchs, and phyllobranchs. Dendrobranch gills, found in the Penaeidae, have two opposite rows of branches on the shaft, and these are split up into secondary branches on the outer convex side. Trichobranch gills, found in Stenopodidea, Astacura and

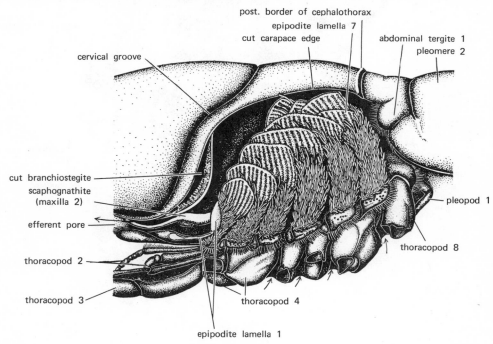

Fig. 13-10. Branchial chamber of *Astacus astacus*, lateral view; 4 cm long. Branchio-stegite removed; arrows indicate the water current. (After Bock.)

Palinura and the primitive Paguroidea, bear tubules on several sides of the shaft (Fig. 13-10). In the lobster each trichobranch may have 10,000 such tubules. In a 9-cm long crayfish, with a shaft 1.2 cm long, there are only 200 tubules. Phyllobranch gills have two rows of subtriangular leaflets attached to the shaft. They are found in most Caridea, most Paguroidea, Galatheoidea, all Hippoidea and Brachyura (Figs. 13-13, 13-14).

The gill shafts stand vertically, extending into the branchial chamber formed by the trunk and carapace walls (Fig. 13-12). In Natantia the ventral edge of the carapace is loose above the coxae, giving the branchial chamber a wide, long gape. Reptantia have the free edge of the branchiostegite (the expanded lateral edge of the carapace) close to the coxae, permitting water to enter posteriorly only between the coxae and from the posterior edge (Fig. 13-13). The branchial chamber is especially tightly closed in the Brachyura, in which the sides of the carapace extend around to the ventral surface and are tightly appressed to the sternites, so that water enters only close to the coxae (as in *Carcinus, Cancer,* and *Maja*) or only in front of the cheliped (Fig. 13-7). In crabs the posterior edge of the carapace is also usually tightly jointed to the trunk. Anteriorly the branchial chamber narrows to a channel that opens close to the mouthparts (Fig. 13-10). The respiratory current is produced by the long exopodite of the second maxilla. These scaphognathites or bailers, extend into the branchial chamber (Fig. 13-11). The scaphognathite is attached in the middle of its longitudinal axis to the shaft of the second maxilla and is thereby divided into an anterior and a posterior portion (Fig. 13-11). The long axis of the bailer lies parallel to the excurrent part of the branchial chamber, but only its anterior part is horizontal; the posterior part is bent dorsally. The bailer rocks on its transverse axis where it is attached to the second maxilla. There may be about 40–250 movements per minute. First the anterior part lowers while the posterior (dorsally bent part) moves forward pushing water out of the channel. Then the bailer turns around its transverse axis to its original position, its posterior part down, the anterior up, closing the expiratory opening. An individual lobster, *Homarus gammarus,* weighing 332 g, pumps 9.6–9.9 liters/hr at 15°C. In aquaria, individuals of many species keep the exopodites of their second and third maxillipeds in constant motion. They direct the expiratory current. Removal of the bailer may cause death. Normally the water is drawn in behind and expelled in front, but in all decapods the direction of flow can be reversed. This is especially true of those Brachyura and Hippoidea which bury themselves in sand; they can take water in through the excurrent opening and move it out at the base of the cheliped or farther posterior.

The path of the water within the branchial chamber can be observed by cutting a window through the side of the carapace and tracing the respiratory water by adding India ink. In *Astacus,* whose bailer normally makes 90 rocking movements per minute at 15°C, the water is forced by the lamellae of the large

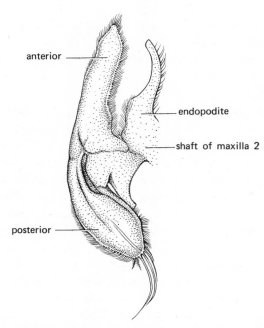

anterior

endopodite

shaft of maxilla 2

posterior

Fig. 13–11. Left scaphognathite (bailer) of *Astacus astacus* **in dorsal view; 1 cm long. The shaft of the 2nd maxilla inserts on the cephalothorax; endites removed. (After Bock.)**

epipodites to rise in the branchial cavity. It is thereby forced through all the branches of the gill on the respective limbs (Figs. 13-10, 13-12). In Brachyura, which usually have only nine pairs of gills and never have epipodite lamellae, the gills converge diagonally in a dorsomedial direction (Figs. 13-13, 13-14). They lie closely appressed to each other and divide the branchial chamber into an upper and a lower tier. The water enters the lower tier between the coxae and edge of carapace, and can reach the anterior excurrent pore only by passing the gills to get into the upper tier.

Dense setae around the incurrent opening in Reptantia protect the gills by straining out detritus. Crabs can close the opening next to the first pereopod with the epipodite of their third maxilliped. In addition the bailer can reverse the water direction in order to clean the gills. In the Brachyura the long seta-bearing permitting the epipodite of the first maxilliped to brush the dorsal side of the gills, while the epipodites of the second and third brush the lower side (Figs. 13-13, 13-14).

CIRCULATORY SYSTEM.   The heart is very short (Fig. 13-9), its muscle layer thick. In the Natantia studied, *Palaemon*, Alpheidae, there are five pairs of ostia, in other groups only three. The anterior aorta is branched several times and widens into a cor frontale (cephalic heart). The posterior aorta supplies species

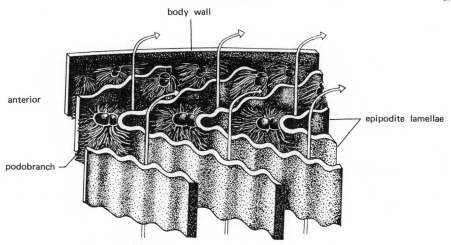

Fig. 13–12. Diagram of a transverse section of the branchial chamber of *Astacus* indicating respiratory currents (arrows). The cross-section of the shafts shows the wide blood channels; the branchiostegite is not shown. (After Bock.)

with long abdomen, with an anterior branch, the gonads, and by successive metameric branches the second to sixth pleopods (Fig. 1-19). Anomura and Brachyura have the arteries in the abdomen specialized.

Three pairs of lateral arteries leave the heart (Fig. 1-19). The most anterior supplies both pairs of antennae, and its proximal branches supply the extrinsic mandibular and the stomach musculature (Fig. 13-9). The median pair branches between the midgut ceca. The third, which becomes median during embryonic development, becomes the desending aorta and enters between the ganglia of the ventral chain to form the ventral subneural artery supplying hemolymph to all the appendages of the cephalothorax except the two pairs of antennae (Figs. 1-21c, 13-8). In *Astacus, Cambarus,* and *Palinurus* the descending aorta originates from a bulbous base of the posterior aorta which is still within the pericardial sinus. Similar blood vessels as in the adults are found in the zoea larvae, which however have more hemolymph circulating through the carapace folds which still are used as an accessory organ of respiration.

The pericardial sacs are found at the posterior end of the cephalothorax, especially in the Brachyura, and are most developed in land crabs (Fig. 13-14). On each side the extension of the pericardial sinus extends into a thin-walled sac which pushes the body wall anteriorly into the branchial chamber. It contains spongy connective tissue penetrated throughout by blood spaces and muscles. In the land crab *Gecarcinus,* before a molt, both sacs store so much liquid that the joint membrane between cephalothorax and abdomen becomes extended. In land crabs they serve for water storage.

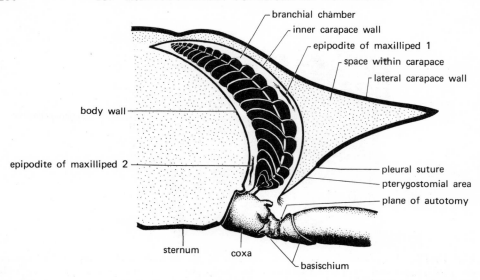

Fig. 13–13.   Cross section through the right half of the cephalothorax of a crab in the area of a walking limb. (After various authors.)

Reproduction

Most decapod are dioecious; a few are protandric hermaphrodites. In the Nephropidae and many crabs, the males are larger than females; in other groups they are smaller. Often there are secondary sexual characters (e.g., fiddler crabs, *Uca*). The testes are one pair of tubes often transversely connected with all kinds of diverticula in larger species. They are located in the thorax above the gut, below the heart (Fig. 1-19). Sometimes they extend into the first or second pleomere (*Homarus*) and in the Paguridae they are entirely within the abdomen. In most decapods each of the two sperm ducts is divided into a glandular portion and a distal muscular ejaculatory duct which opens on the coxae of the last pereopods. Cambarinae, some Paguridae and Brachyura have the openings on papillae.

The ovaries resemble the testes but extend more commonly into the abdomen (Fig. 13-9). The Paguridae and Thalassinidae have the ovary entirely within the abdomen. From each lateral wall originates an oviduct which opens on the coxa of the third pereopods or on the sternum of the third pereomere.

The spermatozoa, like those of most malacostracans, lack a movable tail. They have different shapes in different groups; in *Astacus* they resemble flattened 20-rayed stars. The first one or two pleopods are transformed into gonopods (copulatory appendages) (Figs. 1-18, 13-6), and in the Nephropidae, Astacidae, Brachyura and many Anomura, they function together as a unit. Other groups (e.g., Paguridae, Parastacidae) lack the specialized pleopods, and most shrimp have only the second pair well differentiated.

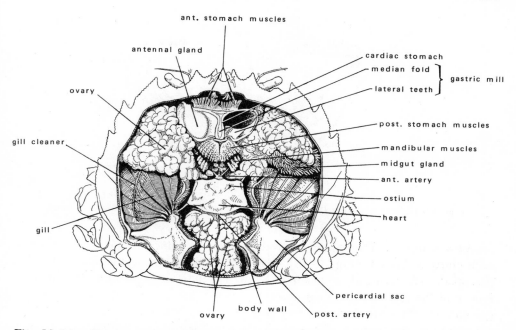

Fig. 13–14.   **Chinese mitten crab, *Eriocheir sinensis*; dorsal dissection, 6 cm wide. The gill cleaner is the epipodite of the first maxilliped; the stomach has been opened on right. (After Panning.)**

In astacine crayfish the sperm mixes with secretions from the sperm duct to produce a viscous white fluid and by bending the body moves from the gonopores to the base of the first pleopods. The rodlike distal portion of the endopodite of the second pleopod is pushed into the first pleopod tube like the plunger of a syringe and the mass is pressed through the tube, producing a 0.5–1-cm long spermatophore. The spermatophore is attached on the sides of the female gonopore. Thus in *Astacus* as in many other decapods the fertilization is believed to be external. In Brachyura the sperm is injected into the gonopore of the female and reaches a seminal receptacle present only in this group. Fertilization is thus internal.

In the cambarine crayfish *Orconectes limosus* mating lasts 2–10 hr. The male turns the female over on her dorsal surface and stands over her, his abdomen bent around outside that of the female. After 10–20 min of inaction, the male rises up and away from the female and settles with the pleopods forward (Fig. 13-15), like the blade of a penknife half open, held by the crossed fifth pereopods anterior to them. The annulus is a depression in the exoskeleton between the coxae of the 4th pereopods. An opening leads into a tubular ectodermal sperm storage organ, which is not homologous with the seminal receptacle. The pleopod tips enter the annulus of the female and as a result of the crossed legs, considerable

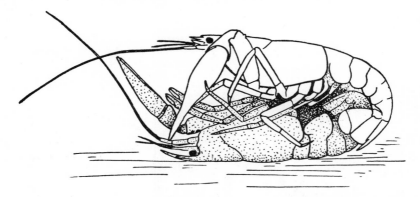

Fig. 13–15. *Orconectes limosus* mating; female stippled. (From Andrews.)

pressure is applied. The hold is secured by hooks or spines on the third article of each third pereopod (Fig. 13-15), fastened to the base of the fourth female pereopods. The terminal papillar part of the vas deferens protrudes into the long groove of the first male pleopod, the sperm is led down to the tip and glides into the annulus. The tip of the second pair of pleopods is applied to the pleopod anterior to it. The sperm is stored within the annulus and fertilization occurs much later, when the eggs are laid.

Males of *Portunus sanguinolentus* behave toward water in which premolt females have been kept as if a premolt female were still present, indicating that a sex-attractant pheromone was released by the premolt female (Ryan, 1966).

A short copulation time and extended precopulatory behavior have been observed in a number of hermit crabs. While copulation in two species of *Anapagurus* (*A. laevis* and *A. chiroacanthus*) was preceded by only a minute or so of mutual appendage stroking, most hermit crabs court for at least several hours before copulation (which usually lasts a few seconds to a minute). About 30 species of diogenid hermits have been observed (*Paguristes, Clibanarius, Calcinus,* and *Diogenes*) and in all the predominant precopulatory activity consists of the males rocking the female back and forth while holding her gastropod shell with his ambulatory legs, and tapping on the hands of her chelipeds with his chelipeds. The pattern of rocking and tapping is distinct for each species. In the pagurid hermits observed (*Pagurus, Anapagurus, Pylopagurus*) the main precopulatory activity is a series of jerking movements made by the male, pulling the female rapidly toward him and pushing her back while holding one of her right ambulatory legs with his minor cheliped.

In a series of observations on mating in *Pagurus bonairensis* and *Paguristes tortugae,* it was found that copulation takes place at various times in the female molt cycle. When the female is soft, copulation is shorter in duration and a pair copulates only once. If the female's exoskeleton is hard, the pair copulates a number of times and the time in copula is significantly longer.

In *Cancer pagurus* the male attends the female for some days prior to her molt. The male rests astride the female and chases off any other males that approach. Similar premolt attendance by males occur in *Carcinus maenas* and *Callinectes sapidus*. The male *Cancer* "helps" push the old exoskeleton off the female and then turns her over onto her back. He uses his chelae to unfold her abdomen and expose the genital openings and then inserts his copulatory organs. Copulation may last 3 hr. The male often guards the female for several days after copulation.

In the grapsid crab, *Hemigrapsus nudus,* the two animals come together and the male grasps the female with his chelae. After some "wrestling" the male lifts the female so that both are perpendicular to the substrate, standing upright on their posterior pereopods. The male then falls back on his carapace and copulation takes place with the female above the male. A similar position is assumed by *Pachygrapsus crassipes.*

Generally terrestrial brachyurans mate on land, using acoustic and visual signals (rather than in water with tactile and chemical signals), and during copulation the female is hard, minimizing injury, and may take a position above the male (Bliss, 1968).

Development

Total cleavage is found only in the primitive Penaeidae, a primitive character also found in *Euphausia*. All possible intermediates are found between total and superficial cleavages in various groups (e.g., *Crangon, Hippolyte,* and *Pagurus*) in which blastomeres have the cell membranes only during the 4- or 8-cell stages. In the 32-cell stage, the cell membranes may not extend to the yolk-rich center of the egg. In most decapods the cleavage is superficial.

Only the Penaeidae hatch as nauplii; most decapods hatch as zoea, some as mysis larvae. In a few, especially as an adaptation to freshwater habitat, the whole development to the decapodid (penultimate) stage takes place in the egg (e.g., crayfish, freshwater crabs, Potamidae, and some Dromiidae). Terrestrial crabs return to the ocean when the larvae are ready to hatch. The number of instars, their appearance and the modification of their appendages are quite diverse, even within groups, resulting in a wide assortment of larvae. In some species, individual animals may have different numbers of stages.

In *Palaemonetes pugio* the rate of development, the time between molts, and also the number of molts depend on the nutrition. Thus larvae of similar size and development may be of different ages and may have passed through a different number of molts. In well-fed animals, pereopods 3–5 appear after two molts; in others only pereopod 3 appears after the second molt, the others following after additional molts.

After the young animal has all metameres and appendages, it is called a post-larva or decapodid. If it is a swimming species, it resembles the adult. In bottom dwellers there often is another metamorphosis after another molt, e.g., the decapodid (megalopa) stage of Brachyura, and the glaucothoe of the Paguridae.

A survey of the different larval periods may be outlined as follows (each period may consist of several stages):

1. Regular anamorphic development, a primitive characteristic, as in Penaeidae: nauplius, metanauplius, protozoea, mysis, decapodid.

2. Irregular anamorphic* development, as in most Natantia: zoea, mysis, decapodid.

3. Irregular anamorphic development, as in Paguridae: zoea, metazoea, decapodid (glaucothoe).

4. Irregular anamorphic development, as in most Brachyura: zoea, metazoea, decapodid (megalopa, Fig. 13-16a).

5. Abbreviated development (rare), as in lobsters: mysis (three stages, of which the last two live on the bottom but nevertheless have exopodites on the pereopods), decapodid.

The swimming megalopa differs from a metazoea by having the carapace depressed, as in the adult, while the metazoea have the carapace laterally compressed (Figs. 2-24, 13-16a). Also the anterior three pairs of thoracopods have transformed into maxillipeds in the megalopa, and the last 5 resemble those of the adult crab. The megalopa swims with pleopods. When walking, the abdomen is folded under, as in the adult.

It is of special interest in the postembryonal development of hermit crabs, is that the asymmetry appears only in the last stage. The swimming decapodid, the glaucothoe, holds on to floating plants. Its abdomen is still segmented, and the pairs of pleopods are still symmetrical. Asymmetry commences on the tail fan and the chelae of the first pereopods. The fourth and fifth pereopods have lagged behind in growth. After the next molt, the final young crustacean emerges. The right pleopods 1-5 and the segmentation of the abdomen have disappeared; the midgut, large sections of the excretory organs and the gonads have moved into the abdomen, which becomes twisted to the right even before being protected by a snail shell, and even if no shells are available.

The average time span of the larval period is 4–5 months in *Sergestes arcticus,* 3 weeks in *Atyaephyra desmaresti,* 4 weeks in *Palaemon adspersus,* 2 months in *Pandalina brevirostris,* 5 weeks in *Crangon crangon,* 3 months in *Palinurus elephas,* 2–3 weeks in *Homarus gammarus,* and 4 weeks in *Carcinus maenas.*

There are various larval adaptations for floating: the carapace spines of Brachyura, and long, antlerlike branched and spined appendages on the carapace and body of the sergestiids. In larvae of polychelids the spherical swelling of the carapace reduces the specific gravity. Very different adaptations are found in

---

* Waterman and Chace (1960) in Waterman, T.H. ed. *The Physiology of Crustacea,* call this metamorphic development, a terminology which does not correspond to usage in other arthropods.

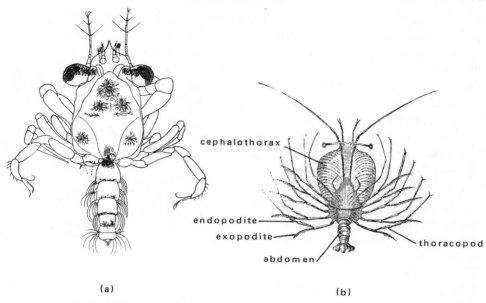

Fig. 13–16. Larval stages: (a) megalopa stage of *Macropipus holsatus* with abdomen extended. The maxillipeds are hidden by the cephalothorax; 2 mm long. (After Williamson from Schellenberg.) (b) Phyllosoma larva in a late (mysis) stage. (After Gerstäcker from Hertwig.)

the first and second pereopods of swimming larvae of *Palinurus, Scyllarus*, and related genera. These hatch as thin phyllosoma larvae (Fig. 13-16b) with a wide transparent body consisting of a head and thoracomeres. The abdomen, segmented in the embryo, is secondarily reduced to a minute cone. The wide, flat carapace extends backward only to the third thoracomere. The phyllosoma larvae generally develop anamorphically, like protozoea larvae (Fig. 2-17). They swim with the thoracic limb exopodites and, in some oceans, grow to 7 cm in 3–8 months before transforming into bottom dwellers. In southern Australian and Japanese waters phyllosomae of *Ibacus* have been found on the exumbrella of medusae. If detached and placed between the medusa tentacles, the phyllosomae were not eaten but climbed back to the exumbrella. Other phyllosomae of 12 different ages have been collected at night in surface plankton, at daytime in 50–100 m depth (Sims, 1966).

AUTOTOMY. If the walking limb of a crab is damaged it usually is immediately autotomized. The crayfish *Astacus*, however, autotomizes only the large chelipeds. The breaking point is a ring shaped region of thin cuticle on the second article, the site of fusion of the basis and ischium (Fig. 13-13). The breakage occurs when the levator of the leg draws the limb up so forcefully that the section distal to the breaking point strikes a coxal projection. Other muscles at the base of the limb do not take part for if these muscles are cut the leg can still be autotomized. The reflex that causes the autotomy is transmitted via the ganglion of the respective limb, as can be demonstrated by cutting the connectives.

That it is actually the leverage that breaks the distal part of the limb can be demonstrated by trying to tear the limb off at the preformed place: autotomy does not occur until weights of 3.5–5 kg are hung on the chela of *Carcinus maenas*. In contrast, the limb was autotomized when weights of only 125–385 g were attached at the tendon of the levator base.

There is virtually no blood lost at the time of autotomy as the former joint membrane between basis and ischium still covers the lumen of the article with openings for nerves and blood vessels only. A major injury to a limb can thus be reduced to a minor one by autotomy.

REGENERATION. In the middle of a wound there is a swelling within which the hypodermis of the remaining stump continues as a tube. In growth it folds twice and, despite limited space in the stump, can form all leg articles. The tissues of the new limb, including its muscles, consist of ectodermal hypodermis cells. The regeneration stump is about as long in a crab as the diameter of the coxa. After the next molt the regenerated leg appears and extends as a normal limb but is shorter than its predecessor. After several molts it reaches its normal length. Regeneration, like molting, is controlled by hormones, the X-organ sinus gland being predominant over the Y-organ. However, if the Y-organ is removed on both sides regeneration does not take place.

Habits

MARINE DECAPODS. Most decapods are marine, and most species inhabit coastal areas of warm oceans. On the coasts, especially in tropical waters, quite a few representatives live in brackish water. In warm countries some inhabit the beach above the high tide line, and some are terrestrial. Only a few families of the Natantia (Atyidae and some Palaemonidae) and Reptantia (Astacidae, Parastacidae, and Aeglidae) a few crabs (Potamidae, Pseudothelphusidae, and Trichodactylidae), and certain species of other families have been successful in freshwater.

In contrast to the Euphausiacea the number of decapods that are permanent pelagic swimmers in the ocean is very small (only some Penaeidea and some Caridea). Most Natantia and some swimming crabs (Portunidae) swim temporarily in search of food, otherwise resting on the bottom. Some species are pelagic, attaching themselves to sargassum seaweed (*Hippolyte acuminatus, Leander tenuicornis,* the small crab *Planes minutus,* etc.). There may also be decapods in floating logs. Large numbers of pelagic forms as well as bottom-dwellers penetrate to great depths, Penaeidae to 5500 m, crabs to 4700 m. Often these are primitive groups. The deep sea decapods tend to be red in color except for burrowers, which are pale. Those that run on mud are spiderlike with long, soft limbs. At depths of more than 1000 m the antennae and often the legs are elongated, increasing the radius of the sense of touch, an adaptation also found in cave animals. The eyes are often reduced.

AMPHIBIOUS AND TERRESTRIAL DECAPODS. In temperate North America and Europe only few decapods leave the water. The widespread *Carcinus maenas* runs on the sand at low tide or digs itself into wet sand. *Ocypode quadrata,* found as far north as Rhode Island, constructs burrows near the high tide line as far as a quarter of a mile from the ocean. Also fiddler crabs, *Uca* are found away from water. The Chinese mitten crab, *Eriocheir sinensis* (Fig. 13-35b) introduced into European freshwaters, may leave the water to walk around a dike or weir, and can live as long as 38 days in wet meadows. Species of the American crayfish *Cambarus* live in burrows in wet meadows, never going into the water. They burrow as deep as 3 m to reach the water level, as the resting chamber at the bottom must be wet.

On tropical coasts at low tides numerous crabs are found on the beach between and on the branches of mangroves or they may even climb cliffs where temperatures reach 37°C. Even beyond the high tide level, in areas reached only by spring tides, many decapods can be seen at daytime or at night. Representatives of some genera go several kilometers inland and only periodically, some in large numbers, come back to the coast for reproduction (*Birgus, Coenobita, Cardisoma,* and *Gecarcinus*).

The main hazard to a crustacean on land is drying up, as crustaceans lack the waxy epicuticular layer in the epicuticle of other land arthropods and have gills. At 78% relative humidity *Gecarcinus* survives only 91 hr, *Cardisoma guanhumi* only 59 hr, and *Ocypode quadrata* only 20 hr. At point of death, they have lost 12.5–22% of their weight, mainly through water loss. At the same time respiration is inhibited when the gills dry, as the diffusion constant of O_2 through a dry chitinous layer is 30 times lower than through water.

Amphibious and terrestrial existence is made possible by resting in moist places during dry periods and limiting activity to nighttime and rainy periods, as well as by some modifications of the respiratory apparatus. *Carcinus maenas* submerge themselves regularly, others burrow to ground water. *Cardisoma* and *Gecarcinus* have especially large pericardial sacs which can store water from a moist substrate and *Birgus* replenishes its water by drinking. Terrestrial hermits (*Coenobita*) carry about a small supply of water in the gastropod shell. The huge pericardial sacs of *Cardisoma* and *Gecarcinus* not only extend far into the branchial chamber but also extend ventroposteriorly to the abdominal tergites and have setae on their tips. Droplets of water placed on these hydrophilic setae will be drawn by capillarity toward the pericardial sacs, so dew and rain can be taken up. The water is believed taken up by the posterior gills (Copeland, 1968).

Terrestrial decapods, except for *Coenobita*, use gills for respiration. The gill surface may be reduced in species that remain out of water for long periods, presumably reducing water loss; such reduction is possible because the O_2 content of air is higher than that of water. The following respiratory adaptations to terrestrial habitat may occur:

1. Using the normal method of obtaining oxygen from the water, terrestrial decapods may obtain respiratory water by burrowing down to ground water.

2. Amphibious crustaceans that respire in the ocean as do marine decapods may increase the oxygen content of the water in the branchial chamber with atmospheric oxygen by a special method. The crab on land holds the trunk up and, using the bailer, expels the water so that it drips from the mouthparts over the ventral surface, reentering the branchial chamber through the opening next to the first or between third and fourth leg coxae, or through the gape at the posterior border. Grooves and depressions guide the water, and humps increase the surface over which the water film flows. The Chinese mitten crab (*Eriocheir sinensis*) in this way can double the oxygen content of its respiratory water. An individual of *Sesarma taeniolatum*, 4 cm wide, using this unusual method of oxygenation, has sufficient water in the branchial chamber to last about 9 hr. After that it has to be replenished by returning to the water or to a burrow that reaches water. The water then enters an opening between the third and fourth pereopod coxae (*Uca, Dotilla,* and *Sesarma*). *Eriocheir* can increase the oxygen content of deoxygenated water by sitting in a puddle with its body extending out.

The same species can also increase the oxygen content of the water between gills by using the bailers to drive air through the branchial chamber. This method is used when most of the liquid has been evaporated.

3. The respiration through the dorsolateral branchial wall can be increased by filling the dorsal area of the respiratory chamber with air. The gills in these Brachyura are not reduced, those in Anomura, however, are reduced in number and size. The respiratory surface of the carapace may be increased by extensions of the blood sinuses as netlike ridges or grape-like humps.

Ghost crabs, *Ocypode* have a space, of varying size in different species, with a septum that separates it from the gills below. Its walls has a network of ridges resembling the roof of the lung in pulmonate snails. At the same time *Ocypode* (except the Peruvian *O. gaudichaudii*) have well-developed gills. They fill the branchial chamber once in a while with ocean water and only its dorsal part functions in aerial respiration. While *O. ceratophthalma* survives only a few days away from water, it survives in ocean water several days without damage.

In the coconut crab, *Birgus latro* (Fig. 13-23) the entire carapace wall of the air-filled branchial chamber is covered by grapelike clusters of warts containing blood sinuses. The respiratory surface is further enlarged by widening the branchial chamber by laterally directed enlargement of the branchiostegites which extend horizontally, then bend and extend toward proximal parts of the limbs. Ventilation takes place in air, by the bailers and by lifting the carapace. Glands of the integument keep the inner surface moist, and the crab can splash water droplets on the inner wall of the branchiostegites with the dorsally directed fifth pereopods. Every 3–4 days at 90% humidity *Birgus* drinks freshwater for

5–10 min by dipping in its large chelae, then draining them onto the third maxillipeds which guide the water to the mouth. *Coenobita clypeatus* drinks similarly. The 10 leaflike gills, one on each side, are much reduced in the narrow, ventral gill chamber. Their extirpation does not seem to damage the crab, and as they are unable to obtain sufficient oxygen from water, the crab drowns in the ocean or freshwater within about 5 hr. Except for reproduction, *Birgus* is a terrestrial animal.

In the Gecarcinidae there is an increased density of capillaries on the smooth inner carapace wall, and at the same time the branchiostegites are widened, increasing the surface (e.g., *Gecarcoidea, Cardisoma,* and *Epigrapsus*) and some Grapsidae (including *Geograpsus*), which stay above the water. In aquatic and marine decapods a sinus containing O_2-deficient blood extends from the dorsal stomach sinus along the dorsal border of the branchiostegites, and sends branches into their walls. These branches distribute themselves as capillaries that run to the ventrolateral edge of the branchiostegites where they fuse into thicker lacunae. These in turn combine on the posterior branchiostegite border into a dorso-posterior sinus that carries oxygenated blood to the heart. In the genera mentioned above there are usually three such capillary systems between the afferent and efferent canals. The capillaries of the afferent vessels unite into several wide sinuses and these divide into a second capillary net that opens into a second sinus. This sinus in turn splits up into a third capillary net before going into the efferent sinus.

The gill number is not reduced, but their surface is, and ridgelike adaptations prevent them from sticking together. *Cardisoma* probably moistens the branchial chamber with ground water in its burrow. Individuals of *Gecarcinus* and *Gecarcoidea* may live kilometers from the beach and have only a few deep burrows. They maintain high humidity in the branchial chamber probably by evaporation from the pericardial sac. They are air breathers, ventilating by raising the branchiostegites.

The postembryonal development by *Birgus* and some Gecarcinidae may include a stepwise widening of the branchial chamber by positive allometric growth, which permits independence from moist retreats. Such adaptations parallel their evolution as terrestrial animals.

4. A new respiratory organ, which has only minor importance in marine decapods, is found in the terrestrial hermit crabs, *Coenobita*. It is an area of numerous blood sinuses underneath the transverse grooves of the back of the abdomen. The branchiostegites also have many sinuses. The gills, however, are very small and do not permit respiration in water. *Coenobita* covered by water drowns in 12–24 hr, but in its normal habitat survives removal of all gills and even of the branchiostegites. The abdominal respiration seems to be most important. Like *Birgus, Coenobita* is fully terrestrial except for reproduction.

5. The water eliminated by the antennal glands may flow back into the branchial chamber.

SHALLOW WATER DECAPODS.   Many Reptantia are shallow water animals that, except during the larval period, probably do not move more than 30–100 m from their resting places. Exceptions are the American lobster, which may move as much as 6 km daily and *Panulirus argus,* which makes long seasonal migrations moving in long single file. Many species, during the course of their post-larval development, descend deeper into the ocean (e.g., *Cancer pagurus*). Developing land crabs and land hermit crabs take the opposite course, moving from the ocean toward the shore before becoming terrestrial. The Chinese mitten crabs, *Eriocheir,* migrate up rivers, but during reproductive periods move back into the ocean, as do the terrestrial crabs. Shrimps and crabs have been known to follow their food supply. Pelagic Natantia have diurnal vertical migrations, rising closer to the surface at night.

PROTECTION.   The large chelae are used for defense by many species; large decapods, such as lobsters, coconut crabs (*Birgus*) and strong individuals of *Cancer,* can cut a man's finger to the bone. *Birgus* (Fig. 13-23) uses the chelae dexterously to escape from boxes made of wood 3 cm thick. Some species of *Alpheus* (pistol shrimp, Fig. 13-18c) have one of their chelae as long as the body. Its immovable finger has a groove that ends in a depression at the base. A projection on the movable finger is pushed into the depression when the chela is closed quickly, forcing a jet of water through the groove anteriorly (Johnson *et al.* 1947). The snap may be heard in all parts of the laboratory in which the shrimp tank is kept. This water pistol is used in defense of territories, in battles with other males and also for catching prey. As numerous individuals of a species live close together, a newcomer causes excitement and one crab after another snaps, making a noise resembling explosions in a brush fire. Whether the sounds, so noticeable to man, have significance to the crabs, is not known. The sight of the open chela is believed to be an effective agonistic stimulus.

The palinurids, which have no chelae, extend the long spiny second antennae toward the attacker. Some crabs (*Lybia, Polydectus*) hold actinians armed with nematocysts in their chelae. *Lybia* used the anterior chelae exclusively for carrying actinians and holds food with the second pair of pereopods. The hermit crab *Diogenes edwardsii* has an actinian sitting on its chela.

On strong stimulation many crabs rear up in a defensive position with limbs spread and chelae raised, making the animal appear larger. In a number of crabs, the chelae are brightly colored, emphasizing them in display positions.

The strong exoskeletons of most Reptantia and spination of Lithodidae serve as a defense against enemies. Many species escape by rapid flight: crabs of littoral habits by running rapidly, long-tailed decapods by shooting backward in the water, repeatedly clapping the abdomen. As they move backward *Panulirus* lobsters produce loud stridulatory sounds by movements of the antennae.

Of great importance, of course, is resemblance to the background, an effective way to hide from predators that find prey visually. Some crabs (*Dromia*) play dead, thereby escaping the attention of predators that rely on vision of movement.

Many species can change their coloration to match the background. Those living on green algae may be green; those on red algae or *Lithothamnion* (e.g., a *Pisa* species) are red. Commensals resemble their host. Sometimes the resemblance extends to the sculpturing, as in crabs that live on corals; their yellow and red carapaces have wide colored depressions resembling eroded corals. Some species of *Hippolyte* seek algae with coloration resembling their own. *Pagurus pygmaeus* lives only in rock piles with epibiotic algae that match its own color.

Some very slow-moving crabs adopt camouflage. Members of the family Majidae cut pieces of surrounding sea weed, sponges, hydrozoan, bryozoan, and ascidian colonies and attach them to hooked setae on their dorsum or appendages. If the masking is related to thigmotaxis, it is often linked with mechanical or chemical protection. Many crabs of the family Dorippidae tend to grasp objects with the fourth and fifth pereopods, and hold them above their back. Most individuals carry mollusk shells like a fan above themselves. If the crab pulls its legs in, it is covered completely by the shells and is even protected mechanically. Some Dromiidae use their two last, dorsally directed walking limbs, to hold over their backs various sponges, ascidian colonies, or seaweed. On being offered paper, modeling clay, ascidians, *Alcyonium,* or sponges, they prefer the last, especially the yellow *Suberites.* Climbing on *Suberites* the crab slowly turns in a circle, using its large chelae to cut a piece of sponge about as large as its carapace. The piece is dropped but is retrieved, hollowed out on one side, and held above the carapace. The crab grows faster than the sponge, and after a few molts the crab has to search for a new sponge and cut a new "suit." *Suberites* not only hides the crab but also offers chemical protection. In an aquarium *Octopus* will attack the crab only if *Suberites* has been removed. In an aquarium *Dromia* will cut paper or cellophane and hold it if no sponge is available. That transparent material is not rejected indicates that this response is strictly thigmotactic.

Many decapods with a thin cuticle, especially Natantia, but also some crabs, have chromatophores that permit them to match the coloration, the lightness or darkness of the background. In *Hippolyte* formation of new and loss of old chromatophores has been observed. *Crangon crangon* can match black, white, gray, yellowish, and reddish substrates; *Hippolyte varians* matches green, brown, and red algae. Some *Palaemonetes* species can be white, gray, black, red, yellow, blue, or green, having a large number of pigments (red, yellow, blue and white).

Burrowing into a substrate protects a crustacean as does the habit of hiding among algae, common among the Natantia and many Reptantia, the animals being thigmotactic at least during the day. Natantia dig shallow, almost horizontal burrows into sand or mud. *Crangon,* by fanning itself with the pleopods and with beats of the abdomen throws sand up; settling into the depression it lets itself be covered by the slowly settling sandgrains. Sometimes it also shoves sand over its back with its second antennae. Only antennae and eyes extend. Similarly, swimming crabs dig themselves in. In contrast, some species withdraw backward into the substrate at an angle of 45° or more (Hippidae, *Corystes*). The use of the chelipeds assures adequate water currents (see below under Calappidae). Oxystomata have special channels in their cuticle which connect the tip of the carapace with the inspiratory opening. *Corystes,* which burrows much deeper into the sand, forms a respiratory tube to the surface kept open by the long second antennae, whose opposite sides have long rows of dense setae, bordering the water tube. The water is led to the mouth in a complicated way via the base of the antennae between rostrum, epistome, third maxilliped and mandibles, entering the anterior opening of the branchial chamber, finally to pass out at the ventral borders of the branchiostegites. In buried *Emerita* and *Albunea* also, the respiratory current is reversed, passing from anterior to posterior; under these conditions the first antennae from the respiratory tube.

Burrows and tubes serve not only as hiding places but also provide mechanical protection. The crab directs its chelae toward intruders, while its body is protected by the substrate.

Lobsters and other Crustacea of rocky coasts live in cavities or caves, and crayfish often in horizontal depressions below banks. Other decapods dig a burrow which may also serve other needs. Fiddler crabs *Uca* are thus protected from being washed away by waves and are provided with respiratory water. Other crustaceans, such as *Upogebia* and the tropical *Thalassina,* which lives to a depth of 1.5 m, use their tunnels also in obtaining food.

Many thigmotactic decapods live in tubes, channels and shells of worms, mollusks, or crabs that have died. The hermit crabs shelter their soft abdomens in gastropod shells which they always carry.

Members of the primitive family Pylochelidae, in which the body is bilaterally symmetrical, push themselves backwards into slits in rocks, holes or sponges, or sometimes into such mollusk shells as *Dentalium* that can be carried. Because the abdomen is straight they rarely use a coiled mollusk shell.

The abdomen of hermit crabs becomes asymmetrical after the first postlarval molt, making it necessary for most of them to use shells twisted to the right. On the right side of the abdomen, the pleopods have been completely lost. The left pleopods (Figs. 13-23, 13-24) fan water for respiration into the shell and also hold the eggs. If one gives *Dardanus arrosor* a shell made of glass, one can see that it holds on by pressing the modified setae of the anteriorly directed uropods against the smooth wall of the shell. By contracting longitudinal muscles the crab widens its abdomen, pressing against the housing, and the two posterior pereopod pairs are jammed against the opening of the shell. These provisions prevent the animal from being pulled out of its shell, and only the second and third pereopods are used for walking. The first chelipeds have transformed in diverse ways in different species so that one alone or both together can close the aperture of the shell tightly after the crab has withdrawn. The growing crab has to replace its shell at intervals. The crab first investigates an empty shell with its chelipeds and then pushes itself into the new shell. A crab may find an empty shell, but frequently it must fight a fellow hermit to obtain a new shell. These fights consist of a series of species specific movements by the attacking animal. The aggressor rotates, shakes and taps the defending crab. The defender, after a number of these movements, comes out of its shell and the pair exchanges shells. *Cancellus* sp., which live in rocks, have similar fights. Species that choose a shell with a slit-shaped mouth (e.g., *Conus* and *Strombus*) generally have a wide, flat carapace.

SYMBIOSIS. Even living animals are used as resting places. Many shrimp such as some species of *Hippolyte,* as well as many Majidae hide among bryozoan or hydrozoan colonies between branches of corals, among the tentacles of antho-zoans and holothurians, and between the spines of sea urchins. Others inhabit hollow spaces within the host as in the cavities of diverse sponges (species of *Synalpheus,* several Pontoniinae, *Spongicola*) in the gill chambers of sessile tunicates (some species of *Pinnotheres* and *Pontonia*), in the water vascular system of holothurians and the mantle cavity of mollusks (some species of *Pin-notheres* and *Anchistus*). At the same time the crustacean may compete for food with its host. Feeding on the host is known in only very few sponge-inhabiting decapods. Reactions of the host to the symbiont, such as changes in growth, are very rare but have been observed in stony corals inhabited by Hapalocarcinidae (Brachyura) and *Paratrypton* (Natantia). The females of *Hapalocarcinus* (Fig. 13-36) are 1 cm long, the males much smaller. They settle in a fork of a seriatoporid and then are surrounded by coral growth, forming a coral "gall." The crustacean is usually completely enclosed except for one or more pores for the feeding and respiratory currents. Other species of Hapalocarcinidae sit in cavities as deep as 7 cm in various corals (Fize and Serène, 1957).

A number of cleaning shrimps (*Periclimenes*) set up their cleaning station about a large sea anemony, gaining protection from the anemone's nematocysts.

The dorsal surface and often the chelae of many hard-shelled Reptantia pro-vide colonization sites for hydrozoa, barnacles, ascidians, *Sabellaria* and other epizootics. Though such commensals occur sometimes in large numbers, they seem to have little effect one way or another on the host, except that the growth hides the crab from visual predators. In some cases mutualism has arisen between crab and symbiont, for instance with the sponge *Suberites* (see above under Protection) and some Hydrozoa and sea anemones (see also Chapters 3–5 of Vol. 1). Often the symbiont sits on a shell carried by the crab. *Diogenes edwardsii* carries the sea anemone *Sagartia paguri* on the smooth surface of the left chela in such a way that when the crab withdraws into its snail shell, it closes off the opening with its large chela, on the outside of which the anemone *Sagartia* is attached. Symbiosis of Brachyura with sea anemones is known; sea anemones commonly live on the back of *Hepatus chiliensis* or may be carried in its claws by *Lybia.* In an aquarium a sea anemone has been observed to move from a stone onto the limb of *Hepatus* and from there to its back. Unlike other hermit crabs, *Paguropsis typica,* about 4 cm long, uses its chelate fourth pereopods to grasp the edge of a disc-shaped colony of zoantharia, *Mamillifera,* and pull it like a pair of trousers over its abdomen. While walking the crab holds its zoantharians with specialized styletlike uropods that hook into the underside of the colony. Despite the strong poison of the nematocysts, hermit crabs often feed on acontia with their food, and apparently become immune to the poison. The immunity can be demonstrated by injecting *Carcinus* with a lethal dose of nematocyst poison along with a large amount of hemolymph from a hermit crab symbiont. The hemolymph protects the crab from the effects of the venom. While hermit crabs covered by actinians may not be attacked by an octopus in an aquarium, they may be attacked by a shark (*Squalus*), which rapidly bites off the parts of *Pagurus bernhardus* and *P. prideauxi* that extend from the snailshell.

The hermit crab *Pagurus* has been observed to insert its chela deep into the gastral cavity of the sea anemone *Calliactis* to fish out a recently swallowed piece of meat. The crabs *Dardanus arrosor* and *D. callidus,* by special patterns of tapping and stroking, cause an anemone to loosen its hold on the substrate. After placing the anemone on its gastropod shell the crab taps it in a different pattern and the anemone attaches.

LOCOMOTION.   Natantia swim with their pleopods, the limbs of each pair beating together, successive ones in metachronal motion. Skin divers have been able to determine an average speed of almost 1 km/hr in *Palaemon northropi*. The shrimps keep their balance by paddling the pleopods to the sides and steer up or down by flexing and stretching the abdomen.

Some swimming crabs, the Portunidae, have the distal two articles of the fifth pereopods flattened into swimming appendages, the proximal ones much shortened. The limb thus is only a short levered paddle (Fig. 13-32b). The Portunidae are skilled swimmers and can zigzag through the water. Even a slowly moving species of *Macropipus* observed, 3–5 cm wide, swims at 1 m/sec. Some hermit crabs (*Iridopagurus*) can swim for short periods.

The Natantia and other long-tailed decapods can shoot backward by suddenly folding the abdomen under with the tail fan spread, an escape reflex made possible by the rapidly conducting giant fibers. An *Orconectes* species observed by a succession of beats, may rapidly zigzag backwards 1–1.5 m. Members of the family Hippidae always swim backward.

Reptantia, which have the trunk longer than wide, Astacura, Palinura, most Anomura, the wide *Lithodes,* and some crabs (*Maja* and *Pisa*) normally walk forward.*

---

* From the literature it is not clear whether all crabs or only some run sideways. Dr. R. Bott and Prof. A. Panning gave the above information which was confirmed by moving pictures of the gait of *Carcinus* and *Macropodia* (Manton, 1953, *Symp. Soc. Exper. Biol.* 7:339–376). *Lithodes* was observed on the deck of a ship to run forward and backward but not sideways. Large numbers of soldier crabs (*Mictyris*), in formation, walk forward rather than sidways over tidal flats (Bliss, 1968).

The leg rhythm is most easily shown by indicating the pereopods on which the animal touches the ground for a moment, while the others are raised.

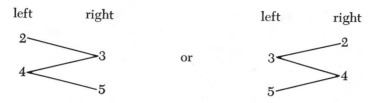

The gait thus resembles that of spiders and insects. The chelipeds usually do not take part in walking, and in hermit crabs the 4th and 5th pereopods are also not used. The rhythm is controlled alone by the appendage ganglia and is changed if a limb is autotomized.

All decapods can run forward. But the many Brachyura in which the cephalothorax is as wide as long and that have large chelipeds, run sideways when moving rapidly.

At times then, the gait is the same as that of other arthropods; when limbs 2 and 4 pull by flexing, limbs 3 and 5 push by stretching. But often it differs. If the circumesophageal connectives are cut, the decapod can run forward but not to the side, indicating that lateral movement is controlled by the supraesophageal ganglion. An *Ocypode* 4 cm wide can run sideways 1.6 m/sec on land; *Carcinus maenas*, 4.5 cm wide, 1 m/sec under water.

SENSES. MECHANORECEPTORS. *1.* Cuticular sensillae. Simple innervated setae, cuticular touch sensillae, are present but have been little studied. Groups of setae distributed over the whole body of the lobster *Homarus gammarus* have sensory neurons that respond to air and water currents and pressure waves. These organs consist of a longitudinal oval sclerite, jointed in the floor of a cuticular pit and bearing setae in a fanlike arrangement.

Certain sclerites have a thick chitinous rod in the middle of the group of setae. In *Homarus gammarus* the rod is bent by water currents of 0.25 cm/sec stimulating the three to five sensory cells with a spike potential. Perhaps this constitutes a sense of touch for distant objects. Similar pit sensory organs have been found in *Procambarus*.

In the lobster pit organs that lack the chitinized rod are stimulated by pressure waves. The stimulus (pressure of 0.4 dyn/cm^2) must touch the side of the setal fan. Experimentally, stimuli were provided by dipping an object into the aquarium, touching an elastic closure membrane on the aquarium surface, and by the passing of a fish.

The effect of water pressure has been investigated only in species of *Galathea*, *Macropipus*, and *Carcinus*. With increase of hydrostatic pressure all three left the floor of the vessel and surfaced. It appears that they sense the pressure. The receptors have not been located.

2. Statocysts. The statocyst on the base of each first antenna, like the utriculus section of the vertebrate inner ear, is an organ of awareness of rotation and position. A cuticle-lined, liquid-filled sphere, it develops from an ectodermal invagination that never completely closed. It contains a statolith and two different types of sensory hairs which have been examined in *Homarus, Carcinus,* and *Maja.* Some reach the statolith, and some are out of reach of the statolith. The sensory cells send spontaneous impulses and control muscle tone.

If a decapod turns around its longitudinal axis, setae touched by the statolith on each side of the body are bent differently and the sensory cells change the level of their tonic impulses. The response will be a tendency to use the legs to turn back into normal position. The eyestalks, even if the eyes are extirpated, take part in the movement, there being a linear relationship between the angle of their turning and the angle of statolith setae, an adaptation that preserves the field of vision.

Proof of the role of the statolith hairs can be provided experimentally. If the hairs in a decapod in horizontal position are bent on one side with a fine wire, the animal immediately adjusts its position to correct the apparent change in position.

A molting *Crangon crangon* uses its chelae to place sand grains in the empty statocyst to form a new statolith. If, instead of sand grains, only nickel particles are present, the metal is used for the statolith. If later a magnet is used to displace the metallic statoliths dorsally, *Crangon* will turn its body by 180°.

It can be shown in *Crangon* that the statocyst organs not only receive information about position but also function as a control. As the shrimp swims down by moving the body axis, the statocyst organs are stimulated, and transmit action potentials in the afferent nerves to the central nervous system resulting in efferent activation of the locomotor organs. If the statocyst is extirpated on one side, so that the stimulus is only of half strength, the downward path becomes steep: with a single statocyst the "desired" path is obtained with a much larger angle of descent. If both statocysts are removed it is impossible for the shrimp to maintain a steady direction for any length of time.

The sensory hairs beyond the reach of the statolith signal body turns also those in the horizontal plane. The stimulus, bending of the sensory hairs, is caused by the relative inertia of the statocyst liquid. These organs respond only to change of speed in turning, and are very directional, their tonic impulse frequency being increased by bending posteriorly or ventrally, decreased by bending in the opposite direction.

Vibration receptors have been demonstrated in the American lobster by recording the action potentials of the statocyst nerve. But the statocyst organs respond only to vibrations of the substrate, not to water or air vibration. The capability of the vibration sense is known in *Uca.* The males emerge from cavities 30 cm deep if a crab of the same size quietly moves by at the ground surface. They remain in their burrows if the passing crab runs or is pulled on a thread. Also species of *Uca* observed respond to drumming of the large chela. The receptors are not known.

3. Chordotonal organs. These proprioceptive organs are found especially in the hinges of limbs. The organs of the propodus-dactyl hinge of the Brachyura studied, are very sensitive to vibrations up to 1000 Hz. The adequate stimulus is the stretching of the connective tissues within which the organs are positioned. The connective tissue tendons extend from a muscle across the hinge to the cuticle of the next leg article. Flexing or extending of the articles cause the sense cells to respond with phasic impulses (producing spike potentials when stimulated), quiescence of the hinge stimulates tonic receptors (producing continuous nerve impulses). The structures (scolopidia) that form the organ resemble those of insects, as many as 40 lying within the connective tissue. Each scolopidium consists of two sensory cells the receptors of which together penetrate an individual scolopale cell, in the cuticular hollow rod that lies in the longitudinal axis of the tendon.

4. Stretch receptors. These organs send signals to the ventral nerve cord on the movement of the articulation between the somites of the trunk. In the abdomen they are in metameric arrangement (but absent in the abdomen of Brachyura). They generally consist of a multipolar sensory cell, the many dendrites of which distribute themselves on a specific thin bundle of dorsal longitudinal muscles, innervated by a motor axon. The dendrites are connected with one or two efferent inhibitory nerves. An adequate stimulation arises from the stretching of the muscle bundle that results when the following trunk segment is bent ventrally. Similar organs are found in the body articulation of the appendages.

5. Specific auditory organs have not been found; however a spiny lobster held suspended in water responded to artificially produced sounds that did not vibrate the substrate. It responded by stridulation even when its statocysts had been removed.

CHEMORECEPTORS. Chemosensory organs (setae on the first antennae, chelae or tarsi of pereopods, also perhaps in the branchial chamber wall) can be separated from those as yet incompletely studied sense organs of the mouthparts. These organs respond to diverse substances. *Carcinus maenas* running over moist filter paper will take to the mouthparts a piece that has been dipped in dilute acetic acid but then rejects it. While the distal articles of the pereopods and mouthparts bear chemosensory organs for contact, the esthetascs of the first antennae serve as sense organs for the scrutiny of distant objects. Their capability is increased by rheotactic behavior of the animal and by the water movements, due to the constant motion of the maxillipeds, sometimes also of the pleopods and the respiratory currents. The chemosensory importance of the inspiratory current has been noted for *Pagurus*. This current, aided by the scaphognathites, sucks in water from greater distances and various directions, producing olfactory channels that bring in information from the surroundings. The water is examined within the respiratory chamber by sensory pores, the excurrent water by the first antennae. As in many mammals and aquatic snails, the olfactory and respiratory currents are closely connected. This relationship can be demonstrated by suspending an eosin-filled tube in an aquarium. The color runs out, forming a red disc on the floor, and this dye slowly diffuses in various directions except for the sector toward which a hermit crab turns its back. Here a colored streak moves toward the crab and enters near the posterior edge of the carapace. If the dye has been combined with a positive attractant such as the scent of meat, the hermit crab will search the surroundings of the colored streak. In summer a hidden piece of food 20–35 cm away is discovered in 6 min, 40 cm away in 10 min, 50 cm away in 14 min. Symbiotic sea anemones are noticed from a distance of 35 cm.

Compared to marine crabs the esthetascs of *Coenobita* have a different dendrite arrangement and rather than hairs with a large surface, are pegs resembling the olfactory pegs of insect antennae, presumbably an adaptation to terrestrial habits (Ghiradella *et al.* 1968).

Electrophysiological studies have only been made of the sensory hairs of the distal articles of the pereopods of some species of *Cambarus, Carcinus* and *Cancer*. These hairs are not stimulated by moistening with sugars, alcohol or proteins but are stimulated by contact with certain amino acids, especially *l*-glutamic acid (threshold value 5×10^{-5}). *Cancer antennarius* is most sensitive to α amino acids. Amino acids, of course, are more soluble than proteins and diffuse easier. Signals could also be detected from the first antennae of *Panulirus argus* after stimulation with glutamic acid, probably coming from esthetascs, sensory processes, each of which contains the dendrites of about 300 sensory cells.

PHOTORECEPTORS.   Color vision has been demonstrated by optomotor experiments in *Crangon, Palaemon,* hermit crabs and *Carcinus* and has been assumed for *Hippolyte,* which matches itself to the substrate. The species of *Ocypode* studied orients its body axis and direction of movement at a certain angle toward the plane of vibration of polarized light, both under water and on land. Fifty flickers per second can be separated by some species of *Cambarus* and *Pagurus,* more than 150 by some *Ocypode* and *Uca.* Some beach crabs of the genera *Uca* and *Ocypode,* only 3–4 cm long, have as many as 12,000 ommatidia in each eye, a 30 cm long lobster, 14,000. The ommatidial angle in these beach crabs is about 2–3.6°, but vision of movement is well developed. The flight distance from a walking person is about 10–25 m.

The ability of such eyes to recognize females of the species has been investigated experimentally for the fiddler crab *Uca tangeri,* the males of which wave the large cheliped in three different intensities depending on the degree of excitation. The lowest intensity is used if they see any shape 5×3 cm moved steadily and horizontally 65–150 cm from their burrow. If the object approaches within 35–65 cm the next intensity of waving is reached only if the object has a downward bent projection on each side (a crab with at least one pair of limbs, or a three-dimensional model in the shape of an M). Approaching still closer to the male in the entrance of its burrow, a steady horizontal movement is no longer sufficient to release the greatest waving intensity; the crab or the model has to be moved rapidly to and fro.

Hermit crabs will also approach immovable objects. *Emerita* orients itself to the shore by sight.

The eyestalk movement of decapods does not result in the area of greatest resolution facing the object but occurs under the following conditions: if the animal is turned around its longitudinal axis it compensates for the shift in field of view. Moving objects are not only followed by the second antennae but also by the eyestalks, which always turn a little behind the stimulation, so that the movement across the ommatidia is almost stopped. In this way the central nervous system obtains information on the speed of the object viewed. (See also the discussion on mandibulate eyes, Chapter 16, Vol. 2.)

The collaboration of different sense organs increases the exactness of information passed on by each one. This is true for the movement of the eyestalk as well as the participation of various organs in the orientation in space: statocysts, proprioceptors, the eyes (which after extirpation of the statocyst take over the equilibrium function in prawns). (For greater detail about the physiology of the sense organs see Waterman, Vol. 2, 1961.)

BIOLUMINESCENCE. Luminescent organs are not common among decapods but are known from about 20 species, most belonging to the Natantia, and one a reptant, *Polycheles phosphoreus*. Species of *Sergestes* may have more than 160 organs distributed over the body, each with a reflector, most with a lens. All flash at the same time or rapidly one after another, greenish yellow in color. Luminous secretions are known from different pandalids. William Beebe observed repeatedly from his deep sea sphere how a large bathypelagic prawn (Oplophoridae), on hitting the window, gave off luminous material. This material was either uniformly bright or weakly luminescent with dozens of bright stars and spots. Only for a moment could the red prawn be seen, then it would disappear while the luminescent cloud lingered. Probably integumental glands produce the liquid.

STRIDULATION. A large number of crabs (e.g., *Portunus, Uca musica, Ocypode*, and *Matuta*), some Natantia (*Metapenaeopsis*) and Axiidae, and many terrestrial hermit crabs and some spiny lobsters have stridulatory organs consisting of a sclerotized ridge, against which a row of grains or parallel ridges is rubbed.

Crabs have a stridulatory area usually on the inner surface of the chela or the movable finger of the chelipeds which is rubbed against a ridge of the ischium of the same limb or against the ventral side of the body. Spiny lobsters have a row of parallel ridges on the median side of the 2nd article of the second antennae and rub these by bending the antennae down toward the back and up against the edge of the cephalic exoskeleton. Spiny lobsters have been observed to respond to stridulation by others of the same species; an animal in hiding responded to the stridulation of a *Palinurus elephas* caught by a diver, and an excited *P. elephas* responded to sound produced by rubbing the finger against moist glass. Both times the sound could not have been transmitted by floor vibrations, but only through the water. Spiny lobsters (*Panulirus argus*) produce a loud rattle, lasting 0.1 sec at 0.04–9 kHz, which can be picked up by a hydrophone submerged in the aquarium. They also make a slower rattle when caught by a diver lasting ¼ sec and consisting of 5–6 tones, 0.5–3.3 kHz. The rattling is made only in daylight when the crustaceans sit in the corners of their containers, otherwise showing no activity. Sounds were heard 76 times within 1.5 hr in a tank containing 24–36 animals. At night when the animals ran about *P. argus* was silent (Moulton, 1957). Animals recorded in nature produce sounds mainly at night (Hazlett and Winn, 1962).

Ghost crabs, *Ocypode*, produce cricketlike or frog (*Rana*)-like calls. Stridulation was observed when one animal was forced into a cavity already inhabited by another. The explosive sound of *Alpheus* is described in the section on protection above and humming of *Homarus americanus* is described below in the systematic section. Defending individuals of *Coenobita clypeatus* stridulate loudly during shell fights.

FEEDING.   Those feeding on large particles are not specialized but take whatever they find. Predators are also scavengers and often take plant parts. Plants may predominate in food of fresh water and terrestrial decapods (*Cambarus*, Coenobitidae, *Ocypode*, *Eriocheir*, *Potamon*, and *Cardisoma*). In the laboratory *Gecarcinus lateralis* needs to be fed eggshells to harden its cuticle after a molt (Copeland, 1968). Portunidae are scavengers and predators.

Most species handle the prey with the chelae. Some, for instance the large spider crab *Loxorhynchus grandis* (Carapace to 26 cm long, 21 cm wide), use their chelipeds to tear living *Octopus* and large starfish. Lobsters crack mollusk shells and *Carcinus*. Even the small chelae of the second and third pereopods assist in pulling prey apart. Pieces are handed from the chelipeds to the third maxillipeds or the maxillipeds reach out for them. Crayfish pass food to the mouthparts with one of the chelipeds. The ischia of the third maxillipeds open laterally like doors, the mouthparts separate, and the food is taken between the mandibles (Fig. 1-11). The width of the third maxillipeds obscures observation of further handling of food.

Large pieces of meat have been found between the strong mandibles of *Carcinus* and *Pagurus*. The pieces of meat hanging out from the mandibles is pressed firmly between the ischia of the second and third maxillipeds. With a strong ventral pull, they pull pieces loose. The mandibles then open and their palps push the meat into the pharynx by running along the biting edges. The medially directed setae of both pairs of maxillae as well as those of the first maxilliped form a grate in front of and below the mandibles, preventing loss of the food. Then the second and third maxillipeds again push the piece of meat torn loose between the mandibles and the sequence is repeated. There is no real chewing. The chewing stomach thus does not contain small particles but only strands of meat. Probably feeding is similar in other decapods that feel on large food items. *Galathea,* however, can use its mandibles to cut pieces out of a worm.

The caridean and stenopodidean shrimps which groom fishes, and *Callianassa* and *Thalassina,* which burrow into the substrate, pick up large as well as small particles.

PLANKTON FEEDERS. Microphagus decapods take plankton out of the water or take the surface layer of the substrate, minute detritus, dead animals, and microscopic animals and plants, and separate these from the sand particles by a system resembling shaking of sieves.

Emerita (Fig. 13-26a) strains the wave back-wash through a filter formed by the second antennae held as a V in front of the body. On one side the antennae have two longitudinal rows of long, dense, branched filtering setae. *Emerita* digs into the sand backwards at an angle of 45°, the anterior facing the sea and when the breakers of the incoming wave run back down to the ocean, the second antennae are extended and the water passes through their "net," leaving strained plankton behind. Rapidly each antenna is beaten toward the mouthparts between the wide doorlike mera of the 3rd maxillipeds. The mouthparts below the third maxilliped pick up the material and take it to the pharynx. The exact method has not been reported. The mandibles are weak. The main food is probably phyto-plankton and fine detrital particles. Within the stomach diatoms. Radiolaria and Foraminifera have been found. Bacteria may also be used as food, indicating the fine mesh of the net (Efford, 1966). Some species of *Emerita,* using fanlike movements of the antennae, can collect plankton in quiet waters.

Other decapods produce their own filter currents. The tube-dwelling *Upogebia pugettensis,* 8 cm long, uses the long setal fans on its pleopods to drive water into the U-shaped burrow. The first two pereopods have filtering fringes on their facing sides. Sitting near the burrow entrance; *Upogebia* holds the two pairs of limbs parallel, across the current. The filtrate is brushed out by the third maxillipeds, which in turn are combed by the second maxillipeds, from which particles reach the mouth. *Callianassa affinis* collects plankton and detritus in a similar manner. Symbiotic *Pinnixa,* which "sublet" in *Callianassa* burrows, use their maxillipeds as filters. Sometimes they live in tubes of echiurids (*Urechis*).

Pisidia longicornis collects plankton by extending the third maxillipeds laterally and fanning out the feathered setae of the last four endopodite articles, forming a concave net. The net is turned around an angle of 30° toward the middle, and is bent toward the mouth at the same time. During these movements of the third maxilliped, considerable amounts of water pass through, leaving the floating particles behind. As the third maxilliped bends toward the mouth the second maxilliped stretches out and combs each seta toward the mouthparts. In the meantime the third maxilliped of the other side has completed its collecting movement and is combed out. The coral inhabitant *Hapalocarcinus* feeds in a similar fashion.

In several genera of hermit crabs (*Diogenes* and *Paguristes*) the specialized feathery second antennae are swept through the water and the collected particles are transferred to the maxillipeds.

Feeding on the microscopic material of the substrate is done very differently. Fiddler crabs and their relatives grab pieces of mud with their chelipeds and take them to the mouthparts. *Pagurus bernhardus* sweeps the floor with its small cheliped and the Atyidae sweep the plant covering with their chelae. *Galathea* uses the third maxillipeds as a broom.

Galathea takes the collected particles to the mouthparts as does *Porcellana*. It brushes up diatoms, etc., with its third maxillipeds stretched out far to the side. Then as the third maxilliped is bent to the mouth, the second maxilliped stretches out and presses its distal article against the third maxillipeds. Then, as the second maxilliped is bent, its setae comb out the setae of the third toward the mouth.

While hermit crabs and *Galathea* may at times take up larger food particles, *Uca* and its relatives mainly strain mud. Feeding has been observed most carefully in Indian species of *Uca*. These small fiddler crabs, 1.3–2.6 cm in carapace width, take up moist mud while moving forward or in very shallow water 3–5 cm deep, and about twice each second take some mud with the left and the right tip of the chela and push it under the operculate basal parts of the third maxillipeds (Fig. 13-7). The space between these maxilliped plates and the mouth is filled with respiratory water coming from the branchial chamber running down the sides of the vertically held body. Within this space, by moving the first and second maxillipeds to and fro, the mud is liquified and, as in a filter sieve, the fine organic particles float and are caught in the setae of the second maxilliped. Other particles settle on the horizontal spoonlike setae, the cavity of which is directed anteriorly, or on the horizontal spatulate setae of the first maxilliped. How they get from here to the mouth has not been observed. The heavy, inorganic particles sink rapidly, bend the spoon-shaped setae down and pass to the base of the third maxilliped. Here the material remaining from 6 to 16 mud clumps is collected into a ball, taken up by the chelae and discarded. In India, where the feeding was observed, the discarded mud contains on the average 30–40 mg N per gram dry weight. Unfiltered mud containing granules of 0.5–0.3 mm averages 10 mg less N for each gram dry weight. The N content of the filtrate corresponds to 63 mg of protein. The material collected by filtering may fill the stomach within 12–40 min. Corresponding to the size of the particles, there is no gastric mill but in its place parallel, dentate or comblike ridges. Except when the animal is sitting in shallow water feeding is combined with respiration.

The same feeding method is used by other mud crabs, including *Dotilla* and *Scopimera,* and at times by some ocypodids.

Lopholithodes and some hermit crabs, which also catch larger prey, do not place the substrate material directly into the mouthparts but stir it up with the chelae. *Dardanus venosus* and *Petrochirus diogenes* use both the chelae and the walking legs to stir up the substrate. *Dardanus arrosor* uses the third maxillipeds and then filters the organic material from the suspension. For filtering *Lopho-lithodes* uses its maxillipeds, which are covered by feathered setae.

Pinnotheres (Fig. 13-34c) lives symbiotically in clams and uses its setose chelae to scrape the ventral edge of the mussel's gills, carrying the mucus containing plankton to its own mouth.

EXCRETION. In the water surrounding decapods are found ammonia compounds, amino nitrogen and small quantities of urea and uric acid. The most common execretory substance, ammonia, is found only in small quantities in urine, if at all, and is excreted even if the nephropore is closed; the gills act as an excretory organ.

The land crab *Cardisoma guanhumi* may accumulate 0.2–15.9% of its dry weight as uric acid, perhaps discarded at times of molt. The crabs are surprisingly tolerant of NH_3 in their blood and stomach, storing 2–3 days supply and eliminating it when entering the water (Gifford, 1968) presumably through gills.

Emptying of the antennal glands results from the pressure of the fluid when the pore is opened; it is accomplished without the aid of muscles. The urine excreted by a marine decapod equals 0.1–6% of the body weight per day. Experimental closure of the nephropore resulted in death of the experimental animals in 1–2 days. The function of the gland is incompletely known. Most recent research has been centered around osmoregulation rather than N excretion. Antennal glands separated from living crayfish, belonging to the genus *Orconectes*, have been examined by micropunctures of the various parts: sacculus, labyrinth (an egg-shaped knot of thin green interconnecting channels), the white winding nephroduct, and the bladder. By ultrafiltration, substances with molecular weights up to 20,000 reach the lumen of the sacculus and form the primary urine. Further evidence for this interpretation are the structure of the wall of the sacculus (Fig. 1-19), and also entrance of radioactive inulin (mol wt 5000) into the sacculus. This resembles the flow from vertebrate glomeruli into Bowman's capsule, where the pressure in the blood vessels is known to be sufficient for ultrafiltration. In crustaceans the blood pressures within the sacculus and adjoining sinus are not known. Polysaccharides, with mol wt to 20,000, behave like inulin; those with molecular weight from 60,000–90,000 are excreted only in small part. Human serum albumin (67,000 mol wt) injected into the blood enters the lumen of the sacculus, as can be demonstrated by marking it with fluorescein. However, injected mammal globulin (about 180,000 mol wt) does not enter urine. That water resorption begins in the proximal parts of the antennal glands is indicated by the greater concentration of inulin in the urine than in the blood, which is known not to be resorbed. Na^+ ions also are resorbed in the labyrinth. Their concentration and that of K^+ ions sinks most rapidly in the distal part of the gland especially in the sacculus. The same has been stated for the chloride content of a species of *Orconectes*, while in a species of *Astacus* it sinks during the passage through the white nephroduct. The strong resorption, of course, is of great importance in the osmoregulation of fresh water crustaceans. Some species of *Homarus* lack the white nephroduct (Riegel, 1963, 1965; Kirschner and Wagner, 1965).

OSMOREGULATION.   The constancy of the hemolymph is easiest to maintain in marine animals in which the blood and surrounding media are more or less isotonic. But even in stenohaline decapods there has to be regulation. The ion content of the blood in decapods does not correspond to that of the ocean. Its concentration is controlled by the antennal glands and perhaps gills. The excreted urine, unlike the blood, often contains a significantly greater concentration of Mg and sulfates, and a lower concentration of K and Ca.

Stenohaline species are tied to ocean water with salt concentration of 34–38‰. In brackish water the osmotic concentration of their blood and water is equal, leading to death of the stenohaline individual. Euryhaline species can penetrate into the littoral, brackish water and even into fresh water. Individuals of *Carcinus maenas* can live in water of different salt concentrations: 35‰ in the North Sea or 8‰ in the Baltic at Darss. *Eriocheir* can live for years in freshwater. This is possible through intense osmoregulation, by which the osmotic value of the blood remains independent of the surrounding water, as in all euryhalines. Such intense osmoregulation depends on the combination of genetically fixed reduction of surface permeability of euryhalines, regulation of the volume entering water by the antennal glands, uptake of salts through the gills, and regulation of osmotic pressures of tissues. Some factors and some examples of osmoregulation are discussed below.

Table 13-1 shows changes in the osmotic values of the blood of crabs from concentrated Mediterranean water (of about 42‰ salt = Δ—2.29°C) after the water was diluted with rainwater to 24‰ (Δ—1.33°C).

A comparison of the permeability of the cuticle of the stenohaline *Cancer pagurus* and the euryhaline *Carcinus maenas* can be made by attaching to each a vessel that opens below into the surface of the carapace. Fifty hours after filling the vessels with 0.75% solution of NaI the blood of *Cancer* has 48 μg/cm^3 I, that of *Carcinus* only 9 μg/cm^3 I.

The urine production of *Carcinus maenas* in normal ocean water is 3.6% of the body weight per day, with little Na loss. But in 40% ocean water, urine production rises to 30% of the body weight daily and the urine contains 20% of the total Na excreted, the remainder being lost through the body surface, prob-

Table 13-1. Comparison of osmotic pressure of stenohaline and euryhaline Crustaceans (After Schwabe)

Lowering of freezing point	Δ start of experiment, °C	After 12 hr, °C	After 36 hr, °C
Maja verrucosa, stenohaline	−2.29	−1.38	−1.33
Carcinus mediterraneus, euryhaline	−2.29	−1.89	−1.82

ably the gills. The gills may replace the salt against the osmotic pressure. Isolated gills of *Eriocheir sinensis* lying in running freshwater took up N+ from NaCl solution of 8 mM/liter (Shaw, 1961).

If an *Eriocheir sinensis* from freshwater is transferred to ocean water, there follows an increase in the concentration of amino acids in its muscle tissues, and in turn an increase in the osmotic value of the blood. In *Carcinus* the concentration of amino acids in the blood approaches 60% of the osmotic value of muscle tissues.

In hyperosmotic euryhaline decapods (e.g., *Carcinus maenas*) the blood is hypertonic in brackish water, in normal seawater, isotonic.

As the urine is isotonic with the blood, the antennal glands regulate not the salts, but the fluid volume.

The hyper-hypoosmotic euryhalines are hypertonic in brackish water, hypotonic in ocean water. To this group belong the prawns of shallow waters (e.g., *Crangon crangon* and *Palaemonetes varians*). They also secrete urine that is isotonic with the blood.

In land crabs which spend many hours daily out of water, the blood is especially constant in osmotic values. The osmotic value of the blood expressed as per cent of normal sea water (34.3‰ salts) is indicated in Table 13-2.

Uca crenulata often lives in lagoons with high salt content. But the water in their burrows may have a concentration of only 17% ocean water, or may be completely fresh after rain. Salt and water are taken up by drinking through the stomach as well as by the gills. The regulation takes place in the gills and antennal glands, which save water by giving off urine that is hypertonic toward the blood. Also in *Ocypode* and *Gecarcinus* water is resorbed by the antennal gland when on land or in high salt concentration water.

The extremely hyperosmotic crayfish *Astacus pallipes* can take only traces of electrolytes from the water and is endangered by having water enter their bodies by losing salts. Experiments with Na^{24} show that *Astacus astacus* takes in and gives off equal amounts of Na+ (Bryan, 1960). The excretion of Na+ is 94% outside of the antennal gland despite low permeability of the integument. Only slight lowering of blood Na+ will increase its uptake. *Astacus pallipes* can still

Table 13–2. Osmotic value of blood of land crabs in various salt concentrations (After Gross)

	Ocean water concentration				Duration of experiment, hr
	50%	100%	150%	168%	
Ocypode ceratophthalmus	79.9	87.6	—	93.2	24
Uca cernulata	84.6	90.7	94.7	—	48
Gecarcinus lateralis	87.8	95.0	103	—	8–10

take up Na+ out of 1/10 normal Na concentration of freshwater (Na+ —0.03 mM/liter). The antennal gland resorbs salts so well that the urine has only 1/10 the osmotic value of the blood. *Astacus astacus* can survive for weeks in brackish water but excretes smaller amounts of more concentrated urine (Table 13-3).

Completely euryhaline crabs are rare. *Eriocheir sinensis* can live in salt-, brackish, and freshwater. Their excellent osmoregulation depends on the ability of the gills to take up ions against solution gradients, low permeability, and high tolerance of the tissues against osmotic pressure. Daily output of urine, which is isotonic with the blood in freshwater, is 3.6% of the body weight.

Classification

Starred (*) names are those on the *Official List of Generic Names in Zoology*, on the *Official List of Family Names in Zoology* or on the *Official List of Specific Names in Zoology*. The spelling and application of these names cannot be changed.

SUBORDER NATANTIA

More than 2000 species are included in the suborder Natantia. The body is almost always laterally compressed, the integument only slightly mineralized. The rostrum is rarely absent. The second antenna has a large scale, its exopodite. Pleopods are swimmerets. The first abdominal somite is not shorter than the others. The pereopods are thin and rarely have exopodites. These are swimming crustaceans with a number of primitive characters. A key to the caridean and stenopodidean shrimp families and genera is found in Holthuis (1955).

Section Penaeidea

The first to third pereopods are of equal size and are almost always chelate. The gills are dendrobranchia. The abdomen is only slightly bent ventrally. The eggs, abandoned in the ocean (except those of *Lucifer*), probably develop by total cleavage in almost all species and hatch as nauplii, metanauplii, or protozoeae. The anamorphic development is regular (Fig. 2–17). Some species are large. The whole group appears primitive.

Table 13–3. Changes is osmotic pressure (lowering of freezing point) of the fresh water crayfish *Astacus* and the euryhaline *Eriocheir* (After Herrmann and Scholles)

Freezing point depression	Fresh water, Δ°C	Blood, Δ°C	Urine, Δ°C	Ocean water, Δ°C	Blood, Δ°C	Urine, Δ°C
Astacus	−0.02	−0.82	−0.09	19‰ −1.02	−1.14	−0.83
Eriocheir	−0.02	−1.2	−1.23	100‰ −1.85	−1.85	−1.91

Penaeidae.* All 5 pereopods are well developed, often with exopodites (Fig. 13–17). They swim or lie on the sand. Many species are benthic. *Penaeus** is benthic, littoral and sublittoral, some species live in brackish and in freshwater. The omnivorous white shrimp, *P. setiferus** (Fig. 13–17) occurs along the east coast of North America from North Carolina to the Gulf of Mexico; it grows up to 18 cm long. The young live in bays in shallow water with low salt content and as they approach maturity move out into the ocean into areas of higher salt concentration and depth. The annual harvest for human consumption of this and related species is 100,000 tons. Others are *Plesiopenaeus** *edwardsianus*, up to 31 cm, *Trachypenaeus** and *Penaeopsis.**

Sergestidae.* At least the fifth but also often the fourth pereopod is, absent or vestigial. Although pelagic and bathypelagic, they may enter brackish waters. The larvae are spinose. *Sergestes,** to 11 cm long, may have luminous organs, *Lucifer,** 1 cm long, includes elongate species, the head one-quarter its total length; they lack gills and luminescent organs. *Lucifer faxoni*, to 100 m depth, is found along the North American and West African Atlantic coasts.

Section Caridea

This group is also called Eucyphidea (Fig. 1–4). The first and second pereopods are chelate or subchelate while the third never has chelae. The epimeres of the first pleomere are covered by the second. The abdomen is usually bent strongly from the third pleomere. The gills are phyllobranchiae. The eggs are cemented to the pleopods and are carried until the zoeae hatch. The anamorphic development is irregular (Fig. 2–19). As in other groups only the most important families are discussed.

Pasiphaeoidea. Only the family Pasiphaeidae* is included. The rostrum is small or absent (Fig. 13–18). The mandibular palp is absent or with one or two articles. The first two pairs of pereopods have slender chelae with pectinate fingers. The fourth is usually smallest. *Pasiphaea** (Fig. 13–18) is pelagic or bathypelagic. *Leptochela**

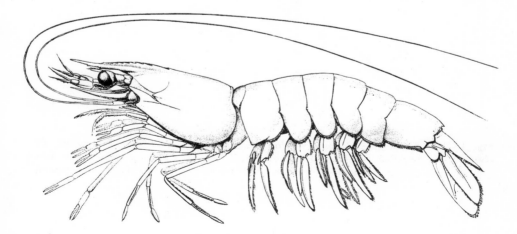

Fig. 13–17. Penaeid shrimp, *Penaeus setiferus*; about 12 cm long. (After Rathbun from Williams.)

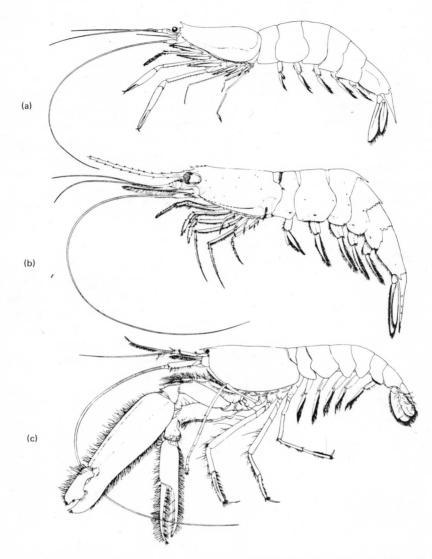

Fig. 13-18.   Caridean shrimps: (a) *Pasiphaea tarda*, Pasiphaeidae; 7 cm long. (From Kemp.) (b) *Systellaspis debilis*, Oplophoridae; 7.7 cm long with rostrum. (From Kemp.) (c) *Alpheus glaber*, Alpheidae; 4 cm long. (From Kemp.) (d) *Pandalus propinguus*,

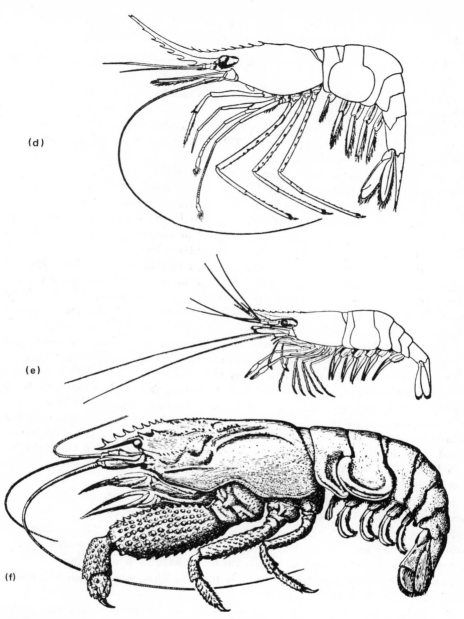

(d)

(e)

(f)

Pandalidae; 8 cm long. (From Smith.) (e) *Palaemonetes intermedius*, Palaemonidae; 3 cm long. (From Rathbun.) (f) *Atya crassa*, Atyidae; 14.5 cm long. (After Bouvier from Balss.)

serratorbita, 21 mm long, is found in coastal waters from the Carolinas to the West Indies.

Oplophoroidea. This group includes the three families Atyidae, Oplophoridae and Nematocarcinidae.

Atyidae.* These freshwater prawns are found almost exclusively in warmer countries, with two genera in brackish water. The wide chelate fingers of the first and second pereopods cannot be used for cutting (Fig. 13–18f). Above their tip usually insert fans of setae whose flat surface is parallel to the article of the movable finger. These setal fans are absent from the chelipeds of the West Indian *Xiphocaris.** If the chela is opened the fan spreads; if closed, the setae converge. Some species may reach 20 cm in length. All are detritus feeders. Many hatch as zoea larvae, others undergo direct development within the egg. *Atyaephyra* desmarestii* is 1.6–2.7 cm long excluding the 0.5 cm long rostrum. It is found in southern Europe, North Africa and Asia Minor and in recent decades has spread from France into Belgium, the Netherlands, and northwestern Germany. It lives in schools in aquatic vegetation along the banks especially among *Potamogeton* and *Myriophyllum.* The transparent animals are hard to see. With a beat of the abdomen they can shoot back 50 cm to disappear among the vegetation. In fall and winter large groups are found among decaying vegetation. With the four chelate limbs individuals grasp such growth as green algae, chrysomonadids, diatoms, and ciliates, and the stomach content indicates that winter food consists of rotting plants or sometimes dead members of the species. Underneath the setal fan there are several sharp, hooked claws used for tearing and perhaps combat. In Central Europe females carry eggs from May to July. Six larval stages have been observed in North Africa. *Syncaris,** 2–5 cm, has been found in California streams, *Palaemonias* ganteri,* 1–2 cm, in waters of Mammoth Cave, Kentucky. *Xiphocaris** is found in fresh waters of the West Indies, *Paratya** in the Far East, Australia, New Caledonia and New Zealand. Three endemic genera with 13 species occur in Lake Tanganyika in Africa.

Oplophoridae.* These are mostly bathypelagic. *Systellaspis** (Fig. 13–18b) has luminescent organs. *Acanthephyra,** up to 18 cm long, is bathypelagic.

Nematocarcinidae.* *Nematocarcinus** is found in more than 200 m depth. The second antennae are twice as long as the body and the long three posterior pairs of pereopods are almost as long, an adaptation for walking on a soft bottom.

Pandaloidea. This superfamily contains three families, the Pandalidae, Thalassocarididae,* and the Physetocarididae.

Pandalidae. The chelae of the first pereopod are microscopic or absent and the carpus of the second chelipeds is two to multiarticulate. *Pandalus,** (Fig. 13–18d) has a large proportion of each population in some species protandric hermaphrodites; *P. montagui,** 6–7 cm long, occurs in moderately deep water of the North Atlantic, feeding on gammarids, mysidaceans, cumaceans, amphipods, polychaetes, and detritus. At night it is colorless or yellowish, at daytime, red or green spotted or banded. *P. danae* is the "coonstriped" shrimp from Alaska to San Francisco, usually in deep water, sometimes in tide pools. Young are all males, becoming females when about 2 years old. *Pandalina** is another genus. *Pantomus*parvulus,** 3 cm long is found from the Carolinas to the West Indies at depth of 150–500 m.

Alpheoidea. The second pereopods are almost always slender, with the carpus divided, while the first is heavier.

Alpheidae.° Snapping Shrimp. The chela of the first pereopod is strong, usually larger on one side than the other. *Athanas*° is found in the eastern Atlantic and Indo-West Pacific areas. *Alpheus,*°† pistol shrimp (Fig. 13–18c), with some species reaching 3.5 cm, may have the chela on one side 2.5 cm long, the entire pereopod is as long as the body. The hand is so large that in molting it is not pulled out of the skin, but the exuvia peels off the chela. If the large chela is lost, *A. californiensis* molts in 5 weeks and has the other smaller chela transformed into a "pistol" although the animal has not grown. *Alpheus californiensis,* 5 cm long, like some other *Alpheus* species, "shoots" with a loud pop if a glass rod is brought near the entrance of its burrow. The noise results when the shrimp snaps the fingers of the chelae together. Pairs of *Alpheus* live in a single burrow. In search of food, the Californian species extends its long antennae out of the burrow. When a small fish approaches, the shrimp crawls slowly out of its burrow and "shoots." The water pressure wave stuns the fish, which is grasped with the chelae and pulled into the burrow to be torn apart, but shared with other pistol shrimp. Snapping while held in the hand produces a stinging sensation. The life span of the shrimp is 8 years. Other species and species of related genera live in large numbers in the bottom, some (especially some species of the genus *Synalpheus*) in sponges. *Alpheus frontalis,* found in the Indo-West Pacific live under porous stones and dead algal blocks. Often pairs observed at the Gulf of Aquaba at Eilat, inhabit burrows 1.5–2 cm in diameter, 15–40 cm long. The burrow is made by glueing together algae, especially *Oscillaria* but also filiform *Ceramium*. The algae continue to grow, as much as 5 cm within 3 weeks, and serve as a source of food, the crab cutting off pieces from the inner wall with its second or third maxillipeds. Tears 4 cm long are cemented together within 1.5 min by the last maxillipeds with a special cement. Only algal remains have been found in feces. In the Red Sea and also at Palau, species of *Alpheus* have been observed to be mutual symbionts with two species of gobies. The fishes rest and hide in the burrows of *Alpheus* and communicate visually perceived warnings through their movements to the prawns, transmitted via antennal contact or water movements. The fishes do not object to the contact with the crustaceans; the prawns depend on the gobies for an extra measure of safety (Magnus, 1967).

Hippolytidae.° The chelae of the first and second pereopods are about equal on both sides, never very large. *Hippolyte*° *varians,*° 3 cm long, of the European Atlantic coasts, lives in large schools in littoral algae to 60 m depth, rarely to 240 m. At daytime it rests on vegetation, at night swims about, the heartbeat increasing from 150/min at daytime to 240/min at night, and the animal becomes translucent blueish or greenish, the red pigment contracting, the blue expanding. Collections indicated that green individuals are collected from green *Ulva,* setose brown individuals from brown algae, red individuals from red algae. In experiments most individuals preferred a uniformly colored background. The color change took place if there was no choice of background. Young *Hippolyte,* if placed on a different colored substrate, changed color within 1–2 days; old individuals change within a week or do not adapt. In green

† Called *Crangon* in older American literature, a name now invalid for this group; *Opinion 334, Int. Comm. Zool. Nomenclature.*

individuals subjected to weak illumination the red pigment spreads and they become brown. Members of this species may be protandric hermaphrodites. Other species are found on North American coasts. *Hippolyte californiensis,* 38 mm long, is found on the North American Pacific coast in mud among ell grass. *Hippolysmata° grabhami* "cleans" fish. *Spirontocaris°* has representatives in the northern Atlantic and northern Pacific.

Processidae.° The rostrum lies parallel to the dorsal carapace surface and its tip usually asymmetrically bifid. The first limb pair is usually asymmetrically chelate, with a simple claw on one side. *Processa°* is found in practically all tropical and temperate seas of the world to depths of 360 m.

Palaemonoidea. This superfamily contains the most common littoral marine-, brackish-, and freshwater prawns.

Palaemonidae.° The first pereopods have small chelae (Fig. 13–18e). *Palaemonetes° varians,°* of European Atlantic coasts, averages 3.7 cm, rarely to 6 cm long. They are active swimmers of shallow brackish water, almost colorless and completely transparent. Zoeae hatch from the 100–450 eggs. They feed mostly on *Neomysis vulgaris. Palaemonetes antennarius,* which is related, lays fewer and larger eggs and lives in freshwater in southern Europe and Turkey. Larvae with five-articled pairs of pereopods hatch from the eggs. Some North American species are found in subterranean waters of caves, others in fresh-, brackish-, and saltwater. Species of the freshwater genus *Macrobrachium,°* which has an hepatic spine on the lateral surface of the carapace, may reach 24 cm in length. *Palaemon° adspersus°* (=*P. squilla*) of the Baltic is sometimes harvested and canned.

The subfamily Pontoniinae has many commensals. *Typton°* is found in sponges and may feed occasionally on their host. *Paratypton°* forms galls in madreporarian corals. In the Indo-West Pacific area, the bivalve *Pinna* is inhabited by the watchman prawn *Anchistus custos.* Numerous individuals can be found within the bivalve, removing mucus with detritus from the gills of their host (Johnson and Liang, 1966). *Chernocaris placunae,* 12 mm long, off Singapore, filters out food particles from the mantle cavity of its host the bivalve *Placuna* (Johnson, 1967). Other genera are *Pontonia* and *Periclimenes.*

Crangonoidea. This superfamily contains two families.

Crangonidae.° The first pereopods have a strong subchela, stronger than the second. Shrimps of the species, *Crangon° crangon,°* (Fig. 1–4) are found along all European coasts; males average 4.5 cm in length, females 7–9 cm. In the Baltic, much smaller shrimps are found in huge numbers and are used for livestock feed as well as for human consumption. In winter they move to deeper waters, and in spring move back into the tidal zones; the optimal temperature seems to be 20°C, the optimal salt content 28‰. The animals remain buried in sand during the day, becoming active at night, when they feed on *Nereis, Corophium,* amphipods, mysids, small mollusks, algae, and detritus. With low tide the animals move into deeper water with the current, returning with high tide. *Crangon* moves to deeper, quiet waters for reproduction. In copulation the male turns the female on her back and, belly to belly, deposits two sperm packages next to the female gonopores. A few hours later the female deposits 3000–4000 eggs which are fertilized externally and are attached with mucus to the pleopods. Breeding occurs two to three times annually, once or twice during summer, and once in November. A female may reach the age of 3 years and produce 20,000 descendants. The species provides food for many kinds of fishes. In the North Sea larvae hatch in 4 weeks at 16°, in 3 months at 6°. Normally they molt 30–35 times. The five larval stages drift 5 weeks with the plankton. Young animals 5 mm long swim toward the coast into shallows to become bottom dwellers, reaching 25 mm in length. They become sexually mature within a year, the females 55 mm long, the males 40 mm. The American Atlantic and North Pacific species, *Crangon septemspinosa,* is also found on the bottom above sand from the low tide line to 100 m or deeper. It is believed that it has a life span of only 1 year and produces one brood a year; others think it lives 3 years. *Crangon* has been used as an experimental animal in the study of crustacean color change. *Pontophilus** is especially abundant in north temperate seas.

Glyphocrangonidae.* This family differs from Crangonidae by having the carpus of the second pereopods multiarticulate. The only genus *Glyphocrangon,** with more than 30 species, about 10 cm long, pantropical and subtropical at depths of 300–3800 m. The species are heavily built, strongly sculptured benthic forms.

Section Stenopodidea

The carapace is not laterally compressed. The epimeres of the first pleomeres more or less cover the second pleomeres laterally. The first to third pereopods are chelate, those of the third much longer than the others (Fig. 13–19). The gills are trichobranchiae. The eggs are carried on the pleopods.

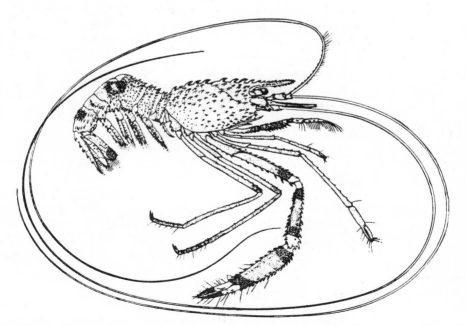

Fig. 13–19. Stenopodid shrimp, *Stenopus scutellatus,* **Stenopodidae; 3.5 cm long. (From Limbaugh et al.)**

Stenopodidae. *Spongicola venusta,* 2.2 cm long, live in pairs enclosed within the glass sponges, *Euplectella* and *Hyalonema. Stenopus* hispidus,* 3.5 cm long, from tropical waters, has been observed in aquaria to clean coral fish, *Dascyllus* and *Balistapus.* The prawn stood on an organ coral and the fish swam backward between its chelae; the prawn used its chelae to feel over the epidermis of the fish and once in a while brought the chelae to the mouthparts.

SUBORDER REPTANTIA

More than 6500 species are included. The body is usually dorsoventrally flattened; the integument, at least of the carapace, is usually strongly mineralized. The rostrum is often reduced or absent. Also the antennal scale is usually reduced or absent. The pleopods are not used primarily for swimming and are often reduced. The first pleomere is shorter than the following ones. The pereopods are strong, the first strongest with an especially large chela. All are bottom-dwellers that carry their eggs attached to the pleopods. Usually the larvae hatch at least as a zoea. There is an irregular anamorphic development.

Section Palinura, Spiny Lobsters

Included are about 130 species whose carapace is depressed with a strong lateral keel, or is cylindrical. The anterior border is usually wide and truncate, the rostrum always small or absent. The sides of the carapace are fused to the episome. The anterior four pereopods are usually almost cylindrical and have chelae only rarely but usually have subchelae. The abdomen is long, straight, and symmetrical with well-developed lateral keels and a wide tail fan. The exopodites of the uropods are not divided by a transverse seam (Fig. 13–20). There are many trichobranchiae. The Section Palinura and the next, Astacura, are often combined in the group Macrura.

Eryonoidea. The first article of the second antennae is not fused to the episome. First to fourth pereopods are chelate, the first being the largest. The carapace is depressed and wide, the telson triangular. Of postembryonal development, only mysis stages, at least 3.5 mm in length, are known; the younger stages are still unknown. The eggs have a diameter of only 0.6–1.5 mm. The bathypelagic larvae (500–5000 m depth), described under the name eryoneicus, have a large, spherical, spiny carapace. In younger ones the abdomen is short and narrow, and only in older ones, 7 cm or more in length, is there the impression of a distinct tagma. This ancient group was first known from fossils only.

Polychelidae.* These are depressed, blind deep-sea dwellers, mainly carrion feeders, that live buried in mud (Fig. 13–20a). *Willemoesia* is found at depths of 2000–4500 m, *Polycheles* typhlops, to 10 cm, in the Mediterranean at depths of 200–2000 m.

Scyllaroidea. The first article of the second antennae is fused to the episome. The first to fourth pereopods lack chelae, and the first is only rarely larger than the others. The telson is rectagular. The gills are trichobranchiae. The larvae pass through a phyllosoma stage.

Palinuridae,* Spiny Lobsters. The carapace is almost cylindrical. Most are very large, edible decapods. Some by making a vertical movement with the second antennae can produce a loud rattling noise. Divers have observed *Panulirus interruptus* to rattle while battling for territories, and others of the same species respond by withdrawing;

fish do not seem to react. *Palinurus* elephas**, the European spiny lobster, up to 45 cm long, may weigh as much as 8 kg. It is highly valued as food in the eastern Atlantic and Mediterranean. It lives on rocky substrate grown over by plants of the littoral and sublittoral and feeds mainly on bivalves and snails, carrion, and also on serpent stars *Ophiothrix*, up to 20–30 at a feeding. Other echinoderms are also used as food. Prey is located by the first antennae, probably chemically. The second antennae, longer than the body, are used to chase off other spiny lobsters and fish while feeding. *Panulirus* argus*, the West Indian Spiny Lobster, Sea Crawfish (Fig. 13–20b) to 45 cm and longer, are found from Bermuda and North Carolina to Brazil harvested in southern Florida and the West Indies. They are found on reefs and among rocks at depths of 2–10 or 25 m at most. Mating season is from March through July, and spawning from March through June in Florida. In mating the male places a spermatophore on the sternum of the usually hard female. The female turns on her back for spawning and the eggs pass over the spermatophore. The eggs hatch in a month and the female may mate and spawn again. A 23 cm long female produces half a million eggs, a 38 cm female 2.5 million, fewer at the second spawning. The last phyllosoma larval stage changes after about 6 months into a postlarva and grows to about 5 cm when a year old. Thereafter it grows about 2.5–4 cm a year at first. In Bermuda tagged individuals homed when transplanted. Most tagged individuals at Key West moved little, but some moved considerable distances. *Panulirus interruptus*, the California spiny lobster, weighing up to 5 kg, is found from southern California to Manzanillo, Mexico, and is of considerable economic importance. Omnivorous scavengers, they are at times caught in tide pools; in winter they live to a depth of 40 m among rocks and in crevices. They are caught in lobster pots baited with decayed fish or abalone at 4–15 m, in winter to 40 m in depth. About 50,000 eggs are attached to the swimmerets and phyllosoma larvae hatch in about 70 days. The larvae swim for several months before settling when about 23 mm long. These spiny lobsters make a noise when molested and ward off predators with their large spiny antennae. Sheephead, jewfish, and octopus are predators. *Jasus,** up to 60 cm long, lives in south temperate waters. *Jasus* edwardsii* is a "crayfish" of economic importance to New Zealand. The eggs hatch as small round naupliosoma after being carried by the female for 6 weeks. The naupliosoma swims with the setose second antennae, the first three pairs of thoracopods are folded ventrally. Within minutes after birth or at most an hour, the naupliosoma molts to a phyllosoma. The first phyllosoma molts after 19–25 days and the total larval duration is expected to be 9–12 months. The phyllosomas are benthic, feeding on polychaetes. Phyllosomas of Australian *Jasus* have been observed to feed on hydromedusae (Batham, 1967).

Scyllaridae,* Spanish Lobsters or Shovelnose Lobsters. The carapace is flattened, with a sharp lateral edge separating dorsal and ventral sides. The very short second antennae are strangely modified: they have a basal article fused to the epistome, three free stalk articles, the first of these formed by the fusion of the second and third articles of the peduncle, and one single larger flagellar article. The next to last stalk article and the flagellar articles are leaf-shaped, flat and very wide, toothed around the margin (Fig. 13–20c). Nothing is known about its adaptive significance and little is known about their life history. *Scyllarus** (Fig. 13–20c) lives in the upper littoral, often on coral reefs, as deep as 400 m; *S. arctus,** to 12 cm, is found on rocks in the

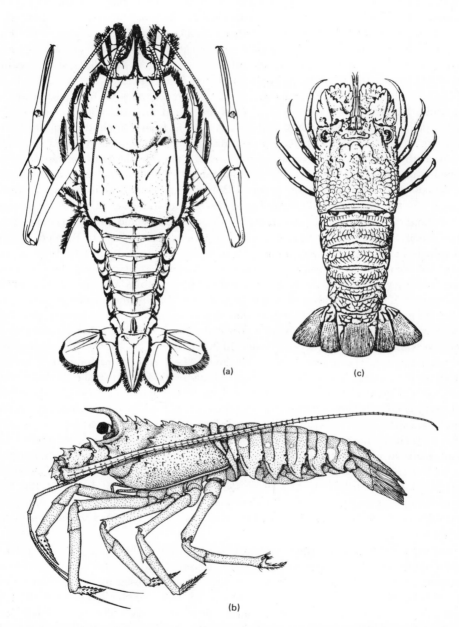

Fig. 13–20. Palinuran decapods: (a) *Polycheles sculptus*, Polychelidae; 14.5 cm long. (From Smith.) (b) Spiny lobster, *Panulirus argus*, Palinuridae; 45 cm long. (From Williams.) (c) Spanish lobster, *Scyllarus chacei*, Scyllaridae; 5.4 cm long. (From Bouvier.)

Mediterranean and eastern Atlantic. Its habits are unknown. It is edible. *Scyllarus chacei,* 5 cm, and S. *nearctus,* 5 cm, are found in the western Atlantic from North Carolina to Brazil. *Scyllarides*° *nodifer,* 12 cm, is found from North Carolina to Bermuda and Cuba at depths between 3 and 73 m. *Ibacus*° is another genus.

Section Astacura

Included are more than 700 species. The carapace is cylindrical, not fused to the epistome. The first, second, and third pereopods are chelate. The abdomen is straight, long, and symmetrical with well-developed, lateral, tergite edges and a wide tail fan. The distal parts of the uropod exopodite are separated by a transverse seam (Figs. 13–6, 13–21). There are many trichobranchiae. The animals hatch in the mysis stage or almost fully developed.

Nephropoidea. This superfamily always has the three anterior pairs of pereopods chelate whereas the third pair is never distinctly so in the Thalassinoidea.

Nephropidae° (=Homaridae), Lobsters. Lobsters have the last thoracic sternite fused to the one anterior. *Homarus*° has the chela on one side much larger than that of the other side and has blunt knobs on its cutting surface. *Homarus gammarus,*° the European lobster, 30–50 cm long, has a preference for rocky bottoms. Except for migrating to deeper water in winter, it prefers shallow water. After its nocturnal outing for food, it returns to its home crevice in rocks or under stones. It feeds on mollusks, which are opened with the cracking chelae, and also on carrion. Like spiny lobsters and other "long-tailed" decapods, a beat of the abdomen shoots the animal backward. In Heligoland eggs are deposited from late July until September. Females 25 cm long carry 8000 eggs on the pleopods; those 31 cm long carry 17,500 eggs; those 37 cm long, 32,000 eggs, each 1.7 mm in diameter. After 11–12 months, in July of the following year, the young hatch as beautiful blue mysis larvae, about 7 mm long, with yellow or red borders. The mysis feed on plankton and molt after 4–5 days. Two additional mysis stages follow. The last does not swim but lies on the bottom; its pereopod exopodites reduced. The postlarva, 1.5 cm long, has a lobster shape. During its first days it swims with the pleopods, later to stay longer and longer on the bottom; it molts eight times before winter, reaching a length of 3 cm. The following summer it molts about five times, growing to 6 cm; in the third summer it molts only four times, reaching 11.5 cm. The female matures in the 6th year, at a length of 23–25 cm, and thereafter molts only every other year. Molting takes 10–20 min, hardening 3–4 weeks. During the entire period the animal stays in its crevice. Old lobsters are covered with symbionts, including barnacles, and molt only rarely. *Homarus americanus,* American lobster (Fig. 13–21a) found off New England and the Canadian Atlantic coast, may reach a length of more than 60 cm and a weight of 22 kg; the average length of adults is 25 cm at a weight of 0.5–1 kg. The life span may be 50 years. American lobsters are nocturnal, feeding on mollusks, carrion, and vegetation. In Nova Scotia 8000 marked individuals were later caught at an average distance of 13.5 km. The longest annual distance traversed is 46 km. Specimens weighing 0.5–1.3 kg produce body vibrations and noise when grasped or alarmed. The sound, lasting 0.1–0.5 sec and having a frequency of 100–130 Hz, cannot be heard but registers on a hydrophone. It is made internally, probably by contraction of the 2nd antennae muscles over the stomach mill. *Nephrops*°

*norvegicus,** the Norway lobster, with females 17 cm, males 22 cm long, lives on soft bottoms of European Atlantic shores and the Mediterranean at depths from 40–800 m. It has narrow chelae.

Astacidae,* Crayfish. The last thoracic sternite is free. The epipodites of the thoracic limbs are V-shaped, folded lamellae (Fig. 13–9). All are freshwater animals; a few are found in brackish water in the Caspian and Black Seas as well as on the Pacific coast of the United States. *Astacus** *astacus,** the European crayfish (Fig. 13–9) has males averaging 16–25 cm long; females are each 12 or rarely 18 cm. They inhabit streams, brooks, ditches, lakes, and ponds. At daytime they hide in burrows under overhanging banks, and the animals also overwinter in these burrows. At night omnivorous *Astacus* goes in search of food. Like all Astacura and Palinura, *Astacus* can move forward, sideways, or backward. By clapping the abdomen, a crayfish can shoot back. The mating season is October and November in Central Europe. The female produces eggs in late November or December; she stands high on all five pairs of pereopods the abdomen held under, and fills the space between thorax and abdomen with mucus from glands in abdominal sternites and pleopods. On contact with water, the mucus forms an elastic skin. She then turns on her back, holding her bent abdomen with the thoracopods. Soon eggs move from the gonopore into the mucus, which in

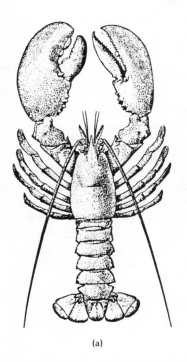

(a)

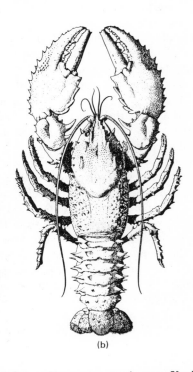

(b)

Fig. 13–21. Astacuran decapods: (a) American lobster, *Homarus americanus*, Nephropidae; male, ca. 40 cm long. (From Rathbun.) (b) Australian crayfish, *Euastacus armatus*, Parastacidae; 30 cm long. (From E. Clark, 1941, in Mem. Natl. Mus. Melbourne, Vol. 12, by permission.)

the meantime has dissolved the spermatophores. Rhythmic movements push the eggs back and they are fertilized. The eggs, covered by mucus, adhere to the pleopod setae; their hardening cover and weight draw the eggs out on a stalk. An 8 cm female produced 70 eggs, 2–3 mm in diameter; a 10 cm female produces 240 eggs. However, only about 20 hatch in a postlarval stage after more than 6 months. At first each baby remains attached to the mother's pleopods with its chelae, its carapace blown up by yolk in the midgut glands. After 10–15 days it molts for the first time; as the yolk is used up the body elongates. The little animal remains another 3 days with the mother until its carapace mineralizes and then goes off by itself. Three to 4 weeks later it molts a second time, after which the uropods appear. The animal now looks like an adult but is only 11.6 mm long. Older literature indicates that during the first year they molt eight times, during the second year five times, during the third year twice. Males mature when 3 years old, the females a year later, when 7.5 cm long. The males then continue to molt, twice annually, the females only once, after the young hatch. The life span is 20 years. Some time before molting, *Astacus* stops feeding, its carapace softens and the gastroliths increase in size. While molting the crayfish lies on its side. The exoskeleton breaks transversely between carapace and first abdominal tergite. The crayfish then bends and stretches, while the cephalothorax slips posteriorly and upward out of the old exuvium, drawing eyestalks, antennae, and all appendages out of the old skin. The blood empties out of the wide chelae, and these collapse to pass through the narrow basal articles; sometimes the old exuvium tears. After the anterior exuvium has been shed, the skin of the abdomen is pulled off with a jerk. The exuvia contain ectodermal organs, fore and hindgut endoskeleton, and gills. The freshly molted crayfish rests 10 days while the carapace hardens.

The crayfish disease of the 1870's destroyed most of the central European crayfish. The causative phycomycete, *Aphanomyces astaci*, which lives in the hypodermis and endocuticle of the joint membranes and in the nervous system, has fruiting bodies that emerge from the joint membranes after the host's death. At present *A. astacus* has only a scattered distribution. *Astacus leptodactylus* is found in eastern Europe and has been introduced into western European waters. *Austropotamobius torrentium,* is found in Central Europe, and *A. pallipes* in southwestern Europe, France and the British Isles, all sensitive to the disease. Close relatives of *Astacus* are found on the North American Pacific slope, about seven species of *Pacifastacus* from British Columbia to California.

Most North American crayfish belong to a different subfamily, the Cambarinae.* Members of this subfamily lack gills on the last thoracic somite; females have the gonopore in an annulus ventralis, males have the first pleopod modified and terminating in two or more distinct parts. The common eastern North American *Orconectes* * *limosus*, resistant to the disease, was introduced into Europe, where it spread rapidly. Now abundant, it is found to be much less edible than the European crayfish, and may compete with *Astacus astacus*, preventing its recovery. *Orconectes limosus,* * (=*Cambarus affinis*) (Fig.13–15), from 11–12 cm, occurs from Maine to Virginia; it digs no burrow but inhabits muddy, vegetated areas and may be active in daytime. In winter it withdraws into deeper water, 3–4 m, but continues feeding. The animals are omnivorous, but mainly vegetarian during summer. In Europe unlike *Astacus,* 5-month-old animals 5–6 cm long, mate in September (see section on Reproduction). The eggs are fertilized while passing the annulus. The eggs are held by a mucous basket below the curled abdomen. The mucus breaks after some hours and is removed. The eggs are attached to the pleopods. The eggs are laid in spring, as in *Astacus.* After 5–8 weeks (at about 14°C), 100 larvae hatch from 200 to 600 eggs. The larvae, 4 mm long, are not as far developed as those of *Astacus* on hatching. They attach to the eggstalk with their chelae, and only after the first molt do they resemble the first instar *Astacus.* The second molt occurs in 48 hr. After the third molt uropods appear and the 8 mm long animals gradually leave the mother. In an aquarium young remained in the third stage for 18 days. While remaining translucent, they changed in color from reddish to green. During the next 3 months they grew to 4–6 cm. When 2 years old, they were 7.5–8 cm long. Females mate when as small as 5 cm, but only 2-year old females produced eggs. There are 66 species of *Orconectes* in the eastern United States and Canada. *Cambarus°* has about 57 species in the same area. Some species make burrows in wet meadows, others burrow only during dry season, making a chimney around the burrow 30 cm high, 30–35 cm in diameter. In digging the chelae are opened and pushed together into the mud. The piece pulled out is pressed against the underside and carried to the entrance. The burrow is left in search of food and a mate. Burrowing crayfish may become a nuisance in moist fields by interfering with crops. Some species live in cave waters. Their habits and life histories differ: the mating season may be spring or fall. After mating, males of cambarine crayfish molt, and their first pleopods change to resemble those of the adult instar. After a later molt they again assume the adult form. There are 111 species of *Procambarus.* Some *Procambarus°* live as long as 6–7 years. Most species die after two to four molts. *Cambarellus°* is most common in Mexico. Several cambarine species are intermediate hosts to the human lung fluke, *Paragonimus westermani.* The cercariae enter a crayfish and metacercariae invade the tissues to infect many mammals that feed on crayfish. Branchiobdellid oligochaetes are common commensals.

Parastacidae.° The last thoracomere is ventrally and laterally movable with the one anterior to it. The epipodites of the pereopods are not V-shaped (Fig. 13–21b). Species are found only in Madagascar, New Guinea, Australia, New Zealand, and South America. *Euastacus serratus,* the edible Murray lobster of Australia, up to 50 cm long, is the largest freshwater crustacean. *Engaeus°* of Tasmania lives in a burrow in moist soil and drowns in water. *Parastacus°* is found in mainly temperate South America and *Astacoides°* in Madagascar, *Cherax°* in New Guinea and Australia, *Paranephrops°* in New Zealand.

Thalassinoidea, Mudshrimp. This superfamily has a membranous to firm exoskeleton. The carapace is compressed. The distal articles of the second and third pereopods often are markedly flattened. The abdomen is large and symmetrical when extended, sometimes with well-developed pleura (Fig. 13–22). The uropods are used for swimming. There are trichobranchiae.

Axiidae.° The carapace has a flattened rostrum. The uropod is split into two parts (Fig. 13–22a). Many members live on coral reefs. *Axius°* lives in sand or mud, in crevices or sometimes as symbionts in glass sponges; *A. serratus* is found off the American Atlantic coast, rarely in less than 20 m.

Thalassinidae.° *Thalassina°* is found in tropical Indopacific. *Thalassina anomala°* is associated with mangroves in the tidal zone and above, where it digs burrows 1.5 m deep that turn at the ground water level and proceed an additional 1.5 m. From the excavated soil a turret 0.5–1.5 m high and to 4 m in diameter at the base is constructed. Rarely, neighboring animals share a single turret which may be 2 m high, 10 m in diameter. A pest in rice fields in the Malay peninsula, its burrows permit saltwater to enter fields at high tide. The turrets may be so abundant as to impede walking. Formerly it was believed that the shrimp feed on rice; it has now been demonstrated that they feed on animal matter. *Thalassina* digs with the first pairs of pereopods and transports the mud with the third maxillipeds. They apparently rarely leave their burrows but search for food by burrowing further.

Callianassidae.* The uropods lack articulations; the third to fifth pleopods are swimmerets (Fig. 13–22b). *Upogebia* affinis is found from Long Island Sound to South Carolina, living in burrows in mud between the tide marks. *Upogebia pugettensis*, 8 cm long, and found on the North American Pacific coast, digs U or Y-shaped tunnel systems as deep as 0.5 m into the substrate. The horizontal part of the burrows, 0.6–1.5 m long, have enlargements every 15 cm that permit the animal to turn. Often more than two burrows lead to the water above. Fanning the pleopods, which are fringed by setae, brings respiratory and food water currents into the structure. *Upogebia*

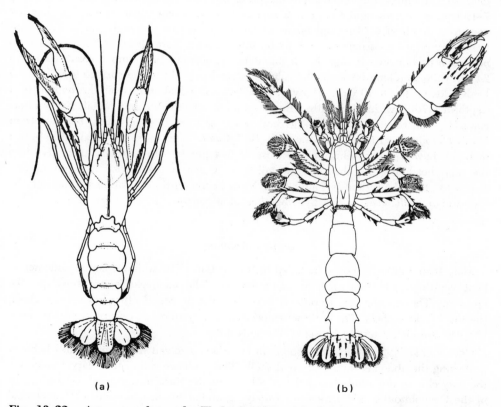

(a)   (b)

Fig. 13–22. Astacuran decapods, Thalassinoidea: (a) *Calocaris macandreae*, Axiidae; 4 cm. (From Rathbun after Bouvier.) (b) *Callianassa goniophthalma*, Callianassidae; 11 cm long. (From Rathbun.)

deltaura, 12–15 cm long, of northern Europe, digs burrows 1–70 m into the coastal mud, where it feeds on polychaetes, mussels, and detritus. In *Callianassa,*° the first thoracic limb on one side has a much larger chela than on the other (Fig. 13–22b). *Callianassa tyrrhena* of European coasts, 4–5 cm long, digs a burrow only 20 cm deep, and the lower horizontal part of the burrow is constantly changed. The earth thrown out produces a mud fountain and settles as a cone around the opening. By bending its body hairpinlike, *Callianassa* can change the direction of its movement. Sometimes the animal leaves its burrow and swims about rapidly, the large chela extended. *Callianassa* feeds on mud with diatoms and other small organisms from the mud but also on large worms such as *Nereis.* It digs by forcing the cephalothorax and the first two pair of walking limbs into the mud, then pushing the mud to the side with the chelae. Deeper in the substrate the first and second pereopods work as a pick ax. Backing up, the crab carries loosened sand in a scoop formed by holding together the ischium and merus of the third maxillipeds, and the mud is deposited at the mouth of the burrow. Sometimes the loose material is mixed with mucus streaming from the area of the mouth and is plastered against the burrow walls by the chelae, the flat propodus of the third pereopod smoothing the wall as a trowel. *Callianassa californiensis,* 7–10 cm long, digs a similar but much more elaborate burrow system starting with a slanting 7–10 cm long burrow and ending in a chamber. At least one burrow from this chamber goes to a depth of 0.75 m and opens under water so that the crab can drive a continuous respiratory current through its burrow system. Normally *C. californiensis* does not leave its burrow. If dug out, it immediately burrows back into the substrate. In 24 hr, while searching for food, the volume of sand brought to the surface in 0.5–1 cm³ loads adds up to 20–50 cm³. Many neighboring individuals turn over the bottom to a depth of 0.75 m within 8 months. But the tunnels of different individuals never interconnect. The related *C. major* digs burrows to 2.1 m deep in sandy beaches from North Carolina to Florida and the U.S. Gulf Coast. The burrows, lined with brown material, have openings clustered as close as 3–4 per ft². In captivity they dive into the sand head first but then burrow tail first. *Callianassa atlantica* lives on muddy shores to 10 m depth and is found from Nova Scotia to South Carolina. Pairs of the Californian *Callianassa affinis* live together in a burrow, as do pairs of some other species.

Section Anomura

More than 1400 species are included in this section. The well-developed abdomen is bent ventrally, and sometimes is asymmetrical. The carapace is not fused to the epistome. The last thoracic sternite is free. The first pereopod often has a large chela, the third lacks a chela. The fifth pereopod differs in position and shape from the third. Uropods are almost always present. The gills differ.

Coenobitoidea. Two separate anomuran lines have evolved the hermit crab's habit of protecting the abdomen in a gastropod shell. The evidence comes from larval anatomy, the only clue in the adults being that the bases of the third maxillipeds are contiguous in the Coenobitoidea while they are widely separated in the Paguroidea (MacDonald *et al.,* 1957). The Coenobitoidea are believed related to the Upogebiinae (Callianassidae).

Pylochelidae. The abdomen is straight and symmetrical, and its well-developed tergites (Fig. 13–23b) are covered by a sponge or a hollow piece of mangrove or bamboo. There are trichobranchiae present. *Pylocheles* is found in moderate depths in tropical and subtropical waters.

Diogenidae. This family is much like the other family of coiled hermit crabs (Paguridae) in superficial morphology; the Diogenidae differ, however, in having chelae of the chelipeds equal in size or with the left larger than the right. Only *Petrochirus* (Fig. 13–22c) has the right chela larger. Other differences are in internal and external anatomy, in larvae, and in distribution, the Diogenidae being restricted to tropical or subtropical latitudes while the Paguridae are more common in colder waters. *Dardanus°* *venosus* is found from the Carolinas to Brazil. *Paguropsis* is found along the Malayan coast. Some species of *Clibanarius°* inhabit brackish water. *Cancellus* has an elongated abdomen; it inhabits holes in rocks and corals or glass sponges, closing the hole with chelae and legs. *Diogenes°* and five other genera are found in the Mediterranean and along African coasts. *Petrochirus diogenes,* with carapace to 75 mm (Fig. 13–23c) occurring from North Carolina to Brazil, harbors symbionts: in its shell, the crab *Porcellana sayana* and on the shell, *Crepidula plana,* bryozoans, and tubicolous worms.

Coenobitidae. Land Hermit Crabs. The abdomen is either soft and spirals in adaptation to the snail shell or the abdomen is straight, wide and subsymmetrical, bent down and with mineralized dorsal plates. These are terrestrial animals that visit the ocean only during reproductive periods. The adaptations of the respiratory system are described above under decapod habits. *Coenobita* are small, carapace 1–5 cm long, inhabiting mangroves and coral reefs above the high tide mark. Adults can move inland. The symmetrical glaucothoe, barely 5 mm long, can swim and crawl. *Coenobita jousseaumei* lives from 5–25 m above the high tide line and between the tide marks of the Red Sea and old coral reefs that continue into the desert. As there rarely is dew or rain, the young, 5 mm long, are restricted to the spray area. Old crabs dig as deep as 13 cm into the sand during daytime or crawl into reef cavities. All fill their shells with water at least once a day during a 10–15 min trip to the edge of the water. They feed on Sargassum weeds, plant remains, and carrion found while moving along the shore, or adult crabs may feed in wadies. They remain within 80–100 m of their resting places. *Coenobita clypeatus* is common on the Florida keys and the Caribbean Islands.

The coconut crab, *Birgus latro* (Fig. 13–23a), up to 32 cm long, folds its abdomen, apparently symmetrical with solid tergites, under the cephalothorax. It inhabits a hole in coral rock or at the foot of a tree and anchors itself with the folded abdomen if one tries to remove it by force. Sensitive to vibrations, the animal moves rapidly into its hiding place if disturbed to reappear only after hours. Large individuals have a chela as large as the fist of a child; with it they can easily cut a branch 3 cm thick. From its hiding place *Birgus* moves only about 100 m in search of food. As it walks around obstructions at some distance, it must be able to see them. The first antennae are of importance as olfactory organs while searching for food. It feeds on carrion, sick members of its own species, and will chase the land crab *Cardisoma* to its hideout, reach in to tear off the chela extended in defense, and feed on it. It also takes the fatty fruit of *Pandanus, Canarium,* sago, and coconut palms. It can climb as high as 20 m up trees, descending backward. To eat a coconut, *Birgus* is believed to remove the outer coat of fibers and then press the large, strong chela into the germinal pore; only very large individuals can open up the coconut itself. No recent observations are available since animals are wary (E.S. Reese, personal communication). On uninhabited islands *Birgus* is diurnal, on inhabited ones, nocturnal. Its postembryonal development indicates the close relationship to *Coenobita.* The symmetrical glaucothoes, 3.5 mm long, pick up shells and emerge from the ocean in large swarms, or sometimes

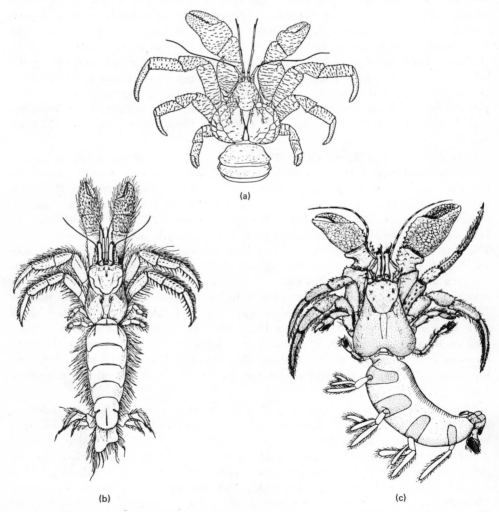

(a)

(b)   (c)

Fig. 13–23. Anomuran decapods, Coenobitoidea. (a) Coconut crab, *Birgus latro,* Coenobitidae; dark brown in color, carapace to 15 cm long. (After Alcock from Balss.) (b) *Mixtopagurus gilli,* Pylochelidae; 5 cm long. (From Benedict.) (c) *Petrochirus diogenes,* Diogenidae; female removed from its snail shell, carapace to 4.4 cm long. (After Provenzano from Williams.)

find their shells on land. The glaucothoes metamorphose in 21–28 days (Reese, 1968). The young crabs wear their shells and live on land in cracks in the soil. The abdomen is now asymmetrical. After half a year, 10–12 mm long, the crab has outgrown its shell and abandons it. The crab's abdomen now becomes almost symmetrical again. The fourth legs, short up to now and used to hold the shell, grow rapidly and take part in locomotion. At the same time the branchiostegites widen. The female spawns on land, attaching the eggs to her pleopods, and the fifth pereopods moisten them with glandular secretions and water. Later she migrates back to the ocean, as far as 3–4 km, and while holding on to a stone shakes off the zoea larvae that have developed inside the eggs. She returns immediately to the land. *Birgus* does not otherwise go into the water; it is the hermit crab best adapted to terrestrial life.

Lomisidae. *Lomis* is believed a primitive offshoot from coenobitid ancestors. Pilgrim (1965) corrects many published errors and points out the relationship of *Lomis* to *Mixtopagurus* in the Pylochelidae.

Paguroidea. This superfamily is believed to have evolved separately from the Coenobitoidea hermit crabs and its closer relationship to the Galatheidae is seen in the larval structure and the wide separation in the adult of the basis of the third maxillipeds (MacDonald *et al.*, 1957).

Paguridae,* Hermit Crabs. The cheliped chelae are subequal or the right one is larger (Fig. 13–24a). The abdomen is soft, asymmetrical, usually wound to the right and covered by a gastropod shell (Fig. 13–24a, b). Rarely, if no gastropod shell is used as cover, the abdomen may be secondarily asymmetrical. Pleopods are found normally only on the left, functioning to provide respiratory currents and as an attachment place for the eggs. The right uropod is smaller than the left. Both deep-water and coastal inhabitants are very active, continuously running about. The larvae, symbiosis and feeding habits have been mentioned in the respective paragraphs of the chapter. *Tylaspis* is a deep-sea form, reduced in size; *T. anomala* lives at 4300 m without a mollusk shell. *Pylopagurus* close the aperture of their shells with an opercular chela. In several species the shell becomes overgrown with a sponge or a bryozoan colony. The crab clears the entrance with its chelae. *Xylopagurus rectus,* occurring in the West Indies at depths of 300–400 m, inserts its elongated abdomen into hollow wooden tubes and closes the posterior opening with the calcified 6th abdominal tergum anterior to the telson. *Pagurus**† *bernhardus* of northern Europe has a carapace up to 35 mm long; larger ones carry *Buccinum* shells often covered with *Hydractinia echinata* or sometimes by *Calliactis parasitica*. Sometimes the sponge *Ficulina ficus* grows around the shell and as it grows enlarges the shell. *Nereis furcata* is found inside the shell as a symbiont. The worm emerges when the crab feeds and participates in the meal, consisting of shovelled up detritus, polychaetes, mollusks, echinoderms, crustaceans. Between 12,000 and 15,000 eggs are produced, and the females emerge part way from their protective shells to swing the eggs, carried by the pleopods, in the water. The first zoea is about 3.5 mm long, the metazoea 8 mm, the glaucothoe only 4 mm. *Pagurus bernhardus* examines the following properties of a mollusk shell to determine its acceptability: optical contrast against the background, movability and

---

† The name *Eupagurus* is an objective synonym of *Pagurus* and is listed on the *Official List of Rejected Names in Zoology*.

weight, existence of a sufficiently large cavity free of obstructions. The fit of the shell is tried out with the abdomen. Some of the 100 species of *Pagurus* are characterized by having the chela on one side with dactylus and propodus transformed into an operculum that closes the aperture of the shell, while the small chela of the other side is pulled back into the house. *Pagurus prideauxi* of the European coast has a carapace 0.8–1.2 cm long. *Pagurus longicarpus* is a common hermit crab along western Atlantic shores, from Nova Scotia to Texas. It feeds on benthic diatoms and detritus, including small particles of animal material. The breeding season is from May to early fall in Massachusetts. It has been used at times as an experimental animal. *Pagurus polli-caris*, the large hermit crab with a carapace 31 mm long, is found from Cape Cod to Texas. Numerous species are found on the North American west coast. *Porcellano-pagurus*, in deeper areas of the littoral of the Indo-Pacific, uses clam shells or *Patella* shells as protection. Their dorsum is flat, the abdomen shortened, and the pleopods are moved dorsally.

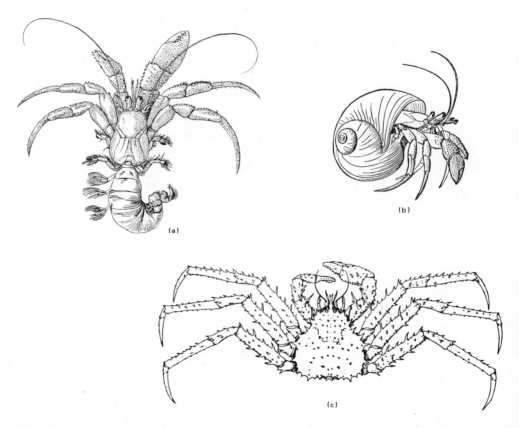

Fig. 13–24.   Anomuran decapods, Paguroidea. (a) Hermit crab, *Pagurus bernhardus*, Paguridae, removed from its snail shell; carapace 3 cm long. (After Calman from Schellenberg.) Hermit crab, *Pagurus* sp., in shell. (From Rathbun.) (c) *Lithodes maia*, Lithodidae; carapace 16 cm long. (From Rathbun.)

Lithodidae.* Members of this family have the carapace heavily mineralized; they resemble true crabs (Fig. 13–24c). The abdomen, except in the primitive Hapalogastrinae, is tightly appressed under the cephalothorax. The abdominal tergites are not uniform plates but consist of several pieces, sometimes so small that the back of the abdomen between the larger lateral plates appears membranous with many tiny sclerites. In females these plates are asymmetrical. There are no uropods. The gills are phyllobranchiae. The morphology of both adults and larvae resembles that of the Paguridae, so that their common origin with that family is not in doubt. The Hapalogastrinae resemble them most. Lithodidae differ from Brachyura by the following characters: the asymmetrical abdomen, the scale on the 2nd article of the second antennae, and the 5th pereopods, which are short and may be hidden in the branchial chamber. All Lithodidae are cold water animals.

Hapalogastrinae contains primitive genera in which the abdomen is not folded tightly against the underside and the dorsal side of the second pleomere can still be seen from above. The dorsal plates of the third to fifth pleomeres are soft and, like the ventral side, unsegmented. *Oedignathus* and *Hapalogaster* are found in the North Pacific.

Lithodinae contains crablike forms. *Lithodes** has a triangular, mineralized carapace with dorsal tubercles and spines. *Lithodes maia* has a carapace to 14.5 cm long and is brick red with darker spines (Fig. 13–24c). It is found in Maine and further north and gets into lobster traps; *L. aequispina*, up to 9 kg, is canned on the West coast. *Lopholithodes foraminatus*, with a carapace 9 cm long, digs itself backward into sand. These animals dig up mud with the large chelae and filter the suspension with the feathered setae of the large maxillipeds. Shells of *Nucula* and *Arca* found while digging are broken by the strong mandibles, crushed and eaten. Species of *Paralithodes* feed on corals in part, judging by fragments found in the feces; *P. camtschatica*, the king crab of northern Pacific, is used as food.

Galatheoidea. The abdomen has well-developed symmetrical tergites, more or less arched. The carapace is not fused to the epistome. The first pereopod has well-developed symmetrical chelae (Fig. 13–25). The small fifth pereopod may be hidden in the branchial chamber. Uropods are always present. The gills are usually phyllobranchiae. Almost all are marine.

Galatheidae.* The abdomen is folded ventrally but can be used to swim backward. The third maxilliped has an epipodite. Metazoeae have long spines. *Galathea* *strigosa*,* to 10 cm long, occurs along European coasts in the littoral to 600 m; they hide at daytime in rock crevices or under stones but they do not dig. They walk slowly but can shoot backward by clapping the abdomen. All are nocturnal, feeding on large and small organisms. *Galathea rostrata*, at 20–100 m depth, with males 18 mm long, occurs along the North Atlantic west coast to Yucatan. *Munida*,* found to a depth of 2000 m, *M. gregaria* is at times found in huge numbers in the open ocean near Tierra del Fuego making the ocean appear red. *Pleuroncodes planipes* swarms in similar numbers off the southern California coast. *Munidopsis** (Fig. 13–25a) is another genus.

Aeglidae. Unlike other Galatheoidea, the gills are trichobranchiae. The genus *Aegla* is the only truly freshwater anomuran and has almost 20 species throughout temperate South America (Fig. 13–25b).

Chirostylidae. The abdomen, though folded ventrally, is visible behind the thorax.
The antennae have a peduncle of five articles. The arthrobranch gills are on the
thorax resembling pleurobranchs. The family includes about 70 species throughout the
world in moderate to considerable depth, living on gorgonians. Genera are *Chirostylus*
and *Uroptychus.*

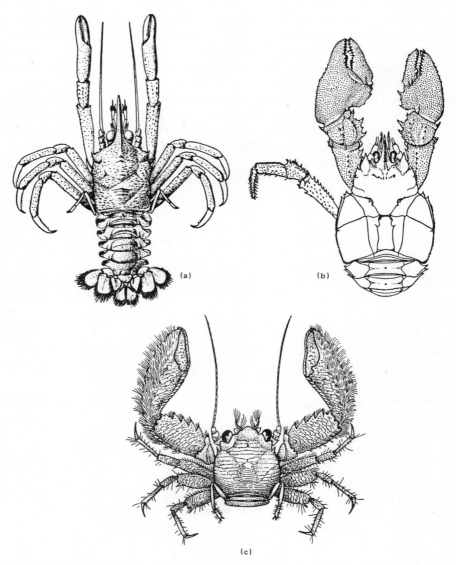

Fig. 13–25.   Anomuran decapods, Galatheoidea: (a) *Munidopsis curvirostra,* **Galatheidae,
3 cm long. (From Rathbun.) (b)** *Aegla parana,* **Aeglidae; most limbs removed, carapace
4 cm long. (From Schmitt.) (c)** *Petrolisthes galathinus,* **Porcellanidae; carapace, 8 mm
long. (From Boone.)**

Porcellanidae,* Porcelain Crabs. The thin abdomen is folded under the crablike cephalothorax (Fig. 13–25c). The presence of uropods and the fifth pereopods, which may lie in the branchial chamber, separate the group readily from the true crabs. These are small, mostly littoral species, that occur under stones or in crevices. Like brachyurans they walk sideways. All are filter feeders (see feeding above). Some species of *Petrolisthes** (Fig. 13–25c) are found on coral reefs, others under rocks in the tidal areas; *P. galathinus,* with a carapace 11–17 mm long, occurs from the Carolinas to Brazil; *P. cinctipes,* to 21 mm, is common under stones on rocky coasts from British Columbia to California. It is relatively flat. The flat-topped crab *P. eriomerus,* to 15 mm, occurs in quiet waters from British Columbia to Lower California. *Pisidia** *longicornis,* with a carapace 6–8 mm long, occurs in eastern Atlantic waters under stones or among sponges or colonies of tube worms: *Porcellana** *sigsbeiana,* 24 mm long, inhabits the edge of the continental shelf from Massachusetts to Virgin Islands and Yucatan. *Pachycheles,* with several species reaching 7 mm, is found in many oceans. *Pachycheles rudis,* the big clawed porcelain crab, to 17 mm, is found intertidally under stones and among *Chama* shells from Alaska to Lower California; *P. pubescens,* to 14 mm, is found in littoral waters from Puget Sound to San Francisco. Other species are found in the Indo-West Pacific. *Polyonyx* species are commensals of the polychaete *Chaetopterus,* straining out small particles carried into the food tube. *Pseudoporcellanella* is a symbiont of sea pens; its anatomy is modified for the habitat.

Hippoidea, Sand Crabs. The body is symmetrical and almost cylindrical, somewhat flattened, especially in Albuneidae. The abdomen is folded ventrally (Fig. 13–26). The carapace is not fused to the epistome. The first pereopods lack chelae in Hippidae, but they are present in Albuneidae. The fifth is small, a grooming appendage in the branchial chamber. All bury themselves in sand and are tropical, subtropical, and a few extend into temperate waters.

Hippidae.* The very long telson is lancet shaped. *Emerita** *talpoida,** up to 3 cm long (Fig. 13–25a), lives in the tidal zone from Cape Cod to Yucatan and filters the water that runs back to the ocean after the waves break. The filtering is done with two rows of featherlike setae on the second antennae. Held V-shaped, these are rapidly extended and retracted. For filtering, the crab digs itself backward into the sand at an angle of 45° with the second to fourth pereopods, facing seaward, back to the returning current. The food intake has been described in the section on feeding. When the water flows out during low tide, some individuals remain, digging themselves 8–15 cm deep into the slowly drying sand until the next high tide. Most emerge rapidly, simultaneously from the sand, let themselves be washed down, and dig in several meters away to continue feeding. As the tide changes the population shifts its location on the beach. With high tide the sand crabs let themselves be washed in. The crabs also turn their backs to an artificial current. If there is no water movement, *Emerita* turns its back to a wide, dark, cardboard disc, indicating that orientation may be optical. *Emerita* swims exclusively with the uropods and only backward, the abdomen held tightly against the cephalothorax. *Emerita analoga* is found on the west coast of North America. Tiny neotenic males (2.5 mm) may be attached to a pereopod of large females (38 mm) of *Emerita* (Efford, 1967), when larger these males (and also those of the albuneid genus *Lepidopa*) change into females (Efford, personal communication).

Albuneidae.* The first pereopods have a subchela (Fig. 13–26b). The telson is short. *Albunea* *gibbesii*, with a carapace 16–20 mm long, occurs from the Carolinas to southern Brazil from the extreme low tide mark to 70 m. It digs backward into sand, as does *Emerita*. It is probably a scavenger and minor predator as are all Albuneidae and Hippidae except *Emerita*. *Albunea paretii* is found from the Carolinas to northern Brazil and West Africa. *Lepidopa* *websteri*, 12 mm, is known from the Carolinas to Mississippi. The first antennae may be up to 4.5 times as long as the carapace and form a respiratory tube for the animal when buried in sand. A recently described Australian genus *Stemenops* has extremely long eyestalks, over 1.2 times the length of the carapace.

Section Brachyura, Crabs

There are more than 4500 species of crabs. This is the largest and most specialized group. The nervous and respiratory systems have been discussed in the respective paragraphs. The carapace usually has a keel all around separating dorsal from ventral surface and is anteriorly fused to the epistome (Fig. 13–7). The antennae usually are very small and lack the scale, or the scale is immovably fused to the second article. The abdomen is short and thin, symmetrical, tightly folded under the cephalothorax

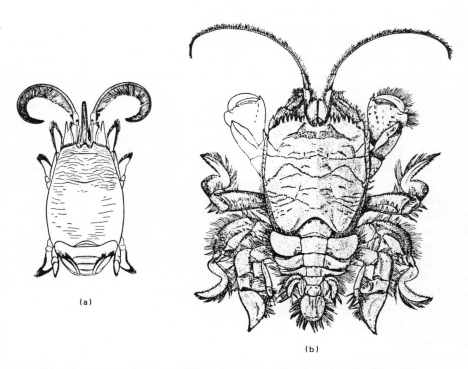

(a)

(b)

Fig. 13–26. Anomuran decapods, Hippoidea: (a) *Emerita talpoida*, Hippidae; carapace 3.4 cm long. (From Rathbun.) (b) *Albunea carabus*, Albuneidae; 1.7 cm long. (From Monod.)

and is not used in locomotion. The chela of the first pereopod is usually large; the third pereopod is never chelate. Uropods are absent and males also lack the third to fifth pleopods. The gills are phyllobranchiae. The zoea and metazoea have long spines. There is a megalopa stage. Most crabs run sideways. (Size measurements given here refer to length and width of the carapace only.)

SUBSECTION DROMIACEA

The carapace is rarely wider than long and often is hairy, as are the appendages. The mouth frame is rectangular. Only the first pereopods have large chelae. The fifth pereopods, often the fourth, are hinged to the trunk dorsally; usually they are thin and

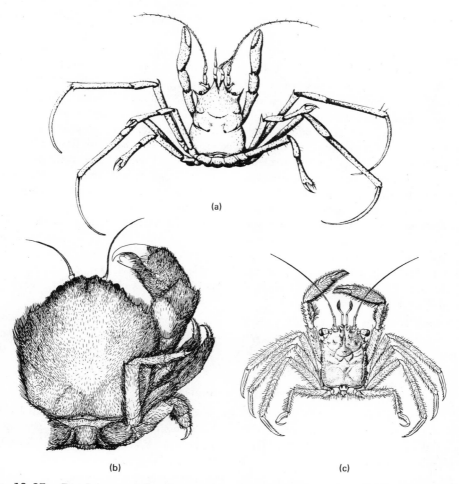

(a)

(b)   (c)

Fig. 13–27. Brachyuran crabs, Dromiacea: (a) *Homolodromia paradoxa*, Homolodro-miidae; 1.8 cm long. (From Milne-Edwards.) (b) *Dromidia antillensis*, Dromiidae; left legs not shown; carapace 3.3 cm long. (From Williams.) (c) *Homola barbata*, Homo-lidae; carapace 3 cm long. (From Williams, after Smith.)

small (Fig. 13–27), rarely reduced and have small chelae or subchelae. The chelae or subchelae usually serve to hold foreign bodies, camouflaging the animal. In both sexes the gonopores are coxal. Usually they have large numbers of phyllobranchiae, unlike other Brachyura. The Dromiacea are the most primitive Brachyura. The zoea lacks the long dorsal spine and the deep dilation of the telson. The metazoea has exopodites on the third maxillipeds; *Dromia* has them also on the first pereopods. Both larvae and megalopas have well-developed uropods absent from the larvae of other Brachyura. Because of these ontogenetic differences, some authors prefer to remove this group from the Brachyura. Serological experiments suggest close relationship with the Scyllaroidea. All species are marine, many occurring in the deep ocean.

Homolodromiidae. These include primitive genera with 21 pairs of gills on each side; some form trichobranchiae, some are intermediate between tricho- and phyllobranchiae. The third maxilliped is almost limb-like. *Homolodromia* (Fig. 13–27a) is found in the deep ocean.

Dromiidae.* The third maxilliped is wide at its base, operculate, and covers the oral region. On each side are 14–16 phyllobranchiae. Oblique sternal grooves in females are the external evidence of internal tubes, part of the reproductive system. *Dromia* *personata* has a highly domed carapace, 5 cm long, almost spherical and, like the appendages, covered with dense short hair; it lives at depths of 20–100 m. Like other masked crabs, it moves very slowly. The sponge crab *Dromia erythropus*, found in the West Indies, cuts out a piece of sponge, settles into the cavity, and covers itself with the sponge fragment; the habit is shared with *D. personata* and other species of *Dromia*. The West Indian sponge crab *Dromidia antillensis*, 32 mm long, 31 mm wide (Fig. 13–27b) occurs from Cape Hatteras to northern Brazil from shore to 340 m. It is usually found carrying ascidians, sponges or zoanthoid polyps. Larvae hatched from berried females in the laboratory had six zoeal stages and a megalopa. The zoeal stages were passed within 17–18 days, the megalopa 12 days feeding on brine shrimp nauplii (Rice and Provenzano, 1966). *Conchoecetes*, like *Hypoconcha*, holds a bivalve shell over its carapace. *Hypoconcha arcuata*, 24 × 24 mm, occurs from the Carolinas to Brazil as deep as 44 m; it clings so tightly to its clam shell that removing the shell is difficult.

Homolidae. The eyes are not covered by the carapace (Fig. 13–27c). The female sternum lacks longitudinal grooves. *Homola* *barbata* (Fig. 13–27c) 30 mm long, occurs in the Mediterranean, Portugal to South Africa, and Massachusetts to the Caribbean at depths of 15–180 m.

SUBSECTION GYMNOPLEURA

The anterior thoracic sterna are broad, the posterior ones narrow. The branchiostegites are reduced, exposing in part the lateral body somites. The last pair of pereopods is dorsal in position (Fig. 13–28). The female gonopores are on the coxae. The antennal sternum is of triangular shape. There are eight branchiae on each side.

Raninidae. The carapace is elongate but does not cover the abdomen (Fig. 13–28). The first four to five pleomeres are dorsally exposed. Both first and second antennal pairs are large. The hand of the chela is flat with the immovable finger bent, allowing the movable finger to close against the anterior border of hand. Most species are

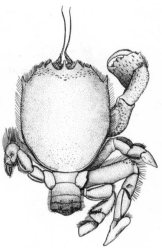

Fig. 13–28. Brachyuran crabs, Gymnopleura. *Ranilia muricata*, Raninidae; 1st to 4th pereopods on right are shown, 5th on left; carapace 2.8 cm long. (From Williams.)

tropical. *Ranilia muricata* (Fig. 13–28) carapace to 39 mm long, is found on sand bottoms 3–30 m from North Carolina to Bahamas (Williams, 1965).

SUBSECTION OXYSTOMATA

The mouth region extends to the anterior border of the carapace and is usually triangularly pointed. Excurrent water passes along a channel that extends to the very tip of the oral region; the roof of the channel is a groove and its floor is the concave dorsal surface of the endopodite of the first maxilliped. Incurrent water runs into a channel, bordered by setae, at the anterior of the carapace. The groove leads to the incurrent opening of the branchial cavity, and this assures the burrowing animal a continuous water current. The fifth pereopod is sometimes modified. Females lack the first pleopod, the first pleomere is usually covered by the carapace. The female gonopore is often on the sternum, rarely on the coxae. There are six to nine trichobranchiae on each side. All are marine.

*Dorippidae.** The carapace is short exposing the second and third abdominal somites from above (Fig. 13–29a). The fourth and fifth pereopods are small, dorsally directed, and have hooks at their tips. The second antennae are long. *Dorippe*lanata,* with a carapace as long as 26 mm, holds a bivalve shell, an ascidian, or the head of a fish above its body with its posterior pereopods. It turns this shield toward potential predators without turning its body. *Ethusa** (Fig. 13–29a) also uses a bivalve shell as a shield.

*Calappidae,** Box Crabs. In German these are called "shame-faced" crabs because they hide their faces behind huge hands, the flattened chelae (Fig. 13–29b). The modified chelae are used to open gastropod shells. The crab rolls a shell found over several times frequently inserting the dactyl into the aperture before positioning it so as to be able to break it. The shell is rolled toward the large tooth on the dactyl by the other chela and the legs, and successive pieces of shell are broken out by pinching the shell between dactyl tooth and two fitting protuberances, exposing the enclosed mollusk or hermit crab (Shoup, 1968).

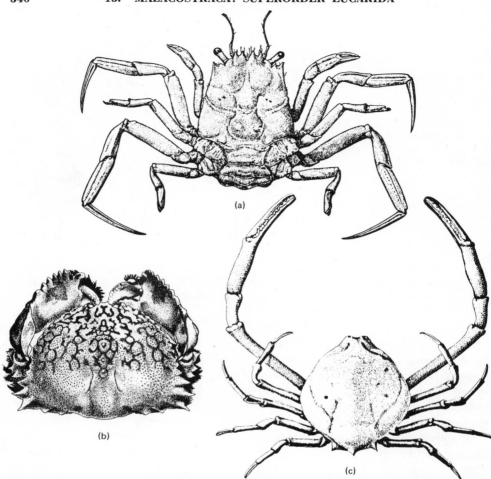

Fig. 13–29.   Brachyuran crabs, Oxystomata: (a) *Ethusa mascarone*, Dorippidae; male, carapace 1 cm long. (From Monod.) (b) *Calappa ocellata*, Calappidae; male, carapace 9 cm long. (From Williams.) (c) *Myropsis quinquespinosa*, Leucosiidae; female, carapace 5 cm long. (From Milne-Edwards and Bouvier.)

When the crab is buried, the vertically flattened chelae enclose a sand-free, water-filled space also bordered by the ventral side of the body. At the same time they cover the water channel on the median part of the branchiostegites that conducts respiratory water from the anterior of the carapace to the openings of the branchial chamber next to the coxae of the large chelipeds. The respiratory water can thus be expelled by the scaphognathites into the median excurrent channel. The current is so strong that it stirs up the sand. The wide carapace covers the abdomen completely. All limbs are directed laterally. The second antennae are short. *Calappa*° digs itself in by pushing its flat chelae against the sand, pushing the body diagonally backward into the sand. At the same time the walking limbs help move the body in this direction. *Calappa* runs as if on stilts, holding the carapace vertically. *Calappa granulata*,° up to 7 cm, 8 cm wide behind, occurs in the Mediterranean and East Atlantic at depths from 30 to 150 m. *Calappa flammea,* 13 cm wide, is found from the surface to 80 m deep along Cape Hatteras, the Gulf of Mexico and the Bahamas, usually buried in the sand. *Calappa ocellata* (Fig. 13–29b), 14 cm wide, occurs from Cape Hatteras to Pernambuco in shallow water to 28 m. *Hepatus epheliticus,* the calico crab, 6.7 cm wide, occurs from Chesapeake Bay to Central America.

*Leucosiidae,** Purse Crabs. The afferent respiratory channels are along the sides of the mouth region and are covered by the exopodites of the third maxilliped. The entrance of the branchial chamber is anterior of the cheliped coxa. The carapace completely covers the abdomen (Fig. 13–29). All pereopods are held out to the sides. The second antennae are short. *Leucosia** is found in the Indo-Pacific region. *Ebalia** *cranchii,* up to 1.1 cm wide, occurs in the North Sea from 20 to 60 m; *E. cariosa,* 1.5 cm wide, from the Carolinas to Brazil, feigns death, closely resembling pebbles. *Speloeophorus** is represented by several species on the southeastern United States coast. *Ilia** and *Iliacantha** are other genera.

SUBSECTION BRACHYGNATHA

The mouth region is quadrangular, covered by the third maxilliped. The fifth pereopods are not modified, rarely dorsal or small. Females lack the first pleopod. The female gonopore is on the sternum.

INFRASUBSECTION OXYRHYNCHA

The carapace is narrow, usually ending anteriorly in one or two sharp points, the pseudorostrum. The body outline is more or less triangular (Fig. 13–30). Orbits are not completely closed. The excurrent opening of the branchial chamber is on the lateral border of the mouth region.

*Majidae,** Spider Crabs. The carapace is longer than wide, the chelipeds usually not much longer than other pereopods (Fig. 13–30). Body and appendages are usually covered by hooked setae used to hold camouflage. All are marine. *Inachus** *phalangium,** to 1.8 cm long, near Heligoland in the North Sea, slowly climbs about the alga-covered bottom, feeding on bivalves, crustaceans, echinoids, polychaetes, and hydroids. *Macrocheira kaempferi,* carapace to 45 cm wide, cheliped span up to 3.6 m, is found on the continental shelf between 50–300 m off Japan. *Macropodia** *rostrata,** to 1.8 cm long, in the eastern Atlantic, in algal meadows of littoral and sublittoral waters, slowly climbs about on long legs, often hiding under the tentacles of *Anemonia sulcata. Hyas araneus,* 8.5–11 cm long, of the Northern Atlantic and Baltic, slowly moves about on algae. *H. coarctatus,* the toad crab, is found on muddy bottoms in the North Atlantic. *H. lyratus* occurs north of Pudget Sound in the Pacific. *Pisa,** 1.3–6 cm long, is found in the Mediterranean and eastern Atlantic. *Maja** *squinado,** 8–11 cm, rarely 18 cm long, occurs along the southwest coast of Europe and in the Mediterranean. It does not molt after reaching maturity, and is unable to autotomize limbs or to regenerate them. The 1- to 2-year-old males mate with recently molted females. Off the English coast, the eggs are deposited about 6 months after mating and are carried for 9 months attached to the pleopods. In July in water 1.5 m deep, mature males and preadult females were observed to form a group of about 60 individuals, 1 m in diameter and 0.5 m high. During the next month 20 individuals joined; within the next 3 weeks the females molted, soft animals moving to the center of the group. In mid-August mating pairs were observed in the group, and the group slowly separated in September. The armored old males on the periphery seem to protect the molting females from attack by an octopus, or other predator. *Maja squinado* grazes algae, Hydrozoa, and Bryozoa. They never attack larger animals. *Huenia*, 2.8 cm wide, is found in the tropical Indo-Pacific. *Loxorhynchus grandis*, the sheep crab, found in deeper waters off the California coast, masks itself when small to 7.5 cm wide; when larger the instinct to mask is lost. It may grow to 19 cm in length, more than 3 kg in weight; *L. crispatus*, the moss crab, to 11 cm long, is an inactive crab in subtidal zones off California. *Libinia emarginata*, 6–10 cm long, (Fig. 13–30a) is found from shore to 60 m deep from Nova Scotia to the Gulf of Mexico. It is most common in July. Juveniles have been seen attached to the bell of the scyphozoan, *Stomolophus*. *Libinia dubia*, up to 10 cm, from Cape Cod to Cuba and Texas, lives most of the year in deep estuarine waters. Young individuals 3–37 mm are found on the subumbrella or genital pits of the jellyfish *Stomolophus meleagris*. *Pelia mutica*, 5–13 mm long, with bright red patches on the carapace, is found from Cape Cod to the West Indies on gravelly bottoms of bays among ascidians, hydroids and sponges on wharf piles. Individuals are often covered by sponges, making their recognition difficult. They have also been reported from *Chaetopterus* tubes. *Podochela riisei*, 21–23 mm long, from the Carolinas to Brazil, is found among hydroids in pilings. *Pugettia producta*, the kelp crab, 10 cm long, is common from Alaska to California at a depth of 20 m. *Scyra acutifrons*, 3.5 cm long, covered by sponges and hydroids, is a common intertidal spider crab from Alaska to San Diego.

*Parthenopidae.** The cheliped is much longer than all other pereopods but not very movable. The movable digit is not parallel when closed but is held at an angle to the immovable one (Fig. 13–30b). The carapace lacks hooked setae. *Parthenope* serrata, 28 mm wide, from the Carolinas to Brazil, occurs from shallows to 120 m deep on various bottoms. *Parthenope pourtalesii*, 47 mm wide, chelipeds 122 mm (Fig. 13–30b), from Massachusetts to the West Indies, occurs on sand or sandy mud bottoms. *Heterocrypta* granulata,* the Pentagon crab, 18–21 mm wide, from Massachusetts to the West Indies, is found on shelly substrate, its angular outline resembling the background. From Monterey south, *H. occidentalis* is found at very low tides; 26 mm wide to 17 mm long, it is pinkish mottled with lighter spots.

INFRASUBSECTION BRACHYRHYNCHA

The body is oval, circular or square. The carapace is wide anteriorly, usually wider than long, or at least as wide. There is little or no rostrum (Fig. 13–32).

*Corystidae.** The carapace is longer than wide with incomplete orbits (Fig. 13–31a). *Corystes* cassivelaunus,* masked crab, 2–3.6 cm long, is a predator in the North Sea.

*Atelecyclidae.** The carapace is usually not markedly longer than broad, generally with complete orbits (Fig. 13–31b). *Atelecyclus** with two species occurs in the eastern Atlantic.

*Thiidae.** The carapace is almost circular, grooves hardly or not indicated. *Thia* scutellata (=*T. polita*), up to 2.2 cm long, occurs along European coasts, digging itself backward into the sand with only the antennae showing.

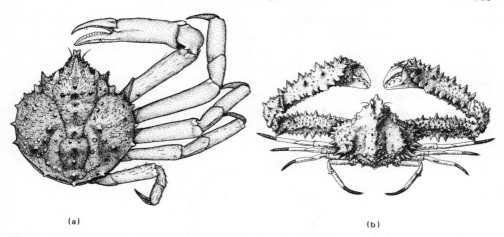

Fig. 13–30. Brachyuran crabs, Oxyrhyncha: (a) Spider crab, *Libinia emarginata*, Majidae; male, left legs not shown; carapace 9 cm long. (c) *Parthenope pourtalesii*, Parthenopidae; female, carapace 3.1 cm long. (Both from Williams.)

Cancridae. The carapace is broadly oval or hexagonal. The last pairs of pereopods are walking legs (Fig. 13–32a). The second antennae fold lengthwise. The flagellum of the first antenna is hairy. *Cancer* irroratus, the rock crab, to 13.5 cm wide, occurs from Labrador to the Carolinas in shallow water north, deeper south; unlike *C. borealis*, it has the anterolateral teeth of the carapace with edges granulate, yellow dotted with purplish. *Cancer borealis*, the Jonah crab or northern crab, to 15.4 cm wide, from Nova Scotia to Tortugas, has the carapace edge denticulate; it is yellowish on the ventral surface, red dorsally. *Cancer* pagurus,* European edible crab, has a carapace up to 30 cm wide. Found on rocky substrates and, with the incoming tide, upper intertidal areas of the European Atlantic coast; it is a predator of fish, crustaceans, mollusks, and echinoderms. When sexually mature it measures 11–13 cm wide. Life history dates have been recorded from the British coasts. The time between mating and spawning is 12–14 months. During this time the spermatophores are in the seminal receptacle. From October to January females move to deeper water to spawn, but the supply of sperm is not exhausted and another batch of eggs is produced the following year without another mating. Females, 14.6 cm wide, produce 500–750,000 eggs; 15 cm wide, one million eggs; 19.5 cm wide, 3 million eggs. The eggs enter the mucus-filled brood chamber between abdomen and sternum and attach to the pleopods. The zoeae hatch after 8 months when the females have moved into warm coastal waters for the summer. At the end of 2 months, the pelagic swimming larvae have developed into minute crabs, 2.5 mm wide. During the first year they molt about eight times and grow to 3 cm; during the second year two times, growing to 4.5 cm; during the third year probably two times, growing to 7 cm; during the fourth year, they molt once in fall, growing to 8.8 cm; and during the fifth year, once in fall, growing to 11 cm. Young animals 2.2 cm wide are found in very shallow waters; animals 5.5 cm wide in the intertidal area; and older animals, 6–11 cm, below the intertidal waters. Large individuals of *Cancer* molt each year in the fall, then mate and move to deeper water for the winter, returning in the spring. Marked males moved less than 9 km along the coast; females often move 37–550 km.

The Pacific North American edible crab, *Cancer magister* (Fig. 13–32a), up to 12.6 cm long, 19.8 cm wide, is found from the Aleutians to Lower California on sandy bottoms some distance from the shore. The predatory crab is trapped, baited with fish, or netted. Unfortunately, crabs are being overfished. Other species of *Cancer* on the Pacific coast are smaller and of less economic importance. Young of *C. gracilis* and the megalopa of this species have been found clinging to the underparts of jellyfish.

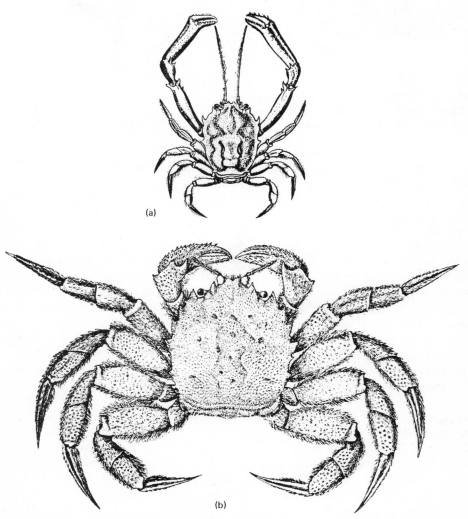

(a)

(b)

Fig. 13–31. Brachyuran crabs, Brachyrhyncha; (a) *Corystes cassivelaunus*, Corystidae; carapace 3.6 cm long. (From Bell.) (b) *Erimacrus isenbeckii*, Atelecyclidae; carapace 10 cm long. (From Benedict.)

Pirimelidae.* The carapace is as long as wide, its surface uneven, with three strong teeth or lobes in front, and anteriolateral margins strongly dentate. There are two genera in eastern Atlantic waters. *Pirimela,** from Norway to the Cape Verde Islands, is found in shallow waters, often among *Zostera* roots. *Sirpus* is a genus of pygmy crabs, 5–7 mm long with two known species, one from the Mediterranean and one from West Africa.

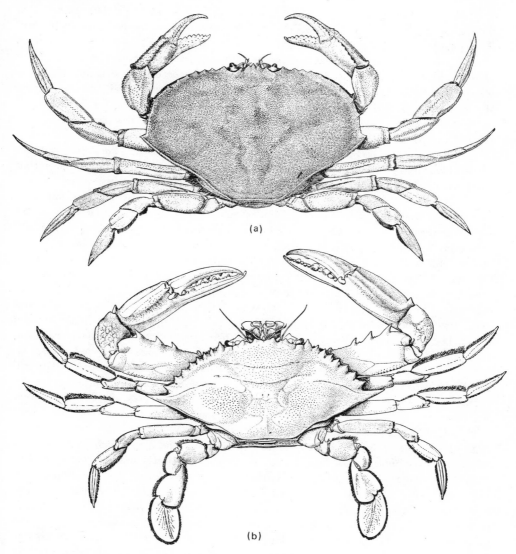

(a)

(b)

Fig. 13–32. Brachyuran crabs, Brachyrhyncha: (a) Pacific edible crab, *Cancer magister*, Cancridae; carapace, 12.5 cm long. (b) Blue crab, *Callinectes sapidus*, Portunidae; carapace 9 cm long. (Both from Rathbun.)

Portunidae, *Swimming Crabs.* The posterior four pairs of pereopods are somewhat compressed, the dactyl of the fifth leg is usually expanded into an oval, leaflike swimming structure (Fig. 13–32b). *Macropipus,* resting on the bottom, observes prey, a fish or *Crangon* swimming above. The crab rises when its own lateral movement cuts across the path of the prey. Rapidly the prey is reached, grasped with the chelipeds, and the crab sinks with it to the bottom. A well fed animal buries itself up to the eyes or head. For digging the flat fifth pereopods are pushed horizontally, like spades, into the sediment, while the bent chelipeds extend forward, moving the body backward into the sand. Just like *Calappa,* the buried animal fits its chelipeds against the curvature of its front end, enclosing a sand-free space with respiratory water. The teeth of the anterior carapace border extend over the water space to the chelae, straining out sand. *Macropipus holsatus,* up to 4 cm long, is one of the common littoral swimming crabs of the North Sea at depths of 0–50 m; it is found from Iceland to the Canaries; *M. puber,* the velvet crab, found from Norway to northwest Africa, has dense hair covering the carapace. *Ovalipes ocellatus,* the lady crab, 7 cm wide, occurs from Canada to the Gulf of Mexico, buried in sand; it is eaten in the South. *Polybius* is semi-pelagic. *Portunus* has long, narrow chelipeds. *Portunis sayi* lives among Sargassum weed and is pelagic, others are littoral or sublittoral. Some large species are of economic importance, such as *P. sanguinolentus* an Indo-West Pacific crab found in Hawaii. *Portunus pelagicus,* which is not pelagic, is of great commercial value in the Indo-West Pacific and through immigration through the Suez Canal, the eastern Mediterranean. *Callinectes* *sapidus,* the blue crab (Fig. 13–32b), up to 20 cm wide and colored blue to green, occurs from Nova Scotia to Uruguay. It is most abundant in the Chesapeake Bay and has been introduced in Holland, Denmark, and in the eastern Mediterranean. Blue crabs are omnivorous. They are of economic importance in North America. The mating period is from May to October in Chesapeake Bay. Females ready for the terminal molt are carried upright under the males. The female is freed during her molt, but afterward the male takes her again, ventral side to ventral side, and using his copulatory stylets, introduces sperm into the seminal receptacles. Copulation lasts several hours, after which the female resumes her upright position but is still carried until hardened. Males mate more than once during their last three intermolts; females mate only once, but the sperm fertilizes several egg masses, the first laid during the following spring. The spawning season is from spring until fall on the North American east coast or winter in Texas, with the peak in June. The total number of eggs laid by one individual is 700,000 to two million. The eggs hatch in 15 days at 26°C, as zoeae which pass through seven instars, sometimes eight, before molting. They transform in 31–49 days after hatching into a megalopa. The megalopa molts into a crab after 6–20 days. The young crab molts 18–20 times before maturity. The first batch of eggs is produced at the approximate age of 2 years, 2–9 months after mating. Late in summer there may be a second spawning and a third at 3 years, the maximum age of the species. The animals are migratory, the little crabs move up into estuaries. Mating occurs in the reduced salinity of estuaries; spawning takes place in deeper water. Males remain permanently in areas of low salinity while females, having returned to deeper water for spawning, remain there for life. Blue crabs are quite tolerant of fresh water and may be found in Florida streams and lakes of low salt concentration. The populations of blue crabs fluctuate. "Soft-shelled" crabs are caught by keeping crabs ready to molt in live boxes and floats, collecting being done by dip or push nets or "scrapers," or bottom dredges. "Trotlines" are used for hard crabs. A rope is baited at intervals of 45 cm with dried fish or meat. The fishing boat moves along the line, lifting it over a roller, and the crabs hanging on to the bait are taken off with a dip net or are pushed off into a bucket. In winter fishing is done only with dredges. Crab pots similar to lobster pots are used in some regions. Blue crabs, being long-lived after their last molt, collect various symbiotic barnacles in their branchial chambers, gills, and on the carapace.

The Indo-Pacific *Podophthalmus*° has long eyestalks. The eyestalks have facilitated experiments on visual neurophysiology allowing single unit recording along the optic pathway. A number of interneurons which respond only to particular types of light change have been found. *Euphylax*° *dovii*° is found from Panama to Chile. *Portumnus*° *latipes*,° to 2.5 cm long, is found in the North Sea. *Carcinus*° *maenas*,° the green crab called shore crab in Great Britain, has females to 5.5 cm wide, males to 8 cm. Its range extends from Nova Scotia to Virginia and from northern Norway to northwest Africa. It is introduced off Victoria, Australia. It is the most common crab in the region of Woods Hole, Massachusetts and in the North Sea and even in the Baltic. Running sideways in shallow water between tides, it travels 2.5–3 km daily and returns at low tide, or at times may be left in the drying intertidal areas or may dig into sand. They may enter brackish waters. The crab is most active at night and at high tide, depending on a daily component of the activity rhythm of 24 hr and a tide component of 12.4 hr. In experiments these components persist in constant twilight and constant temperature in submerged animals or outside of the water. Food consists mainly of mollusks, amphipods, mysids, shrimp, worms, and small fish. Running rapidly sideways and even jumping in the water, the crab may catch swimming prey such as shrimp, *Palaemon*, and small fish. Threatened from above, it raises itself and attacks with the chelipeds or even jumps. That it is not a good swimmer can be deduced from the narrow dactylus of the fifth pereopod. In mating the male turns the freshly molted female over. The female unfolds her abdomen and with it covers the male. The male supports his abdomen on his telson held at a right angle between the female gonopores, introducing his long first pleopods into these. Mating lasts 1–4 days, spawning 1 day. There are four planktonic larval instars and a megalopa 1.15 mm wide. On the Scottish coast a young crab measures 1.8 mm wide; at the end of the first year 1.3 cm; of the second year 4–5 cm; of the third year 5.5–6 cm. Often the crabs are attacked by *Sacculina*. This widely distributed crab is eaten in some countries.

Potamidae° (= *Potamonidae*). Freshwater crabs include numerous species in subtropical and tropical areas (Fig. 13–33a). The family is now often split into the Potamidae, Pseudothelphusidae° and Trichodactylidae.° *Potamon*° *fluviatile*, with males 4.3 cm, females 5 cm wide, inhabits freshwater of Mediterranean countries; it is edible. Some species are found to 2100 m in mountains. Others are found in humid forests of the Caucasus where the precipitation is 1.2–1.6 annually, and terrestrial species occur in the upper Congo in Africa under stones and in litter. South American species of the genera *Pseudothelphusa*° and *Valdivia*° also are terrestrial. As in crayfish the young are born in an advanced state of development. There are no planktonic stages. *Potamon berardii* has been observed to have only deciduous leaf litter in its stomach, but young animals, less than 1 cm wide, have one-third aquatic insects. They aim a stream of water at an intruder if disturbed. *Trichodactylus* petropolitanus is found in Brazil.

Xanthidae, Mud Crabs. These crabs are oval to hexagonal, broad in front. There are many species, some on coral reefs. Three or four species of *Xantho* are found in European waters. *Eriphia gonagra,* 44 mm wide, from the Carolinas to Patagonia, occupies various habitats, including coral reefs. *Menippe mercenaria,* the stone crab (Fig. 13–33b), 12.7 cm wide, occurs from the Carolinas to Yucatan and the West Indies. The young, 1.2 cm wide, move from deep water among shell fragments to shallow waters among oysters and pilings. When full grown they move to a shoal and make burrows 18 cm in diameter and 30–50 cm deep. They spawn in spring. Females have been observed to molt, immediately mate after spawning in the laboratory, and

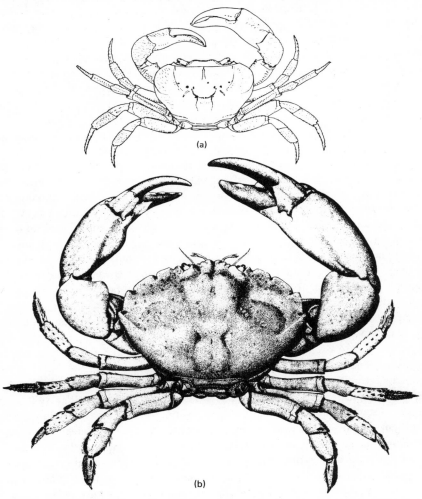

(a)

(b)

Fig. 13–33. Brachyuran crabs, Brachyrhyncha: (a) *Potamonautes johnstoni,* Potamidae; carapace 4 cm long. (From Chace.) (b) Stone crab, *Menippe mercenaria,* Xanthidae; carapace 8.5 cm long. (After Rathbun from Williams.)

spawn again immediately after the previous eggs have hatched. Females have been observed with six egg masses in 69 days, each with 0.5–1 million eggs. They can tolerate reduced salinity. This crab has been observed to prey on small and large oysters, consuming as many as 219 oysters per crab per year. *Pseudocarcinus* gigas** of the south coast of Australia, to 40 cm wide, is edible. *Lybia* uses actinians to defend itself. *Pilumnus** is a widespread genus with numerous species on the southeast coast of North America; *P. hirtellus,** 2 cm wide, in the North Sea has been observed to feed on the clam *Tellina.* Other genera are *Heteropanope** and *Rhithropanopeus.** Several xanthids may be poisonous if eaten: *Zosimus* aeneus** from the Ryukyu and Amami islands in the Pacific (although edible in some areas) and *Atergatis* floridus* from the Indo Pacific. The poison is found in the exoskeleton and is similar in action to the neurotoxin saxitoxin from the protistan *Gonyaulax,* which contaminates some mollusks (Konosu *et al.,* 1968).

Goneplacidae. The members resemble Xanthidae but the body is usually rectangular, the sides more or less parallel (Fig. 13–34a). All are bottom dwellers. *Goneplax* hirsuta,* to 19 mm long, 29 mm wide (Fig. 13–34a) is found at 20–40 m a depth from the Carolinas to Brazil. *Euryplax* nitida,** to 25 mm wide is found in shallow water to 25 m from Carolinas to Texas and West Indies.

Geryonidae. Geryon has a hexagonal, laterally dentate carapace with a quadridentate front (Fig. 13–34b). It has long compressed legs. The larger of the two species from the western Atlantic attains a carapace width of more than 16 cm and it may eventually prove to be of economic importance as food. *Geryon* tridens,** to 7.4 cm wide, occurs on northern European coasts at depths of 450–1800 m; *G. longipes* is found in the Mediterranean.

Pinnotheridae. The carapace is membranous, the anteriolateral margins entire. Orbits and eyestalks are small (Fig. 13–34c). Species of these small crabs live within bivalves, ascidians, and worm tubes; others burrow. *Pinnotheres** includes more than 120 commensal species; *P. pisum,* pea crab, with males to 1 cm wide, females to 1.8 cm, lives free in the mantle of various bivalves of the eastern Atlantic and Mediterranean. The first crab stage following the megalopa has long swimming hairs on the walking legs and searches for a host. It molts within the host to become a soft-shelled, second instar crab that cannot swim and does not leave the host. The next stage is hard shelled and has swimming hairs; at times they remain free-living, the males searching for females. After mating the males (of *P. ostreum*) die. The immature mated females molt again into a soft stage that, like later stages, remains in the clam. In *P. ostreum* (Fig. 13–34c) the seventh female stage matures and spawns. *Pinnotheres pisum* produces only about 6000 eggs. *Pinnotheres ostreum,* the oyster crab, 4–15 mm, occurs from Massachusetts to Brazil. Chiefly found in oysters, it invades small spat; two crabs were found in an oyster only 4.2 mm long. The crab feeds on food strings picked from the mantle of the oyster with its chelipeds. The crabs harm their hosts by causing gill erosion. In the mussel crab, *P. maculatus,* 14 mm wide, from Massachusetts to Argentina, the males may survive the hard stage; *P. puggettenis* is found in tunicates of Puget Sound; *Pinnixa* chaetopterana,* the parchment worm crab, 14 mm wide, from Massachusetts to Brazil, lives commensally with *Chaetopterus* and *Amphitrite,* rarely outside the tubes. The crabs are thigmotactic and will enter glass tubes left lying on the sand. Within tubes the crabs brace themselves with all

legs. The respiratory current of the crab is weak and only mouthparts, eyes and antennae are cleaned. If the tube of the host is too small to leave, the crab bites
through it.

    Ocypodidae.° Orbits, their outer walls indistinct, occupy the entire anterior border
of carapace. The crabs are amphibious, living in littoral waters and estuaries. Ghost
crabs, *Ocypode,*° inhabitants of sandy beaches, include 20 species in warm waters, one,
O. cursor, in the Mediterranean. *Ocypode quadrata,* 44 mm long, 50 mm wide, occurs
from Rhode Island to Est. Santa Catarina, Brazil. In late spring, females carry eggs.
At intervals they unfold the abdomen, turn upside down and rotate the eggs and force
water through the egg mass. The young have vertical burrows near the water. Older
crabs make their burrows farther away from the water, at an angle 45° away from
shore with a vertical shaft almost to the surface; burrows higher on the beach may lack
the vertical branch. The burrows reach the water level. Burrowing is done at daytime,
the excavated sand spread out near the mouth of its burrow. At noon the burrows may
be plugged. In the evening the whole population appears on the beach to feed as
scavengers or predators. They enter water only to moisten gills or when disturbed, but

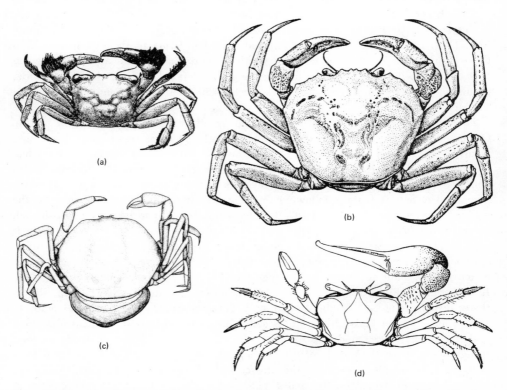

(a)

(b)

(c)

(d)

Fig. 13–34.   **Brachyuran crabs, Brachyrhyncha: (a)** *Goneplax hirsuta,* **Goneplacidae;
carapace 1.9 cm long. (After Borradaile from Williams.) (b)** *Geryon affinis,* **Geryonidae;
carapace 13 cm long. (From Monod.) (c) Oyster crab,** *Pinnotheres ostreum,* **Pinnotheridae; carapace 1.1 cm long. (From Williams.) (d) Fiddler crab,** *Uca pugilator,*
Ocypodidae; male, carapace 1.7 cm long. (From Rathbun.)

usually wait above the water for a wave to wet them. The crabs usually run sideways but at times slowly backward or forward. The eyes are sensitive to changes of light intensity, but the crabs do not avoid bright light and may hide when the light is turned off or if an object on the beach moves. In the Indo-Pacific region eastward to Hawaii, in mangroves, tropical *O. ceratophthalmus** uses the chelipeds to turn leaves and branches and to snatch up flies that alight on the underside in the shadow. The ommatidia of *Ocypode* nearly surround the vertical eyestalk and each cylindrical compound eye views 360° in the horizontal plane. Approached within 15–20 m by a human, adult crabs flee to their burrows; even if the mound surrounding the burrow is leveled, the crab recognizes the site. Laboratory experiments indicate that they orient by polarized skylight. Males of *O. saratan* of the Red Sea shores erect a sand pyramid 16 cm high, 40 cm from the mating burrow. The pyramid is connected to the burrow by a tamped path. The pyramid attracts females, which then follow the path into the burrow. The male uses vibration signals only when the female is close to the burrow. The pyramids also stimulate mature males to territorial defense and to build their own pyramids. The sand pyramids spaced at least 1.35 m apart rather than the animal itself guides the females to the burrow where the male waits. The male stays 4–8 days during which time he does not feed. After feeding he returns to conquer a new pyramid area. After the mating period in April, the males dig simple resting burrows near the water line like the females.

Uca,* fiddler crabs (Fig. 13–34d), with about 65 species, are found on the coasts of most warm and tropical oceans. Fiddler crabs are small, usually only 0.8–3.5 cm wide and some Indo-Pacific species are colorful; coral red, bright green, and golden yellow or light blue. The males have one very large cheliped, often accounting for half the total weight of the animal. Most species live close to the tide zones, usually in open sunny areas, rarely among mangroves or other plants, but always in quiet places where the flood tide brings up detritus and deposits it in a thin layer. Most populations densely occupy the empty tidal areas and shallow pools with as many as 50 individuals per m². Depending on their size they build vertical burrows 12–50 cm deep to ground water; above the high tide line burrows may be 1 m deep. The animals moisten their gills in the burrow at any time but during flooding respire air, as the water does not fill the covered burrows. In burrowing the crab presses into the mud several pereopods on one side, bends the legs, and lifts up a clump of mud, embraced even by the small cheliped. With the free pereopods *Uca* carries the mud 10–15 cm before depositing it. The hole is made deeper and deeper. Some species make a chimney around the burrow entrance. Half an hour before high tide, members of the Indian species, *U. vocans* (=*U. marionis*), close their burrows from inside with fresh mud. They reemerge at low tide. The significance of this behavior can be understood by placing a fiddler crab in the water during high tide. Rolled around by the waves, the fiddler makes corkscrew movements of the body, attempting to burrow backward into the bottom. During low tide the crabs enter their burrows if a crow flies over at 3–8 m or if a man approaches. In the presence of man *Uca tangeri* has a flight distance of 25–30 m; at 6–10 m, the crab disappears into its burrow. The eye position and visual field are as in *Ocypode*. By color marking it has been demonstrated that individuals of *U. vocans* stay close to their burrows for only 30 min after the water has left. After this period some individuals move as far as 66 m and flee if need be into the nearest burrow. Such wandering crabs build a new burrow before high tide as far as 40 m from the old one. *Uca annulipes* behaves the same way on mud flats, while in rocky areas, grass or mangroves, it may use the same burrows for considerable periods. Mud flat fiddler crabs feed by going from the burrow in various directions and returning to it, resulting in radiating lines of food remains. Algae, mud, fruit, and fish carcasses are eaten.

The activity rhythm is synchronous with the tides in Indian and American species that have their burrows in areas flooded daily. Four behavior periods follow after emergence from the burrow at low tide. The first is a feeding period. This is followed by a period of claw waving and fighting, in *U. vocans* starting 31–60 min after the beginning of low tide. It is distinguished by fighting between males or females alternating with feeding, running about in groups and claw waving by the males. The male raises the large cheliped in vertical position, then lowers it. In some species the movement is complicated by horizontal stretching before raising the claw, or the small cheliped may be moved at the same time while the legs make "knee bends." The species differ in the frequency of waving, with intervals of ¼, ¾ or 1 sec. In Indian species this waving is courtship. In some species it stimulates the active males as well as the females. In *Uca annulipes* it attracts females, which follow the waving male to its burrow where mating takes place. The males wave when they see one of their species that lacks a large cheliped or holds it so that it cannot be seen. The females are thus recognized by a negative character. This behavior period lasts 1–2 hr. The following period is one of burrow building. At this time the crabs do not wave or fight even when disturbed. Finally, about 1 hour before high tide, there is another feeding period followed by closing of the burrow, so that the two Indian species disappear from the scene at least half an hour before flooding.

Not all *Uca* species have such a regular activity rhythm. The South American *U. maracoani* may exhibit all activity phases within a day or during a half moon phase. The West African *U. tangeri*, also found in the Bay of Cadiz of the Iberian Peninsula lives on solid mud or in dense reeds above the median high water mark and has no typical rhythmic activity. They wander about on the recently flooded areas looking for food until the water returns and do not wave during this time. Building is done in the evening. The waving occurs on harder, higher areas of mud in the morning, independent of the high tides.

Threatening of males and waving are very distinct. Males wave even when no females are close. In *Uca tangeri* the waving intensifies to a lateral stretching of the large cheliped if the female approaches within 35 cm. This species also courts by beating the large cheliped on the ground, especially in reeds where visibility is poor. Here it could be shown that the vibrations are carried along the ground. Other species combine waving and beating.

Dotilla has habits resembling those of *Uca* but does not have an enlarged cheliped and has no distinct waving movement. *Dotilla blanfordi*, 14 mm wide, feeds by shaving off the mud with closed chelae. The shaking sieve feeding method is the reverse of that used by *Uca* so that the unusable particles appear at the tips of the maxillipeds. It withdraws at high tides into vertical channels 5–10 cm deep. In some species of *Macrophthalmus* both eyestalks together are slightly longer than the width of the carapace. A related genus is *Scopimera*.

Palicidae.° Anterolateral margins of the carapace are dentate. The last pair of pereopods is dorsally placed and smaller than the others (Fig. 13–35c). *Palicus*° *alternatus* to 9 mm wide, from 2 to 30 m deep, is found from North Carolina to west coast of Florida; *P. faxoni,* to 11 mm wide (Fig. 13–35c), is found from North Carolina to Yucatan at depths from 16 to 25m.

Mictyridae. The carapace is spherical, orbits indistinct. The rudimentary flagella of the first antennae fold vertically. There are only a few Indo-Pacific species, all belonging to one genus. The soldier crab *Mictyris longicarpus,* about 2 cm wide, inhabits beaches and large numbers in close formation traverse the beaches together. If chased all disappear within seconds into the mud, lying on one side to burrow with the legs of that side only, while slowly turning, penetrating like a corkscrew. The hole is closed from inside. They also bury themselves at high tide, to emerge when the water leaves.

Grapsidae. The carapace is usually square, the front usually wide, sides generally straight or slightly arched, the orbits usually at the anterolateral angles (Fig. 13–35a). The crabs are found on rocky shores, but some are terrestrial, and others live among mangroves or enter freshwater. *Geograpsus* are tropical, venturing 5 m away from the high tide line. The inhalant opening is a pore between third and fourth pereopod coxae. *Grapsus* lives on cliffs in the surf; using spine-tipped legs to hold on; it can run rapidly sideways on almost vertical cliffs to hide in a crack. *Grapsus grapsus* observed on the Galapagos walk slowly forward, flee sideways and can leap from rock to rock and also swim short distances. They pick algae with their chelae, and take meat and small animals as food. The males when threatening raise the body and make ritualized movements with the chelipeds (Kramer, 1967). *Pachygrapsus*° *marmoratus,* 3–3.2 cm wide, is found on Mediterranean coasts as a cliff inhabitant. At 20°C the crab can stay out of water 5 days, at 28°C only 10 hr, at 35°C only 1 hr. In California, *P. crassipes*° lives in rocky areas hiding in cracks and fissures. *Pachygrapsus* feeds on fish carrion, small living animals, and the algal film on cliffs, which is slowly scraped off with alternate cheliped tips and is spooned to the mouth. In captivity they remain out of water 12 hr daily. *Planes*° *minutus,* the Gulf weed crab, 1.1–1.4 cm long, occurs in warm and tropical oceans, only at or near the surface of the open sea, by holding onto turtles, pelagic snails, driftwood, sargassum weeds; most abundant in the Sargasso Sea, it occurs from tropical oceans to the North Sea, Newfoundland, and the Mediterranean, but is absent from the Gulf of Mexico.

Males of the Chinese mitten crab, *Eriocheir sinensis* (Fig. 13–35b), up to 7.5 cm wide, have "fur" on the chelipeds. Native to low-lying areas in China, they were found in Europe for the first time in 1912 in the Aller, a north German river. Perhaps specimens were transported with ballast water in tanks of ships from Chinese harbors and later, when pumped out, entered European rivers. The crab rapidly spread in Germany and by 1928 reached up the Elbe River to Saxony; it now reaches upstream as far as Prague. It was found in the Rhine in 1930, in the Seine in 1943, but the greatest abundance remains in the Elbe and Weser, from which it is spreading over low-lying areas of Germany, Scandinavia, Holland, and Belgium. In the Elbe drainage, 500,000 kg of mitten crabs were fished in 1935. Since that time the infestation has been reduced. The postembryonal development has to take place in the ocean. The habits of the zoeae, 1.7 mm long, are known only from aquarium observations. Feeding on algae, they pass through three further planktonic stages before reaching the megalopa stage within 5 weeks. The last metazoeae and the 5 mm long megalopas are carried upstream by tidal currents. The megalopa is a good swimmer and walker. Within 3 months the metazoeae and megalopas have travelled 100 km. The young crabs remain a year within the tidal areas of rivers but the following spring, at 1¾ years of age and 20–25 mm long, the crabs go upstream in deep water at night, often following each other in long rows. Dams and obstacles are overcome by walking on land and as many as 90,000 crabs a day have been collected in pit traps. By color marking at dams it can be shown that 1–3 km are traversed daily by animals 25–35 mm long. The migration stops by fall when the migrants have moved into ponds and ditches where they remain until mature, others continue to migrate in spring. The mature crabs, 5 years old, begin migrating back to the ocean in mid-July, moving 8–12 km daily. The males return before the females and wait in huge numbers at the outlets of rivers. The ovaries mature as the crabs migrate out into deeper water to spawn in November. The young hatch in late spring. After spawning only once, all females die in shallow waters. The stomach contents indicate that the crabs are mainly vegetarians but consume some animal food such as mollusks, hydroids, and arthropod larvae. Fishes are attacked only at weirs. In feeding the crab runs backward, stirring up mud. In the tidal areas the crabs make burrows into which they withdraw at low tide and during the day. The large numbers of these introduced crabs are tremendously destructive. Caught in

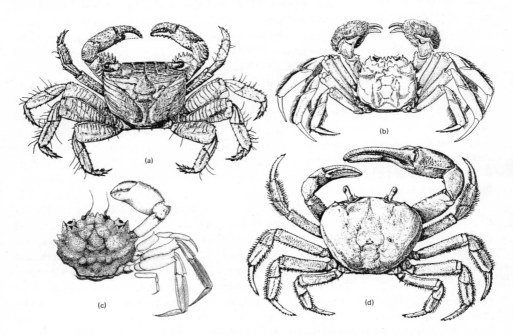

Fig. 13–35.   Brachyuran crabs, Brachyrhyncha: (a) *Goniopsis cruentata*, Grapsidae; male, carapace 3.2 cm long. (From Monod.) (b) Chinese mitten crab, *Eriocheir sinensis*, Grapsidae; male, carapace 7 cm long. (From Schellenberg.) (c) *Palicus faxoni*, Palicidae; male, left legs removed; carapace 1.5 cm long. (From Williams.) (d) Land crab, *Cardisoma armatum*, Gecarcinidae; male, carapace 6.7 cm long. (From Monod.)

fishermen's nets, they damage fishes and tear nets. They also eat bait out of traps. While the Chinese mitten crab is considered a delicacy in China, it has been used in Europe only occasionally for animal fodder. The hard shell makes their use difficult.

To 26 mm wide, *Aratus pisonii* lives on mangroves in America, climbing up branches and chewing leaves. It is found from Florida to Brazil and from Nicaragua to Peru. In Jamaica the crabs breed throughout the year. They mate above the waterlevel, male uppermost. The incubation period is 16 days. The eggs hatch at both full and new moon as prezoeae, causing the female to enter the water head first, to vibrate the abdomen sending clouds of larvae into the water. The crabs mature when about 6 months old and have a life span of 2–3 years (Warner, 1967). *Metaplax* ranges from India eastward to China and Japan.

The only completely freshwater grapsid crabs are four endemic crabs in Jamaica, West Indies, whose larvae even mature in freshwater: *Metopaulias depressus*, the pine crab, the river crab *Sesarma bidentatum* lives in upper reaches of rivers; *S. jarvisi* may be terrestrial and *S. verleyi* inhabits water in caves but leaves the water (S. Peck, personal communication). *Metopaulias depressus*, with the carapace to 20.8 mm wide, is widespread on Jamaica; it lives at elevations up to 900 m in bromeliads that have a pool of water at each leaf base, many individuals on a single plant. They leave the water as scavengers. The size of the eggs of *M. depressus* is larger and their number smaller than marine relatives. A first *zoea* hatches, followed by two *zoea* stages that do not feed and a megalopa that does feed. *Sesarma bidentatum* to 27 mm wide has a first zoea, 1.8 mm, with abundant yolk. It does not feed. There are two zoeal stages within 3 days before the megalopa which does not swim (Hartnoll, 1964); marine *Sesarma* have three to four zoeal stages before the megalopa.

More than 150 species of *Sesarma* are found on coasts of tropical and subtropical oceans. Some species go up streams as far as 1000 m altitude and are terrestrial in undisturbed forests of south east Asiatic islands. The coastal inhabitants dig burrows 3–8 cm wide, to 1.5 m deep, down to ground water at the blind end. The mud chimney around the entrance may be 20 cm high. The crabs go as far as 30 m from their burrows, returning rapidly if alarmed. The flight distance from a walking man is 10–15 m. Some avoid high tide by climbing mangroves, others allow themselves to be covered by water. On the east coast of North America there are two species. *Sesarma reticulatum*, 28 mm wide, occurs from Massachusetts to Texas; they build communal burrows with a male and several females. *Sesarma cinereum*, the wharf crab, 23 mm wide, from Maryland to Venezuela, has been kept over 4 years in captivity. They have been observed to feed on land. In water they are rather inactive, their oxygen consumption being reduced.

Gecarcinidae. Land crabs. These crabs return to the ocean only during reproductive periods. They are found in tropical America, West Africa and the Indo-Pacific. All are omnivorous. *Cardisoma* (Fig. 13–35d) lives in open fields, mangroves, shrubby forests, and coconut plantations, plowing up the ground with its burrows. *Cardisoma guanhumi,* found in Bermuda, West Indies, Florida and rarely to Texas, weighs about 500 g when 11 cm wide. It lives above the tide level and moves as far as 8 km inland. The burrows are 0.3–2 m deep. In shady areas they search for food at any time of day, run toward falling fruit, cut off leaves from branches lying on the ground, and carry them to burrows for feeding. They also take carrion. They rarely go more than 2 m from the burrow entrance, partially due to the territorial behavior of other crabs. In a terrarium they return to water daily for 2 hr. After leaving the water, they rear up vertically, driving the water out of the anterior openings of the branchial chambers. They accept ocean water, brackish- or freshwater, and can survive submerged in running seawater for 6 months. Their blood in ocean water is Δ −1.7°C, in freshwater Δ −1.4°C. The eggs mature in lunar and semilunar periods and are carried by the mother to protozoea or zoea stages. On full-moon nights large numbers of crabs migrate to the ocean to shake off the eggs.

Epigrapsus inhabits the Indo-Pacific, extending eastward to the Society Islands. *Gecarcoidea,* to 7.5 cm long, lives on the coast, inland terraces and central plateau of small southeastern Asiatic islands and large numbers run around at daytime in shrubs. The female enters the ocean only to shake off her brood of zoea larvae. The young may emerge in such numbers as to form a brown-red carpet on the beach. With body length of 12–14 mm, they live in humid spaces between roots and under stones. The branchiostegites do not yet have the width typical for adults. The branchiostegites grow allometrically. When 14–30 mm long the animals come more often out of their hideouts, but do not move far away. Only the adults respire dry air, retiring to their burrows only a few hours each day. *Gecarcinus,* found in tropical America and West Africa, with a carapace to 9 cm wide, often lives far from the ocean and large numbers may migrate together. This is the genus best adapted for terrestrial life; its burrows are only 30–40 cm deep. *Gecarcinus lateralis* is found from Bermuda and southern Florida to French Guiana. In its burrows, at a depth of 38 cm, the noon temperature was 30°C with 95–100% relative humidity, while next to the entrance the temperature was 50°C.

A fruit fly, *Drosophila carcinophila,* whose closest relatives live in deserts on cacti, is the strange symbiont of the West Indian and Bahaman land crab *Gecarcinus ruricola.* The young larvae live in the moist hairs of the excretory groove below the excretory pore together with symbiotic mites and nematodes. Older larvae crawl to the inner posteriolateral surface of the third maxilliped exopodite and where they pupate. The puparia are cemented to the smooth surface. Four to eight flies are active above the carapace feeding on microorganisms (H. Carson, 1967). *Ucides* lives in brackish water among mangroves. Its burrows to 80 cm deep, extend to ground water. *Ucides* may belong to the family Ocypodidae (F.A. Chace Jr., personal communication).

SUBSECTION HAPALOCARCINIDEA

Hapalocarcinidae. The carapace is elongate, convex, and twice as long as wide. The long abdomen does not fold tightly under the cephalothorax; in the female it forms a wide brood pouch for eggs and larvae by fusion of somites (Fig. 13–36). Males are only 1.2 mm long. The larvae are typical brachyuran zoeae. These crabs live in corals, producing water currents through their cavities, and feeding on plankton. Genera of the family are *Hapalocarcinus* (Fig. 13–36), female carapace to 5.5 mm long, *Cryptochirus* and *Troglocarcinus,* all found off southeastern Asia. *Troglocarcinus* is found also in West Atlantic waters; Bermuda, and the West Indies.

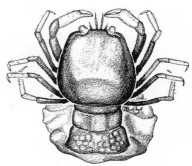

Fig. 13–36. Brachyuran crab, *Hapalocarcinus mar-
supialis*, Hapalocarcinidea; female, carapace 3.2 cm
long. (After Fize and Serène.)

References

Euphausiacea

Bargmann, H.E. 1937. The reproductive system of *Euphausia superba*. *Discovery Rep.*
14: 325–350.

———— 1945. The development and life history, *Euphausia superba*. *Ibid.* 23: 103–176.

Boden, B.P., Johnson, M.W., and Brinton, E. 1955. The Euphausiacea of the North Pacific.
Bull. Scripps Inst. Oceanogr. 6: 287–400.

Brinton, E. 1962. The distribution of Pacific euphausiids. *Ibid.* 8: 51–270.

———— 1966. Euphausiacean phylogeny. *Proc. Symp. Crustacea, Ernakulum* 1: 255–259.

Fisher, L.R. and Goldie, E.H. 1959. Food of *Meganyctiphanes norvegica*. *J. Marine Biol.
Ass.* 38: 291–312.

———— et al. 1955. Vitamin A and carotenoids in Euphausiacea. *Ibid.* 34: 81–100.

Kampa, E.M. 1965. The euphausiid eye. *Vision Res.* 5: 475–481.

———— and Boden, B.P. 1957. Light generation in a sonic-scattering layer. *Deep Sea Res.*
4: 73–92.

Manton, S.M. 1928. Anatomy and habits of the lophogastrid Crustacea. *Trans. Roy. Soc.
Edinburgh* 56: 103–119.

Marr, J.W.S. 1962. The natural history and geography of the Antarctic krill, *Euphausia superba*. *Discovery Rep.* 32: 37–464.

Mauchline, J. 1958. The circulatory system of the crustacean euphausid, *Meganyctiphanes*.
Proc. Roy. Soc. Edinburgh. 67B: 32–41.

———— 1959. The development of the Euphausiacea, especially that of *Meganyctiphanes*.
Proc. Zool. Soc. London 132: 627–639.

———— 1960. The biology of the euphausiid crustacean, *Meganyctiphanes norvegica*.
Proc. Roy. Soc. Edinburgh 67B: 141–179.

———— 1967. Feeding appendages of the Euphausiacea. *J. Zool. London* 153: 1–43.

Ponomareva, L.A. 1966. Euphausiids of the North Pacific, their distribution and ecology.
[Transl. Russian in *Acad. Sci. U.S.S.R. Inst. Oceanol.,* 1963] *Israel Program for Scien-
tific Translations:* 1–54.

Taube, E. 1909. Entwicklungsgeschichte der Euphausiden; die Furchung des Eies bis zur
Gastrulation. *Z. Wiss. Zool.* 92: 427–464.

———— 1915. Entwicklungsgeschichte der Euphausiden; von der Gastrula bis zum
Furciliastadium. *Ibid.* 114: 577–656.

Zimmer, C. and Gruner, H.E. 1956. Euphausiacea *in Bronns Klassen und Ordnungen des
Tierreichs*. Akad. Verl., Geest & Portig, Leipzig 5(1) 6(3): 1–286.

Decapoda

Abrahamczik-Scanzoni, H. 1942. Muskulatur und Innenskeletts der Krabben. *Zool. Jahrb. Abt. Anat.* 67: 293–380.

Alexandrowicz, J.S. 1953. Nervous organs in the pericardial cavity of the decapod Crustacea. *J. Marine Biol. Ass.* 31: 563–580.

———. 1967. Receptor organs in thoracic and abdominal muscles of Crustacea. *Biol. Rev.* 42: 288–326.

Allen, J.A. 1960. The biology of *Crangon allmani. J. Marine Biol. Ass.* 39: 481–508.

———. 1963. The biology of *Pandalus montagui. Ibid.* 43: 665–682.

Altevogt, R. 1957. Biologie, Ökologie und Physiologie indischer Winkerkrabben. *Z. Morphol. Ökol.* 46: 1–110.

———. 1957. Biologie und Ethologie von *Dotilla. Ibid.* 46: 369–388.

———. 1959. Ökologische und ethologische Studien an Winkerkrabbe *Uca tangeri. Ibid.* 48: 123–146.

———. 1959. Rennkrabben. *Natur und Volk* 89: 129–133.

———. 1964. Eine Klopfkode und eine neue Winkfunktion bei *Uca tangeri. Naturwissenschaften* 51: 644–645.

Arvy, L. 1953. Sang et la leucopoïèse chez *Pachygraspsus marmoratus. Ann. Sci. Nat. Zool.* (11) 14: 1–13.

Atkins, D. 1954. Leg deposition in the Brachyuran megalopa when swimming. *J. Marine Biol. Ass.* 33: 627–636.

Balss, H. 1927. Decapoda *in* Kükenthal, W. and Krumbach, T. *Handbuch der Zoologie.* de Gruyter, Berlin 3: 840–1038.

———. *et al.* 1940–1957. Decapoda *in Bronns Klassen und Ordnungen des Tierreichs.* Akad. Verl., Geest & Portig, Leipzig 5(1) 7: 321–480.

Barnwell, F.H. 1966. Daily and tidal patterns of activity in individual fiddler crabs. *Biol. Bull.* 130: 1–16.

———. 1968. The role of rhythmic systems in the adaptation of fiddler crabs to the intertidal zone. *Amer. Zool.* 8: 569–583.

Batham, E.J. 1967. The larval stages and feeding behavior of phyllosoma of the palinurid crayfish *Jasus edwardsii. Trans. Roy. Soc. New Zealand* 9: 53–64.

Bauchau, A. 1966. *La Vie des Crabes.* Lechevalier, Paris.

Bennett, M.F. 1963. The phasing of the cycle of motor activity in the fiddler crab, *Uca pugnax. Z. Vergl. Physiol.* 47: 431–437.

——— and Brown, F.A. 1959. Experimental modification of the lunar rhythm of running activity of the fiddler crab, *Uca pugnax. Biol. Bull.* 117: 404.

Bernhards, H. 1916. Der Bau des Komplexauges von *Astacus fluviatilis. Z. Wiss. Zool.* 116: 649–707.

Bliss, D.E. 1963. The pericardial sacs of terrestrial brachyura, *in* Whittington, H.B. and Rolfe, W.D.I. ed. *Phylogeny and Evolution of Crustacea.* Mus. Comp. Zool., Cambridge, Mass. p. 59–78.

———. 1968. Transition from water to land in decapod crustaceans. *Amer. Zool.* 8: 355–392.

——— and Mantel, L.H. 1968. Adaptations of Crustacea to land. *Ibid.* 8: 673–685.

Bock, F. 1925. Die Respirationsorgane von *Potamobius astacus. Z. Wiss. Zool.* 124: 51–117.

Borchert, H.-M. and Jung, D. 1960. Mitteilung über den Erstfund einer Süsswassergarnele, *Atyaephyra desmaresti. Zool. Beitr.* (N.F.) 5: 365–366.

Bott, R. 1950. Die Flusskrebse Europas (Decapoda, Astacidae). *Abhandl. Senckenbergischen Naturf. Ges.* 483: 1–36.

Bouvier, E.L. 1940. Décapodes marcheurs *in Faune de France,* 37: 1–399.

Brock, F. 1926. Das Verhalten des Einsiedlerkrebses *Pagurus arrosor* während der Suche und Aufnahme der Nahrung. *Z. Morphol. Ökol.* 6: 415–552.

———— 1949. Das Verhalten dekapoder Crustaceen beim Aufbau des Beutefeldes. *Verh. Deutschen Zool. Ges.* 12: 416–426.

Brown, F. et al. 1954. Proof of endogenous rhythm. *J. Cell. Comp. Physiol.* 44: 477–506.

Bryan, G.W. 1960. Sodium regulation in the crayfish *Astacus fluviatilis*. *J. Exp. Biol.* 37: 83–128.

Bückmann, D. and Adelung, D. 1964. Der Einfluss der Umweltfaktoren auf das Wachstum und den Häutungsrhythmus der Strandkrabbe *Carcinides maenas*. *Helgoländer Wiss. Meeresunters.* 10: 91–103.

———— 1965. Der Einfluss der innersekretorischen Organe auf den Häutungsrhythmus von *Carcinus maenas*. *Verh. Deutschen Zool. Ges.* 28: 131–136.

Burger, J.W. 1957. The general form of excretion in the lobster, *Homarus*. *Biol. Bull.* 113: 207–223.

Bush, B. 1965. Proprioreception by chordotonal organs in the joints of *Carcinus maenas* legs. *Comp. Biochem. Physiol.* 14: 185–199.

Cochran, D.M. 1935. The skeletal musculature of the blue crab, *Callinectes sapidus*. *Smithsonian Misc. Coll.* 92: 1–76.

Cohen, M.J. 1960. The response patterns of single receptors in the crustacean statocyst. *Proc. Roy. Soc. London* 152B: 30–49.

Carlisle, D.B. 1959. Moulting hormones in *Palaemon*. *J. Marine Biol. Ass.* 38: 351–359.

———— 1959. On the sexual biology of *Pandalus borealis*. *Ibid.* 38: 381–394; 481–506.

Carson, H. 1967. Association between *Drosophila carcinophila* and its host the land crab *Gecarcinus ruricola*. *Amer. Midland Natur.* 78: 324–343.

Case, J. and Gwilliam, G.F. 1961. Amino acid sensitivity of the dactyl chemoreceptors of *Carcinides maenas*. *Biol. Bull.* 121: 449–455.

Chace, F.A. *et al.* 1959. Malacostraca *in* Edmondson, W.T. ed. *Ward and Whipple Freshwater Biology.* Wiley, New York.

Chaisemartin, C. 1964. Gastrolithes dans l'économie due calcium chez *Astacus pallipes*. *Vie et Milieu* 15: 457–474.

Chassard, C. 1956. Polymorphisme des populations d'*Hippolyte varians* et comportement en fonction de leur adaptation chromatique présente. *Bull. Soc. Zool. France* 81: 413–418.

Chassard-Bouchaud, L. 1965. L'adaption chromatique chez les Natantia. *Cah. Biol. Marine* 6: 469–576.

Choudonneret, J. 1956. Le système nerveux de la région gnathale de l'écrevisse, *Cambarus affinis*. *Ann. Sci. Nat. Zool.* (11) 18: 33–61.

Christensen, A.M. and McDermott, J.J. 1958. Life history and biology of the oyster crab, *Pinnotheres ostreum*. *Biol. Bull.* 114: 146–179.

Claus, C. 1886. Zur Morphologie der Crustaceen. *Arb. Zool. Inst. Wien* 6: 1–108.

Cook, I. 1964. Electrical activity and release of neurosecretory material in crab pericardial organs. *Comp. Biochem. Physiol.* 13: 353–366.

Copeland, D.E. 1968. Fine structure of salt and water uptake in the land crab *Gecarcinus*. *Amer. Zool.* 8: 417–432.

Crane, J. 1957. Basic patterns of display in fiddler crabs. *Zoologica*, New York 42: 69–82.

———— 1958. Aspects of social behavior in fiddler crabs with special reference to *Uca maracoani*. *Ibid.* 43: 113–130.

———— 1967. Combat and ritualization in fiddler crabs. *Ibid.* 52: 49–76.

Cubit, J. 1969. Behavior causing migration and aggregation in *Emerita analoga*. *Ecology* 50: 118–123.

Dall, W. 1965. Composition of *Metapenaeus* and structure of the integument. *Australian J. Marine Freshwater Res.* 16: 13–23.

_____ 1967. Functional anatomy of the digestive tract of *Metapenaeus*. Ibid. 15: 699–714.

Daumer, K., Jander, R. and Waterman, T. 1963. Orientation of the ghost crab *Ocypode* in polarized light. *Z. Vergl. Physiol.* 47: 56–76.

Davis, W. 1968. The neuromuscular basis of lobster swimmeret beating. *J. Exp. Zool.* 168: 363–378.

Dawson, C.E. and Idyll, C.P. 1951. The Florida spiny lobster, *Panulirus argus*. *Univ. Miami, Marine Lab. Tech. Ser.* 2: 1–38.

Degkwitz, E. 1957. Natur der proteolytischen Verdauungsfermente bei verschiedenen Crustaceenarten. *Veröff. Inst. Meeresf. Bremerhaven* 5: 1–13.

Demal, J. 1953. Genèse et différenciation d'hemocytes chez *Palaemon varians*. *Cellule* 56: 85–102.

Demeusy, N. 1957. Respiratory metabolism of the fiddler crab *Uca pugilator* from two different latitudinal populations. *Biol. Bull.* 113: 245–253.

Dijkgraaf, S. 1963. Statocysten als Drehsinnesorgane. *Ergebn. Biol.* 26: 63–65.

Edwards, E. 1966. Mating behaviour of *Cancer pagarus*. *Crustaceana* 10: 23–30.

Efford, I.E. 1966. Feeding in the sand crab *Emerita analoga*. Ibid. 10: 167–182.

_____ 1967. Neoteny in *Emerita*. Ibid. 13: 81–93.

Fage, L. 1927. Le "Stade Natant" de la langouste commune *Palinurus vulgaris*. *Arch. Zool. Expér. Gén.* 67: 32–38.

Fingerman, M. 1957. Relation between position of burrows and tidal rhythm of *Uca*. *Biol. Bull.* 112: 7–20.

_____ 1965. Chromatophores. *Physiol. Rev.* 45: 296–336.

_____ 1966. Neurosecretory control of pigmentary effectors in crustaceans. *Amer. Zool.* 6: 169–179.

_____ and Yamamoto, Y. 1967. Daily rhythm of melanophoric pigment. *Crustaceana* 12: 303–318.

Fish, J.F. 1966. Sound production in the American lobster, *Homarus americanus*. Ibid. 11: 105–106.

Fishelson, L. 1966. Habitat and behavior of *Alpheus frontalis*. Ibid. 11: 98–104.

Forster, G.R. 1951. Biology of the common prawn *Leander serratus*. *J. Marine Biol. Ass.* 30: 333–360.

Garth, J.S. 1958. Brachyura of the Pacific coast of America, Oxyrhyncha. *Allan Hancock Pacific Exped.* 21: 1–854.

Ghiradella, H.T. *et al.* 1968. Structure of aesthetascs in marine and terrestrial decapods. *Amer. Zool.* 8: 603–621.

_____ 1968. Aesthetasc hairs of *Coenobita*. *J. Morphol.* 124: 361–386.

Gifford, C.A. 1962. Biology of the land crab, *Cardisoma guanhumi* in south Florida. *Biol. Bull.* 123: 207–223.

_____ 1968. Accumulation of uric acid in the land crab *Cardisoma guanhumi*. *Amer. Zool.* 8: 521–528.

Gilles, R. 1967. Biochimiques de l'euryhalinie chez les crustacés. *Ann. Soc. Zool. Belgique* 97: 31–34.

Goldsmith, T. and Fernandez, H. 1968. Studies of crustacean spectral sensitivity. *Z. Vergl. Physiol.* 60: 156–175.

Gray, I.E. 1957. A comparative study of the gill area of crabs. *Biol. Bull.* 112: 34–43.

Grobe, J. 1960. Putz-Symbiosen swischen Fischen und Garnelen. *Natur und Museum* 90: 152–157.

Gross, W.J. 1964. Water and salt regulation among aquatic and amphibious crabs. *Biol. Bull.* 127: 447–466.

_____ 1964. Water balance in anomuran land crabs on a dry atoll. Ibid. 126: 54–68.

_____ and Marshall, L.A. 1960. The influence of salinity on the magnesium and water flexes of a crab. *Ibid.* 119: 440–453.

Guinot-Dumortier, D. and Dumortier, B. 1960. La stridulation chez les crabes. *Crustaceana* 2: 117–155.

Gurney, R. 1942. *Larvae of Decapod Crustacea.* Ray Soc., London.

Hagen, H.O. von. 1962. Freilandstudien zur Sexual - und Fortpflanzungsbiologie von *Uca tangeri. Z. Morphol. Ökol.* 51: 611–726.

Haig, J. 1960. Porcellanidae of the Eastern Pacific. *Allan Hancock Pacific. Exped.* 24: 1–440.

Harada, E. 1958. Naupliosoma and newly hatched phyllosoma of *Ibacus ciliatus. Publ. Seto Marine Biol. Lab.* 7: 173–180.

Harms, J.W. 1937. Lebenslauf und Stammesgeschichte von *Birgus latro. Z. Naturwiss.* 71: 1–34.

Hartnoll, R.G. 1964. The freshwater grapsid crabs of Jamaica. *Proc. Linnean Soc. London* 175: 145–169.

Hazlett, B.A. 1966. Social behavior of the Paguridae and Diogenidae of Curaçao. *Stud. Fauna Curaçao and other Caribbean Isl.* 23: 1–143.

_____ 1968. Communicatory effect of body position in *Pagurus bernhardus. Crustaceana* 14: 210–214.

_____ 1968. Sexual behavior of some European hermit crabs. *Pubbl. Staz. Zool. Napoli* 36: 238–252.

_____ 1969. Individual recognition and agonistic behavior in *Pagurus bernhardus. Nature* 222: 268–269.

_____ and Bossert, W.H. 1965. Analysis of aggressive communications of some hermit crabs. *Anim. Behaviour* 13: 357–373.

_____ and Winn, H.G. 1962. Sound production and associated behavior of Bermuda crustaceans. *Crustaceana* 4: 25–38.

Heegard, P. 1966. Larvae of decapod Crustacea, the oceanic penaeids. *Dana Rep.* 67: 1–147.

Heidje, W.D. van der. 1940. *Das Skelett und die Kiemen von Palinurus vulgaris im Vergleich mit denen von Astacus fluviatilis.* Verl. H.D. Paris, Amsterdam.

Herreid, C.F. 1963. Observations on the feeding behaviour of *Cardisoma guanhumi* in southern Florida. *Crustaceana* 5: 176–180.

Herrnkind, W.F. 1968. Adaptive visually directed orientation in *Uca pugilator. Amer. Zool.* 8: 585–598.

Hinsch, G.W. 1968. Reproductive behavior in the spider crab *Libinia emarginata. Biol. Bull.* 135: 273–278.

Hoffmann, C. 1963. Vergleichende Physiologie der mechanischen Sinne. *Fortsch. Zool.* 16: 269–332.

_____ 1964. Bau und Vorkommen von proprioceptiven Sinnesorganes bei den Arthropoden. *Ergebn. Biol.* 27: 1–38.

Holthuis, L.B. 1951, 1952. A general revision of the Palaemonidae of the Americas. *Allan Hancock Found. Occas. Pap.* 11: 1–332; 12: 1–396.

_____ 1955. The Recent genera of the caridean and stenopodidean shrimps with keys for their determination. *Zool. Verh. Leiden* 26: 1–157.

Horridge, G.A. 1966. Optokinetic memory in the crab, *Carcinus. J. Exp. Biol.* 44: 233–245.

_____ 1966. Perception of edges versus areas by the crab *Carcinus. Ibid.* 44: 247–254.

_____ 1966. Optokinetic response of the crab *Carcinus* to a single moving light. *Ibid.* 44: 263–274.

_____ 1966. Direct response of the crab *Carcinus* to the movement of the sun. *Ibid.* 44: 275–283.

_____ 1966. Adaptation and other phenomena in the optokinetic response of the crab *Carcius. Ibid.* 44: 285–295.

Huggins, A. and Munday, K. 1968. Crustacean metabolism. *Adv. Compar. Physiol. Biochem.* 3: 271–378.

Hughes, D. 1966. Behavioural and ecological investigations of *Ocypode ceratophthalmus. J. Zool. London* 150: 129–143.

Inoue, A. *et al.* 1968. A new toxic crab *Atergatis floridus. Toxicon* 6: 119–123.

Janisch, E. 1923. Der Bau des Enddarms von *Astacus fluviatilis. Z. Wiss. Zool.* 121: 1–63.

Johnson, D.S. 1967. Commensal decapod crustaceans from Singapore. *J. Zool. London* 153: 499–526.

_____ and Liang, M. 1966. Biology of a watchman prawn *Anchistus custos* a commensal of the bivalve *Pinna. Ibid.* 150: 433–455.

Johnson, M.W. *et al.* 1947. The role of the snapping shrimp *Crangon* and *Synalpheus* in the production of underwater noise in the sea. *Biol. Bull.* 93: 122–138.

_____ *et al.* 1960. The offshore drift of larvae of the California spiny lobster, *Panulirus interruptus. Rept. California Coop. Oceanic Fish. Invest.* 7: 147–161.

Kampa, E.M., *et al.* 1963. Vision in the lobster, *Homarus vulgaris,* in relation to the structure of its eye. *J. Marine Biol. Ass.* 43: 683–699.

Keim, W. 1915. Das Nervensystem von *Astacus fluviatilis. Z. Wiss. Zool.* 113: 485–545.

Keller, R. 1965. Hormonale Kontrolle des Polysaccharidstoffwechsels bei *Cambarus affinis. Z. Vergl. Physiol.* 51: 49–59.

Kinne, O. 1963. Adaptation, a primary mechanism of evolution. *in* Whittington, H.B. and Rolfe, W.D.I. ed *Phylogeny and Evolution of Crustacea.* Mus. Comp. Zool. Cambridge, Mass. p. 27–50.

Kirschner, L.B. 1966. Physiology of the crayfish antennal gland. *Proc. Symp. Crustacea, Ernakulum* 1: 17–27.

_____ and Wagner, S. 1965. The site and permeability of the filtration locus in the crayfish antennal gland. *J. Exp. Biol.* 43: 385–395.

Kleinholz, L. 1966. Separation and purification of Crustacean eyestalk hormones. *Amer. Zool.* 6: 161–167.

Knight-Jones, E.W. and Quasim, S.Z. 1955. Responses of some marine plankton animals to changes in hydrostatic pressure. *Nature* 175: 941.

Konosu, S. *et al.* 1968. Comparison of crab toxin with saxitoxin and tetrodotoxin. *Toxicon* 6: 113–117.

Korte, R. 1966. Sehvermögen einiger Dekapoden. *Z. Morphol. Ökol.* 58: 1–37.

Kramer, P. 1967. Biologie und Verhalten der Klippenkrabbe *Grapsus grapsus* auf Galapalos. *Z. Tierpsychol.* 24: 385–402.

Kümmel, G. 1964. Das Cölomsäckchen der Antennendrüse von *Cambarus affinis,* eine elektronenmikroskopische Untersuchung. *Zool. Beitr.* (N.F.) 10: 227–252.

Kunze, P. 1967. Histologische Untersuchungen zum Bau des Auges von *Ocypode. Z. Zellforsch.* 82: 466–478.

Laverack, M.S. 1962. Responses of cuticular sense organs of the lobster, *Homarus vulgaris.* Hair-peg organs as water current receptors. *Comp. Biochem. Physiol.* 5: 319–325.

_____ 1962. Responses of the cuticular sense organs of the lobster, *Homarus vulgaris.* Hair-fan organs as pressure receptors. *Ibid.* 6: 137–145.

_____ 1964. The antennular sense organs of *Panulirus argus. Ibid.* 13: 301–321.

_____ and Ardill, D.J. 1965. The innervation of the aesthetasc hairs of *Panulirus argus. Quart. J. Microscop. Sci.* 106: 45–60.

Leersnyder, M. de and Hoestland, H. 1963. La régulation osmotique et la régulation ionique du *Cardisoma. Cah. Biol. Marine* 4: 211–218.

Limbaugh, C. *et al.* 1961. Shrimps that clean fishes. *Bull. Marine Sci. Gulf Caribbean* 11: 237–257.

Lindberg, R.G. 1955. Growth, population dynamics, and field behavior in the spiny lobsters, *Panulirus interruptus*. *U. Calif. Publ. Zool.* 59: 157–248.

Linsenmair, K.E. 1965. Optische Signalisierung der Kopulationshöhle bei der Reiterkrabbe *Ocypode saratan*. *Naturwissenschaften* 52: 256–257.

———— 1967. Konstruktion und Signalfunktion der Sandpyramide der Reiterkrabbe *Ocypode saratan*. *Z. Tierpsychol.* 24: 403–456.

Lockwood, A.P.M. 1962. The osmoregulation of Crustacea. *Biol. Rev.* 37: 257–305.

———— 1967. *Aspects of the Physiology of Crustacea.* Aberdeen Univ. Press.

MacDonald, J.D. *et al.* 1957. Larvae of the British species of *Diogenes, Pagurus, Anapagurus* and *Lithodes*. *Proc. Zool. Soc. London* 128: 209–257.

Magnus, D.B.E. 1961. Zur Ökologie des Landeinsiedlers *Coenobita jousseaumei* und der Krabbe *Ocypode aegyptiaca* am Roten Meer. *Verh. Deutschen Zool. Ges.* 24: 316–329.

———— 1967. Ökologie sedimentbewohnender *Alpheus* - Garnelen des Roten Meeres. *Helgoländer Wiss. Meeresunters.* 15: 506–522.

Mantel, L.H. 1968. The foregut of *Gecarcinus* as an organ of salt and water balance. *Amer. Zool.* 8: 433–442.

Manton, S.M. 1964. Mandibular mechanisms and the evolution of arthropods. *Phil. Trans. Roy. Soc. London* 247B: 1–183.

Mayrat, A. 1958. Le système artériel des Pénéides comparison avec les autres Décapodes et les Mysidacés. *Arch. Zool. Exp. Gén.* 95: 69–78.

Mellon, De F., Jr. 1963. Electrical responses from dually innervated tactile receptors on the thorax of the crayfish. *J. Exp. Biol.* 40: 137–148.

Moulton, J.M. 1957. Sound production in the spiny lobster *Panulirus argus*. *Biol. Bull.* 113: 286–295.

Naylor, E. 1958. Tidal and diurnal rhythms of locomotory activity in *Carcinus maenas*. *J. Exp. Biol.* 35: 602–610.

———— 1960. Locomotory rhythms in *Carcinus maenas* from non-tidal conditions. *Ibid.* 37: 481–488.

———— 1962. Seasonal changes in a population of *Carcinus maenas* in the littoral zone. *J. Animal Ecol.* 31: 601–609.

Nicol, E.A.T. 1932. Feeding habits of the Galatheidea. *J. Marine Biol. Ass.* 18: 87–106.

Niggermann, R. 1968. Biologie und Ökologie des Landeinsiedlerkrebses *Coenobita scaevola* am Roten Meer. *Oecologia* 1: 236–264.

Palmer, J.D. 1963. A persistent diurnal phototactic rhythm in the fiddler crab, *Uca pugnax*. *Biol. Bull.* 123: 507.

Panning, A. 1924. Die Statocyste von *Astacus fluviatilis* und ihre Beziehungen zu dem sie umgebenden Gewebe. *Z. Wiss. Zool.* 123: 305–358.

———— 1952. *Die Chinesische Wollhandkrabbe.* Die Neue Brehm - Bücherei, Akad. Verlagsges., Leipzig.

Pearce, J.B. 1964. On reproduction in *Pinnotheres maculatus*. *Biol. Bull.* 127: 384.

Pearson, J.C. 1939. The early life histories of some American Penaeidae, chiefly the commercial shrimp, *Penaeus setiferus*. *U.S. Dept. Int. Bull. Bureau of Fisheries* 49: 1–73.

Pennak, R.W. 1953. *Fresh-water Invertebrates of the United States.* Ronald, New York.

Peters, N. and Panning, A. 1933. Die Chinesische Wollhandkrabbe *Eriocheir sinensis* in Deutschland. *Zool. Anz. (Erg. Bd.)* 104: 1–180.

———— 1938. Untersuchungen über die Wollhandkrabbe. *Mitt. Hamburger Zool. Mus. Inst.* 47: 1–171.

Pilgrim, R.L.C. 1965. The morphology of *Lomis hirta* and a discussion of its systematic position and phylogeny. *Australian J. Zool.* 13: 545–557.

————— and Wiersma, C.A.G. 1963. Musculature of the abdomen and thorax of *Procambarus clarkii* with notes on the thorax of *Panulirus interruptus* and *Astacus*. *J. Morphol.* 113: 453–487.

Pochon-Masson, J. 1965. Schéma général du spermatozoïde vésiculaire des décapodes. *Compt. Rend. Acad. Sci. Paris* 260: 5093–5098.

————— 1965. L'ultrastructure des épines des spermatozoïde chez les décapodes. *Ibid.* 260: 3762–3764.

————— 1968. L'ultrastructure des spermatozoïdes vésiculaires chez les crustacés décapodes. *Ann. Sci. Nat. Zool.* 10: 1–100.

Powell, B. 1966. Control of the 24 hour rhythm of colour change in juvenile *Carcinus maenas*. *Proc. Roy. Irish. Acad.* 64: 379–399.

Provenzano, A.J. 1961. Larval development of the land hermit crab *Coenobita clypeatus*. *Crustaceana* 4: 207–228.

Rao, K.R. 1968. The pericardial sac of *Ocypode*. *Amer. Zool.* 8: 561–567.

————— 1968. Adaptive color changes during the life history of *Ocypode macrocera*. *Zool. Jahrb. Abt. Physiol.* 74: 274–291.

————— et al. 1967. Physiology of the white chromatophores of *Uca*. *Biol. Bull.* 133: 606–617.

Rathbun, M.J. 1918. The grapsoid crabs of America. *U.S. Natl. Mus. Bull.* 97: 1–461.

————— 1925. The spider crabs of America. *Ibid.* 129: 1–613.

————— 1929. *Canadian Atlantic Fauna. 10. Arthropoda. 10m. Decapoda.* St. Andrews, New Brunswick, Biol. Board Canada.

————— 1930. The cancroid crabs of America of the families Euryalidae, Portunidae, Atelecyclidae, Cancridae and Xanthidae. *U.S. Natl. Mus. Bull.* 152: 1–609.

————— 1937. The oxystomatous and allied crabs of America. *Ibid.* 166: 1–278.

Redmond, J.R. 1955. The respiratory function of the hemocyanin in Crustacea. *J. Cell. Comp. Physiol.* 46: 209–247.

————— 1962. Oxygen-hemocyanin relationships in the land crab, *Cardisoma guanhumi*. *Biol. Bull.* 122: 252–262.

Reese, E.S. 1963. The behavioral mechanisms underlying shell selection by hermit crabs. *Behaviour* 21: 78–126.

————— 1968. Shell use, and adaptation for emigration from the Sea by the coconut crab. *Science* 161: 385–386.

————— and Kinzie, R.A. 1968. The larval development of *Birgus latro* in the laboratory. *Crustaceana* 21: 117–144.

Rice, A.L. and Provenzano, A.J. 1966. Larval development of the West Indian sponge crab *Dromidia antillensis*. *J. Zool. London* 149: 297–319.

Riegel, J.A. 1963. Micropuncture studies of chloride concentration and osmotic pressure in the crayfish antennal gland. *J. Exp. Biol.* 40: 487–492.

————— 1965. Micropuncture studies of the concentrations of sodium, potassium and inulin in the crayfish antennal gland. *Ibid.* 42: 379–384.

Ringel, M. 1924. Zur Morphologie des Vorderdarmes von *Astacus fluviatilis*. *Z. Wiss. Zool.* 123: 498–554.

Ross, D.M. and Sutton, L. 1968. Detachment of sea anemonies by commensal hermit crabs. *Nature* 217: 380–381.

Rudy, P. 1968. Water permeability in selected decapod Crustacea. *Comp. Biochem. Physiol.* 22: 581–589.

Ryan, E.P. 1966. Pheromone in a decapod crustacean. *Science* 151: 340–341.

Salmon, M. and Atsaides, S.P. 1968. Visual and acoustical signalling during courtship by fiddler crabs, *Uca. Amer. Zool.* 8: 623–639.

Salmon, M. and Stout, J.F. 1962. Sexual discrimination and sound production in *Uca pugilator. Zoologica,* New York 47: 15–20.

Schäfer, W. 1954. Form und Funktion der Brachyura-Schere. *Abhandl. Senckenbergischen Naturf. Ges.* 489: 1–65.

Schellenberg, A. 1928. Decapoda *in* Dahl, F. *Die Tierwelt Deutschlands,* G. Fischer, Jena. 10: 1–146.

Schlieper, C. 1950. Temperaturbezogene Regulation des Grundumsatzes bei wechselwarmen Tieren. *Biol. Zentralbl.* 69: 216–226.

———— 1964. Ionale und osmotische Regulation be ästuarlebenden Tieren. *Kieler Meeresf.* 20: 169–178.

Schmidt, W. 1915. Die Muskulatur von *Astacus fluviatilis. Z. Wiss. Zool.* 113: 165–251.

Schmitt, W.L. 1921. The marine decapod Crustacea of California. *Univ. California Publ. Zool.* 23: 1–470.

———— 1965. *Crustaceans.* Univ. Michigan Press, Ann Arbor.

Schöne, H. 1954. Statozystenfunktion und statische Lageorientierung bei dekapoden Krebsen. *Z. Vergl. Physiol.* 36: 241–260.

———— 1957. Kurssteuerung mittels der Statocysten. *Ibid.* 39: 235–240.

———— 1959. Die Lageorientierung mit Statolithenorganen und Augen *Ergebn. Biol.* 21: 161–209.

———— 1963. Minotakische Orientierung nach polarisiertem und unpolarisiertem Licht bei der Mangrovekrabbe *Goniopsis. Z. Vergl. Physiol.* 46: 496–514.

———— 1968. Agonistic and sexual display in aquatic and semiterrestrial brachyuran crabs. *Amer. Zool.* 8: 641–654.

———— and Schöne, H. 1963. Balz und andere Verhaltensweisen der Mangrovekrabbe *Goniopsis cruentata* und des Winkverhalten der eulitoralen Brachyuren. *Z. Tierpsychol.* 20: 641–656.

———— 1967. Integrated function of statocyst and antennular proprioceptive organ in spiny lobster. *Naturwissenschaften* 54: 289.

Schwabe, E. 1933. Osmoregulation verschiedener Krebse. *Z. Vergl. Physiol.* 19: 183–236.

Schwartzkopff, J. 1953. Pulsfrequenzen von Garneelen. *Naturwissenschaften* 40: 609.

———— 1955. Untersuchungen der Herzfrequenz bei Krebsen. *Biol. Zentralbl.* 74: 480–497.

———— 1955. Die Grössenabhängigkeit der Herzfrequenz von Krebsen im Vergleich zu anderen Tiergruppen. *Experientia* 11: 323–325.

Seabrook, W.D. and Nesbitt, H.H.J. 1966. The morphology of the brain of *Orconectes virilis. Canadian J. Zool.* 44: 1–22.

Shaw, J. 1961. Studies on ionic regulation in *Carcinus maenas* and *Eriocheir sinensis. J. Exp. Biol.* 38: 135–162.

Shoup, J.B. 1968. Shell opening by *Calappa. Science* 160: 887–888.

Sims, H.W. 1966. The phyllosoma larvae of the spiny lobster *Palinurellus gundlachi. Crustaceana* 11: 205–215.

———— 1966. Notes on the newly hatched phyllosoma of the sand lobster *Scyllcerus americanus. Ibid.* 11: 288–290.

Snodgrass, R.E. 1952. The sand crab *Emerita talpoida. Smithsonian Misc. Coll.* 117: 1–34.

———— 1952. *A Textbook of Arthropod Anatomy.* Comstock Publ., Ithaca.

Stainer, J.E. *et al.* 1968. The fine structure of the hepatopancreas of *Carcinus maenas. Crustaceana* 14: 56–66.

Stephen, G.C. 1962. Circadian melanophore rhythms of the fiddler crab. *Ann. New York Acad. Sci.* 98: 926–939.

Terao, A. 1929. Embryonic development of the spiny lobster *Panulirus japonicus*. *Japan J. Zool.* 2: 387–449.

Theede, H. 1964. Physiologische Unterschiede bei der Strandkrabbe *Carcinides maenas*. *Kieler Meeresf.* 20: 179–191.

Thomas, L.R. 1963. Phyllosoma larvae associated with medusae. *Nature* 198: 208.

Travis, D.F. 1960. The histology of the gastrolith skeletal tissue complex and the gastrolith in the crayfish *Orconectes virilis*. *Biol. Bull.* 118: 137–149.

————— 1963. The histochemical changes associated with the development of the non-mineralized skeletal components of the gastrolith discs of the crayfish *Orconectes virilis*. *Acta Histochemica* 15: 269–284.

Van Weel, P.B. 1960. Secretion of digestive enzymes by the marine crab, *Thalamita crenata*. *Z. Vergl. Physiol.* 43: 567–577.

————— and Christofferson, J. 1966. Electrophysiological studies in perception in the antennulae of certain crabs. *Physiol. Zool.* 39: 317–325.

Vasserot, J. 1966. Un prédateur d'Echinodermes: *Palinurus*. *Bull. Soc. Zool. France* 90: 365–384.

Veillet, A. *et al.* 1963. Inversion sexuelle et parasitisme par *Bopyrina virbii* chez la *Hippolyte inermis*. *Compt. Rend. Acad. Sci. Paris* 256: 790–791.

Vernberg, F.J. and Costlow, J.D. 1967. Handedness in fiddler crabs. *Crustaceana* 11: 61–64.

Völker, L. 1967. Zue Gehäusewahl des Land Einsiedlerkrebses *Coenobita scaevola* vom Roten Meer. *J. Exp. Biol. Ecol.* 1: 168–190.

Wald, G. 1968. Single and multiple visual systems in Arthropods. *J. Gen. Physiol.* 51: 125–150.

Warner, G.F. 1967. Life history of the mangrove tree crab *Aratus pisoni*. *J. Zool. London* 153: 321–335.

Waterman, T.H., ed. 1960, 1961. *The Physiology of Crustacea*. vols. 1,2. Academic Press, New York.

Weygoldt, P. 1961. Embryologische Untersuchungen an *Palaemonetes varians*. *Zool. Jahrb. Abt. Anat.* 79: 223–294.

Wiersma, C.A.G. 1947. Giant nerve fiber system of the crayfish. *J. Neurophysiol.* 10: 23–38.

————— and Ripley, S.H. 1952. Innervation patterns of crustacean limbs. *Physiol. Comp. Oecol.* 2: 391–405.

Wilder, D.G. 1963. Movements, growth and survival of marked and tagged lobsters. *J. Fish. Res. Board, Canada.* 20: 305–318.

Williams, A.B. 1965. Marine decapod crustaceans of the Carolinas. *U.S. Fish and Wildlife Service. Fishery Bull.* 65: 1–298.

Wilkens, J.L. and Fingerman, M. 1965. Heat tolerance and temperature relationships of the fiddler crab, *Uca pugilator*. *Biol. Bull.* 128: 133–141.

Wright, H.O. 1968. Visual displays in brachyuran crabs. *Amer. Zool.* 8: 655–665.

Zehnder, H. 1934. Die Embryonalentwicklung des Flusskrebses. *Acta Zool.* 15: 261–408.

14.

Superorder Pancarida

Pancarids are minute malacostracans that have the carapace fused to the first thoracic somite and extending posteriorly to form a dorsal brood pouch. The lumen of the brood pouch connects with the lateral branchial chambers. The mandibles have a movable process near the teeth, the lacinia mobilis. The maxillipeds have two endites and an exopodite, endopodite, and epipodite, although either exopodite or endopodite may be absent. The epipodites extend into the branchial chamber of the carapace. There are no nephridia.

ORDER THERMOSBAENACEA

The single order includes six known species, the largest of which is *Monodella halophila,* 4 mm long.

The Thermosbaenacea have the trunk regularly segmented without constriction between thorax and abdomen (Fig. 14-1). The carapace is fused to the first thoracic tergite, and usually forms a roof over the second and third. In the female it is expanded, forming a dorsal brood pouch. The brood pouch opens along the sides into the branchial chamber, which lies between the inner carapace wall and the thin lateral walls of head and thorax. Within this gill chamber the epipodite of the maxillipeds swings back and forth, drawing in water from the posterior and expelling it anteriorly next to the base of the antennae. Blood in lacunae under the thin carapace wall takes up oxygen from the water.

The first thoracomere bears a maxilliped. The remaining seven thoracic somites and six abdominal somites are free except that the last abdominal somite is fused to the large telson in *Thermosbaena.* The pereopods are biramous and lack epipodites. *Thermosbaena* lacks limbs on the last two pereomeres. The uniramous pleopods consist of a single article and are present only on the first two abdominal somites. The sixth pleomere has biramous uropods. There are no eyes. The central nervous system has a normal supraesophageal ganglion and separate metameric pairs of ganglia in the thorax and abdomen. *Thermosbaena* has a cardiac and

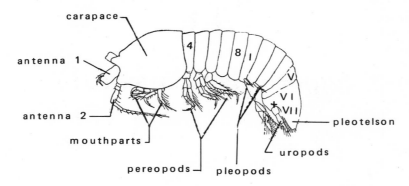

Fig. 14–1. *Thermosbaena mirabilis*, lateral view; 3 mm long. Arabic numerals indicate thoracomeres, Roman numerals, pleomeres. (After Monod from Giesbrecht.)

pyloric stomach. One pair of long ceca originates from the anterior part of the midgut. Corresponding with the position of the respiratory organs, the sac-shaped heart is in the posterior of the cephalothorax. The heart has one pair of ostia and an anterior and posterior aorta. There are paired testes or ovaries in the thorax.

Reproduction

The sternite of the eighth thoracomere of the male bears a pair of flexible penes. Mating has not been observed. The fertilization is internal in *Thermosbaena* and the eggs develop within the ovaries to the 12-cell stage. How the eggs reach the brood pouch has not been observed. In *Monodella* oviducts from the posterior edge of the ovaries open on the sixth thoracic sternite and a second duct leads dorsally to the brood pouch. However, this was reported only once and could not be substantiated by other authors. The 4–10 eggs in the swollen brood pouch are continuously washed by respiratory water. In *Thermosbaena* newly hatched young resemble the adults but are much smaller. The young of *Monodella* pass through a postmarsupial development during which the sixth and seventh pereopods and the pleopods are formed.

Relationships

In having the mandibular lacinia mobilis the Thermosbaenacea resemble the Peracarida. However they lack oostegites. The presence of pereopod exopodites appears primitive.

Habits

Thermosbaenacea live exclusively in interstitial or cavernicolous habitats. *Monodella halophila* was found in the sand of brackish coastal ground water, perhaps the ancestral habitat of the order. Other species have been found in warm springs, 30°C, along the Dead Sea and in subterranean waters having normal temperatures 15°C, in Italy. *Monodella argentarii* was found within a cave, as far as 300 m from the entrance, in complete darkness. Members of this species swim quickly, propelled by their thoracic limbs, and frequently rise to the surface. The vegetal detritus on which they feed completely fills the gut. Small particles are torn with the distal endites of the first maxillae, larger ones with the mandibles. *Thermosbaena* is known only from hot springs, 45°C in Tunis and dies at a water temperature of 30°C. It probably reaches the hot springs through subterranean waters. Soil organisms, blue-green algae, diatoms, bacteria, pollen, and detritus have been found within the gut of *Thermosbaena*. How it feeds is not known. *Thermosbaena* hardly swims, but runs along the sides and bottom of the springs.

Classification

Thermosbaenidae. Monodella, up to 4 mm long, has an elongate body, adapted to dwell in spaces between sand grains. *Monodella argentarii* is found in freshwater in a Tuscan cave, *M. stygicola* in brackish water in a southern Italian cave, *M. halophila* in salty water in a cave on the Dalmatian coast, and *M. relicta* a few meters above the surface of the Dead Sea in a hot salt spring. Recently *M. texana* was discovered in fresh cool water of a cave in Hays County, Texas, 200 km from the ocean, 180 m above sea level. *Thermosbaena mirabilis*, up to 3.4 mm long, has a stout body (Fig. 14–1). The carapace usually covers the third thoracic somite, or the whole thorax when the brood pouch is full. It is found in 34‰ saline, hot springs in Tunisia, North Africa.

References

Barker, D. 1962. A study of *Thermosbaena mirabilis* and its reproduction. *Quart. J. Microscop. Sci.* 103: 261–286.

Fryer, G. 1964. Studies on the functional morphology and feeding mechanism of *Monodella argentarii. Trans. Roy. Soc. Edinburgh* 66: 49–90.

Maguire, B. 1964. Crustacea: a primitive Mediterranean group also occurs in North America. *Science* 146: 931–932.

———. 1965. *Monodella texana,* an extension of the range of the crustacean order Thermosbaenacea. *Crustaceana* 9: 149–154.

Monod, T. 1940. Thermosbaenacea *in Bronns Klassen und Ordnungen des Tierreichs.* Akad. Verl. Geest & Portig, Leipzig. 5(1) 4: 1–24.

Por, F.D. 1962. *Monodella relicta* dans la dépression de la Mer Morte. *Crustaceana* 3: 304–310.

Siewing, R. 1958. Anatomie und Histologie von *Thermosbaena mirabilis. Abhandl. Akad. Wiss. Lit. Mainz, Math. Naturwiss. Kl.* 7: 197–270.

Stella, E. 1959. Riproduzione e lo sviluppo di *Monodella argentarii Riv. Biol.,* Perugia 51: 121–144.

15.

Superorder Peracarida and Order Mysidacea

SUPERORDER PERACARIDA

There are 10200 species of peracarids known. The isopod, *Bathynomus gigan-teus,* up to 27 cm long, is the largest species.

Peracarida are malacostracans that are characterized by the oostegites of the females, inwardly directed plates on the basal article of several or all thoracopods. Together the oostegites form a brood pouch (marsupium). A further character of the group is that the chewing surface of the mandibular endite, between the incisor process and the molar process, has a movable toothed article, the lacinia mobilis, otherwise found only in Pancarida (Vol. 2, Fig. 16-8, p. 307).

Direct development is characteristic of the Peracarida. The characteristic oostegites of the females (Fig. 15-2) may appear only at the molt before spawning, or sometimes earlier in the form of small buds. Their number varies. Though attached to the inner surface of the coxae, these thin folds can be considered epipodites. Claus (1886) observed that in some Mysidacea they first appear on the outer sides of the coxae, moving to the inner side during the course of development. Claus and later Giesbrecht considered both the gills and the oostegites to be epipodites; the oostegites as appendages of the epipodites and not as new structures. All peracarids have a cephalothorax, but often the head is combined with only a minute thoracic somite. Present in primitive groups, which have a shrimplike appearance, the carapace becomes smaller during the course of development in the most specialized groups and the isopods and amphipods lack it.

The Mysidacea probably represent the ancestral group, and they resemble the eucarid Euphausiacea so much that they were formerly combined within one order. Certain primitive characters of the oldest peracarids can be seen in the Lophogastrida, in their seven pleomeres, heart with many metameric arteries, and antennal and maxillary glands (Fig. 15-3). From this group evolved two different lines. In both lines the eyestalk has been lost except in Tanaidacea. In *Tanais* the eyes are located on paired lateral projections, each containing a muscle. Though the seventh pleomere has otherwise disappeared, it can still be seen, with its ganglion but lacking coelomic pouches, in the embryos of *Limnoria* and *Hetero- tanais*. Later in development the borders between the sixth and seventh pleomeres disappear and the ganglia of the seventh fuse with those of the sixth.

The first evolutionary line includes the orders Cumacea, Tanaidacea and Isopoda, characterized by having maxillary glands, paired posterior aortae, a convex folding of the venter of the embryo and by lacking the eighth thoracopods in the first free stage, the manca stage. Furthermore, the carapace and its respira- tory organs display progressive reduction from the least to the most specialized group, the isopods. With the disappearance of the carapace in isopods, the respiratory function is transferred to the pleopods. The sixth pleomere is fused to the telson. There is a dorsal plate with hooks in the cardiac stomach. The gut epithelium appears syncytial. The antennal glands are lost except for vestiges sometimes found in adult Tanaidacea and some Isopoda.

The second line, represented by the amphipods alone, is characterized by: retention of the antennal glands; there is a funnel that, as in the Mysidacea, directs coarse food particles from the pyloric stomach to the gut; the dorsal ceca of the gut are retained; there is a median posterior aorta; the venter of the embryo undergoes concave bending; and the first free stage has all pairs of limbs.

Little is known about the relationship of the Spelaeogriphacea, whose internal anatomy has not been studied.

ORDER MYSIDACEA, OPOSSUM SHRIMPS

There are 450 known species. The largest is *Gnathophausia ingens*, up to 35 cm long.

Most Mysidacea are elongate, 1–3 cm long. The carapace is fused to the thoracic tergites, three at most, forming lateral branchial chambers. The eyes are stalked. The thoracopods have large exopodites. The flattened uropods are leaf-shaped. Fossil mysids are known since the Triassic. Carboniferous fossils previously regarded as Mysidacea are now placed in a separate Superorder, the Eocarida, recently created for certain "mosaic" forms.

Anatomy

The cephalothorax is formed by fusion of the head with no more than three thoracomeres, a characteristic that separates the mysids from the superficially similar euphausiaceans. Posteriorly the carapace roofs over almost all thoracic somites (Fig. 15-1), but usually a dorsal emargination leaves the last one or two thoracic somites visible dorsally (Fig. 15-2). Only in the Lophogastrida do the pleomeres have epimeres (posterior lateral extensions).

The very long sixth pleomere is divided by a transverse groove in the lophogastrids, suggesting that it might consist of two metameres. A seventh pleomere
is developed in the mysid embryo, but fuses with the sixth.

The first antennae have two flagellae. The exopodites of the second antennae
are scalelike (Fig. 15-3). The mouthparts are described below in the section on
feeding. The first thoracopod is shortened and specialized as a maxilliped. It
always has an epipodite which extends laterally into the branchial chamber and
produces the water current (Fig. 15-2). Frequently the second and rarely the
third thoracopods are maxillipeds. The second to eighth thoracopods always have
many-articled exopodites and, in the female, oostegites. The oostegites in most
Mysida are limited to the posterior thoracopods (Fig. 15-2). In *Gastrosaccus* the
pleural plates of the first pleomere are expanded into a pair of lateral lamellae
which take part in the formation of the brood pouch. Epipodite gills are present
only in the Lophogastrida, usually in the shape of two- to four-branched tubular
gills (Fig. 15-3). In the Lophogastrida the pleopods are well-developed biramous
limbs. They are well developed and biramous in many Mysida males, but
vestigial in females. A tail fan is formed by the flattened uropods and the telson
(Fig. 15-1).

The exoskeleton is usually flexible; only rarely is it stiff and mineralized
(*Ceratolepis*). Over the light background color there is a pattern resulting from
star-shaped, branched, dark chromatophores. Deep-sea animals are often red.

SENSE ORGANS.   Sense organs include compound eyes, statocysts, sensory
hairs, and sensory tubules. Statocysts are found only in Mysidae where they occur
on the base of the uropod endopodite. A narrow slit connects the statocysts to the
outside. Inside is a statolith whose organic nucleus is made up of layers of calcium
fluoride to which sensory setae adhere from a number of sensory cells (about 60

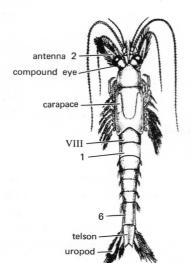

antenna 2
compound eye
carapace
VIII
1
6
telson
uropod

Fig. 15–1. *Mysis relicta*, female, dorsal view; 13
mm long. Roman numerals refer to thoracomeres;
Arabic to pleomeres. Only the exopodites of the
posterior five pereopods are visible. (After Tattersall.)

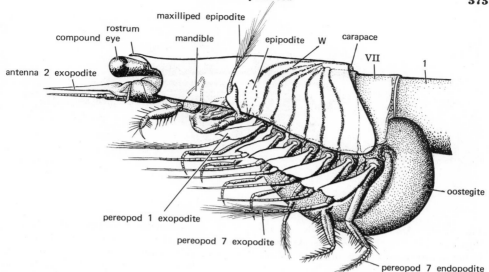

Fig. 15–2. *Praunus flexuosus,* anterior of female with marsupium; lateral view; 7 mm long. The basal article of the mandible, the epipodite of the maxillipeds and the transverse swellings of the body wall can be seen through the transparent carapace wall. Each swelling (w) contains a sinus that runs from the pereopods to the pericardium. Posteriorly the carapace hangs over the pereon. Roman numeral refers to pereomere; Arabic to pleomere. (After Vogt.)

cells in *Praunus*). The compound eyes are usually on a movable stalk. Only some deep sea species (e.g., *Pseudomma* and *Amblyops*) and cave species (e.g., *Spelaeomysis*) have sessile eyes, platelike without any visual components. In other bathypelagic species they are, however, very large and divided into a frontal eye and lateral eye. The frontal eye faces in the direction of the eyestalk and the lateral eye faces at a right angle (Vol. 2, Fig. 16-23, p. 320). In some genera the lateral eye may be smaller than the frontal or, as in *Arachnomysis,* it may completely disappear.

NERVOUS SYSTEM. The nervous system is primitively segmented. All ganglia behind the mandibular ganglion are separated from each other by connectives. If there is a seventh pleomere, the sixth and seventh abdominal ganglia are fused (Fig. 15-3). In marked contrast, the lophogastrids (*Eucopia* and *Gnathophausia*) have the ganglia of the three mouthparts and the first maxilliped fused into a characteristic subesophageal ganglion. In *Mysis* the ganglia of the mouthparts and those of all thoracomeres fuse to a long ribbon that extends through the length of the cephalothorax.

DIGESTIVE SYSTEM. There is a single cecum or a pair of dorsal ceca, and several ventral ceca originate at the anterior end of the midgut (Fig. 15-3). The long gut and cuticularized hindgut have very distinct cells.

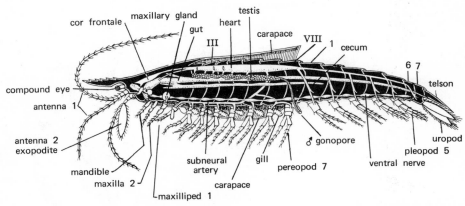

Fig. 15–3.   Anatomy of a male mysidacean of the suborder Lophogastrida. Roman numerals refer to thoracomeres, Arabic to pleomeres. At the base of the second antenna is the nephropore of the antennal gland. The uropods in embryonic development move to the tip of the abdomen, although they belong to abdominal metamere 6. The internal organs of the pereon from dorsal to ventral are heart with ostia, aortae, and 9 pairs of lateral arteries, testes, gut with short dorsal and long ventral ceca, nerve cord, subneural artery. (After Siewing from Remane.)

EXCRETORY SYSTEM.   There are antennal glands, and the lophogastrids, at the base of the second maxillae, also have one pair of small maxillary glands with a very short canal (Fig. 15-3), apparently secondarily reduced. In addition there are nephrocytes mainly clustered in blood lacunae above the eighth thoracopods, and phagocytes in the dorsum of the thorax, in the telson, and in the uropods.

CIRCULATORY SYSTEM.   The circulatory system appears much more primitive in the Lophogastrida than in other peracarids. The heart extends from the second to the eighth thoracomeres. It takes up blood through three pairs of ostia in the anterior part where the sinus coming from the branchial chamber opens into the pericardial sinus (Fig. 15-3). In diastole the blood leaves the heart through an anterior and a posterior aorta and nine pairs of lateral arteries. The first pair, probably cephalic arteries, arises ventral to the origin of the anterior aorta and runs parallel to it, emptying into lacunae in the vicinity of the cardiac stomach. The second to seventh pairs enter ventrally between the gonads, and open into a large lacuna above the gut. The right branch of the eighth, the descending artery, takes part in the formation of the gut sinus; the left branch continues to the ganglion of the sixth thoracomere. Here it continues as the subneural artery, sending branches to all thoracopods (Fig. 15-3). The ninth pair originates close to the posterior aorta and may be a branch of it, descending into the pleopods, as do all following branches of the posterior aorta. The anterior aorta is widened above the cardiac stomach into the cor frontale, a cephalic heart (Fig. 15-4), before branching into cerebral vessels.

Mysis has a specialized circulatory system. The heart has only two ostia left and usually does not extend to the posterior end of the thorax. In addition to the paired cephalic arteries, there are only three single ventral arteries, the two anterior ones supplying the gut sinus. The posterior larger one, the descending artery, divides into 3 branches above the nerve cord. The anterior branch is the subneural artery, the middle one supplies the sixth thoracopods, the posterior supplies the seventh and eighth. There may also be a posterior pair of arteries, perhaps only a branch of the posterior aorta.

The circulation through the lacunae has been studied in living *Mysis*. The well-defined sinus conducts blood from the second to seventh thoracopods into the pericardial sinus (Fig. 15-2). A considerable part of the content of the thoracic lacunae, together with blood returning from the head and the first maxillipeds, reaches a channel along the lower edge of the carapace, which dorsally leads into a lacunar net that covers the inner respiratory wall of the carapace. Oxygenated blood collects in a large sinus which leads into the pericardial sinus (Fig. 15-4).

REPRODUCTIVE SYSTEM. The paired testes are divided into tubules and saccules and open anteriorly into vasa deferentia which in turn open near the base of the 8th thoracopods, on genital papillae of variable length. The ovaries are two uniform tubes, which, in Mysida, are connected at midlength by a narrow bridge.

Reproduction

Mating, observed in *Praunus, Mesopodopsis, Paramysis* and others, is nocturnal and occurs after the female has molted and has a brood pouch. The males pay attention only to females that have molted within the last 12 hr and have mature ova in the oviduct. Presumably they recognize such females by olfactory senses.

The male of *Hemimysis* swims under the female, its venter toward the female, its anterior toward the female's telson. The male holds the female's abdomen with his pereopod endopodites and pushes his long genital papillae between the oostegites. Within 1–2 sec the sperm is sprayed into the brood pouch and the male departs. After 20–30 min the eggs move into the marsupium where they are fertilized. *Mesopodopsis* sheds the sperm into the water between the mating pair and presumably the sperm is transported into the brood pouch by the modified male pleopods.

In Arctic species the females molt after the young hatch, after which the oostegites are reduced. In temperate areas the female, after molting, keeps her marsupium and rears another brood. Thus *Praunus inermis* starts its reproductive period at the end of April and produces up to five broods, one after another, before dying in July or August. The first brood soon begins to reproduce and has five broods during the first summer before dying off in fall, but the other broods over-winter to spawn the following April.

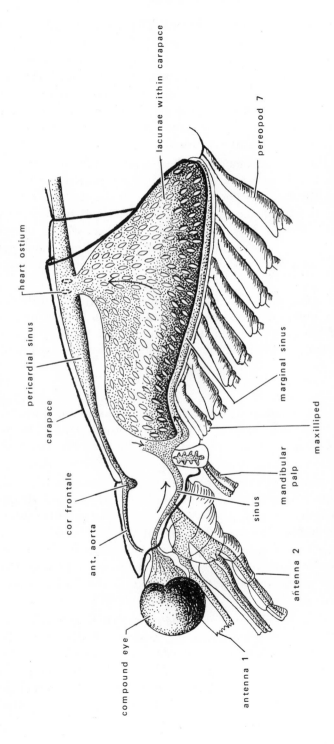

Fig. 15–4. Important blood vessels of a mysid; about 7 mm long. Arrows give direction of blood flow. From the gills the oxygenated blood flows through the lacunae of the carapace fold to the pericardium through ostia into the heart. The ostia are seen through the pericardial wall. (After Delage.)

The number of eggs depends on the size of the mother, which increases between broods. *Mysis relicta,* 13–15 mm long, carries 10–20 eggs; 17–21 mm long, 25–40 eggs. Within the marsupium the number of eggs declines for unknown reasons.

Development

Oostegite movements produce water currents in the brood pouch. The time needed for development differs in different species and also depends on water temperature. *Mysis relicta* spawns in winter in northern Europe; the brood remains in the marsupium for 2–3 months and the young hatch 3–4 mm long. Summer broods develop much faster, those of *Praunus inermis* in barely 2.5 weeks. Development in tropical waters is also faster. The young of *Mysis orientalis* along the coast off Madras at temperatures of 25–29°C leave the brood pouch after only 4 days. The young differ little from the adults. The number of molts passed before maturity is not known.

Mysis relicta, M. mixta and the three species of *Praunus* of northern European waters have a life span of 1–1.5 years, while the Arctic *Mysis oculata* lives for at least 2 years.

Habits

Most Mysidacea are marine, a few live in brackish waters and a very few in freshwater. The female of *Gastrosaccus sanctus* may be amphibious on moist sand beaches. Of interest is the freshwater *Mysis relicta,* related to *M. oculata* and *M. litoralis* of northern oceans, and found in brackish water in the Baltic Sea and in fresh inland lakes not connected with the ocean in Ireland and northern Europe and America. Some of these populations may be relicts from times when the lakes were connected with Arctic oceans; the presence of other populations can be explained by transport south in proglacial lakes formed by damming in front of glaciers during the Pleistocene.

Mysinae, Siriellinae, Gastrosaccinae, Heteromysini, and many Leptomysini are shallow-water animals. Other species are found in deep waters: *Gnathophausia gigas* at depths from 1500–3500 m, *Eucopia* at 2000 m with extremes from 500–6500 m. Most live near the bottom hovering over it. Many are benthic by day and pelagic at night. A few species rest on plants such as *Ulva* and *Zostera* during the day (*Praunus inermis*) or on stones, and cliff ledges (*Neomysis, Siriella jaltensis*). Only a few mysids hide themselves in sand. *Gastrosaccus spinifer* digs by throwing up sand grains with moving exopodites while lying on the sand. *Paramysis pontica,* to 18 mm long, digs ditches in muddy sand with the claws of the first to third pereopods. Digging for long periods of time, they produce open ditches to 5–10 cm within an hour.

Almost all bathypelagic and bottom-dwelling mysids, even those that burrow in sand, rise at night, as is indicated by nocturnal plankton hauls. Perhaps the stimulus is light intensity. The sand dweller, *Gastrosaccus sanctus,* begins to rise in winter at 16 hr (4 pm), in summer at 21 hr (9 pm). The time it takes to rise depends on the speed, depth of water and other factors. Some species rise only 20 m, *Eucopia unguiculata* and *Boreomysis microps* rise 200–600 m. The ascent is most distinct during the spawning periods and may be of importance for the distribution of young leaving the marsupium.

Like many other shallow water crustaceans, littoral species may move to deeper water in fall and return in spring or summer (*Neomysis integer, Mesopodopsis slabberi,* and *Praunus flexuosus*).

Some species swarm. Swarms of *Neomysis integer* may be several km long, one to several meters in diameter. All members within a swarm belong to the same species.

LOCOMOTION. Mysidacea are mainly swimmers. The primitive Lophogastrida swim with their biramous pleopods. Female Mysida having reduced pleopods swim with the exopodites of the thoracopods. They hold these toward the sides and rotate them so that the tip describes an oval (Figs. 15-1, 15-2). The limbs do not alternate rhythmically as might be expected, but move continuously in slightly different phase. They draw water from the sides toward the dorsum so that two strong currents, parallel to the abdomen and some distance from it drive the body forward. Many species hold the body horizontal, dorsum up, while swimming. *Praunus flexuosus* holds the anterior part of the body almost vertical. All can swim up, down, forward, and backward.

If suddenly disturbed, as by a moving shadow, the Mysidacea jerk back by flexing the abdomen and tail fan against the thorax. Bottom-dwellers walk slowly with the endopodites, the exopodites also in constant motion. Members of the genus *Erythrops,* like other crustaceans that live on bottom mud, have long endopodites, making them resemble spiders.

Species of *Mysidium* swim in schools that behave much like schools of fishes. They respond to fish predators by taking refuge among the spines of sea urchins or in the nest caves of pomacentrid fishes. The nesting pomacentrid defends the cave against the predatory fish, but does not feed on the mysids.

COLORATION. Chromatophores, each composed of a cluster of five to nine cells, often are imbedded deep in the connective tissue below the hypodermis. Their shape does not change, nor their number and position, and may be different in different species. In shallow-water species, the background color and light intensity influence distribution of the pigment within the stellate chromatophores. On a black background *Praunus flexuosus* becomes dark, and *P. neglectus,* usually light green and found among green algae, may change to dark olive. In numerous Mysidacea diurnal and nocturnal pigment migration has been observed. *Praunus,* for instance, may be completely transparent at night as the result of clumping of the pigment into small spheres within the centers of the chromatophores.

SENSES. Chemical senses have barely been studied. Except for the observation that *Praunus* notices moving prey with its eyes from distances up to several centimeters, the importance of the head sense organs is known only in regard to vertical migration and body position. The eyes and impulses from the statocysts control the horizontal body position even in the dark. If *Mysidium* is illuminated from the side it turns its body so that its back is toward the light. The angle of turning may be greater than 45° especially if the light is strong, the optic control overriding the equilibrium organs. *Mysidium* orients by polarized light; in clear water it swims transversely to the plane of polarization of the light coming from above.

The schooling of Mysidaceans depends on optic signals during the day and probably sensitivity to swimming currents of neighbors at night.

When the statocysts are removed, *Mysis* does not react to vibrations of the water. *Praunus* and *Schistomysis* respond to increased hydrostatic pressures by swimming toward light (in nature toward the surface). Very slight differences in pressure are detected, suggesting that this sense may be important in moving with the tides. The receptors for pressure have not been localized.

FEEDING. Most Mysidacea are filter feeders, removing fine detritus, mollusk larvae, diatoms, and other plankton out of the water while swimming just above the bottom. Species of *Lophogaster* lack a filter and take larger particles. But all filter feeders may also take larger food when they feed as predators or on carrion. *Praunus, Siriella,*and *Neomysis* have been observed to catch live copepods, cladocerans, amphipods, and small mollusks. *Siriella* drops from above on living *Calanus* or zoeae, grasping them with the claws of the anterior pereopods and putting them into the mouth. *Praunus* detects its prey from a distance of several centimeters, swims toward, and grasps it with the anteriorly directed endopodites of the anterior five thoracopods. Feeding is done while swimming, as in other Mysidacea. The endopodites of the maxillipeds, aided by the mandibulaı palps, move the food under the mandibles and the first maxillae, pressing it against these cutting appendages. *Eucopia,* with strong subchelae (Fig. 15-5), is presumably a

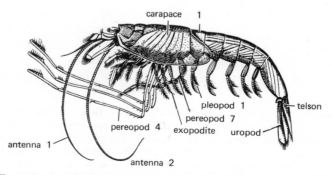

Fig. 15–5. *Eucopia unguiculata,* a predacious lophogastrid from 1,000 to 26,000 m deep; young female about 2.6 cm long; 1, first pleomere. (After Sars from Smith.)

predator. *Hemimysis* has been observed to feed like *Praunus,* on larger detritus, dead amphipods, and similar material.

Food is also filtered while swimming. *Neomysis integer* in a very dense diatom culture of *Nitzschia* filtered about a million cells per hour.

Hemimysis lamornae swims 5–7 cm above the bottom, stirring up the water with the lateral exopodites while standing vertically on the first antennae and antennal scales. The filtering of plankton has been studied in detail in this species. The plankton-containing water is sucked in by movements of the exopodites of the pereopods and moved into the longitudinal ventral groove. The groove, bordered by the coxae on each side, extends from the head to the end of the thorax. As in the Cumacea and Tanaidacea the second maxillae filter the plankton out of the water.

The water is swept into the filter by extending the exopodites out to the sides and moving their tips in an oval in the shape of a cone (Fig. 15-2). Because the tip moves faster than the base, there is stronger pressure near the tip. The water is thus drawn from the region of high pressure to low, from the tip to the base. It enters between the coxae, which face anteriorly toward the ventral groove.

The movement of the exopodite is like that of a spoon stirring sugar into coffee. The spoon is wider than its handle. Like the spoon, the exopodite articles

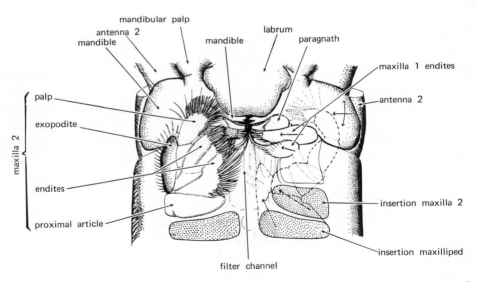

Fig. 15–6. Mouthparts of *Hemimysis lamornae* seen from the venter; 1.3 mm wide. The maxillipeds (thoracopod 1) and left 2nd maxilla are cut off. On the right, dotted lines show the maxilliped with other mouthparts underneath; its setae have only been shown on the proximal endite and the dactyl. On the right, thick arrows indicate movements of the mouthparts; those of the maxilliped and its endites are represented by dotted arrows. The endites of the mandibles are in part under the labrum. The mouth is also underneath the labrum. (After Cannon and Manton.)

in the dorsal or backward part of its path are wider because the setae on each side spread and form a paddle, increasing the pressure. Just as the sugar is stirred up toward the handle, the plankton reaches the base of the exopodites. The water with plankton moves foward in the ventral groove to the second maxillae where the groove widens and is closed anteriorly by the paragnaths.

A row of median setae on the proximal endites of the second maxillae are the filters, directed dorsally toward the midline. The filter combs of both endites form the walls of the filter chamber (Figs. 15-6, 15-7). The ceiling of the chamber is the venter of the head and the floor is formed by the basal articles of the maxillipeds (Fig. 15-7).

From the filter chamber the water can leave only through the two lateral combs. As it flows through these, the plankton is deposited on their ventromedian surfaces. Besides the circling exopodites, the basal article of the flat second maxillae takes part in producing currents. Its concave anterior (dorsal) wall, close behind the posterior (ventral) surface of the first maxillae, regularly beats down and back, producing a low pressure area between the maxillae. This suction draws the exopodite of the second maxillae against the slit. The exopodite presses from outside against the slit between the first and second maxillae, closing the filter chamber laterally (Fig. 15-6). Also water is drawn from the groove through the filter combs toward the space between the two pairs of maxillae. If the second maxilla again moves forward (dorsally) appressing against the first maxilla, water forces the valvelike exopodites aside and spurts laterally from the shrinking space between the maxillae. As in a jet pump this spurting expiratory water in the branchial chamber produces a further pull toward the filter chamber.

By feeding iron saccharate it could be demonstrated that the particles suspended in water stick to the ventral (posterior) side of the filter of the proximal endite of the second maxillae. The particles are combed out by anteriomedian movements of the setae situated below (behind), which are directed dorsomedially and arise from the basal endite of the first maxilliped (Fig. 15-7). Individual long setae

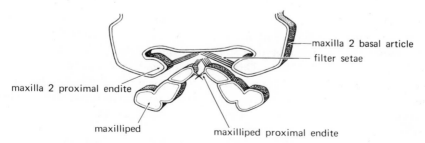

Fig. 15–7. Cross-section through the filter chamber of *Hemimysis lamornae*; 1 mm wide. The inner edge of the maxilliped has a row of dorsal and a row of ventral setae. The ventral setae form the ventral wall of the filter channel; the dorsal setae can comb out the filter setae of the 2nd maxillae. (After Cannon and Manton.)

(directed toward the mouth) of the endites of the second maxillae and the first maxilliped by moving forward brush the particles into the area of the distal articles of the first maxilliped and palps of the first maxillae, bringing them to the biting teeth of the first maxillar endites (Fig. 15-6).

In *Gnathophausia* the food groove of the head is not connected with the longitudinal ventral groove of the thoracomeres. It can be assumed that the filter current is produced only by the 2nd maxillae and the suction of the expiratory current, but this has not been verified in living animals.

RESPIRATION.   Only the Lophogastrida use epipodites (Fig. 15-3), combined with a carapace respiratory surface for respiration. The specialized inner surface of the carapace is the main respiratory area of the Mysida.

In the narrow cavity underneath the carapace, the epipodite of the first maxilliped moves anteriorly and down from a vertical position, pushing the water out of the gape between carapace and maxillae (Fig. 15-2). This rapid stream draws water from the dorsolateral border of the carapace into the branchial chamber. The thin inside wall of the carapace encloses many blood sinuses which take up the oxygen from the water (Fig. 15-4).

ECONOMIC IMPORTANCE.   The large number of individuals and swarms of Mysidacea form an important supply of food for fish. In the Orient mysids are used for human food, sometimes being made into a paste.

Classification

SUBORDER LOPHOGASTRIDA

These are primitive in structure. Thoracopods 2–7 (rarely also 8) have branched epipodites that function as gills (Fig. 15-3). The pleopods are biramous in males and females and are used in swimming. There are seven pairs of oostegites. The cardiac stomach has seven to eight pairs of pyloric food grooves. There are antennal glands and maxillary glands. The circulatory system has a pair of arteries in each thoracomere (Fig. 15-3). All species are marine.

Lophogastridae. Thoracopods 3–8 are alike, without subchelae. The last pleomere, between abdominal somites 6 and 7, has a distinct transverse groove, which in other groups appears only in the embryo. *Lophogaster typicus*, to 22 mm long, is a predator occurring at depths of 35–750 m. *Gnathophausia* found at depths of 500–2600 m has primitive features, and at the base of the exopodites of the 2nd maxillae the opening of a gland that secretes material luminescent on contact with the ocean water. There are no reflectors or lenses. *Gnathophausia ingens* is up to 35 cm long. *Ceratolepis* belongs to this family.

Eucopiidae. The endopodites of the first four pairs of thoracopods are short, with strong subchelae; the fifth to seventh endopodites are long and soft but also have a subchela (Fig. 15-5). *Eucopia*, to 43 mm long, is bathypelagic, usually at depths from 1500–3000 m. The second maxillae lack filter setae.

SUBORDER MYSIDA

Thoracopods lack branched gills. In females the pleopods are usually much reduced and are not used in locomotion. The number of oostegite pairs is usually reduced to two or three. The cardiac stomach has at most three pairs of filter grooves. There are no maxillary glands. Some thoracic arteries are single, and some have been lost. Included are marine, brackish,- and freshwater species.

Petalophthalmidae. There are seven pairs of oostegites. *Petalophthalmus* to 50 mm long is bathypelagic, occurring at depths of 1000 m or more, and is a predator that lacks filter setae on the second maxillae.

Mysidae.[*] There are usually only two to three pairs of oostegites. *Boreomysis,* 17–30 mm long, occurring at depths of 200–1900 m, has seven pairs of oostegites. *Siriella* is 10–20 mm long. *Gastrosaccus* is 10–20 mm long; *G. spinifer,* 21 mm long, occurs as deep as 50 m in the western Baltic. *Erythrops erythrophthalma,* 9–11 mm at 40–275 m, occurs in Arctic waters to Cape Cod; the eyes are red, there is an orange patch on the carapace, white spots on the trunk and yellow on the ventral side. Other genera are *Pseudomma, Amblyops,* and *Euchaetomera,* up to 11 mm long. *Arachnomysis,* 8 mm long, is bathypelagic, occurring 500–3000 mm deep. *Leptomysis* is 13–18 mm long, *Mysidopsis bigelowi,* up to 7.5 mm, is found from the Gulf coast to Cape Cod at 16–50 m; *M. almyra,* up to 9.4 mm occurs at low salinities to 2.0‰ from Florida to Louisiana.

Hemimysis lamornae, up to 10 mm long, is found at depths of 4–90 m. *Mysis*[†] *stenolepis,* up to 30 mm, lives among weeds from the Gulf of St. Lawrence to New Jersey in shallow water; *M. mixta,* up to 25 mm, lives from eastern Canada to New England from the tide zone to 200 m. *Mysis*[†] *oculata,* 27–29 mm long, circumpolar in the Arctic oceans, also occurs in brackish waters. The scale of the second antennae is five times as long as wide. The invagination in the telson is 18–25% of the telson length. A very similar freshwater species, *M. relicta,* is found in lakes, probably a relict from the Pleistocene. It and *M. litoralis* formerly were confused with *M. oculata* of the Arctic Ocean. Of interest are four related species of *Mysis* found in the Caspian Sea. *Mysis relicta* (Fig. 15–1), up to 21–25 mm long, matures when 13 mm long. The antennal scale is only four times as long as wide; the telson invagination is only 8–17% of its length. Once in a while specimens are washed down river and are found in the ocean. The waters inhabited are never warmer than 14°C. They include northern Eurasian and North American inland lakes of glaciated areas, including the Finger Lakes and Great Lakes. It has been found in Lake Superior at a depth of 270 m, but is also found in small lakes. The animals are swimmers, keeping themselves in deep, cold water during summer and moving in winter to shallow places where they reproduce. The 40 young stay 1–3 months in the brood pouch. At night *Mysis relicta* may rise. In northern Germany they do not occur in water with an O_2 concentration less that 4 cm^3/liter; in Wisconsin they were found in water with only 1 cm^3/liter water. The species feeds mainly on detritus stirred up from the bottom, including diatoms and unicellular algae. But remains of *Bosmina, Cyclops* and *Cypris* have been found in the stomachs, indicating that they also feed as predators and on carrion.

[*] The name Mysidae is on the *Official List of Family Names in Zoology.*
[†] The name *Mysis* is on the *Official List of Generic Names in Zoology.*

Eleven species of *Paramysis* are found in the Caspian Sea. *Schistomysis kervellei*, 17 mm long, is found in the western Baltic; *S. ornata*, 19 mm long, has been found in the North Sea. *Praunus flexuosus*, 16–26 mm, is a common euryhaline shallow-water form that enters rivers; although European, it has been found several times in New England waters. *Praunus neglectus*, 14–20 mm long, occurs in shallow water; *P. inermis*, 11–18 mm, occurs in shallow water. *Mesopodopsis slabberi* may reach 15 mm in length. *Neomysis integer*, to 17 mm long, is euryhaline along European coasts but cannot live in freshwater; it has two generations a year in the Baltic but three off Holland. The first generation, with an average of 20 eggs, appears from May to October and lives 4–5.5 months. The second generation, with about 40 eggs in each brood, appears from July to September and lives 10–11 months. The time of reproduction depends on temperature. The first generation can reproduce 1–1.5 months after leaving the brood pouch and produces four to six broods; the winter generation reproduces after 7–8 months and has three or, rarely, four broods. The number of eggs per brood declines during the summer. *Neomysis americana*, to 12 mm, is common from the Gulf of St. Lawrence to Virigina, intertidal to 200 m. *Acanthomysis awatchensis*, up to 15 mm long, is found in brackish water on the Pacific coast of North America and Asia. It is sometimes found in isolated water among Oregon sand dunes or freshwater lakes connected to the coast. *Taphromysis louisianae*, 8 mm long, inhabits small freshwater pools in Louisiana; *T. bowmani* lives in brackish water along the Gulf coast. *Mysidopsis almyra* occurs in brackish waters in Florida and Louisiana. *Teganomysis novaezealandiae* is found in brackish water in New Zealand, *Diamysis americana* lives in freshwater in Surinam and Venezuela.

Several mysids are known from caves. *Heteromysis cotti*, with reduced eyes, occurs on an island of the Canaries. Some members of the genus *Heteromysis* are commensal with invertebrates; *H. actiniae* lives among the tentacles of the sea anemone *Barthelomea annulata*. *Heteromysis harpax* lives within gastropod shells occupied by hermit crabs. Other species are associated with sponges, coral, and brittle stars. One species of *Antromysis* lives in holes of shore crabs above the high tide mark in Costa Rica, presumably in brackish water; another species is found in freshwater in Yucatan. *Troglomysis vjetrenicensis* occurs in a cave in Yugoslavia. *Mysidium* belong to this family.

Lepidomysidae. Members of this family are cave forms that lack a statocyst but may still have vestigial eyes. *Lepidomysis servatus* (=*Lepidops servatus*) is found in Zanzibar in slightly brackish water. *Spelaeomysis bottazzii* is found in caves in southeastern Italy and *S. quinterensis* is found in freshwater in a cave in Tamaulipas, Mexico.

Stygiomysidae. The members of this family are modified for living in caves, having a vermiform body and a reduced carapace. There are no statocysts. *Stygiomysis hydruntina* is found in caves in southeastern Italy in fresh or slightly brackish water and *S. holthuisi*, up to 9 mm, is found on St. Martin, Lesser Antilles.

References

Peracarida

Brooks, H.K. 1962. The Paleozoic Eumalacostraca of North America. *Bull. Am. Paleont.* 44 (202): 163–338.

Claus, C. 1886. Beiträge zur Morphologie der Crustaceen. *Arb. Zool. Inst. Wien* 6: 1–108.

Giesbrecht, W. 1921. Crustacea in Lang A. ed. *Handbuch der Morphologie der Wirbellosen Tiere.* G. Fischer, Jena.

Haffer, K. 1965. Der Kaumagen der Mysidacea im Vergleich zu dem verschiedener Peracarida und Eucarida. *Helgoländer Wiss. Meeresunters.* 12: 156–206.

Siewing, R. 1951. Verwandtschaft swischen Isopoden und Amphipoden. *Zool. Anz.* 147: 166–180.

———— 1956. Morphologie der Malacostraca. *Zool. Jahrb. Abt. Anat.* 75: 39–176.

———— 1963. Malacostracan morphology *in* Whittington, H.B. and Rolfe, W.D.I. ed. *Phylogeny and Evolution of Crustacea.* Mus. Comp. Zool., Cambridge, Mass. pp. 85–103.

Mysidacea

Băcescu, M. 1955. Mysidacea. *Faune Rep. Popul. Romāne* 4(3) 3–122.

Banner, A.H. 1947, 1948. The Mysidacea and Euphausiacea of the northeastern Pacific. *Trans. Roy. Canadian Inst.* 26: 345–399; 27: 65–125.

Bowman, T.E. 1964. *Mysidopsis almyra,* a new estuarine mysid from Louisiana and Florida. *Tulane Stud. Zool.* 12: 15–18.

Cannon, H.G. and Manton, S.M. 1927. The feeding mechanism of a mysid crustacean, *Hemimysis. Trans. Roy. Soc. Edinburgh* 55: 219–253.

Chace, F.A. et al. 1959. Malacostraca *in* Edmondson, W.T. ed. *Ward and Whipple Freshwater Biology.* Wiley, New York.

Clutter, R.I. 1967. Zonation of nearshore mysids. *Ecology* 48: 200–208.

———— 1968. Soziales Verhalten von Spaltfusskrebsen. *Umschau* 68: 78.

———— (in press). *J. Exp. Marine Biol. Ecol.*

Fage, L. 1932. La migration verticale saisonnière des Mysidacés. *Compt. Rend. Acad. Sci. Paris* 194: 313–315.

———— 1933. Pêches planctoniques à la lumière effectuées à Banyuls-sur-mer et à Concarneau. *Arch. Zool. Expér. Gén.* 76: 105–248.

Gordan, J. 1957. A bibliography of the order Mysidacea. *Bull. Amer. Mus. Natur. Hist.* 112: 279–394.

Gordon, I. 1960. *Stygiomysis* from the West Indies. *Bull. Brit. Mus. (Natur. Hist.)* 6(5): 285–326.

Haffer, K. 1965. Der Kaumagen der Mysidacea im Vergleich zu dem verschiedener Peracarida und Eucarida. *Helgoländer Wiss. Meeresunters.* 12: 156–206.

Holmquist, C. 1959. *Marine Glacial Relicts an Account of Investigation on Mysis.* Berlinska Boktryck, Lund.

Hutchinson, G.E. 1967. *A Treatise on Limnology,* vol. 2. Wiley, New York.

Ii, N. 1964. *Fauna Japonica: Mysidae* (Crustacea). Nat. Sci. Mus. Tokyo.

Jander, R. 1962. The swimming plane of the crustacean *Mysidium gracile. Biol. Bull.* 122: 380–390.

———— and Waterman, T.H. 1960. Sensory discrimination between polarized light and light intensity patterns. *J. Cell. Comp. Physiol.* 56: 137–159.

Kinne, O. 1955. *Neomysis vulgaris* eine autökologische Studie. *Biol. Zentralbl.* 74: 160–202.

Labat, R. 1954. Sur l'accouplement et la ponte de *Paramysis nouveli. Bull. Soc. Natur. Toulouse* 89: 406–409.

Lockhead, J.H. 1950. *Heteromysis formosa* in Brown, F.A. ed. *Selected Invertebrate Types.* Wiley, New York.

Manton, S.M. 1928. The embryology of a mysid crustacean, *Hemimysis lamornae. Philos. Trans. Roy. Soc. London.* 216B: 363–463.

——————— 1928. Anatomy and habits of the lophogastrid Crustacea. *Trans. Roy. Soc. Edinburgh* 56: 103–119.

Mauchline, J. 1967. Biology of *Schistomysis spiritus. J. Marine Biol. Ass.* 47: 383–396.

Mayrat, A. 1956. La système artériel de *Praunus flexuosus. Bull. Sta. Oceanog. Salammbo* 53: 44–49.

——————— 1962. Étude au microscope électronique des yeux des Crustacés. *Comp. Rend. Acad. Sci. Paris* 255: 766–768.

Nouvel, H. 1937. L'accouplement chez une espèce de Mysis, *Praunus flexuosis. Compt. Rend. Acad. Sci. Paris* 205: 1184–1186.

——————— 1940. La sexualité d'un Mysidacé, *Heteromysis armoricana. Bull. Inst. Océanograph. Monaco* 789: 1–11.

Pennak, R.W. 1953. *Freshwater Invertebrates of the United States.* Ronald, New York.

Petriconi, V. 1968. Bildung des präantennalen Mesoderm bei *Neomysis. Zool. Jahrb. Abt. Anat.* 85: 579–596.

Rice, A.L. 1961. Responses of certain mysids to changes in hydrostatic pressure. *J. Exp. Biol.* 38: 391–401.

Siewing, R. 1953. Morphologische Untersuchungen an Tanaidaceen und Lophogastriden. *Z. Wiss. Zool.* 157: 333–426.

——————— 1956. Untersuchungen zur Morphologie der Malacostraca. *Zool. Jahrb. Abt. Anat.* 75: 39–176.

Tattersall, W.M. 1951. Mysidacea of the U.S. National Museum. *Bull. U.S. Nat. Mus.* 201: 1–292.

——————— and Tattersall, O.S. 1951. *British Mysidacea.* Ray Soc., London.

Vogt, W. 1932. Die Morphologie und Histologie der Antennendrüse und der thoracalen Athrocytenorgane der Mysideen. *Z. Morphol. Ökol.* 24: 288–318.

——————— 1935. Die Entwicklung der Antennendrüse der Mysideen. *Ibid.* 29: 418–506.

——————— 1935. Über ein Seitenorgan der Mysideen. *Ibid.* 29: 507–510.

Wigley, R.L. 1963. *Praunus flexuosus* in New England. *Crustaceana* 6: 158.

Zimmer, C. 1933. Mysidacea in Grimpe, G. and Wagler, E. *Die Tierwelt der Nord-und Ostsee.* Akad. Verl., Geest & Portig, Leipzig 10(23)4: 70–120.

16.

Orders Cumacea, Spelaeogriphacea, and Tanaidacea

ORDER CUMACEA

Of the 770 species known, the largest is *Diastylis goodsiri,* up to 3.5 cm long.

The Cumacea are small Peracarida, usually only 0.5–1 mm long, with the cephalothorax widened by the bulging carapace and set off distinctly from the long abdomen. Some thoracopods may have exopodites but there never are exopodites on maxillipeds 1 and 2 or on pereopod 5. The pleopods, found only on males, are cylindrical. The uropods have a slender rodlike stalk with branches.

Cumacea are known fossil since the Upper Permian.

Anatomy

This interesting order appears to be intermediate between the Lophogastrida and Tanaidacea. The well-developed carapace fuses dorsally with 3–4, rarely 5–6, anterior thoracic somites. The carapace completely encloses the cephalothorax, producing on each side a large branchial chamber (Figs. 16-1, 16-2, 16-6). Unlike other peracarids, cumaceans have the branchial chamber closed behind, the posterior carapace border being fused along its whole length to the lateral body wall (Fig. 16-6). Ventrally there is only a narrow slit because the lateral wall, as in the crabs, bends inward and extends horizontally toward the maxilliped attachments, forming a ventral wall to the branchial chamber (Figs. 16-6, 16-8c). Anteriorly the carapace narrows, ending with a small projection over the head on each side (Fig. 16-2). These projections fuse to form the pseudorostrum, beneath which open the lateral branchial chambers (Figs. 16-6, 16-8a).

The pereon, depending on the size of the cephalothorax, consists usually of 4–5 metameres, rarely only 2–3, usually with only short posterior lateral portions (epimeres). The much narrower, long and flexible abdomen consists of six rings, which have epimeres only in the males. In some families the last somite is fused to the telson.

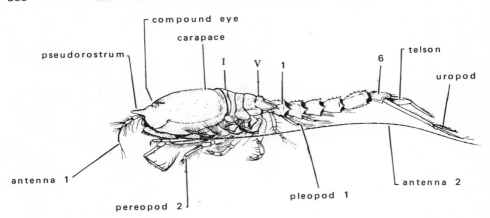

Fig. 16–1. *Diastylis cornuta,* male in lateral view; total length 12 mm. The exopodites are visible above the pereopod endopodites. Roman numerals refer to pereomeres, Arabic to pleomeres. (After Sars from Fage.)

Both pairs of antennae are short in females and young males, never consisting of more than six articles. In males the second antennae become very long during maturation (Fig. 16-1). The mandibles lack palps. The endopodite of each first maxilla bends backward, extending into the chamber under the carapace. Its function is to remove detritus particles from the respiratory water (Fig. 16-8c). The first three thoracopods are maxillipeds, the first one most differentiated, the last least. The short, compact first maxilliped is hooked to its partner by setae. A large epipodite on its coxa extends into the branchial chamber and bears a row of 4 to 12 (or up to 40) fingerlike, rarely foliaceous, gills (Fig. 16-7).

The pereopods are diverse and may have many-articled, setaceous exopodites modified for swimming. The endopodites are not modified and there are no epipodites. Only the males have biramous pleopods, which are two-articled at most. For grooming the abdomen is bent under, and slender uropods are used to groom the cephalothorax.

The exoskeleton is in part soft and flexible, in part strong and mineralized. Only rarely are the Cumacea transparent; usually they are slightly or completely opaque. Their color is white, yellowish or bright yellow, pink, bright red, or olive brown. Brightly banded or spotted species are found, particularly on beaches. The color is either diffuse or due to chromatophores, but nothing is known about the chromatophore function.

NERVOUS SYSTEM.   The protocerebrum, corresponding to the small eyes, is very simple. Its three optic masses cannot be separated with certainty. Also the globuli cells are not well differentiated. The mouthpart ganglia and those of the first maxilliped have moved together leaving only a small space between. The following 13 pairs of ganglia are all well separated but the two ganglia of each somite are almost completely fused in the midline.

SENSE ORGANS. The compound eyes when present are not on stalks but are on the wall of the cephalothorax (Fig. 16-2). They are poorly developed, and almost always fused in the midline of the dorsum into a median organ. Usually each eye has only 3 to 4, at most 11, ommatidia (Fig. 16-3).

DIGESTIVE SYSTEM. The structure of the chewing stomach is intermediate between that of the lophogastrids and that of the Tanaidacea and Isopoda. Behind the stomach there originate one to four pairs of ceca on the venter of the gut (Fig. 16-3). The midgut, like that of isopods, is believed to be syncytial.

EXCRETORY ORGANS. The excretory organs are maxillary glands, which extend posteriorly beyond the stomach. There are also clumps of four to five large glandular cells at the bases of both pairs of antennae, the first maxillae, the third pereopods, and in the blood channels that run to the pericardium from the maxillipeds and the next two limbs.

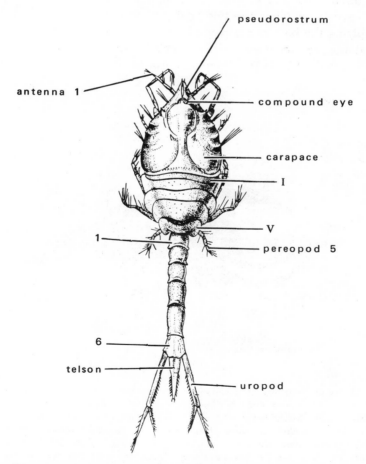

Fig. 16–2. *Diastylis rugosa*, dorsum of female; total length 8 mm. Roman numerals refer to pereomeres, Arabic to pleomeres. (After Sars from Fage.)

RESPIRATORY ORGANS AND CIRCULATORY SYSTEM.   Respiration is localized
at the anterior end of the cephalothorax. Only the first maxilliped has an epipodite
bearing gills. The surface of the carapace facing this epipodite also has a respiratory surface. Matching this localization, the dorsal vessel is sac-shaped and
differentiated into a heart only in the thoracic region (Fig. 16-3). In primitive
genera (e.g., *Diastylis*) it extends backward to the seventh thoracomere, while in
other genera (e.g., *Cumopsis*) it is secondarily shortened posteriorly, ending in
the fifth thoracic somite. It has only a single pair of ostia. The anterior aorta has a
cor frontale, in the region of the stomach. Posteriorly the heart has an unusual
pair of aortas extending to the telson (Fig. 16-3). Perhaps they are homologous
to lateral arteries, which are also present, four pairs in *Diastylis*, only three in
Cumopsis. These extend directly to the pereopods and do not have a branch for an
enteric sinus. The first pair of lateral arteries supplies thoracopods 3–5; of the
remaining three pairs in *Diastylis*, one supplies each of the thoracopods 6–8.
Cumopsis has the last (here the third pair) of lateral arteries forked near their
origin, supplying the posterior pereopods. There is no subneural artery.

The blood escapes from the arteries into unlined lacunae and via the respiratory
organs returns to the heart. The circulation can be observed in animals that have
just molted.

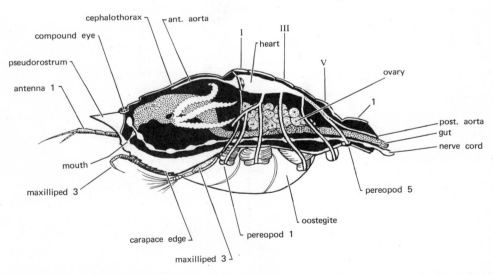

Fig. 16–3.   Internal anatomy of a broodcarrying cumacean female in lateral view. The
legs and abdomen have been cut off. The body wall is transparent. The epipodite of the
first maxilliped lying in the branchial chamber has not been drawn in. A lateral artery
enters each pereopod. Nervous and circulatory systems and digestive ceca are white.
The digestive system except the ceca is stippled. The artery of the 5th pereopod parallels
the posterior aorta for some distance. Roman numerals refer to pereomeres, Arabic to
pleomere. Diagram about 1.5 mm long. (After Sars and Ólze.)

REPRODUCTIVE SYSTEM. The gonads are paired tubes lying in the pereon above or to the sides of the gut. The oviducts open mesally on the coxae of the sixth thoracic limbs. The testes at their posterior end continue as a pair of vasa deferentia opening through two adjacent papillae on the last thoracic sternite. Males and females already differ in late embryonic stages. The adult females are recognized by overlapping oostegites on thoracopods 3–6. The adult males have well developed sense organs, much longer second antennae (Fig. 16-1), long first antennae and the eyes, if present, are larger. Also the males have their locomotor organs better developed, longer exopodites and well-developed pleopods, and have specialized second (or rarely first) pleopods for clasping females. There may also be sexual differences in keels or spines of the exoskeleton.

Reproduction

Mating probably occurs in many species during nocturnal swarming, in which the population rises to a certain height in the water or even to the surface. The animals have only been observed before mating, and it is not known whether mating takes place during or after the maturation molt of the female. Soon after maturation the females produce eggs, which develop in the brood pouch. The number of eggs depends on the size of the mother. Females of *Diastylis rathkei* that are 10 mm long produce 12–49 eggs, females 15 mm long produce 85–102.

Development

The eggs hatch in the brood pouch where the young remain to molt three times. After reaching the manca stage the larvae leave the mother. The animals then have the appearance of the adult but still lack the last pair of pereopods.

On the Swedish west coast, at aquarium temperatures of 10–13°C, the young of *Diastylis rathkei* leave the marsupium after 55–60 days, or after 76 days at 3–8.5°C, after 90–101 days at 3.5°C. When leaving the brood pouch, the young are about 2.3 mm long.

The young grow, molting repeatedly, and on reaching the penultimate instar already have sperm and ova, though presumably they do not mate yet. The secondary sex characters of the male develop gradually after molts and similar gradual development of oostegites has been observed in females of some species.

The postmarsupial life history of *Diastylis rathkei* on the Swedish west coast has been observed. The animal molts six times: there are 2 manca stages lacking the eighth thoracopods, three additional stages that have all appendages, one penultimate stage, and the adult. Female diastylids live 4 years, but most Cumacea live only about 1 year.

Habits

The Cumacea are mainly marine, there being only a few species inhabiting brackish- and freshwater. All spend most of their lives buried in sand or mud substrate rich in food, the individual species each having special requirements.

The population at times may be very dense. In the western Baltic, *Diastylis rathkei* was found in an abundance of 1214 individuals per m² in a yellow mud bottom; in mixed sand and mud with detritus there were only 111 individuals per m². *Diastylopsis dawsoni* has been found at the amazing density of 17 thousand individuals per 0.1 m² at 15 m depth on mud off northern California (R.R. Given, personal communication).

Many species occur at various depths, *Diastylis laevis* from 13–3000 m. Others prefer a narrower range, for instance *Cumopsis goodsiri* from 0–32 m. At night they make vertical migrations to shallow water or to the surface, appearing at times in dense swarms. Most of the individuals collected at night with plankton nets are adult males, but also there are penultimate females and manca stages. It is not known whether vertical migrations occur in all species at all seasons or only during the reproductive periods.

LOCOMOTION.   Burrowing is accomplished by lateral shoveling of the substrate using the last three pairs of pereopods while holding the abdomen bent over (Fig. 16-4). The animal sinks into the hole while enlarging it by flexing and extending the abdomen. The sand thrown up falls on its back, covering it. Sticky mud may be stirred up by beating the abdomen back after it has been bent under. All species, almost without exception, remain buried during the day, protected from many predators. They may move within the mud or sand, but have never been observed to run on the surface.

Otherwise cumaceans move about by swimming. Females swim by paddling motions of the pereopods or by bending and extending the abdomen. Males use their pleopods also. The body is extended, rarely bent up and over the back or down and under. The male of *Pseudocuma longicornis*, 3 mm long, moves 1 m up in 45 sec at 12°C, using four pairs of pereopods. A female of the same size needs twice the time, having only two pairs of pereopods with well-developed exopodites.

SENSES.   The habits of the animals correspond with the development of the eyes in the particular species. All deep-sea species lack eyes; the best development of eyes is found among the species living in the littoral zone. Particularly good eyes are found in species that regularly visit the surface, especially males. However, in general the eyes are poorly developed and have so few ommatidia that they cannot form an image. Such limited vision seems correlated with the habit of living within the substrate. Nothing is known about other senses.

FEEDING.   Feeding has been observed in only a few species. Cumaceans that live in the sand graze off sand grains: *Lamprops fasciata, Iphinoe trispinosa, Pseudocuma,* and *Cumopsis goodsiri.* The anterior of the body projects from the bottom of a pit in the sand. The long first pereopod picks up a sand grain and passes it to the maxillipeds. The maxillipeds hold it, constantly turning it, while the first maxillae and the mandibles brush it with sawlike movements, their setae removing algae and detritus. The cleaned grain is then dropped—*Iphinoe* throws it backward over the head—and the pereopods bring a new grain. *Lamprops* may handle particles as large as its carapace. *Diastylis rathkei* may rasp off parts of free-living plants. *Cumella vulgaris* feeds in the finer substrates, cleaning off grains from 0.15–0.5 mm in diameter.

Diastylis rathkei extends the cephalothorax over the mud surface (Fig. 16-5), and extends the three pairs of maxillipeds, forming a water-filled funnel between the appendages and the ventral body wall. Looking into the funnel from above, one can see the functioning of the mouthparts and the two anterior pairs of maxillipeds. The endopodites of the third maxilliped reach to the side and gather some mud. If microorganisms or organic detritus are present, the mud is brought into the funnel, diluted, taken up with the mouthparts and eaten. However, if the mud consists only of inorganic particles, it is expelled between the basal articles of the first antennae. When the third maxillipeds cannot reach any more mud, the first pereopods or the first antennae participate in collecting food, and a ditch develops slowly around the animal (Fig. 16-4). When the food supply at a particular place is exhausted, *Diastylis* moves a bit farther through the mud. *Leucon nasica* feeds in the same way, but obtains the mud from below its body. How the particles are brought to the mouth cannot readily be observed, but possibly this depends on a sucking or filtering structure, the presence of which has been demonstrated in many sand and mud inhabitants. Only in *Diastylis bradyi* has the physiology of the structure been studied.

This filter structure draws water with floating particles from anterior toward the second maxillae. As in Mysidacea, the cumacean filter has a row of feather-like setae on its proximal endites (Fig. 16-8c). These setae point diagonally toward the midline, forming the lateral walls of a filter chamber, the roof of which is the venter of the head (Fig. 16-8c), and the floor of which is a median extension from the body venter that projects forward between the insertion of the first maxillipeds. This extension is wide at its base and closes off the filter chamber posteriorly (Figs. 16-7, 16-8c). Anteriorly the filter chamber is closed by the distal parts of the V-shaped paragnaths (labium) and by distal endites of the second maxillae.

Fig. 16–4. *Diastylis rathkei*: left, digging in; middle, animal submerged in mud; right, funnel in front of cephalothorax to take in food. The white sector indicates the outline of the cumacean body. The animal is 15 mm long. (After Forsman.)

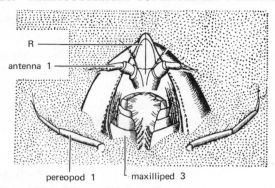

R

antenna 1

pereopod 1 — maxilliped 3

Fig. 16–5. *Diastylis rathkei* feeding; view on the exposed anterior of the animal. The 3rd maxillipeds have feathered, lateral setae which guide the respiratory current into a channel between 3rd and 2nd maxillipeds posteriorly to the inhalant opening of the carapace (see Fig. 16–6). The tips of the 1st pereopods help to obtain food. The venter of the pseudorostrum is closed by the siphon branch (R) of the 1st maxilliped. The part of the animal in view is 2.3 mm wide. (After Zimmer.)

As the 2nd maxillae move rhythmically down and out, their distance from the first maxillae above is increased. Water thus sucked in from below passes through the filter setae, on the outer side of which the filtrate is deposited (Fig. 16-8c). When the maxillae move dorsomesally again, the water that has entered is driven out of the filter chamber through a slit between the first and second maxillae. This lateral valve was previously closed from outside by the exopodite of the second maxillae (Figs. 15-6, 16-8c), but it is opened by increased water pressure. The water passing through reaches the anterior excurrent canal of the branchial chamber, the mesal wall of which is formed by the mouthparts touching one another (Fig. 16-8c).

The rhythmic beat of the epipodites of the first maxilliped within the branchial chamber reinforces the low suction and pressure that operate the valve of the second maxillae. When the pressure in the branchial chamber falls, strong suction develops, drawing the contents of the filter chamber through the maxillary slit. If the pressure of the branchial chamber increases, the suction in the filter chamber is increased, analogous to a water jet pump. The respiratory water never enters the filter chamber, because of the valve action of the exopodite of the second maxillae.

Movements of the endites of the first maxillipeds also contribute to producing the currents (Fig. 16-8c). Alternating with the endites of the 2nd maxillae, they comb out the filtrate on the under side of the filter setae with their row of erect setae. The endites of the maxillipeds (Fig. 16-8c) beat dorsomedially toward the anterior and up, pushing the particles toward the proximal endites of the first maxillae. The setae of the endites move the food to the first maxillae. In biting, the setae of the first maxillae move the food between the paragnaths toward the mandibles.

The feeding of the *Campylaspis* group probably differs. Among the stomach contents of these animals have been found Foraminifera and crustacean limb fragments. The strongly toothed pars incisiva and a stylet-shaped pars molaris on the mandibles bear evidence of predatory habits. No filtering structures are present.

RESPIRATION. Respiration takes place on the inner surface of the carapace and the gills, which arise from the posterior branch of the long epipodites of the first maxillipeds (Fig. 16-7). The whole epipodite is within the branchial chamber and its anterior (siphon) branch reaches under the pseudorostrum where it closes the exit of the chamber (Figs. 16-6, 16-7). While the epipodite rocks on its transverse axis, which runs through its attachment to the protopodite of the first maxilliped, the water current is propelled along and is forced out under the pseudorostrum. Fresh water is sucked in through a ventral slit between the carapace edge and the point of insertion of the second and third maxillipeds (Fig. 16-6). Feathered setae of the third maxilliped hold back large particles, preventing them from plugging the respiratory ducts (Fig. 16-5).

The gill branch of the epipodites divides the branchial chamber into upper and lower chambers. The branch rocks up and down around its base 40 times per minute (Fig. 16-8c). When it rises, the upper chamber contracts and the water

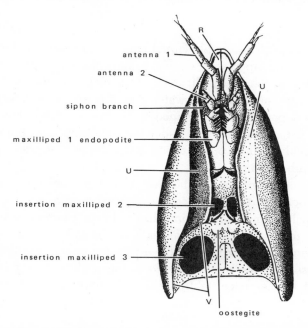

Fig. 16–6. **Ventral surface of cephalothorax of a female *Diastylis rathkei*; 4 mm long. Maxillipeds 2 and 3 have been removed at their insertion. Arrows behind insertion of maxilliped 2 indicate inhalant opening into branchial chamber. The siphon branch belongs to the first maxilliped. A line of fusion (V) closes the branchial chamber behind. R, ventral side of pseudorostrum; U, mediodorsal fold of carapace edge. (After Zimmer.)**

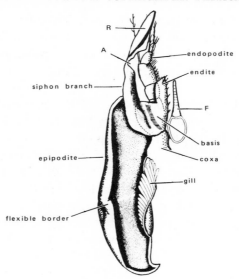

Fig. 16–7. First maxilliped of *Diastylis bradyi* with its appendages as seen from the anterior (dorsal) ; 2 mm long. A, notch in siphon branch of the maxilliped, which fits around basal articles of antenna; F, median extension, the floor of the filter channel; R, distal portion, which forms the floor valve of the pseudorostral chamber. (After Dennell.)

is put under pressure. The flexible lateral edges and posterior edge of the canoe-shaped epipodite are appressed closely against the carapace wall, preventing water from entering the lower chamber (Fig. 16-7). As the posterior carapace border is fused to the body, the water can only move anteriorly. The water presses the distal portion of the epipodite (siphon branch) down (Fig. 16-7), and forces out the stream of water under the pseudorostrum. Immediately after-ward the triangular tip recoils into its original position, closing the anterior respiratory chamber again.

By the raising of the gill branch the lower chamber is enlarged and water enters through a slit between the carapace and the insertion of the second and third maxillipeds. On lowering the gill branch, the lower chamber contracts and the enclosed water moves the soft lateral walls of the gill branches away from the carapace, allowing the water to enter the expanding upper chamber and reach the gills. The pressure at the same time closes the respiratory slit by pushing the flexible edge of the carapace tightly against the maxillipeds (Fig. 16-8c).

Classification

A key to the seven families is given by Jones (1963).

Bodotriidae. There is no free telson and the pleopods have a process on the outer edge of the inner ramus. *Bodotria scorpioides,* 5–7 mm long, occurs at depths from 3–

120 m, *Cumopsis goodsiri,* 3–6 mm long, from 0–32 m, and *Iphinoe trispinosa,* 9–10 mm long, from 4–150 m, along the European coast. *Cyclaspis varians* of the Atlantic North American coast inhabits brackish water; *C. nubila* of the Pacific coast does not.

Leuconiidae. There is no free telson and the pleopods lack an external process on the inner ramus. *Leucon americanus,* 4–5 mm long, is in brackish waters from Woods Hole to Chesapeake Bay. *Eudorellopsis deformis,* 4–5 mm long, from 2–100 m deep is found from the Arctic to Long Island. *Eudorella monodon,* 5 mm long, is found on the Gulf Coast; *E. pacifica* is abundant on the Pacific coast of North America from Vancouver to southern California, from 40–200 m deep.

Nannastacidae. There is no free telson and the male lacks pleopods. *Campylaspis canaliculata,* 4 mm, occurs from 3–212 m deep off California. *Cumella* is included in this family.

Lampropidae. There is a free telson and the three pairs of pleopods if present have an external process on the inner ramus. European *Lamprops fasciata,* 5–9 mm long, is found from 4–70 m deep; *L. korroensis,* which occurs along the coast of Kamchatka, is believed to live in fresh water; *L. quadriplicata,* 8–9 mm, occurs from Newfoundland to New England at 2–150 m, and also on the Kamchatka coast, the Kuriles, Vancouver, B.C. and the Alaskan Arctic.

Pseudocumidae. The telson is free but small and the two pairs of small pleopods lack an external process on the inner ramus. The uropod endopodite has one article. Members of the only genus *Pseudocuma* are found in the Volga delta; they are apparently unable to live upriver in freshwater. Others are found on the Atlantic coasts of Europe and Africa and the Black Sea.

Diastylidae. The telson is free and the pleopods if present lack an external process on the inner ramus. *Diastylis rathkei,* 13–18 mm long, rarely to 24 mm, occurring from 5–800 m deep, is widely distributed. In the Baltic the young leave the mother in late January or February, mature the following November and mate. The females molt again at the end of April, 8–10 weeks after the young have left, and now lack oostegites. But after the next molt, in November, they again have oostegites. Females live 4 years, producing one brood each year, while males die in their first year, after mating. *Diastylis bradyi,* 10–12 mm long, from 4–375 m deep, and *D. rugosa,* 8–9 mm long, are both European. In the western Atlantic are found *D. abbreviata, D. polita,* 12–14 mm long, *D. quadrispinosa,* 11 mm long, occurring from 4–400 m and often common, and *D. sculpta,* 9–10 mm long, found from the tide zone to 400 m deep. *Diastylis goodsiri* of the Arctic Ocean is 25 mm long; *D. californica,* 12 mm long, occurs 5–100 m deep off the California coast. *Oxyurostylis smithi,* 6–7 mm long, is found from Maine to Louisiana from the surface to 41 m deep; *O. pacifica,* 7 mm long, occurs on the California coast from 3–77 m deep. *Diastylopsis dawsoni,* 9–15 mm long, is found at times in high densities from Alaska to Central California.

Ceratocumidae. This family has a small separate telson and five pairs of pleopods with an external process on the inner ramus of the adult male. The only species is *Ceratocuma horrida* to 4 mm long, which has been found at 99 m depth in the Irish Sea and at 805 m off Cape Natal, South Africa.

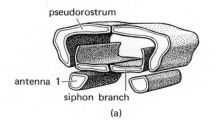

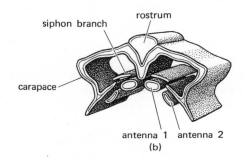

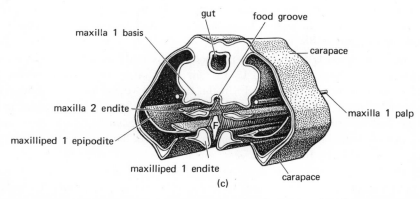

Fig. 16–8. Cross section through the branchial chamber and filter chamber of *Diastylis bradyi*. (After sections from Dennell.) (a) Cross section through anterior of branchial chamber and pseudorostrum, in which the siphon branch forms the floor valve; 0.1 mm wide. (b) Cross section near the base of the second antenna; 0.2 mm wide. Here the anterior of the body divides the branchial chamber into a left and right chamber of the carapace whose floors are formed by the carapace fold and the siphon branch, which closes against the carapace fold and the antennae. The respiratory water is under pressure and must flow anteriorly into the pseudorostrum. (c) Cross section of the filter chamber at the level of the base of the first maxilla; 0.3 mm wide. The floor of the respiratory chamber is formed by the carapace fold, the first maxilliped and the median extension (F). The mesal walls of both branchial chambers are formed by both pairs of maxillae and the maxillipeds. In the section above, the distance between these appendages has been exaggerated. The palp of the first maxilla keeps the branchial chamber clean.

ORDER SPELAEOGRIPHACEA

Only 1 species, to 7.5 mm long, is known.

Spelaeogriphacea are peracarids whose carapace is fused to the first thoracic somite, enclosing on each side a branchial chamber. All pereomeres and pleomeres have posterior lateral extensions, epimeres. The anterior three pereopods have an exopodite with two articles. The first four pairs of pleopods are also biramous and have wide swimmerets, their edges bordered by setae.*

Anatomy

The cephalothorax includes the head and first thoracic somite. The carapace has a triangular rostrum and loosely covers the first pereomere. On the sides the carapace hangs down forming the walls of the respiratory chambers, in each of which lies a stalked, spoon-shaped epipodite of the maxilliped (Fig. 16-9). Viewed from above, the trunk has parallel sides, somewhat dorsoventrally flattened. The telson is not fused to a somite. The stalked eyes are nonfunctional, represented by a short unpigmented lobe. The mouthparts, including one pair of maxillipeds, resemble those of Tanaidacea and Isopoda. The exopodites of the anterior three pereopods are setose. The posterior pereopods except for the last have reduced exopodites. The endopodites are walking legs. The anterior four pairs of pleopods are wide swimmerets that, on a protopodite consisting of a single article, bear setose exopodites and endopodites. The fifth pair of pleopods is reduced, the uropods are long. Females have oostegites on the five anterior pereopods, the first of them being very small. The internal anatomy of this recently discovered group is incompletely known.

Reproduction

During mating the male clasps the female with his modified first antennae. The juvenile stages have not yet been observed.

Relationships

The presence of oostegites forming a ventral brood pouch indicates that Spelaeogriphacea are Peracarida, but their position within the group is uncertain. Their respiratory system indicates that they are relatively specialized Peracarida. The Mysidacea must be regarded as primitive and related to the eucaridan line because epipodial respiration appears to be the primitive condition in Peracarida. The respiratory system of the Spelaeogriphacea seems to indicate a phylogenetic position for the group branching off the isopod line close to the Tanaidacea which have a similar respiratory system.

* Dr. J.R. Grindley, Director of the Port Elizabeth Museum, South Africa, who read the initial English draft on Spelaeogriphacea, generously made available his yet unpublished observations.

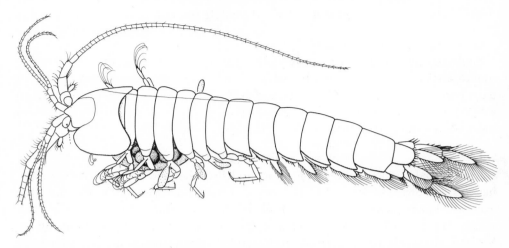

Fig. 16–9. *Spelaeogriphus lepidops*, female with eggs; 7 mm long. (Courtesy of Dr. I. Gordon and British Museum, Natural History.)

Habits

Spelaeogriphus creeps about using its pereopods or swims with active wave motions of the whole body.

The exopodites on pereopods 4–6 and rarely 7, are reduced and modified to serve as gills. A cuplike epipodite on the base of the maxilliped is apparently also respiratory in function. The exopodites on the anterior three pairs of pereopods are segmented and fringed with setae. Their constant rhythmic motion, repeated several times per second, produces a respiratory current that flows backward over the gills. The flow of hemolymph through the gills is clearly visible through the transparent cuticle.

The animals appear to feed on streamborne particles of detritus, and they often gather in sandy hollows in the stream bed where eddies result in the accumulation of detritus. Detritus particles are manipulated by the mouthparts and there is no filter-feeding mechanism.

Classification

Spelaeogriphus lepidops is found only in a stream, at +10°C in Bats Cave on Table Mountain near Cape Town, South Africa.

ORDER TANAIDACEA

Of the 300 to 400 known species of Tanaidacea, the largest is *Herpotanais kirkegaardi* which attains a length of 2.5 cm.

The Tanaidacea are peracarids that have the carapace fused to thoracomeres 1 and 2, enclosing a branchial chamber on each side. The first thoracopod bears a maxilliped, the second a chela, often very large.

Tanaids occur as fossils since the upper Permian.

Anatomy

The cylindrical trunk, usually only 2–5 mm long, is generally weakly flattened dorsoventrally and lacks distinct epimeres. The cephalothorax consists of the head fused to the first two thoracomeres. Its dorsum is a short carapace (Figs. 16-10, 16-11, 16-13). The long pereon consists of six free somites.* The short pleon has its last, sixth somite fused to the telson to form a pleotelson (Figs. 16-11, 16-12, 16-13).

The antennae and mouthparts are diverse in structure (see section on Classification, below). As in the isopods each mandible is hinged to the lateral wall of the head by two condyles. The first maxillae have a posteriorly directed endopodite lying within the branchial chamber in some groups. The first thoracic limb is differentiated into a maxilliped bearing an epipodite that, as in the Cumacea, also extends into the branchial chamber (Fig. 16-14). This branch is very large and concave in *Apseudes* (Fig. 16-15); in *Tanais* it is platelike and has respiratory epithelium, and in *Leptognathia* it is rod-shaped. Adult males of some families have reduced mouthparts.

The second thoracopod, the cheliped, is usually very large and armed with a chela or subchela (Fig. 16-13). The six pereopods, are uniramous walking legs (Fig. 16-13). But there are indications that the ancestors of the Tanaidacea had exopodites on all pereopods, suggesting close relationship with the Mysidacea and Cumacea. The Kalliapseudidae frequently have exopodites with three articles on the cheliped and first pereopods. The last manca stage of *Kalliapseudes,* has exopodites of four articles on the last three pereopods. *Apseudes* often has the first pereopod specialized for burrowing (Fig. 16-13) and the distal articles are wide and flattened. In Tanaidae the first three pairs of pereopods are specialized for spinning, the dactylus and claw are elongated and on its tip is an opening through which "silk" emerges produced by glands in the pereon. In Paratanaidae only the first one is so modified. Females develop large oostegites on the first to fourth pereopods in most species or on the fourth pereopod alone in others, in *Tanais* there is a small oostegite on the cheliped (see Classification, below). The pleopods consist of a protopodite of two articles and foliaceous branches of one or

---

* By convention specialists of the group consider the second thoracomere, which is not free, to be the first pereomere, and its appendage, the cheliped, to be the first pereopod. Here we follow Lauterbach's (1970) suggestion and consider the appendages of the fused first two thoracomeres to be the maxilliped and the cheliped; the third thoracomere, which is the first free one, is regarded as the first pereomere and bears the first pereopods.

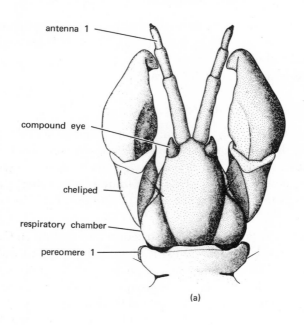

(a)

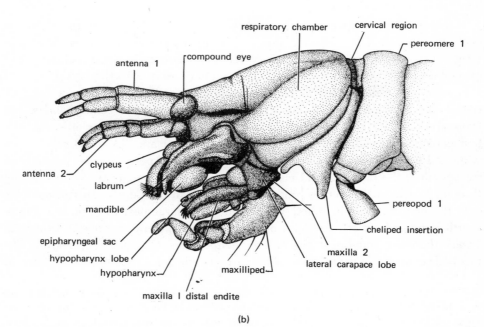

(b)

Fig. 16–10. *Tanais cavolinii*, cephalothorax: (a) Dorsal view of male; (b) lateral view of female. Most setae have been removed. The carapace of the male is about 0.5 mm wide. (From Lauterbach, courtesy of K. E. Lauterbach.)

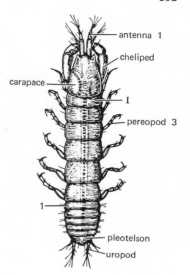

Fig. 16-11. *Heterotanais oerstedi*, female in dorsal view; 2 mm long. Roman numeral refers to pereomere; Arabic to pleomere. (After Sars from Nierstrasz, Schuurmans-Stekhoven.)

two articles, their edges lined with setae. The pleopods may be absent in females. They have no respiratory epithelium. The rodshaped endopodites of the uropods are usually much longer than the exopodites.

The fairly thick integument is usually whitish or yellowish, sometimes with darker pigment spots or borders.

SENSE ORGANS AND NERVOUS SYSTEM. Besides sensory setae and esthetascs, eyes are often present. The eyes may be on cone-shaped projections from the head (Figs. 16-10, 16-14). In *Tanais* the reduced eyestalk has a movable insertion on the head capsule, and one levator muscle. There are only a few ommatidia: in *Heterotanais*, 12 in each eye; in *Apseudes*, which lacks crystalline cones, only 8.

The supraesophageal ganglion lacks globuli and antennal glomeruli and its optic center is only weakly developed. In *Apseudes*, the somatic ganglia form a long chain. Only the first four, belonging to the mouthparts and the maxillipeds, have short connectives; the others have long ones. There is a rudimentary seventh ganglion in the abdomen. In *Heterotanais* and *Tanaissus*, however, the ganglia of the first five postoral appendages and those of the abdomen lie very close to each other, separated only slightly by constrictions, in contrast to the six pairs of pereon ganglia. The seventh abdominal ganglion is absent.

DIGESTIVE SYSTEM. The chewing stomach of *Apseudes* is more complicated, and resembles that of lophogastrids. In *Heterotanais* and especially in *Tanaissus* the chewing stomach is reduced and its folds partly fused. It is intermediate in *Tanais*. *Apseudes* has two pairs of ventral midgut ceca, while the other two genera have only one pair. The epithelium of the long gut, like that of isopods, appears to be syncytial, and has a cuticularized intima. The ectodermal hindgut in the pleotelson opens at the end of the telson.

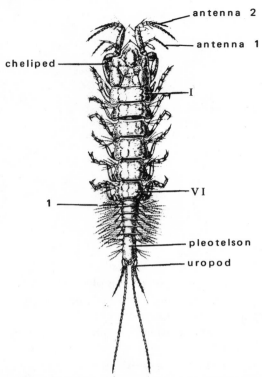

Fig. 16–12. *Apseudes talpa*, female in dorsal view; 6 mm long. Roman numerals refer to pereomeres; Arabic to pleomere. (After Sars from Nierstrasz, Schuurmans-Stekhoven.)

EXCRETORY SYSTEM.   In addition to nephrocytes, there are paired maxillary glands at the base of the second maxillae. Embryos of *Apseudes* and *Tanais* have a rudimentary gland in the second antennae.

CIRCULATORY SYSTEM.   In *Apseudes* and *Heterotanais* the heart extends the length of the pereon; in *Tanaissus* it is short, ending in the third pereomere. It lies in a pericardial sinus with a well-developed muscular pericardial membrane, to which well developed transverse lateral muscles are fused. The heart usually has a pair of ostia in the second and third pereomeres. The right ostium of the anterior pair closes during postembryonic development of *Apseudes*.

Anteriorly there is a cephalic aorta that widens in the area of the chewing stomach to form the cor frontale. It supplies the brain, labrum and antennae with blood. Besides the usual single pair of lateral arteries there is an additional pair in *Heterotanais* and at least three in *Apseudes*. All lateral arteries open near the large gut lacunae and midgut ceca.

The postmandibular appendages receive the blood from a longitudinal lacuna. As in Mysidacea, the blood returning from the limbs flows back to the pericardial sinus in podopericardial channels lined by membranes.

RESPIRATORY ORGANS. The inner surface of the carapace fold, the lateral wall of the cephalothorax facing the branchial (or respiratory) chamber, and in *Tanais* the epipodites of the maxillipeds, function in respiration. Underneath these walls the sinus coming from the head, antennae and mandibles divides into a net of capillary lacunae which continues straight into the pericardial sinus at the posterior border of the carapace. The oxygen enriched blood is thus taken up mainly by the first ostia and for the most part is returned to the head. The posterior region receives hardly any oxygenated blood.

REPRODUCTIVE ORGANS. The paired tube-shaped ovaries extend the length of the pereon and open through an oviduct to the outside at each side of the fourth free sternite. The testes are much shorter. Their vasa deferentia widen into seminal vesicles. The male gonopores are always paired and lie on a single median cone (Monokonophora) or a pair of cones (Dikonophora) on the last, the eighth thoracic sternite.

Reproduction

Mature males can be distinguished from females by their better developed sense organs: larger eyes (when present), a larger number and longer esthetascs, and also by their more developed pleopods. The male of *Heterotanais oerstedi* uses his much larger chelae for tearing open the dwelling-tube of the female prior to mating.

Heterotanais oerstedi is a potential hermaphrodite. In this species there are two kinds of males, primary males that develop from the second neuter stage after a molt and secondary males that develop from females when they molt after brooding. This molt is accompanied by reduction of the ovary. There are also gonochoristic females. If a female lives in an aquarium with a male it will not change sex. How the male influences the sex determination is not known. Larvae raised with an adult male develop into females, those kept with adult females become males. Larvae raised in isolation become either gonochoristic females or females that later become males (protogynous).

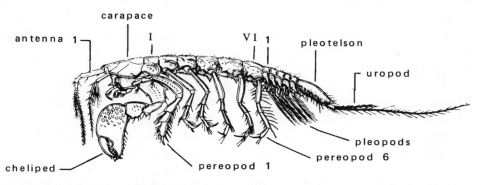

Fig. 16–13. *Apseudes spinosus*, male in lateral view; 12 mm long. Roman numerals refer to pereomeres; Arabic to pleomere. (After Sars from Claus, Grobben.)

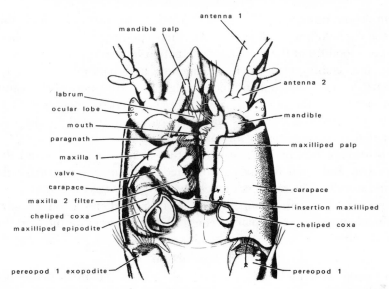

Fig. 16–14.   Cephalothorax of *Apseudes talpa*, venter; 2 mm long. At left the carapace wall has been removed, exposing branchial chamber. The right maxilliped and both chelipeds have also been cut off. The valve is the epipodite of the maxilliped. Lower arrow on right, inflowing water; arrow right center, outflowing water. (After Dennell.)

Apseudes talpa is also suspected of having females that turn into males. The antarctic *Apseudes spectabilis* has concurrently mature ova and sperm in its gonads thus being a simultaneous hermaphrodite.

The male of *Heterotanais* enters the dwelling tube of the female, who is receptive only if she has undergone the "maturation" molt and has become adult, but has not produced eggs yet. In the narrow tube there is prenuptial display, the partners running back and forth for as long as 11 hours. Finally they take a position parallel to each other, their venters adjacent, and the female above the male extends her oostegites. The sperm are released in 2–3 sec and the marsupium closed. Immediately the eggs are laid, about 6–16 in 20 min. The female, with her chelae, now takes the male by the pleotelson or the pleopods and drives him out of the tube. Afterward both ends of the tube are spun shut and the tube walls are strengthened.

Development

Heterotanais females move the oostegites rhythmically one after another, rolling the eggs anteriorly and posteriorly, supplying them with freshwater. If the eggs get caught, a pereopod reaches into the marsupium and pushes them along. At 22°C the embryo hatches after 13 days with undeveloped appendage segmentation. The first manca stage leaves the marsupium and feeds on diatoms; after a few days it breaks out of the mother's tube, spinning a branch tube which the mother separates off by repairing the hole in her own tube. In the second manca stage pleopods appear as well as the last pair of pereopods, which reach final development after the next molt when the animal enters the neuter 1 stage. From the neuter 1 stage development may follow two directions: the animal may molt into neuter 2 and molt again into a primary male; or the animal may molt into a female with small oostegites and then molt into a female with a complete marsupium. This female may molt again to become a male or molt to become a female with small oostegites.

Habits

This group is mainly marine, although some species enter brackish water, and *Tanais stanfordi* enters rivers in Brazil and Argentina and has been found in freshwater lakes in the Kuril Islands. The Tanaidacea are found along coasts as well as in the deep ocean down to 8200 m. Most stay on the bottom and hide in the mud or among plants, colonial animals or debris, although *Kalliapseudes* has been reported from the plankton. *Heterotanais* swims with the pleopods, but soon burrows into mud with its chelipeds. Females of *H. oerstedi* can bury themselves in 1 min, males which have larger chelae, still more rapidly.

Heterotanais oerstedi lives individually in spun tubes that lie parallel to the surface in the sediment. In spinning the tube, the third to fifth thoracopods move alternately to and fro. At the same time from the tips of their claws flows a mucus secretion produced in glands that lie in the corresponding trunk somites. The secretion is pulled into threads that stick to the surroundings. When the claws are withdrawn, diatoms become stuck to their tips and outside on the threads. The animal also picks up detritus with the chelae and spins it into the tube walls. Females deposit fecal columns they produce regularly every 2.5 min, parallel to each other on the lower half of the tube. The ends of the tube remain open. The tube can be extended by pushing the cephalothorax against the walls, sometimes aided by the chelae. Holes are covered up or sometimes a new tube is made.

FEEDING. When feeding, *Heterotanais* extends its cephalothorax and pereon from the tube. It takes up clumps of detritus and brings them to the maxillipeds which pass the material to the mouth. Large pieces are broken up by the chelae. Microorganisms in the detritus are probably of main importance as food. Diatoms are sucked into the tube by the respiratory currents produced by the pleopods, and stick to the walls. They are taken by the first manca stage as their first food. At times females pick up small nematodes and tear them with their chelae. Adult males do not feed.

Apseudes, like primitive peracarids, takes in a small part of its food by filter-feeding. The paired filters are near the midline of the body and, as in the Cumacea, have the shape of a slanted roof which covers the filter channel (Figs. 16-14, 16-16). Both slanted areas consist of a row of dorsomedially directed filter setae, standing on the anterior (dorsal) edge of the proximal endites of the second maxillae. The second maxillae are embraced proximally and behind (ventrally) by the anteriorly (dorsally) concave maxilliped. The maxilliped endites, which run parallel to the midline, are connected to each other by hooked coupling setae, forming the floor of the filter chamber. Posteriorly the filter channel is closed by the paired maxillipeds. Water can therefore enter only from the anterior. Suction is produced in two ways: (1) by moving the maxillipeds rhythmically posteriorly, increasing their distance from the second maxillae and the suction chamber, and (2) by the respiratory current which comes through a slit between the first maxillae and the posterior second maxillae, diagonally, producing a suction on the content of the filter chamber from outside and behind. The respiratory water drawn out of the filter chamber by the expiratory stream leaves its filtrate on the sides of the maxillae filter setae and enters the branchial chambers (Fig. 16-16). Then the dorsal setae of the maxilliped endites comb out the filtrate and take it to the mouth (Fig. 16-16).

RESPIRATION.   The respiratory surface is the integument of the carapace wall facing the branchial chamber (Fig. 16-15). In *Apseudes* the chamber is closed anteriorly by the thick base of the mandible, laterally by both pairs of maxillae, and posteriorly by the swollen base of the cheliped. If one brings iron saccharate into the surroundings of *Apseudes,* one can see the powder disappear at the posterior carapace border and reappear near the base of the maxilliped under the lateral edge of the carapace (Fig. 16-14). The posteriorly directed epipodite of the maxilliped, hemispherical in shape with the concave side against the swollen base of the cheliped coxa (Fig. 16-15), causes the water to move. The epipodite moves constantly to and fro around a diagonal axis. If it moves anteriolaterally, up, its distance from the coxa is increased, causing water to be pulled up from behind (Fig. 16-15). At the same time, its setose, lappet-shaped anterior valve is pulled back, opening a slit between the bases of the first and second maxillae that leads into the filter channel (Figs. 16-15, 16-16). Thus the suction also pulls water from the filter chamber into the branchial chamber. The return of the epipodite is posteriomedial and down. The concavity approaches the swollen base of the cheliped, forcing the water out. At the same time the rising water pressure presses the lappet-shaped branch valve against the maxillary slit, closing it. The water flowing out cannot enter the filter chamber or the carapace opening because the ventral border of the epipodite presses tightly against the base of the first coxa. The water thus escapes ventrally in front of the first leg coxa and reaches the narrow anterior part of the branchial chamber. As this is closed by the mandible base, the water has to escape near the base of the maxilliped (Fig. 16-14, upper arrow). The next anteriolateral, dorsal movement of the epipodite sucks water up out of the filter chamber and from the surroundings (Figs. 16-14, lower arrow, 16-15). The small exopodites of the first and second pereopods also push water toward the inhalant opening, aiding in the uptake of water (Fig. 16-14). The endopodite of the first maxilla reaches into the branchial chamber to remove particles that have adhered to respiratory surfaces.

In *Tanais* the respiratory chamber is closed anteriorly by the fusion of the carapace to the wall of the cephalothorax and posteriorly by a large sidepiece of the chelipeds and the folded rim of the carapace. The respiratory current is reversed in Tanaidae: water enters anteriorly and leaves through a dorsal opening behind the cephalothorax. Thus the respiratory current and the feeding current within the animal's tube go the same direction, unlike that of Apseudidae and *Heterotanais,* an adaptation to living in a tube (Lauterbach, 1970).

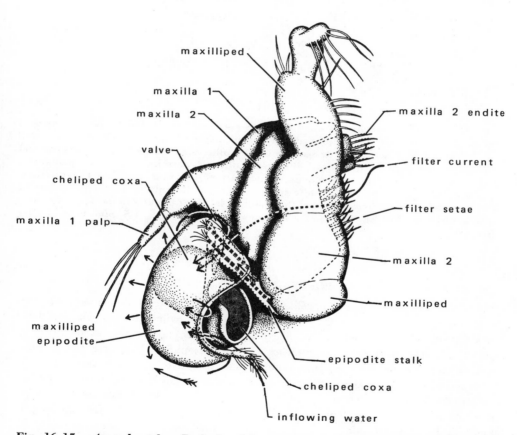

Fig. 16–15. *Apseudes talpa.* Both the right maxillae and the maxilliped viewed from below; diagram showing the start of suction of the maxillipedal epipodites. Parts of maxillae 1 and 2 dorsal to the maxilliped are drawn in as dotted lines. The valve is the epipodite of the maxilliped; the filter setae stand on the proximal endite of maxilla 2. The arrows with tails show the movement of the epipodite, those without tails, water currents. Currents that run over the maxilliped (under, in the diagram) are dotted; the cheliped has been cut off at the coxa. Maxilliped, 1 mm long. (After Dennell.)

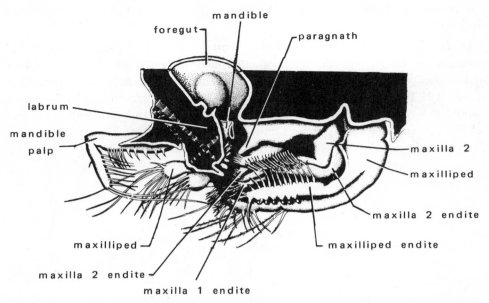

Fig. 16-16. Median section through the lower portion of the cephalothorax of *Apseudes talpa*. Endopodite of the maxilliped has dorsally directed combing-out setae and coupling hooks that link it with its partner on the other side of the body. The proximal endites of the 2nd maxillae bear filter setae directed dorsally toward the paragnath. Below the paragnath one can see the distal part of the first maxilla. The grinding (molar) surface and below it (white spots in Figure) the cutting (incisor) surface of each mandible meet directly under the mouth. In the foregut can be seen the constriction caused by the mandibular adductor. Cephalothorax 1.5 mm long. (After Dennell.)

Classification

SUBORDER MONOKONOPHORA

The first antennae have two flagella; the second antennae usually have an exopodite, a scale consisting of one article. Each mandible has a palp. The first maxillae have two endites each. The cheliped and first pereopods each may have an exopodite (Fig. 16–14). There are 4 pairs of oostegites in the marsupium. The male has a single median genital cone. This appears to be the more primitive of the two suborders.

Apseudidae. Neither the first pereopod nor any other has a sense organ on the last article. Included are *Apseudes spinosus* (Fig. 16–13), up to 13 mm long; *A. talpa* (Fig. 16–12), up to 6 mm long, found on European coasts; *A. garthi*, 1.9 mm long, is found in the Gulf of California. *A. sibogae*, an abyssal species, reaches a length of 21 mm. Other representative genera in this poorly understood family are *Pagurapseudes*, an inhabitant of gastropod shells, and *Sphyrapus*.

Kalliapseudidae. The base of the last article (dactylus) of at least the 1st pereopod bears a sense organ with long tubules or the whole article has been replaced by the sense organ. This family has a cosmopolitan distribution in tropical and subtropical waters. It has been found intertidally and subtidally in mud, sandy ooze, fine and coarse sand and gravel; among ascidians, shells and worm tubes; and on mangroves. It is the only family of the order in which filter feeding is apparently the main means for obtaining food. The genera in the Kalliapseudidae are *Kalliapseudes, Hemikalliapseudes,* and *Psammokalliapseudes.*

SUBORDER DIKONOPHORA

The first antennae have only one flagellum. The second antennae lack an exopodite (scale). The mandibles do not bear palps. The epistome, labrum and mandibles have shifted anteriorly to a position typical of the isopods and amphipods. Pereopods lack exopodites. Males have a pair of genital cones. The brood pouch is composed of one or four pairs of oostegites. The group has secondarily reduced appendages, stomach, and circulatory system. The differences in the cephalothorax probably result from the change from filter feeding to feeding on larger particles.

Neotanaidae. Each first antenna is composed of seven or eight articles, the second has nine articles. In the female, each first maxilla has two endites and the second maxillae are well developed. Males lack first maxillae, and their mandibles and second maxillae are reduced, but maxillipeds are present, each with a fairly well-developed palp. Four pairs of oostegites are borne on the first to fourth pereopods. The Neotanaidae are inhabitants of the deep ocean, from about 300–8200 m. *Neotanais* and *Herpotanais,* the latter up to 2.5 cm long, are included. *Neotanais americanus* is a species found off the eastern coast of North America.

Paratanaidae. The flagellum of the first antenna is more or less reduced. The second antennae never have more than six articles. The first maxillae have only one endite and a palp with one article. The second maxillae are vestigial. At most the males retain the palp of the first maxilla and the maxillipeds with their epignaths. The marsupium consists of one or of four pairs of oostegites on the first to fourth pereopods; if only one pair is present it is on the fourth. One of the best known members of this family, *Heterotanais oerstedi,* about 0.6–10 mm in length (Fig. 16–11), is found in shallow European waters, where it makes mucus tubes, attached to the substrate. It walks with its pereopods. Pleopods are used for swimming and also to circulate water in the tubes. Males occur only during 3–4 months of the year and can be found outside their tubes in the daytime. They lose their mouthparts when they become mature and probably live only for a short time after mating. The ovary contains 10–20 eggs, but there are fewer in the marsupium. They are laid following a molt, only in the presence of a male. If no male is present they are resorbed. The manca stage that leaves the marsupium is 0.6 mm long and molts after 10–11 days. The female molts after the young have left, losing her oostegites. After the following molt oostegites reappear, and the female may breed again. *Heterotanais groenlandicus,* 2 mm in length, has been recorded from mud and fine sand at 10 m depth from the coast of British Columbia. *Leptognathia longiremis,* up to 4 mm long, has been collected from 3–100 m in sandy mud off British Columbia. It is also found on northern European coasts. Many members of the Paratanaidae occur at abyssal or hadal depths. Other genera representative of this large and taxonomically very poorly known family are *Cryptocope, Leptochelia, Nototanais, Paratanais, Pseudotanais, Paranarthrura,* and *Typhlotanais.*

Tanaidae. This family differs from the preceding by having the palp of the first maxilla two-articled, and the males do not have their mouthparts reduced. Females have only one pair of oostegites, on the fourth pereopods. *Tanais cavolinii,* up to 5 mm long (Fig. 16–10), is found on both sides of the North Atlantic in shallow water among vegetation and corals. *Tanais stanfordi* is found in freshwater lakes on Kunashir in the Kuril Islands and in Brazilian and Argentine rivers. *Zeuxo robustus* has been found between the carapace scales of the Atlantic loggerhead turtle, *Caretta caretta. Pancolus* is included in this family.

References

Cumacea

Băescu, M. 1951. Cumacea. *Fauna Rep. Popul. Române* 4 (1): 3–91.

Calman, W.T. 1912. Cumacea in the collection of the U.S. Natl. Mus. *Proc. U.S. Natl. Mus.* 41: 603–676.

Dennell, R. 1934. The feeding mechanism of the Cumacean, *Diastylis bradyi. Trans. Roy. Soc. Edinburgh* 58: 125–141.

Dixon, A.Y. 1944. The biology of *Cumopsis goodsiri* and some other Cumaceans in relation to their environment. *J. Marine Biol. Ass.* 26: 61–71.

Fage, L. 1951. Cumacés. *Faune de France* 54: 1–136.

Forsman, B. 1938. Die Cumaceen des Skageraks. *Zool. Bidr. Uppsala* 18: 1–161.

Jones, N.S. 1963. New Zealand Cumacea. *Mem. New Zealand Oceanogr. Inst.* 23: 1–79.

Mensy, J.J. 1963. La glande androgène chez deux Crustacés Péracarides: *Paramysis nouveli* et *Eocuma dollfusi. Compt. Rend. Acad. Sci. Paris* 256: 5425–5428.

Oelze, A. 1931. Anatomie von *Diastylis rathkei. Zool. Jahrb. Abt. Anat.* 54: 235–294.

Siewing, R. 1952. Morphologische Untersuchungen an Cumaceen. *Zool. Jahrb. Abt. Anat.* 72: 522–559.

———. 1953. Morphologische Untersuchungen an Tanaidaceen und Lophogastriden. *Z. Wiss. Zool.* 157: 333–426.

Stebbing, T.R.R. 1913. Cumacea. *Das Tierreich* 39: 1–210.

Wieser, W. 1956. Factors influencing the choice of substratum in *Cumella vulgaris. Limnol. Oceanogr.* 14: 274–285.

Zimmer, C. 1933. Cumacea *in* Grimpe, G. and Wagler, E. *Die Tierwelt der Nord-und Ostsee.* Akad. Verl., Geest & Portig, Leipzig 10.

———. 1936. California Cumacea. *Proc. U.S. Natl. Mus.* 83: 423–439.

———. 1941. Cumacea *in Bronns Klassen und Ordnungen des Tierreichs.* Akad. Verl., Geest & Portig, Leipzig 5(1) 4(5): 1–222.

Spelaeogriphacea

Gordon, I. 1957. *Spelaeogriphus,* a new cavernicolous crustacean from South Africa. *Bull. Brit. Mus., Natur. Hist., Zool.* 5: 31–47.

Tanaidacea

Bückle-Ramírez, L.F. 1965. Untersuchungen über die Biologie von *Heterotanais oerstedi. Z. Morphol. Ökol.* 55: 714–782.

Dennell, R. 1937. Feeding mechanism of *Apseudes talpa* and the evolution of the peracaridan feeding mechanism. *Trans. Roy. Soc. Edinburgh* 59: 57–78.

Forsman, B. 1956. The invertebrate fauna of the Baltic. *Ark. Zool.* (2)9: 389–419.

Lang, K. 1949. The systematics and synonymies of the Tanaidacea. *Ibid.* 42A(18): 1–14.

———— 1953. *Apseudes hermaphroditicus*, a hermaphroditic Tanaide from the Antarctic. *Ibid.* (2)4: 341–350.

———— 1953. The marsupial development of the Tanaidacea. *Ibid.* (2)4: 409–422.

———— 1956. Kalliapseudidae, a new family of Tanaidacea *in* Wingstrand, K.G. ed. *Bertil Hanström, Zool. Pap. in Honor of his 65th birthday*, Lund pp. 205–225.

———— 1956. Neotanaide, with some remarks on the phylogeny of the Tanaidacea. *Ark. Zool.* (2)9: 469–475.

———— 1957. Tanaidacea from Canada and Alaska. Contributions de Département des Pêcheries, Quebec, No. 52.

———— 1958. Protogynie bei zwei Tanaidaceen-Arten. *Ark. Zool.* (2)11: 535–540.

———— 1967. Taxonomische und phylogenetische Untersuchungen über die Tanaidaceen. *Ibid.* (2)19: 343–368.

———— 1968. Deep-Sea Tanaidacea. *Galathea Rep.* 9: 23–210.

Lauterbach, K.E. 1970. Der Cephalothorax von *Tanais cavolinii*. *Zool. Jahrb. Abt. Anat.* 87 (in press).

Menzies, R.J. 1953. The apseudid Chelifera of the eastern tropical and north temperate Pacific Ocean. *Bull. Mus. Comp. Zool.* 107: 443–496.

Miller, Milton A. 1940. The isopod Crustacea of the Hawaiian Islands (Chelifera and Valvifera). *Occ. Pap. Bishop Mus. Honolulu* 15: 295–361.

Moers-Messmer, W. von. 1936. *Das Marsupium der Amphipoden und Tanaidaceen.* Inaug. Diss. Friedrich-Wilhelms-Universität Berlin. Konrad Triltsch, Berlin.

Nierstrasz, J.F. and Stekhoven, J.H.S. 1930. Anisopoda *in* Grimpe, G. and Wagler, E. *Die Tierwelt der Nord-und Ostsee.* Akad. Verl., Geest & Portig, Leipzig 10: 134–167.

Richardson, H. 1905. A monograph on the isopods of North America. *Bull. U.S. Natl. Mus.* 54: 1–727.

Scholl, G. 1963. Embryologische Untersuchungen an Tanaidaceen. *Zool. Jahrb. Abt. Anat.* 80: 500–554.

Siewing, R. 1953. Morphologische Untersuchungen an Tanaidaceen und Lophogastriden. *Z. Wiss. Zool.* 157: 333–426.

Wolff, T. 1956. Crustacea Tanaidacea from depths exceeding 6000 m. *Galathea Rep.* 2: 187–241.

Zimmer, C. 1926. Tanaidacea *in* Kükenthal, W. and Krumbach, T., *Handbuch der Zoologie,* de Gruyter, Berlin 3(1)10: 683–696.

17.

Order Isopoda

There are more than 4,000 species of isopods known. *Bathynomus giganteus,* which may reach a length of 35 cm, is the largest.

Isopods are peracarids that have the head fused to the first thoracic somite, rarely also the second, forming a cephalothorax. There is no carapace. The thoracic limbs are cylindrical legs; the pleopods are flattened.

This interesting and very diverse order is the only crustacean group that, in addition to many marine and freshwater inhabitants, has evolved numerous true terrestrial species, in all stages of development independent of the marine or freshwater environment. Isopods are almost always dorsoventrally flattened. In only a few families is the trunk cylindrical (Anthuridae) or laterally compressed, amphipod-like (Phreatoicidea) (Figs. 17-6, 17-15). Some females among the parasitic Epicaridea lose their crustacean shape when sexually mature, so that only careful study of postembryological development makes it possible to relate them to Crustacea and Isopoda (Fig. 17-2). Isopods are known as fossils since the Permian.

Anatomy

The pereon, the largest body division, is made up of seven somites (rarely only six), as the first (at times the second also) of the eight thoracomeres is fused to the head, forming a cephalothorax. Its tergites are usually quite distinct, only rarely fused with each other or the abdomen (some Asellota, Gnathiidea, Fig. 17-24). The abdomen, the pleon, is noticeably short and its segmentation is often variable as some (in Sphaeromatidae, Idoteidae) or nearly all tergites (in Asellota) are fused with the telson into a uniform dorsal plate (Fig. 17-28). But almost always the last pleomere, the sixth, fuses with the triangular or semicircular telson to form the pleotelson. Exceptions are the Anthuridea. In many species the trunk is widened, shieldlike, and plates, the epimera, project laterally above the leg attachments (Fig. 17-1). The epimeres rarely are transformed sternites or tergites, but usually are leg coxae transformed into plates (Fig. 17-27), which fuse more or less indistinguishably to the tergites (Fig. 17-1). Only among the Asellota and almost all Phreatoicidea do the leg coxae remain movably articulated with the body (Fig. 17-6).

The first antennae are nearly always uniramous. They are very short in terrestrial isopods (Oniscidea), barely visible under the cephalothorax edge (Fig. 17-3). In general, except for the second antennae, all appendages used as feelers are shorter in terrestrial isopods than in aquatic species. The mandibles, as in insects, have anterior and posterior insertions on the head and work pincerlike against each other, their endites bent at a right angle to the midline of the body (Vol. 2, Fig. 16-8c, p. 302). Often a palp with three articles is present. The plate-like first maxillae often have strong teeth but lack palps. They are always stronger than the second maxillae, which also lacks palps (Fig. 17-5). The maxillipeds often have flat proximal articles and lie parallel to one another on the posterior (ventral) surface of the other mouthparts, their straight median edges touching along the midline (Figs. 17-7, 17-21). The mouthparts therefore are enclosed posteriorly and below by the maxillipeds, like the labium of insects. Parasitic isopods may have the mouthparts pointed, more or less transformed into stylets

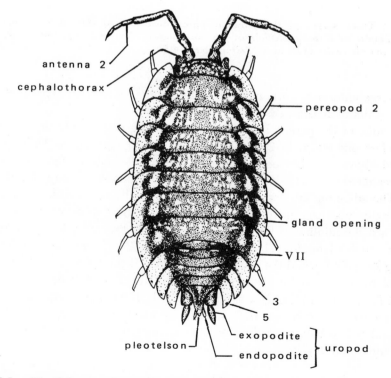

Fig. 17-1. Wood louse, *Oniscus asellus*; 15 mm long. Roman numerals refer to pereomeres, Arabic to pleomeres. (From Paulmier after Van Name.)

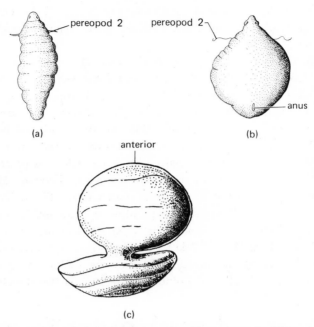

(a)   (b)

(c)

Fig. 17–2.   Three stages of the parasitic epicaridean isopod, *Liriopsis*: a, b, Young parasitic females of *L. monophthalma*; 1.6 mm long. c, Female *L. pygmaea* with brood, free in the mantle cavity of its host (*Peltogaster*) ; 4.5 mm long. (After Sars and Caullery.)

that converge anteriorly, forming a projecting cone (Fig. 17-4). The Gnathiidae have two pairs of maxillipeds.

Each somite of the pereon has one pair of walking limbs. Only the Gnathiidae lack the last one (Fig. 17-24). The pereopods are uniramous. The coxae, as mentioned above, can be recognized as a leg article only in the Asellota and most Phreatoicidea; usually they are modified as widened epimeres fused to the trunk. (The adjoining articles after the coxa are usually transverse, at right angle to the longitudinal axis, in many, particularly terrestrial species, bent V-shaped back toward the midline of the body.) The basis is directed toward the middle, the other articles laterally. This arrangement (Fig. 17-7) prevents the leg from extending laterally beyond the protective epimeres. If the leg articles are diagonal instead of transverse, as in the Anthuridea and the Phreatoicidea, the trunk is often not flattened, and the legs, as in amphipods, form two groups, the anterior group directed anteriorly and the posterior group directed posteriorly. This leg position, related to habits, results from a late rotation that has been observed in the postembryonal development of *Anilocra*.

The pereopods are specialized as walking legs, as gnathopods, or as swimmerets. While in some groups (e.g., Oniscidea) all seven are walking legs, in other families some of the anterior pairs have become gnathopods, and in the larvae

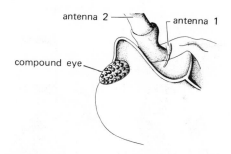

Fig. 17–3. Cephalothorax of an oniscid, *Cylisticus convexus*, in anteriodorsal view. First antenna, 0.4 mm long. (After Leydig.)

of Epicaridea, all have transformed (Fig. 17-12). The end of the gnathopod pereopod may be a strongly bent distal article (dactyl) a subchela or a chela in which the dactyl can work against the widened propodus, the previous article (Fig. 17-12), or rarely the propodus against the carpus.

The pleopods have foliaceous branches, at times divided by a transverse groove into two articles. In most isopods these branches are so wide that they cannot lie next to each other, so the exopodite lies in front of the endopodite (Figs. 17-7, 17-8). As the pleopods lie posteriorly against the venter, the outer branch covers the inner one (Figs. 17-7, 17-8). Sometimes one or two pairs of pleopods are modified to form an operculum for all posterior abdominal appendages. This is true of the first pair in the Anthuridea, the second pair of the Parasellidae females, the third pair in Asellidae and Stenetriidae, and the fourth exopodite of the Serolidae. The second pleopod of the male is almost always transformed into a gonopod, its inner branch with a rodlike, often grooved, appendage lateral to the gonopore (or gonopores) in position. In the Oniscidea and Parasellota this appendage is missing and the endopodites themselves are pointed. In most terrestrial isopods the first pleopod of the male is also cylindrical and takes part in sperm transfer.

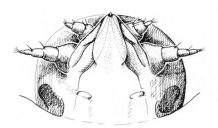

Fig. 17–4. Mouth cone of first larval stage of *Cancricepon elegans*. Outside of the mouth cone there are one pair of maxillae and one pair of maxillipeds, but it is not known whether the 1st or 2nd maxillae are retained. Much enlarged. (After Bonnier.)

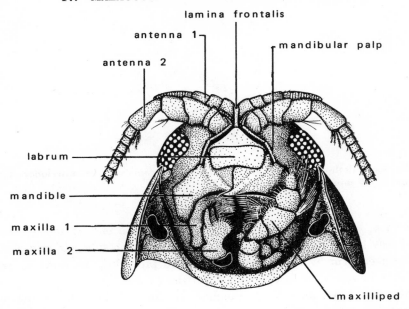

Fig. 17–5.   Mouthparts of the predacious isopod, *Cirolana borealis*; cephalothorax and first pereomere as seen from below; 5 mm long. (After Gruner.)

It is characteristic of isopods that the pleopods have become gills. They can be recognized by the soft exoskeleton, increased surface due to folding (Sphaeromatidae), brushlike attachments (*Bathynomus*), or branching (some Epicaridea). Usually these specializations are found on the inner branch and do not affect all pleopod pairs. The pleopods of terrestrial isopods are specialized for terrestrial respiration (discussed under the section Respiration). Differences in the uropods, the last pair of appendages, are useful in classification. They are lost in some females of the Epicaridea and nearly always in males. In the

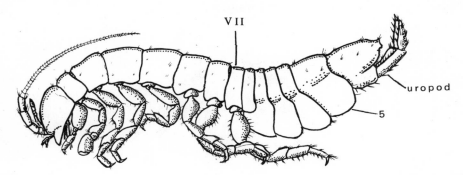

Fig. 17–6.   *Neophreatoicus assimilis*; 12 mm long. The long, ventral epimeres of the 2nd–5th abdominal somites make the round trunk look like the laterally flattened body of an amphipod. Last pereomere, VII; fifth pleomere, 5. (After Chilton.)

Flabellifera, Gnathiidae, and Anthuridea they have retained the basic caridoid form, being flattened and widened, and in conjunction with the pleotelson, forming a tail fan (Fig. 17-26). In Asellota, Phreatoicidea, and Oniscidea they are cylindrical or ribbonlike, and may be very reduced (Fig. 17-7). In the Valvifera they are ventral, their stem and the inner branch together forming a long plate that can be closed over the long pleopods like a door (Fig. 17-9).

The exoskeleton is leatherlike and flexible in most species. In some marine species it is mineralized (Sphaeromatidea, Serolidae, and Parasellota), as it is in terrestrial Oniscidea. In these, epicuticle, pigment layer, calcareous layer, and lamellate zone can be differentiated. The parasites however are usually soft. In color, isopods may be gray or brown with markings. *Jaera* and many species of

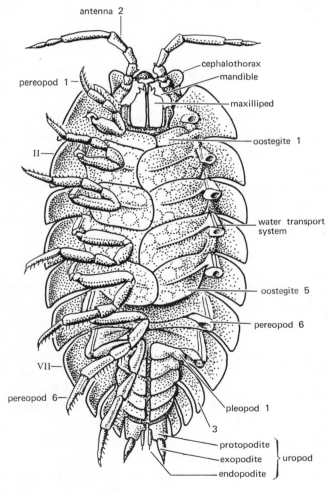

Fig. 17–7. **Venter of eggcarrying female oniscid, *Porcellio scaber*; 12 mm long. The left walking legs (on the right) have been cut off. Roman numerals refer to pereomeres, Arabic to pleomere. (Modified after Van Name.)**

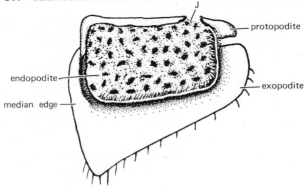

Fig. 17–8.   Third right pleopod of the oniscid _Porcellio scaber_ in dorsal view; 1.6 mm wide. J, insertion of protopodite on the pleosternite; the insertion is lateral on the sternite. (After Unwin.)

Idotea are especially well marked, the pattern determined by melanophores and leucophores (see below under Neurosecretion).

The distribution of chromatophores does not change during the life span of _Idotea,_ but individuals have different patterns and many distinct patterns appear convergently in different species. As the contrasting black-white animals are exclusively females, and the young of a single litter display only a few patterns, it may be assumed that the pattern is genetic. Inheritance of the pattern has been demonstrated in crossings of _Idotea balthica_ ($=I.$ _tricuspidata_). Also in _Sphaeroma serratum_ five main patterns are inherited as four independent alleles, involving yellow and red chromatophores in addition to brown and white. Each population of _Sphaeroma serratum_ displays the five patterns in a characteristic percentage, which remains constant over many years.

In addition to chromatophores, such factors as blood, gut contents, or internal organs may influence the coloration (_Idotea chelipes, I. metallica,_ and many Gnathiidae). Species living in the dark usually lack color.

SENSE ORGANS.   There are sensory setae and sensory tubules on the antennae. Statocysts are found in many genera of the Anthuridae at the base of the telson (Fig. 17-15). The compound eyes are flat on the cephalothorax, not stalked (Fig. 17-5). The eyes of cave inhabitants are more or less reduced, and in deep-sea families they are absent. While _Bathynomus_ has 3,000 ommatidia, the eyes of the small predacious Aeginae have about 400. True terrestrial isopods, _Porcellio_ and _Oniscus,_ have only 20–30 ommatidia, with their circular lenses domed, separated from each other, quite different from the angular corneas of the other malacostracans (Fig. 17-3).

NERVOUS SYSTEM.   The simple protocerebrum lacks globuli. Further reduction of the brain is found in Oniscidea that live far from the ocean; these have few ommatidia and short antennae, adaptations to their secluded habits. In some Oniscidea the optic masses are simple, containing only one crossing of fibers. The deutocerebrum is also weakly developed, its olfactory lobe and antennal glomeruli vestigial. In contrast, the large complex glomeruli of the tritocerebrum, connected to the second antennae, are well developed, much more so than in other peracarids.

Mouthparts and maxilliped ganglia are usually fused into a subesophageal ganglion, though their segmentation is usually still recognizable. The other seven ganglia of the pereon form a cord with connectives (Fig. 17-10). In contrast, the abdominal ganglia have more or less moved together, especially in the Oniscidea. The specialized females of the Epicaridea often have a much reduced nervous system.

DIGESTIVE SYSTEM. The ectodermal foregut is a specialized pump in blood-sucking species; in others, it is a chewing or filtering stomach, reduced in terrestrial isopods. At the anterior of the pereon the long gut starts. In the terrestrial isopods studied, as in *Astacus*, the gut is ectodermal, and its ectodermal origin is also indicated by the presence of a cuticle. Up to three pairs, at most, of tube-shaped, endodermal midgut ceca hang from its origin at the stomach, their length differing in different families.

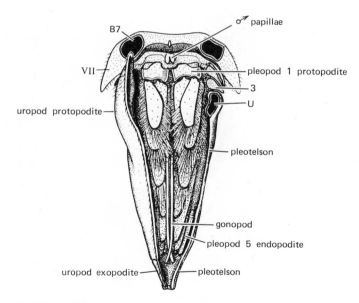

Fig. 17-9. Posterior of male *Saduria entomon*, in ventral view; 10 mm long. The 7th pereopod has been cut off (B7). On the right the lamellate uropod has been cut off at its basal joint (U); at the left it has been turned on its joint to show the long pleopods, which are normally covered. The exopodite of the long uropod reaches to the tip of the long pleotelson and, like a door, can close over the pleopods. The pleotelson is ventrally concave. Only its lateral border shows. The gonopods are specialized median parts of the endopodite of the 2nd pleopod called appendix masculinum. Papillae guide the sperm toward the gonopods. Epimere of 7th pereomere, VII; epimere of the 3rd pleomere, 3. (After Gruner.)

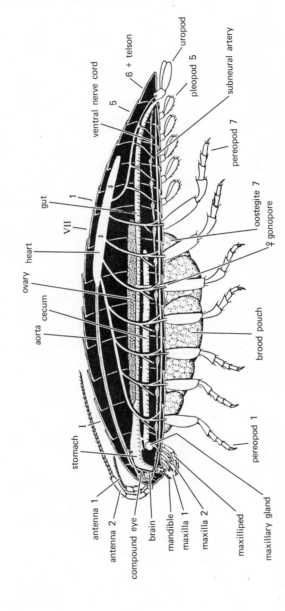

Fig. 17-10. **Anatomy of a primitive isopod. The female gonopore is on the sternite of the 6th thoracomere. The 2nd maxilla has the pore of the maxillary gland at its base, behind it is the maxilliped, which is the first thoracopod. The subneural artery is supplied by median branches of the anterior aorta. Roman numerals indicate pereomeres; Arabic, pleomeres. (From Gruner after Siewing.)**

EXCRETORY ORGANS. There are nephrocytes and a pair of maxillary glands with the nephroduct longer in freshwater isopods (and in *Anilocra*) than in the marine species examined.

CIRCULATORY SYSTEM. The heart is separated from the body cavity by segmental transverse (alar) muscles of the pericardial septum. Its position in the abdomen is close to the respiratory organs (Fig. 17-10). It has one pair of ostia in terrestrial isopods, two pairs in freshwater and marine isopods and in *Ligia*. In some species the ostia are asymmetrical. The ostia receive blood from the pericardial sinus. A cor frontale is present.

From the anterior of the heart an aorta goes anteriorly. According to Silén, a pair of arteries clings to the aorta. The next pair of lateral arteries forms branches to the first four pairs of pereopods, while three separate pairs of arteries go to the posterior pereopods (Fig. 17-10). The posterior aorta gives off five branches in the abdomen that do not enter pleopods. The arteries to appendages branch to supply other organs, including the nerve cord, which receives a subneural artery that goes into the cephalothorax and combines there with a branch of the anterior aorta (Figs. 1-21, 17-10).

As it leaves arterial branches in the pereon, the blood moves posteriorly toward the abdomen, via two routes. One portion passes through unlined lacunae within a central bloodsinus that surrounds the gut, and another portion through a pair of ventrolateral lacunae, close to the exoskeleton, that also receive blood from the pereopods. Within the abdomen the blood flows medially into the pleopods, and finally leaves on the sides, moving toward the dorsal pericardial sinus.

Specializations are found in the terrestrial isopods, in which the heart, corresponding to the localization of respiratory organs, is thicker and longer than in marine species. The proportion of heart to body length is at least 0.4 in the Oniscidea; in marine isopods and *Asellus* it is at most 0.25. Also the heart here has only one pair of ostia (except in *Ligia*) and the arterial branching is simpler. In contrast, the lacunae are much better defined by septa. Protopodites of the pleopods are connected to the pericardial sinus by vertical membranous septa forming a kind of channel that direct the blood (compare to Fig. 18-3 of an amphipod).

REPRODUCTIVE SYSTEM. The gonad primordium in both sexes of young Porcellionidae and *Sphaeroma* which have just left the marsupium consists of a tube on each side from which originates in each of the second to seventh pereomere lateral mesodermal cords. After the third postmarsupial molt, the three anterior cords of the male become testis tubes, the fourth and fifth become suspensory cords, the sixth vas deferens. The original paired tubes become seminal vesicles.

In females, however, the longitudinal tube becomes the ovary, the fourth lateral cords in the sixth thoracomere the oviducts, the other cords become suspensory filaments. The same primordium thus may develop into male or female gonads.

The male opening at the end of the thorax is on a median pair of papillae, fused in the Gnathiidae and most Oniscidea into a single median cone. The papillae, which may be finger-shaped or lance-shaped, conduct the sperm to the gonopods; there is no penis (Fig. 17-9).

The paired tubular ovaries are in the pereon. They are so swollen in each somite in the Epicaridea that they fill up the entire body cavity, an adaptation to parasitism. The oviducts which start out from the ovaries in the sixth thoracomere open separately in a median gonopore next to the sixth thoracopods (Fig. 17-10). The sclerotized lining of the distal part of the oviducts indicates an ectodermal origin. This part is widened or has lateral evaginations forming seminal receptacles.

Oostegites are found medially on the coxae of the pereopods. In families that have flat coxae, the oostegites seem attached to the sternum (Fig. 17-7). In young animals they may be short appendages, slowly increasing in size. In the molt following mating they increase tremendously, covering the midventer and forming a brood pouch. The molt which produces complete oostegites is called the parturial molt. In some families oostegites are found only after this molt just before spawning. Usually the oostegites are only on the first to fifth or first to fourth pereopods, rarely on all seven (Figs. 17-7, 17-10). Some Arcturidae have only one pair.

Many Sphaeromatinae do not carry their eggs in the marsupium (though it is present) but in 1–4, usually paired, hypodermal pockets formed by invaginations of the intersternal membranes into the body cavity, and opening by narrow slits into the marsupium. *Tylos* and *Helleria* have a similar, but median, brood pouch that extends from the border between the fifth and sixth pereomeres to the abdomen. This structure may be an adaptation to the habit of curling up to form a sphere, although it is not present in all species that form balls. In parasitic isopods (Cryptoniscidae) there are similar pockets in place of the marsupium. Dajidae have the oostegites replaced by large ventrally bent folds of the trunk.

Secondary sexual characters and formation of copulatory organs depend on a hormone of the androgenic glands, as has been shown in *Oniscus, Porcellio* and *Armadillidium*. These ribbon-shaped glands are close to the testes, and were only recently distinguished from the testicular tissue. In experiments they have been accidentally transplanted along with the testes, resulting in erroneous interpretation of the experiments. The oostegite formation depends on an ovarian hormone, as has been demonstrated by extirpation experiments in *Asellus, Armadillidium,* and *Porcellio.* (See also Chapter 1.)

Reproduction

Most isopods are dioecious. Some parasites (Cymothoidea and Cryptoniscina) are protandric hermaphrodites. In the genus *Trichoniscus,* parthenogenesis is known. *Trichoniscus pusillus,* besides a diploid bisexual race, also has a parthenogenetic triploid race ($3n = 24$) found farther north, especially in relatively dry habitats. But even in this parthenogenetic race, males are found once in a while: 1.6% in 4600 in southern France, fewer or none farther north. In general monogenic females come from unfertilized eggs. Parthenogenetic females are also found in diploid bisexual races of *Trichoniscus*, *Armadillidium vulgare* and *Cylisticus convexus*. There is only a single equal maturation division in oogenesis, no reduction division, and only one polar body. In these races there is no fertilization after mating.

Usually sexual differences are small. Noticeable differences are found in Gnathiidea, the males of which have large mandibular pincers, and among the Epicaridea in which the females become specialized while growing to large size, the males remaining dwarfed in comparison (Figs. 17-19, 17-24). The sex is usually determined genetically but in Bopyridae the sex organs develop late from presumably hermaphroditic primordia. Young bopyrids, attaching as first arrivals on a decapod crustacean, become female; later arrivals attach on the female and become males. If a female of *Stegophryxus* just attached is taken off and transplanted into the marsupium of another female the transplanted one becomes male. If an attached male is removed from a female of *Ione thoracica* and placed on a crustacean gill, it attaches and transforms into a female.

Mating is accomplished with the second pleopod, in terrestrial isopods with both anterior pairs of pleopods. The male of *Porcellio scaber* recognizes the mature female only after touching it with the distal articles of the second antennae. If these are amputated it cannot find females. In mating it perches transversely on the back of the female and bends its pleon under that of the female, bringing its anterior pleopods to a female gonopore. Then it turns 180° horizontally and transfers sperm to the gonopore on the other side.

Asellus aquaticus mates in almost the same way, but the male can mate only during the interval between the molts of the anterior and posterior parts of the female. In the parturial molt, while losing the skin behind the fourth pereomeres, the female gonopores extend three times their length and almost double in width. Ten hours later (at 18–22°C) the anterior molts, producing oostegites that block access to the gonopores. The period of availability is short but at 18–22°C the females molt every 3 weeks. The male approaches a female before she molts and follows or rides for days. Mating is similar in *Helleria*. Two to ten hours after mating, females of *Asellus*, *Jaera*, and *Sphaeroma* produce eggs into the new brood pouch.

In terrestrial isopods the female undergoes a parturial molt only after mating. Exceptions have been observed in *Ligidium*, *Trichoniscus*, *Philoscia* and *Oniscus asellus*, in which the unfertilized eggs succumbed. In *Porcellio dilatatus*, the molt 3 or more weeks after mating is a parturial molt. At the same time, the oocytes grew from 150 μm diameter to 400+ μm. In a group of 105 females, one mating sufficed for half of them, afterward isolated, to bring on parturial molts within the next year. The other half underwent parturial molts in the following year. During the course of 2–3 years, 81 hand-raised virgin females underwent only regular molts, not parturial molts.

The number of eggs differs in different species and individuals. As might be expected, it is very large in parasites, 2450 in *Cymothoa oestrum.* Large wood lice produce more eggs than small ones: *Oniscus asellus,* 13 mm long, produces 47; a female 17 mm long had 83 in the marsupium. Temperature may also influence number of eggs. *Porcellio* during the summer may have 85 embryos, in fall hardly more than 30. In many species the number within the marsupium decreases. Large *Asellus aquaticus* produce 150 eggs but only 80 offspring leave the marsupium; in *Idotea* only four to six leave out of 30–40 eggs; in *Jaera marina,* three or four or rarely 12 from 15–16 eggs.

Females circulate a current of fresh respiratory water through the brood pouch. From the maxilliped coxae a posteriorly directed median projection (homologous to an oostegite) extends into the brood chamber, and often helps to produce the current. Its movement directs water toward the head (Asellidae, Idoteidae, Cymothoidae, and Bopyridae). The oostegites may be raised, permitting water to enter. Terrestrial isopods fill the brood chamber with a fluid derived in part from their capillary water (see Respiration, below) or probably from the cotyledons, metameric fingerlike evaginations of the sternites. They carry the young in a built-in aquarium and therefore, unlike land crabs, are independent of ponds, streams or oceans. Movements of oostegites and pleopod exopodites produce a current in the fluid. That there is no nutrient fluid in the marsupium can be concluded from a few observations that eggs of *Armadillidium nasutum* and *A. vulgare* deposited in water develop normally.

Development

Only a few isopods (e.g., *Paragnathia*) are viviparous. The young in all species remain in the marsupium for a while after hatching. They differ from adults in lacking the seventh pereopods and the secondary sexual characters. In *Sphaeroma* and *Jaera* there are three such manca stages in succession. In terrestrial *Porcellio* the last thoracic legs appear on reaching the juvenile stage after the third molt; at the same time the tracheal lungs develop. Depending on the temperature it takes *Porcellio scaber* 49–102 days to reach the juvenile stage. Of this period, 35 days may be embryonic development, 21 days for the first manca stage, 16 of which are spent in the marsupium, and 23 days each for the second and third manca stages.

Larger differences between the stages are found in parasitic isopods. Young Cymothoidae have larger eyes and antennae than the adult, and the pereopods are setose walking legs, not modified as holdfasts. The manca stage of *Gnathia,* in contrast to the adult, swims and is armed with biting mouthparts. It attaches to a fish and takes up so much blood that the pereon swells, obscuring the segmenta-

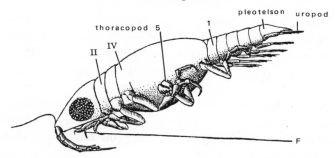

Fig. 17–11. Larval stage of *Gnathia,* parasitic on a fish fin (F); 2–4 mm long. The blood taken in has stretched the intersegmental membranes between the posterior thoracomeres. Roman numerals refer to thoracomeres, Arabic to pleomeres. The first and second thoracomeres are part of the cephalothorax, and the third is the first pereomere. (After Monod.)

tion (Fig. 17-11). The fed parasite sinks to the bottom and molts only once to become a bottom-dwelling adult, with mouthparts unusable for further feeding (Fig. 17-24).

The young Epicaridea change so drastically in metamorphosis that the young are called larvae. They leave the mother as pelagic epicaridia (Fig. 17-12). The 0.2–0.35 mm long, highly domed animals have the same segmentation as other manca stages. However, they swim with the long setose second antennae and the setose, usually biramous, cylindrical pleopods. The six pereopods are modified as holdfasts. The epicaridium feeds on stored yolk for two weeks. Its fate has been observed only in *Portunion kossmanni,* but is probably the same in other species. The *Portunion* epicaridium attacks a pelagic copepod and transforms into

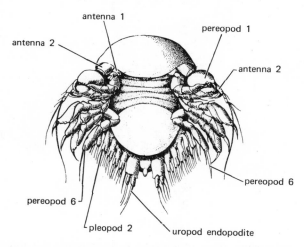

Fig. 17–12. Epicaridium stage (manca stage) of *Clypeoniscus meinerti*; 0.16 mm long. (After Giard and Bonnier from Nierstrasz and Bender.)

the parasitic microniscium. This, after several molts, loses its swimming setae, sensory setae of the second antennae and pleopods, and the segmentation of the appendages becomes reduced. Later it molts into a swimming cryptoniscium, 0.5–2 mm long (Fig. 17-13). This pelagic stage is known for many species. It swims with its setose 2nd antennae and pleopods. Most of the seven pleopods are holdfasts. As in the epicaridium, the tips of the styletlike mandibles extend from the mouth cone. This stage attaches to the final host, where it molts to become adult. This final metamorphosis may be drastic in the females. Within each family members may take either of two paths.

The cryptoniscium of *Bopyrina* chooses a decapod crustacean as its final host. If the decapod is already parasitized by this species of *Bopyrina,* the cryptoniscium attaches to the parasite, molts and becomes an adult male (Fig. 17-24). While it hardly changes in size or shape, the eyes, antennae, and pleopods are reduced to

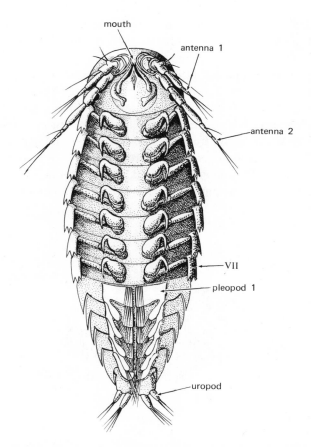

Fig. 17–13. Cryptoniscium stage of *Portunion kossmanni*; 0.3 mm long. Behind the mouth can be seen the styletlike mandibles and the first maxillae which show through the integument. VII, 7th pereomere. (After Giard and Bonnier.)

vestiges; however, holdfasts are not reduced. If the host is not already parasitized, the cryptoniscium undergoes a number of molts, in the course of which the trunk, gonads and marsupium grow extensively, and transforms into a female. In its final form it is far from a typical isopod. It may even become asymmetrical and lose pereopods (Fig. 17-29).

The cryptoniscium of *Cryptoniscina* attacks members of various crustacean orders. Without molting it transforms into a male. Later eggcells appear in its testes and then, during the course of a number of molts, the isopod-like male transforms into a female. The female loses its isopod resemblance, becoming a sac without appendages (Fig. 17-14). In this group, therefore, are only old females and young males. This protandric hermaphroditism can be successful only if the host has several parasites.

The number of postmarsupial molts differs in different species: only one in *Gnathia;* four or five in *Idotea chelipes,* six in the males; 10 to 13 in central European Oniscidea. In the species studied, the body behind the fourth pereon somite molts first, the skin tearing transversely between the fourth and fifth somites. Depending on external conditions or on the species, the anterior part may molt after a few hours or after a few days by rhythmic stretching and contraction of the trunk. Division of ecdysis into two parts is related to mating and spawning.

Oniscus asellus live 3–4 years, *Aega* at least 2¼ years; *Asellus aquaticus* at room temperatures live 1 year, under natural cooler conditions probably longer. Cryptoniscids live half a year at most.

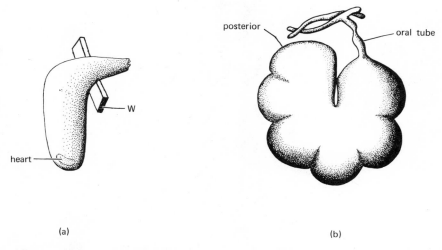

(a)   (b)

Fig. 17–14. Female of *Danalia curvata*: (a) young animal, 1.2 mm long; (b) female with brood pouch, 8 mm long if straightened. W, wall of branchial chamber of host. (After Caullery.)

Habits

The habits of isopods are not well known. Most marine species live on the bottom, as is suggested by the flattened body, so reminiscent of trilobites and horseshoe crabs. But many predacious and parasitic species occasionally leave the bottom and a few species can be collected in nets in the pelagic region, presumably while searching for food (e.g., *Xenuraega, Munnopsis,* (Fig. 17-16) *Eurydice,* and young Epicaridea). The ocean is inhabited from the beach to a depth of 10,000 m (Wolff, 1956). Many marine species can tolerate dilution of the water and live in brackish water: *Jaera marina, Cyathura* sp. (Fig. 17-15) and *Saduria entomon.* In *Saduria entomon* the pleotelson length depends on the salt content. A few families of marine isopods have representatives in freshwater.

Among the Asellidae, certain freshwater species may be found in brackish water, as may some of the peculiar terrestrial Australian and south African Phreatoicidea (Fig. 17-6). Caves and subterranean waters are inhabited by white, eyeless species, some of which belong, surprisingly, to the marine families Sphaeromatidae and Cymothoidae (Thienemann, 1950). However in those animals (esp. *Asellus*) examined carefully by sectioning, vestiges of eye tissue were found. That these vestiges may differ in amount within a population is probably accounted for by balanced polymorphism, a selection in pleiotropic genes that also control eye characters, or perhaps introgression through hybridization. ·

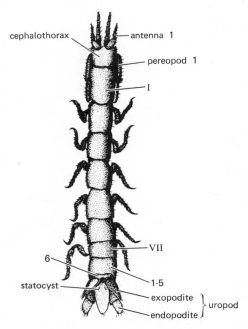

Fig. 17–15. *Cyathura carinata;* 15 mm long. Roman numerals refer to pereomeres, Arabic to pleomeres. (After Gruner.)

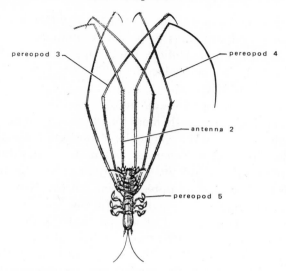

Fig. 17–16. Male of *Munnopsis typica*, with long uniramous uropods; 16 mm long. (After Sars from Nierstrasz.)

The terrestrial isopods illustrate well how a marine group can become terrestrial. Almost all belong to the suborder Oniscidea. They have invaded land directly from the ocean without first passing through freshwater. The various stages from halophilic beach-dwellers to denizens of marshes and moist terrestrial habitats to desert inhabitants are to be found. Internal fertilization, brood pouches, and walking habits are preadaptations to terrestrial habits. The halophilic inhabitants of rocky beaches (*Ligia*) as well as inhabitants of wet terrestrial habitats (*Ligidium*) still escape into the water. In experiments, both survived in aerated water (*Ligia*, of course, in salt water) for more than 80 days. They could even feed under water. Found in nonmarshy but still moist environments, under stones, in manure, and cellars, are *Oniscus, Cylisticus,* and some species of *Trachelipus* and *Porcellio. Trachelipus rathkei* may be kept for almost 2 months in freshwater, *Porcellio scaber* and *Oniscus* for 1 month, *Cylisticus* for only a few hours. Unfortunately there is no indication of the water temperatures at which these experiments were made, a variable of importance to metabolism and respiration. For comparison, the experiments should have been made at the same temperature. Aeration of the water is also important: *Ligia italica* will die in 200 ml saltwater at 28°C in about 9 hr if it is not aerated.

In mesic forest-edge habitats are found members of the genus *Armadillidium* that can live only 10–24 hr under water. In arid areas *Hemilepistus* and some species of *Porcellio* are found.

ADAPTATION TO TERRESTRIAL HABITATS. Different genera exhibit a stepwise adaptation to terrestrial habitats. The adaptations can be broken down into several separate requirements: protection against water loss, resistance to temperature extremes, and structures to permit respiration.

1. Water economy. The epicuticle, unlike that of insects, lacks a waxy layer; there are lipids only in the endocuticle. The exoskeleton is therefore water-permeable in both directions, mineralization being of minor importance in water conservation. The water loss through transpiration in some individuals immediately after a molt was only slightly more than between molts. *Venezillo arizonicus* may have a water barrier in the epicuticle; chloroform but not abrasion increased transpiration in experiments. *Venezillo* rests under stones during the day at an air temperature of 45° and relative humidity of only 15% and may roll up. Its transpiration is the lowest of all isopods examined (Warburg, 1967).

Transpiration of Oniscidea brought into a low humidity chamber is greatest during the first half hour, then declines, probably as the result of lowering of body temperature (Warburg, 1967) and perhaps also due to increased osmotic pressure of the blood (Table 17-1). The water loss through the cuticle is so great that *Oniscus asellus* and *Porcellio scaber* can live only a limited time at 86% relative humidity at 20°C, while *Armadillidium* can survive longer, as would be expected judging by the drier habitat in which it lives. Most transpiration is through the venter which is weakly sclerotized and bears the pleopods. *Oniscus asellus* at 90% relative humidity and 14–18°C succumbs much faster if they have to stay on wire screen than when they can protect their venter by keeping it close to the substrate.

In an extreme example, the desert isopod *Hemilepistus* can survive 5 hr on sand at 32–40°C, with an air temperature of 28°C and 30% relative humidity.

Water can be taken up through the integument from the air by terrestrial isopods brought from low humidities into an atmosphere of at least 98% relative humidity, as has been demonstrated by careful weighing of *Porcellio scaber*. Freshly killed *Porcellio* still show this uptake, but not as distinctly.

Water regulation of Oniscidea seems a simple matter of avoiding transpiration loss by moving into microclimates of high humidities. During the day they remain in a cool refuge with saturated air, under stones, bark, rotting logs, their venter

Table 17-1. **Water loss through transpiration of *Oniscus asellus* at 60% relative humidity and increasing temperature. This is calculated in mg, and to 1 cm² surface per hour (After Edney)**

Time, min	Temperature, °C	Water loss in mg
0–30	17	4.73 ± 0.14
30–60	21	3.18 ± 0.16
60–90	24	2.82 ± 0.12

close to the surface. When at dusk and night they move about after food, they lose only limited amounts of their fluids, which can be replaced in their refuges and from moist food.

Oniscidea avoid staying permanently in saturated air. After 21 hr in saturated air, 18–30% of the *Porcellio* in choice experiments moved to a place of 75% relative humidity, perhaps because the short maxillary nephridia give off only small amounts of fluid.

Behavior is probably the most important factor in the water economy (Warburg, 1968), drinking being resorted to in extreme cases. *Ligia,* by drinking through the hindgut, can replace the water lost in transpiration. The desert isopod *Hemilepistus,* which can tolerate 5 hr of drying, takes refuge during the day in a desert microhabitat and is able to imbibe water during the cool hours of early morning or late night when the air is saturated and there is dew. *Oniscus asellus* and others may become more thigmotactic in dry than in moist air. The odor of each species (*Oniscus asellus, Porcellio scaber* and *Armadillidium vulgare*) is attractive to others of the same species especially to dry individuals, guiding individuals to more humid microhabitats and aggregations (Kuenen and Nooteboom, 1963).

2. Protection against heat. Burrowing and transpiration provide protection against heat. Like many other desert animals, *Hemilepistus* burrows during the day; several together may penetrate to 30 cm depth. They push with their front ends, scratch with the anterior pereopods and brush the sand to the side with the posterior ones. *Oniscus* also digs in if the temperature rises to 25°C and the soil is moist sand. Burrowing is done as by *Hemilepistus,* but a diagonal, then horizontal, burrow covering the body suffices. Burrowing *Oniscus* and *Porcellio* also can be found under stones.

Lowering of temperatures by transpiration is known to occur in a number of terrestrial isopods and can be observed in living individuals. For instance, at an air temperature of 21°C, *Ligia* can remain on the surface of rocks that are sunheated to 36°C, and the body temperature remains at 28°C, 8° below that of the rocks. Other wood lice also, even in sunshine, can maintain body temperatures lower than the temperature of the substrate; *Oniscus,* 4–5°C; *Armadillidium,* 4°C; *Porcellio,* 2–3°C. Dry or dead wood lice take on the same temperatures as the substrate. *Venezillo arizonicus* can stand the high desert temperatures and can cool itself only little by transpiration.

Further, isopods are able to adapt to higher temperatures. If *Porcellio laevis* or *Armadillidium vulgare* collected at low temperatures are kept for 2 weeks at 30°C they will survive for 30 min when placed in 41°C chambers, while animals that have not been heat adapted will die. Also, the oxygen requirements and the number of heart beats of heat adapted animals rises less in heat than in those not so adapted (Table 17-2).

Table 17-2. Heart rate in temperature adapted *Porcellio laevis*. Numbers refer to heart beats per minute (After Edney, 1964)

	Individuals kept for 2 weeks at		
	10°C	20°C	30°C
Transplanted to 10°C	55	44.3	39.4
Transplanted to 20°C	129	117.5	95.2
Transplanted to 30°C	237	201	171

3. Protection against cold. Except in tropics, terrestrial isopods need protection from severe cold. In a cold winter only the individuals of *Porcellio scaber* hiding in the center of a refuse pile survived, those on the periphery died. These hazards exist for marine animals in the tide zones only.

4. Respiratory adaptations to terrestrial habitats are discussed under Respiration below.

5. In terrestrial isopods, the osmotic pressure of the blood is lower than in marine isopods. The osmotic value depends on humidity, season, and proximity of the next molt. *Porcellio* in dry air has a hemolymph concentration of Δ —1.34°C (of lowering of freezing point), in humid air in summer Δ — 1.54°C, in winter Δ — 1.68°C. *Ligia*, however, has the osmotic value of 50–100% ocean water. If kept on sand saturated with ocean water (Δ — 1.98°C), the osmotic pressure varies between Δ — 1.98 to —2.33°C lowering of freezing point.

The nitrogen content of excretory products of *Oniscus* is very low (0.3 mg N/10 g per day) compared to *Ligia* (1.3 mg N/10 g per day) and *Asellus* and marine isopods (2.0 mg N/10 g per day). *Oniscus* and *Porcellio* give off 50% of the nonprotein nitrogen as NH_3 which escaped in part (10–30%) as gas. *Ligia* excretes 70–80% as NH_3.

Besides these adaptations, the Oniscidea have some other special characters that do not appear directly related to terrestrial habits. For instance, the number of ommatidia has become reduced. *Ligia* and fast predacious marine isopods have 500–800, *Ligidium* 120, *Oniscus*, *Porcellio*, and *Armadillidium* only 20–30. Another character is the reduction in size of the first antennae and in number of articles of the distal flagellum of the second antennae. *Ligia* has more than 12 articles, *Ligidium* 10–15, *Oniscus* 3, and *Porcellio* and *Armadillidium* only 2. There is also a reduction of the maxilliped endopodites and reductions of the corresponding sections of the brain. Since, except for reduction of the antennae, the fresh water *Asellus* shows similar reductions, they can be related to the availability of food and nocturnal habits. With life in darkness comes the loss of pigment movement: only *Ligia* and *Tylos* have isolated chromatophores whose pigment can contract and expand. In all other Oniscidea the chromatophores have fused to form a continuous net.

PROTECTION. Active defense is of little importance to isopods. Male Gnathi-idea and some Cirolanidae (e.g., *Conilera*) can bite and the Aegidae defend themselves by biting with the piercing mouthparts. As defense against spiders, the Oniscidea have repugnatorial glands that open on the lateral border of epimeres and uropods. The epimeral glands have few cells and several open together in a tiny sieve plate on the border of each epimere (Fig. 17-1). In a number of experiments using different stimuli, the isopods secreted only if bitten on the soft venter by a spider. The secretion is repellant to predators, which after touching it, immediately abandon the prey and clean their chelicerae. However, spiders will readily eat wood lice that have been cut open, avoiding contact with the secretion, while they would not touch flies smeared with it (Tretzel, 1961). Of course many spiders such as *Dysdera* prey on wood lice despite the secretions. Perhaps the extreme length of *Dysdera's* long chelicerae is a modification for killing without approaching closely.

Uropodal glands open individually in 50–60 long narrow channels on the tip of the protopodite and the sides of the uropod exopodites of terrestrial isopods. They also are repugnatorial. *Porcellio scaber* touched by the spider *Coelotes* on a web ran off, *Porcellio's* secretion forming long viscid threads behind. The spiders did not follow and if its legs got stuck to the threads, these were removed with difficulty using the chelicerae.

Passive protection is bestowed upon many marine isopods by their color resemblance to the shade and sometimes to the color of the substrate. A few isopods mask themselves with sponges. Burrowers, such as *Saduria*, are protected by their burrows just below the surface. Some sphaeromatids tunnel with their mandibles in tufa, soft sandstone or wood (*Limnoria*). Other isopods crawl among worm tubes or barnacles. *Zenobiana* hides itself in a tube made of plant fragments, as do caddis fly nymphs. Isopods of many species derive mechanical protection from the epimeres by pressing them against the substrate and drawing in all legs. The exoskeleton is most protective in species that can roll up into a sphere, the pill-bugs.

The ability to ball up has evolved convergently among the Sphaeromatidae as well as in many Oniscidea resulting in various stages of perfection. In terrestrial isopods the curling is activated by muscles that are also present in noncurling species, with only a difference in their development. The ventral, longitudinal muscles of the pereon and pleon shorten the venter. At the same time the diagonal dorsoventral muscles of the pereon pull the sternites in. The body arches, venter concave, providing room for the legs and, in species that curl up completely, for the second antennae. In highly specialized species the edges of the anterior and posterior parts of the body fit against each other. There is a straight ridge on the cephalothorax in front of which the head is truncate; the posterior end is also truncate. The uropods not extending beyond the pleotelson, fill the space between pleotelson and the epimeres of the fifth abdominal somite (Fig. 17-17).

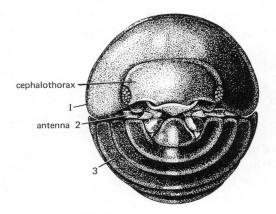

Fig. 17-17. Pill bug, *Armadillidium vulgare***; 6 mm diameter. The animal is not completely rolled up, permitting the 2nd antennae to be seen; the antennae are completely covered when cephalothorax and abdomen meet. I, first pereon tergite; 3, tergite of 3rd pleomere. (After a photograph by Wächtler.)**

Marine isopods of the genera *Ligia, Idotea,* and *Sphaeroma* can match the background by clumping or extending pigments within star-shaped chromatophores, a reaction that is controlled by neurosecretions from the cephalothorax and the nerve cord (See Chapter 1). Twenty-four-hour rhythms of color change have been reported for several species of *Ligia* and *Idotea.* In a light intensity of 150 foot candles (1600 lux) *Ligia occidentalis* shows a clear relationship between the degrees of pigment spread in the chromatophores and the light reflected from the substrate. At day time *Ligia* running around on dark rocks has the melanin expanded, at night contracted. An endogenous rhythm that works also in constant darkness in the laboratory controls this adaptation. In constant illumination it is immediately suppressed. The leucophores of *Idotea* have a similar light influenced rhythm that continues for a while in constant light. *Idotea* resembles the substrate in shade and often in coloration not due to chromatophores, but to gut contents showing through the transparent integument. The blood and coelomic fluid take the color of the food. *Idotea chelipes* feeding on green algae, usually are green despite the distribution of the pigment; on brown or red algae they are brown or red, and match the background so well that the collector has difficulty finding them. If the food is changed, within 10–15 days there is a corresponding color change. But *Idotea* does not search for a matching background.

LOCOMOTION.   Most isopods walk with the pereopods; a few can make considerable speed on the beach (50 cm/sec). Most marine isopods can also swim; some cannot, including all adult Gnathiidae and *Jaera.* Many freshwater isopods, the Phreatoicidea and *Asellus,* cannot. Even members of one genus (e.g., *Sphaeroma*) may differ as to swimming ability. Young *Gnathia* and many Cirolanidae are very rapid swimmers, shooting through the water to catch living prey. Other good swimmers are *Eurydice,* young Epicaridea, and Valvifera. The pleopods serve as swimmerets. Whether all or just some pleopods take part in swimming in different species has not been studied. Certainly not all pleopods take part in young Gnathiidea, Epicaridea, *Idotea, Cyathura,* and *Limnoria.* However, those Asellota that can swim, especially Parasellidae, paddle only with the pereopods not with the pleopods. *Sphaeroma* and *Saduria* turn their venters to the surface, *Idotea* swims dorsum up.

SENSES. Terrestrial isopods are generally thigmotactic, probably an adaptation against water loss. *Oniscus* and *Porcellio* do not respond to the turning of a seesaw on which they sit. But members of both genera, and *Ligia* also, can find their way out of water in the dark presumably due to negative geotaxis. Of 64 *Ligia* carried inland, 50 immediately returned to the ocean perhaps due to positive geotaxis, perhaps due to orientation from blue areas in the cloudy sky. A sense of gravity is shown by *Cyathura,* which digs only vertically downward if the statocysts are undamaged.

A sensitivity to humidity is found in all terrestrial isopods and also in *Ligia.* If they have the choice in the dark of two wide open chambers at a time (97–100%, 77–76% or 34% relative humidity), they will stay most of the time in the most humid chamber. Here they rest and walk slower than in the one with dry air, where they run about rapidly until they find the chamber with higher humidity. *Ligia italica* is able to differentiate between chambers of 97% and 100% relative humidity. The receptors are believed to be on the underside of the lateral plates (Jans and Ross, 1963).

As can be demonstrated by extirpation experiments, the chemical sense in wood lice is localized in the distal articles, the flagellum, of the second antennae. Removal of the small first antennae has no effect. Males of *Porcellio scaber* can recognize females only after touching them with the flagella of the second antennae. Experiments indicate that out of 207 *Porcellio scaber,* 119 individuals avoided their own smell as well as that of their fecal matter if kept for 24 hr in air of 100% humidity. In a dish with its bottom moistened half with water, half with 2% glucose solution, 77% of individuals of *P. scaber* chose the water.

Also individuals of *Armadillidium, Oniscus,* and *Porcellio* react to different temperatures. In a circular terrarium divided into sections with temperature differences in steps from 4–30°C, all with saturated air, representatives displayed temperature preferences that differed for each species. The receptors are not known.

Eyes of the predatory fish isopods, each with 500–800 ommatidia almost certainly can see movements. In wood lice the biological importance of eyes is probably the ability to avoid light. The ommatidial angles are very large, in *Oniscus asellus* about 13.7°, in *Porcellio scaber* about 20°; also the number of ommatidia is only 19–23 in these species.

Astronomical orientation has been demonstrated for *Idotea balthica* and for the wet beach inhabitant, *Tylos*. The marine coastal *Idotea balthica basteri*, if its resting place dries, takes a path toward water at a 90° angle to the coast line. Each population depending on the direction of the coast line has different orientation. The 90° path toward water is kept as the animals use the sun as a compass and are able to compensate for the changing of solar azimuth. When the sun is reflected in the mirror the isopods will change their direction of movement 180°. But even if the sun is hidden most individuals move directly to the water.

Tylos, the peculiar beach isopod, also uses its eyes, each with about 400 ommatidia, for astronomical orientation. On dry sand the animal moves toward the ocean, on very wet sand or in water, it moves toward land. Using mirrors it can be shown that the animal uses the sun for orientation. The angle toward the sun shifts continuously but is accounted for by an internal clock. At night *Tylos* orients by the moon. Although *Tylos* can also orient in the shade, polarized light vision could not be demonstrated. Under a polarized light screen the animal becomes disoriented. No polarization plane was found that was preferred statistically.

FEEDING. The freshwater isopods (*Asellus*) feed on decaying plants and detritus. Most terrestrial isopods feed on fallen leaves. Species of *Oniscus* and *Porcellio* that live in large populations are of importance to humus formation. Their excretory matter may accumulate in layers that are further degraded by Collembola. Some *Sphaeroma* and the gribble, *Limnoria*, feed on wood.

Mixed feeders are found among marine isopods. *Idotea, Jaera,* some *Sphaeroma* and *Ligia* feed on epiphytic algae, especially brown algae, but will also take carrion. *Idotea chelipes* in addition feeds on Bryozoa, Hydrozoa polyps and on worms. *Serolis* takes mostly detritus, but will also feed on meat.

Many isopods search the bottom for food and take up small animals and carrion. Some *Cirolana* feed mostly on carrion, other species are also predators. Most *Cirolana* take injured prey or prey caught in nets. There are two reports of sturgeon (1.25–1.80 m long) caught in nets at Rovinj, Istria being eaten, skin and all, down to the skeleton during the night by *Cirolana hirtipes*. *Cirolana* and *Eurydice* commonly even attack swimmers, and they may start to chew between the toes of an angler standing in water. *Conilera* in swarms attacks fish and may attack weakened decapods.

In an aquarium *Saduria entomon* attacks *Chironomus* larvae or *Asellus* that they have touched with one pair or all three pairs of gnathopods. Running around on sand, they notice buried *Chironomus* and pushing the cephalothorax into the substrate, pick up the insect with the chelipeds. The prey is eaten within the burrow or sometimes pulled to the surface. They never use the gnathopods to tear prey but only to hold it until the mouthparts take over. At times *Saduria* buries itself except for the antennae. If a moving *Chironomus* larva touches the antennae, *Saduria* rises up to capture the prey. Other isopods such as Arcturidae

(Fig. 17-18) wait in ambush. *Astacilla*, with the three posterior pereopods, clings to a sea pen *Funiculina*, bryozoan colonies, or *Laminaria* and, sometimes rocking, extends the anterior part of the body into the water. As soon as a small crustacean approaches, the isopod shoots toward it, propelled by the long anterior two pairs of setose pleopods. The prey is taken with the two long chelate antennae and pressed to the spiny four anterior pairs of pereopods, which form a basket below the mouth. Then *Astacilla* immediately returns to its ambush place. These observations are old, and new ones are desirable.

Adaptations of blood sucking and parasitic isopods intergrade with those of the predators, such as the Cirolanidae. Aegidae are intermediate, having styletlike mouthparts and three pairs of pereopods modified as holdfast structures with which the isopod can attach to fish while it sucks blood. Cymothoidae have all pereopods transformed into holdfasts and they live in part sucking blood between gills or on the skin and partly probably as commensals in the mouth of fishes. These large isopods may attack swimmers. The manca stage of Gnathiidae also feeds on fish blood. It takes up so much blood that the segmentation of the posterior thoracomeres disappears (Fig. 17-11). The large amount of food is stored in the digestive ceca and permits the adult to live and reproduce without feeding again.

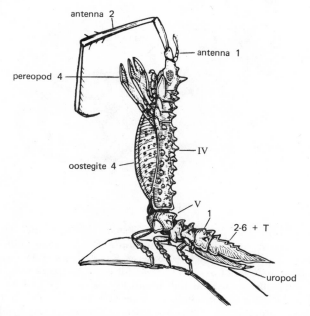

Fig. 17–18. Predacious isopod, *Astacilla pusilla*, in ambush position; 12 mm long. The long 4th pereomere divides the body: the anterior is specialized for predation, the posterior as a holdfast and for swimming. The oostegites of the first 3 pairs of pereopods are minute, only the 4th are large and functional. Roman numerals refer to pereomeres, Arabic to pleomeres; 2–6+T, pleotelson. (After Sars from Nierstrasz and Schuurmans-Stekhoven.)

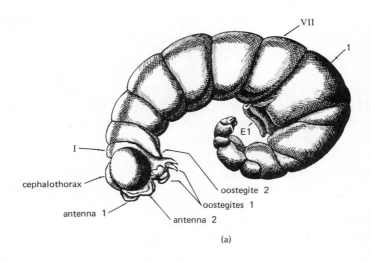

(a)

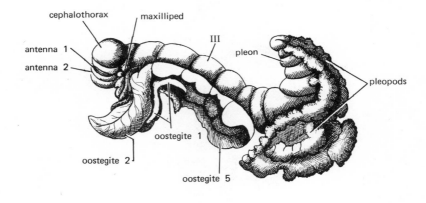

(b)

Fig. 17–19. Development of female *Portunion maenadis* in branchial chamber of the crab, *Carcinus maenas*, from an isopod-like swimming larval stage (Fig. 17–13). Roman numerals refer to pereomeres, Arabic to pleomeres. E1, epimere of 1st pleomere, (Adapted after Giard and Bonnier.) (a) Loss of legs and filling of gut ceca through several molts results in a caterpillar-like stage with indications of oostegites 1 and 2; 3 mm long. (b) After successive molts the oostegites enlarge, especially those of the 2nd pereopods, and grow anteriorly, surrounding the first oostegites. The abdomen bends back and its appendages become expanded gills with swollen and frilled edges that provide a large surface; 6 mm long. (c) Sexually mature female with marsupium, inside of which the ribbonshaped first oostegite is found. The attachment site of the oostegites has moved dorsally and the oostegites have fused to each other, completely covering the thorax. The animal lies in a deep invagination of the branchial chamber wall (K) forming a sac and hanging into the body cavity of the host. The mouth of the invagination (M) re-

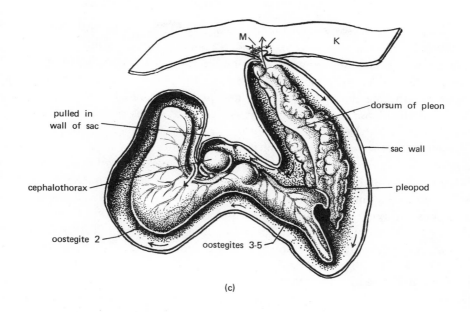

(c)

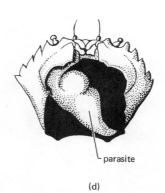

parasite

(d)

mains open and through it the isopod draws respiratory water from the branchial chamber of the host. The current, produced by raising and lowering pleopods and pulling in the ventral pereon wall, flows between isopod and sac toward the head (arrows) where, close to the cephalothorax from above, it enters the caplike elongation of the marsupium (arrows). The current then flows over the eggs within the marsupium on the venter of the isopod and passes toward the abdomen, pushed by pumping movements of thoracic sternites and peristaltic motions of the first ribbonshaped oostegites. Here it now enters a channel formed by the sternites of the abdomen and the pleopods and gets to the posterior end. The posterior end extends into the mouth of the sac (M) and spurts the CO_2 enriched water into the respiratory chamber of the host. Food is taken in through the funnelshaped, pulled-in wall of the sac. Female parasite is 20 mm long. (d) Position of the parasite in its sac within the body cavity of its host, *Carcinus maenas*. Part of the carapace has been cut off. The body cavity is black; sac is 32 mm long.

Parasites of crustaceans are found among the Epicaridea. All attack only crustaceans and the females lose their locomotor organs while the ovaries, marsupium, and egg number grow enormously. Bopyrina pierce the decapod host with the mandibles, which have a spoonlike tip with sharp teeth that extend from a mouth cone formed by labrum and paragnaths (Fig. 17-4). Only a few species sit outside on the host; most attach with pereopods in protected positions on the abdomen of hermit crabs, or in the branchial chamber of a host. Entoniscidae push into the soft inner wall of branchial chambers of young Brachyura and sink into the depression, which gets deeper and deeper and will eventually close above the parasite. A single pore remains through which respiratory water and males can enter, and the spawn leave (Fig. 17-19). These females lose their pereopods and body segmentation. Such reduction is found also in Cryptoniscina, which parasitize crustaceans, the parasite becoming an egg-filled sac, lacking appendages and losing all resemblance to crustaceans (Fig. 17-14).

Feeding is difficult to observe as the maxillipeds, which meet in the midline, cover up other mouthparts. On the posterior (ventral) edge of the endites of all four pairs of mouthparts of *Idotea* there are cutting teeth (mandibles) or cutting spines (first maxillae and maxillipeds, Fig. 17-20). The first maxillae and maxillipeds are used to scrape algae held by the first pereopods. The mandibles crush or cut pieces off. Dorsally directed transport spines on the anterior (dorsal) edge of the endites push the particles toward the mouth when the endites work

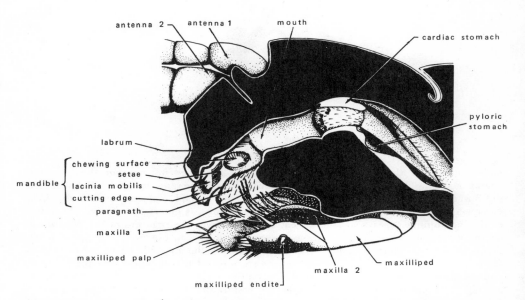

Fig. 17–20. Median section through the head of *Idotea baltica*, showing the transport setae directed toward the mouth; 2.5 mm long. The endite of the maxilliped has hook-shaped setae, which couple it to its opposite partner. (After Naylor.)

against each other (Fig. 17-20). The predacious Cirolanidae have huge spines on their first maxillae, which probably take part in cutting up meat (Fig. 17-5) while the small second maxillae are split distally into two fingers, and can be used only to transport the food.

The blood-sucking isopods have a mouth cone made up variously of labrum, paragnaths, maxillipeds and, the cymothoids, the endites of the second maxillae. From the tip of the cone extend the mandibles and frequently the first maxillae also. In Aegidae and Cymothoidae the mouth cone is vertical. Out of it, at an angle, extends the pointed chewing edge of the mandibles, which can make a pinching movment as the basal article has two articulating condyles (Fig. 17-21). An opening in the host is snipped as with tips of scissors, then enlarged by sawing with the toothed cylindrical first maxillae, which are pushed forward and back by bending and stretching both basal articles. The mouth cone becomes a sucking tube and only the blood flowing into it serves as food.

The anteriorly directed mouth cone of the blood sucking *Gnathia* manca stages is different, though it also encloses cylindrical mandibles and first maxillae. The larva swims toward a fish, hooks the long sickle-shaped claw on the tip of each first pereopods into a fish fin and with these strong appendages pulls the mouth cone into the skin. The mandibles, which barely move, are thus pushed in until

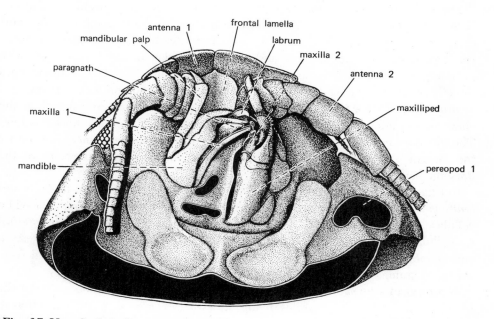

Fig. 17–21. Cephalothorax of *Aega psora*, as seen in ventrolateral view, showing piercing mouthparts. On the left, the right 2nd maxillae and maxilliped have been removed from their attachment to show the mandible and first maxilla lying underneath. The first maxilla is stylet shaped. Cephalothorax is 8 mm wide. (After Gruner.)

their barbs hook fast. During feeding, the isopod stays attached to the host with these two appendage pairs only (Fig. 17-11). Then the first maxillae are rapidly drawn in and out several times, five such movements followed by a pause. They penetrate to blood vessels. After a while, the first pereopods take hold more anteriorly and push the mouth cone deeper into the host.

Most Epicaridea bore into the integument of the host with stylet-shaped converging mandibles, the tips of which have a spoonshaped enlargement with teeth on the edges. Usually the first maxillae are absent or rudimentary. But the second maxillae have been lost and maxillipeds do not take part in feeding (Fig. 17-4). *Danalia* (Cryptoniscidae) are hyperparasites that attack the parasitic roots of *Sacculina* from the integument of the crab. The mouthparts are completely reduced in the female, and food is taken through a tube that forms in the oral region and, probably aided by enzymes, penetrates the crab's integument, growing until it meets the rootlets of *Sacculina*. The pulsating foregut can be seen within the tube, each beat lasting about a minute (Fig. 17-14).

In blood- and lymph-sucking *Aega*, as also the manca stage of *Gnathia*, the pharynx is equipped at one or two places with strong muscles that expand the lumen to form a pump. The filter stomach almost lacks the ridges and channels present in most Malacostraca. The blood passes directly into the gut, so much widened in Aegidae and *Gnathia* as to take up the full width of the body cavity in the pereon, a modification that permits the intake of large amounts of blood at one time. Later the contents are shifted anteriorly through the stomach into the short ceca where digestion takes place. Female Entoniscidae have a much enlarged ectodermal stomach, its anterior and posterior portion are wide and suck with pulsating movements. The connecting part is crescent-shaped. The blood goes directly into the two long roomy ceca. The long gut is blind and lacks an anus.

DIGESTION. In *Ligia oceanica* the preferred food, *Fucus,* is pressed in the chewing and filtering stomach and the fluid part enters the cuticle-lined hindgut. The gut does not secrete enzymes. Breakdown results from enzymes secreted by the midgut glands and mixed with the mashed food by the movement of the dorsomedian fold, as in *Oniscus.*

In *Ligia* the pH of the midgut glands is 6, that of the the the gut, 6.5. In *Ligia* and *Oniscus* absorption takes place in the cecal walls and the walls of the gut. While cellulase, proteinases, lipase and esterase have been found in *Oniscus asellus,* no cellulase has been found in *Porcellio.* The woodborer, *Limnoria,* digests a considerable amount of wood, as indicated by the dry weight of food and fecal matter. Lignin, however, is not attacked.

OSMOREGULATION. In *Asellus aquaticus* the freezing point depression of the blood is $\Delta - 0.55°C$. Species living in estuaries are of particular interest. *Cyathura polita,* on the Atlantic coast of North America, lives in water with salt content varying between 0.1–37‰, depending on the shallowness of the water. In water of 1–28‰ the blood is hypertonic; at 29–42‰, however, it is hypotonic against the surrounding medium. The oxygen taken up within a given time is the same in 1‰ as in 37‰ as also in the Chinese mitten crab *Eriocheir*. *Saduria entomon*, in the Baltic Sea of Europe, has hypertonic blood in water with a freezing point depression of Δ — 0.41°C and 3.6 g Cl/l (about 7‰ salt). If the salt content of the water is raised above 15 g Cl/l, the blood becomes isotonic. The animal still can live in dilute brackish water, 0.9 g Cl/l, while a freshwater race has the blood concentration the same as in brackish water inhabitants accustomed to water of 14 mg Cl/l. Osmoregulation of *Ligia* and terrestrial isopods has been discussed above with their adaptation to terrestrial habitats.

RESPIRATION.   Marine and freshwater isopods use the vascularized pleopods both as swimmerets and for respiration. When resting, the pleopods are turned posteriorly, almost flat against the venter of the abdomen (Figs. 17-7, 17-9), so that their anterior surface is ventral and their morphological posterior has a dorsal position. They are ventilated by beating forward and back, the speed of movement depending on several variables. But to give an approximate indication: at 14°C *Asellus aquaticus* makes 18–167 beats/min, *Idotea tridens* 39–199, *Ligia italica* 64–120.

Of interest is the transformation of this respiratory organ for terrestrial life in the Oniscidea. It is surprising that in most north temperate wood lice the endopodites of the pleopods continue to work as gills just as they do in aquatic species. Like the gills of *Asellus*, they are hollow pouches that have the thin cuticle and high hypodermis characteristic of the respiratory epithelia of crustaceans that take oxygen out of water (Fig. 17-22). Even in inhabitants of fairly arid regions (e.g., *Armadillidium*) the endopodites are covered by a film of liquid, evaporation of which is prevented by the anterior exopodites lying below. These exopodites are dorsally hollowed out, forming a capsule that the endopodites fit into. The fit is tighter in inhabitants of arid regions (e.g., *Armadillidium*) than in marsh inhabitants (e.g., *Ligidium*).

Experiments in replacement of evaporated liquid have all been made under laboratory conditions, none in nature. It was found that the evaporated fluid is replaced by two possible methods. Almost all central European Oniscidea, except for *Armadillo* form a tube by placing the left and right cylindrical endopodites of the uropods together. If the uropods dip into dew or raindrops, capillarity draws water up to reach the base of the pleopods (Fig. 17-8). Fanning pleopod movements, otherwise rare in terrestrial isopods, spread the water over the endopodites.

The second method of water replacement is through a water transport system found in *Oniscus, Porcellium, Porcellio,* and *Armadillidium*, but not in isopods (Ligiidae, Trichoniscidae, *Armadillo, Tylos,* and *Helleria*) of very wet habitats. Every drop of water that falls on the back of the isopod is channeled to the pleopods. If one places a drop of red colored water on the second free tergite, it moves to the slit between second and third epimeres and then appears on the venter. Then a thread of red liquid moves from leg insertion to leg insertion and to the pleopods. The pleopods make movements that conduct the fluid to all endopodites. If there is abundant water, the endopodites of the uropods which lie against each other form a capillary tube that removes the surplus water. The lateral water channel on the underside of the pereon, examined under the microscope, turns out to be a longitudinal groove, bordered by many spiny and finger-like projections, and running from one walking leg insertion to the next on each side (Fig. 17-7).

Besides gill respiration, direct uptake of oxygen from the air is employed. In *Ligia, Ligidium,* and *Oniscus* the anterior (ventral) wall of the exopodites has a much thinner cuticle than the dorsal side. Also there is a very thin hypodermis, difficult to see and comparable to that of insect tracheae in that it is permeable to oxygen. The dryer the habitat the greater the importance of taking oxygen directly from the air. In *Oniscus* and *Philoscia* the lateral part of the pleopod exopodites is thin and deeply grooved on the posterior wall. This part is set off by a fold against the thicker median part of the exopodite (Fig. 17-22), resulting in wide spaces between the successive thin lateral parts of the exopodites, which are filled with air. Ventilation takes place as the abdomen is bent: bending up sucks air in, bending down removes air. This can be observed most easily in an atmosphere rich in CO_2. If the wood louse is injected with reduced indigo, the differentiated lateral exopodite parts become blue, leaving no doubt that oxygen is taken up through the surface enlarged by grooves.

More specialized respiratory organs are found in *Porcellium, Cylisticus, Porcellio, Armadillidium,* and *Armadillo.* In these genera there is a thinly lined air sac within the pleopod exopodites from which leave thin, branching, blind tubules, that go into the blood filled space of the exopodite (Fig. 17-23). The sac develops from a hypodermal invagination on the posterior (dorsal) side of the exopodite. At its origin the invagination remains open as the spiracle. The thin respiratory epithelium here has moved toward the inside. While the surface has been increased, there are adaptations against drying up. This tracheal lung, called pseudotracheae, has evolved convergently with the tracheae of insects and spiders. In *Cylisticus, Porcellium,* and *Trachelipus* all five pairs of exopodites have such pseudotracheae, from outside resembling chalky white cushions. They are spread over one-fifth to one-fourth of the exopodite. *Porcellio* and *Armadillidium,* inhabitants of dry areas, have pseudotracheae on the two anterior exopodites only, but they are larger, being spread over at least half the interior of the exopodite with their branches. Protection against water entering from the capillary duct keeping the endopodites moist, are the lateral position of the spiracle and an area around the spiracle which is water repellant. Air is replenished in part by diffusion and by the movements of the abdomen.

Beach isopods of the genera *Tylos* and *Helleria* have 1–11 separate pseudotracheae in each exopodite of the second to fifth pleopods. Each pseudotrachea has spiracles centrally located on the anterior (ventral) surface of the exopodite. Endopodites are present.

It can readily be demonstrated that the drier the habitat, the more important the pseudotracheal respiration. If isopods from dry habitats are submerged in water, they are unable to get sufficient oxygen from the water through the endopodites. If the slightly specialized pleopod exopodites are removed from *Oniscus,* it can continue to live in moist surroundings, getting oxygen through the endopodite gills. If in *Porcellio* all four pseudotrachea are filled with oil, the oxygen consumption of the isopod sinks to about one-third and the animal dies, although it can get rid of its CO_2.

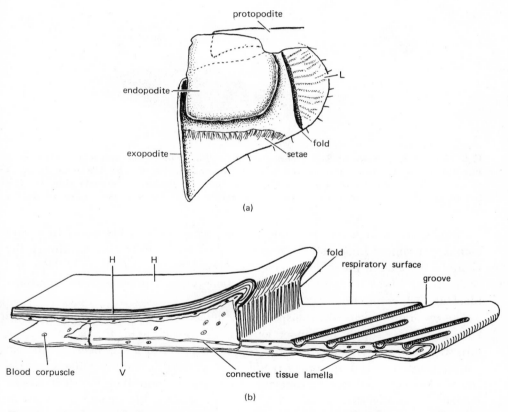

Fig. 17–22. **Fifth pleopod of wood louse *Oniscus asellus.* (After Unwin.) (a) Posterior (dorsal) view; 1.4 mm wide. The lateral respiratory surface (L) is separated by a fold. On the left is the median edge. Foreign material coming with the water spurted out by the anus, or taken up by uropod endopodites, is kept out by the setae. (b) Cross section through lateral half of pleopod exopodite. H, posterior (dorsal) wall; V, anterior (ventral) wall. The connective tissue lamella divides the blood sinus into 2 chambers. The grooves increase the respiratory surface, separated by a fold from the median part.**

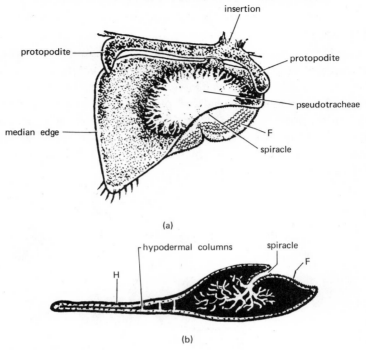

(a)

(b)

Fig. 17–23. Exopodite of 2nd pleopod of *Porcellio scaber*. The spiracle, the opening into the pseudotracheae, lies behind a water repellent area, F. (a) Posterior (dorsal) view; 1.3 mm wide. (After Unwin.) (b) Cross section. Posterior (dorsal) side, H. The hypodermal columns connect anterior and posterior walls; 1.3 mm wide. (From Unwin.)

Respiration through the integument is important only in moist air. In a very humid atmosphere *Ligia* and *Oniscus* can take 50% of their oxygen through the ventral exoskeleton, *Porcellio scaber* and *Armadillidium vulgare* only 34% and 26%. To arrive at these figures, the pleopods were painted with an emulsion, blocking their respiration.

The respiratory oxygen is transported by hemocyanin. While marine Crustacea have a limitless supply of copper ions for the hemocyanin, this is not true of terrestrial animals. In studies of the copper metabolism it has been found that marine isopods picked up copper readily and excreted it readily, while terrestrial isopods have evolved storage capacity and means of controlling loss. The copper content of the digestive glands of terrestrial isopods is higher and less variable than of their marine relatives; intertidal species are intermediate. Also the hemocyanin content becomes less variable with increasingly terrestrial habits. Since it is difficult to get sufficient copper for herbivorous isopods (and amphipods) from terrestrial food materials in which copper is bound to complex organic compounds, microorganisms aid in breaking down organic compounds in poorly digested fecal material which is then eaten again (Wieser, 1966, 1968).

Classification

The classification followed is that of Gruner (1965, 1966). Unless stated otherwise, the species mentioned are marine.

SUBORDER GNATHIIDEA

The cephalothorax consists of the head and two thoracomeres, the second of which is dorsally limited by a suture. The small appendages of the second thoracomere have become an operculum for the mouthparts. The following five thoracomeres (3–7) are much larger and have epimeres. The last, the eighth thoracomere has been reduced and lacks appendages. It thus appears that there are only five pairs of pereomeres. Adult females have lost their mouthparts except for the maxillipeds. Males have also large forcepslike mandibles, which have moved to the wide anterior border of the cephalothorax (Fig. 17–24). The cephalothorax of the male is much more sclerotized than the rest of the integument. Females lack this development of the anterior. The uropods are flattened forming a tail fan with the pleotelson.

Gnathiidea live in burrows that the male digs in mud, or in cracks between rocks, corals, tubeworms, barnacles, sponges. Adults do not swim and do not feed. The foregut is narrow, and the empty gut ends blindly in *Paragnathia*. The midgut glands are filled with fish blood obtained by the larva, a fish parasite. Usually 2–5 mm long, the largest may reach 17 mm.

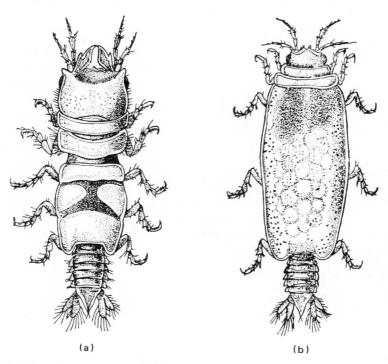

(a)   (b)

**Fig. 17–24.   Gnathiidean isopod, *Gnathia dentata*, in dorsal view; 4 mm long. (a) Male.
(b) Female. (After Monod and Sars from Gruner.)**

Gnathiidae. Gnathia oxyurea is 2.4–5 mm long. *Gnathia cerina* is found on the east coast of North America. *Paragnathia* is viviparous, the young developing completely within the ovaries. The marsupium is vestigial. Males 3.8 mm long excavate diagonal tubes 15–25 cm long into vertical mud banks. The entrance of the tube is only 1 mm wide and can be closed by the head of the male. Below, the burrow widens into an oval chamber, 4–5 mm in diameter, housing up to 10 females. The larvae leave the burrow, swimming off to attach to fishes as temporary parasites. There are three larval stages.

SUBORDER MICROCERBERIDEA

The mouthparts are modified for chewing. Only the first pair of pereopods are subchelate, the dactyl with two claws. The endopodites of the second male pleopods have two articles. These are minute interstitial isopods, some species have been found in caves. Like many other interstitial animals, they have the trunk greatly elongated but are of dwarf size (Fig. 17–25).

Microcerberidae. There are more than 16 species of *Microcerberus; M. abbotti,* to 1.2 mm long and 0.09 mm wide, inhabits fine sand in subterranean coastal waters in California. Other species are found around the Mediterranean, southern European caves, West Indies, Mexico, Brazil, Africa to India, all in subterranean waters.

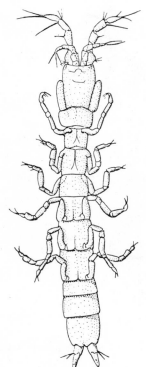

Fig. 17–25.   Microcerberidean isopod, *Microcerberus pau-liani*, in dorsal view; 0.7 mm long. (From Delamare Debout-teville by permission of Hermann Paris.)

SUBORDER ANTHURIDEA

The body is elongated, almost cylindrical. The pereopods, inserted on the sides, are oriented vertically, not transversely as in other isopods (Fig. 17–15). The last, the sixth pleomere is separate from the telson. Mouthparts may be chewing or piercing. The first pereopod has a subchela and is larger and stronger than the others; the second and third usually have also a subchela. In some genera the first pair of pleopods is large and wide, forming a cover over the others. The base of the telson often has statocysts.

Habits of most are almost unknown. Some species live in worm tubes, where they can close the tube with the wide telson and spread uropods. Other species live in cracks between worm tubes and barnacles, barely wider than the diameter of the body. In these the telson is pointed. Usually the tail fan is small. Others live in mud (e.g., *Cyathura*). All walk, swimming only short distances. *Paranthura* has been found attached to and parasitizing fish. Otherwise nothing is known about food habits. One species is found in freshwater springs on Curaçao, Lesser Antilles, another on Réunion.

The only family, Anthuridae has most species in the southern hemisphere. Species with chewing mouthparts have paired statocysts if any. *Anthura* has a large tail fan; *A. gracilis,* to 11 mm long, occurs in the Mediterranean to British coasts. *Cyathura carinata* (Fig. 17–15), to 27 mm long, is found in the North Sea and Baltic under stones between shells, barnacles decaying algae. The animal is euryhaline and can live in water with 2.1‰ salt. During daytime *C. carinata* lives in a vertical burrow in mud, leaving it at night to run on the surface of mud in search of food. Only rarely does it swim short distances, dorsum and anterior up, posterior hanging down. The walking legs are spread when swimming, and only the pleopods beat. The food habits are not known. If *C. carinata* falls on mud it digs in immediately, vertically down, head end first. If the container is turned 90° on its side, the isopod will continue to dig vertically, making a 90° angle. This is done only if the statocysts are not damaged. Loss of statocysts however has no effect on swimming ability and ability to leave the mud. Females of *C. carinata* even those with marsupium and embryos, become males in winter. There are also primary males.

Cyathura polita, to 18 mm, is common on the eastern and Gulf coasts of North America from Maine to Louisiana in brackish waters with currents. They inhabit simple unlined tubes or worm burrows, preferably having some sand and plant debris on the bottom; in northern New England they live in clay. *Cyathura munda* is found on the west coast, *C. milloti* in a spring on Réunion Island. *Crurigens fontanus* of New Zealand is freshwater hypogean. Species that have piercing mouthparts and have no statocysts or only one are: *Calathura branchiata* found on the eastern coast of North America, and *Paranthura elegans* found on the California coast.

SUBORDER FLABELLIFERA

The body is usually flattened. The abdomen is either segmented or in part fused, its sixth pleomere is part of the pleotelson. Usually the mouthparts are chewing, in some piercing. Pereopod coxae are flattened, and form epimeres that usually have a suture toward the trunk (Fig. 17–26). The pereopods, except in Serolidae, lack subchelae, but some or all may be strong prehensile clasping limbs. The uropods are usually flattened and form a tail fan with the pleotelson. Flabellifera are predators, parasites, or feed on vegetation. Many species swim.

The first three families form a series from carrion feeders and predators to temporary and permanent parasites that are only slightly modified.

Cirolanidae. Members of the family have chewing mouthparts and none of the pereopods are prehensile. These are carrion feeders or predators (Fig. 17–5). The pereopods are walking legs. Usually marine, but many representatives are found in caves and ground water, a few in freshwater. The northern European *Cirolana borealis,* 27–33 mm long, normally feeds on fish and crustacean carrion, but will attack diseased fish or fish caught in nets or weirs. It has been reported from the North American Atlantic coast. *Cirolana concharum* is a scavenger from Nova Scotia to South Carolina; *C. harfordi* occurs on the North American Pacific coast; *C. browni* is believed to inhabit freshwater in Cuba. Some cirolanids inhabit American caves. *Cirolanides texensis* inhabits freshwater springs and caverns in the western United States from Wyoming to Texas. *Eurydice pulchra* of northern Europe, to 7 mm long with males 4.2 mm, feeds on injured animals and when burrowing seems to dive into the substrate, the pleopods stirring up sand. *Bathynomus gigantea,* to 35 cm long, is found in the Caribbean.

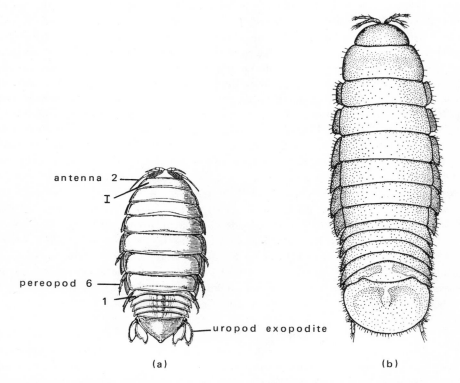

(a)   (b)

Fig. 17–26.   Flabelliferan isopods: (a) *Aega psora,* 4 cm long. (After Sars from Nierstrasz and Schuurmans-Stekhoven.) (b) Woodboring *Limnoria lignorum*; 4.5 mm long. (After Gruner.)

Aegidae. There are piercing mouthparts (Fig. 17–21). The first three pereopod pairs are prehensile. These active swimmers prey on fish, piercing where the skin is thin and sucking blood. Some remain on the host for some time. The salve bug, *Aega psora* (Fig. 17–26), lives to 900 m depth, and attacks *Gadus, Anarhichas, Cottus,* and *Hippoglossus.* It is found in the North Sea and from Greenland to the Mississippi. *Rocinela belliceps* is found on the Pacific coast. *Xenuraega* is another genus.

Cymothoidae. These have piercing mouthparts, and at least the first to sixth pereopods are prehensile. They are permanent parasites that attach to fishes, rarely to cephalopods, and suck blood. They prefer gills, mouth and gill cavities but also skin. A number reach 20 mm in length, some 30–35 mm. Most are marine, a few occurring in freshwater. They will readily bite the hand of the collector. *Cymothoa* are protandrous hermaphrodites. *Livoneca ovalis* is found on the Atlantic and Gulf coasts, *L. vulgaris,* and *L. californica* on the Pacific. *Nerocila californica* is found on the Pacific North American coast, *N. munda* on the Atlantic. Other genera are *Anilocra* and *Meinertia.*

Limnoriidae. The body is flattened and the animals can ball up. The abdomen is segmented but the last pleomere is fused to the telson. The mouthparts are chewing. The uropods have clawlike, vestigial exopodites. They are plant feeders. Wood-boring gribbles, *Limnoria* (Fig. 17–26), include about 20 species that feed on algae, driftwood, and wood of docks as long as it remains submerged. The isopod uses its sharp mandibles to cut off the fine splinters it feeds on, consuming large quantities. Some *Limnoria* in burrowing follow the spring growth of an annual ring into the inner parts of the log just above the surface. The burrow is often winding, punctured by a series of windows, which provide respiratory water. In some places small secondary burrows, inhabited by young animals, branch off, all remaining close to the surface. The animals sit at the ends of burrows. In heavily infested wood there are 12–15 burrows per 1 cm³, usually of twice the diameter of the trunk of the isopod. A dense network of burrows permits the isopods to bore 1–2 cm deep. Between the burrows remain lamellae, but in heavily infested wood the lamellae may be broken down by waves. As the diameter of the wood is reduced, the isopods can penetrate deeper, doing enormous damage. All the wood particles pass through the gut, serving as food. The presence of cellulase has been demonstrated in the midgut glands. *Limnoria* can utilize hemicelluloses; it cannot survive on sterile wood. Wood is attacked only after it has been in the ocean for some weeks, by which time the surface has mycelial growths of ascomycetes and Fungi Imperfecti which are eaten along with the wood; conidia are found in the fecal matter. At times submarine cables have been attacked, and their rubber insulation penetrated. Practically mature and mature *Limnoria* can be seen walking outside the burrows, or swimming rapidly, dorsum down. Five pairs of pleopods with long swimming setae serve as swimmerets. The rapid swimming movement causes the animal to take an erratic course and distances of barely 1 m can be traversed. As there is no pelagic larval stage, the distribution of *Limnoria* is probably caused by driftwood, wooden ships and ocean currents. *Limnoria lignorum,* up to 5 mm long, is a large widespread species of the northern hemisphere, in the Atlantic as far south as Rhode Island, in the Pacific, south to California. In summer it carries 8–10, rarely 30 young in the marsupium, in fall only 1–4. During growth some eggs die.

They may tolerate salt content down to 15‰. *Limnoria tripunctata*, found in warmer water from Rhode Island to Venezuela, from California to Mexico, enters pressure-creosoted wood shortly after it is submerged in the ocean, going deep inside where creosote has not penetrated. (The sensitive larvae of shipworms cannot enter creosoted wood.) *Limnoria tripunctata* burrows with the wood grain, and outer wood may have fewer animals than inner wood (Menzies, 1957). Species of *Limnoria* are carried around by man; *L. pfefferi* occurs in the Indian Ocean and Florida, *L. saseboensis*, Japan and Florida.

Sphaeromidae (= Sphaeromatidae). The body is convex and can ball up. The abdomen has only three rings; the first ring is concealed by the seventh pereomere. The second consists of two fused somites while the third is the pleotelson. The inner branch of the uropods is immovable, but the outer moves if present. The animals feed on vegetable matter and have chewing mouthparts. A number of species of this otherwise marine group are found in subterranean waters. *Sphaeroma* swims venter up. Some species bore into wood, moving at a right angle to the surface. The European *S. hookeri*, 10 mm long survives in the brackish Baltic. On brackish water coasts one finds many individuals on the dry beach. In the laboratory the animals have been kept successfully without water on a substrate of 8–30‰ salt and feeding them on decaying *Fucus*. If the substrate is saturated by freshwater or ocean water of 60‰ salt, they do not survive. At 15°C the eggs develop within 45 days. The young have three manca and six juvenile stages. During the last the primordia of oostegites are visible at the bases of the second to fourth pereopods. In the molt after the next, the parturial molt, the marsupium becomes complete. A week after the young have left, the mother molts into a stage that lacks a marsupium. But the marsupium reappears after the next molt which may occur in the same season or the following spring. If males are not present at the parturial molt the animals do not spawn and the eggs are resorbed in the ovary. But the female molts eventually as if it had produced young. They swim with the venter up. On the east coast of North America south to Florida *S. quadridentatum*, 8 mm long, is found under stones and among algae along shores; *S. pentodon* in California may burrow into and destroy sandstone; *S. destructor* is found on the Gulf coast, *S. quoyana* of New Zealand has been found destructive to soft stone employed in port installations, and may enter concrete. *Sphaeroma terebrans* a cosmotropical species that bores in wood has been found in freshwater in the southern United States. *Cymodoce japonica*, to 17 mm long, has been introduced along the North American Pacific coast, probably with Japanese oysters. *Exosphaeroma amplicauda* is found on the California coast; *E. thermophilum* is found in warm springs in New Mexico. *Gnorimosphaeroma oregonensis* is found in brackish water on the Pacific coast, and has also been found in freshwater. *Caecosphaeroma burgundum* is found in European cave waters. It has a low oxygen consumption compared to marine relatives and only 4–15 eggs in the marsupium, less than marine relatives.

Serolidae. Serolids have a wide, flat, disc-shaped body. Two thoracomeres are partly fused into the cephalothorax. The last pereomere is not seen dorsally on all species. The fifth and sixth pleomeres are frequently dorsally fused with the pleotelson. There are chewing mouthparts. Females have the first, males the first and second pereopods with subchela; the others are walking legs. The first three pleopods are swimmerets. The uropods have one or two movable flat branches. *Serolis* may reach about 6 mm long; the only genus in the group, its species are mostly found in the southern hemisphere.

SUBORDER VALVIFERA

The attachment of the uropods has moved to the venter. The very long uropods are lamellate and can be folded toward the midline of the body (Fig. 17–9), protecting the pleopods. While swimming the uropods are kept open. The pleomeres are more or less fused, the last is always a part of the pleotelson. There are chewing mouthparts. The pereopod coxae are flattened as epimeres and are often fused to the trunk without suture. The pleotelson is often elongated.

Idoteidae. The body is dorsoventrally flattened. The pereomeres are all of about the same length (Fig. 17–27). The pereopods are usually walking legs, the anterior three pairs with subchelae. Individuals of *Idotea* appear in large numbers especially among brown algae in the sublittoral and in *Zostera* stands. Omnivorous *Idotea balthica*, to 3 cm long, is cosmopolitan and euryhaline; the lowest salt concentration it can tolerate is 3.5‰. Other common European species are *I. chelipes*, up to 1.5 cm long, found in shallow brackish waters of salt concentrations to 4.5‰, and *I. granulosa*, up to 2.6 cm, which can be found at 6‰ concentration. *Idotea phosphorica*, up to 2 cm, is found among weeds of the New England coast; *I. metallica*, up to 1.8 cm, is cosmopolitan, found swimming or on vegetation. Numerous species of the genus are found along the North American Pacific coast. *Ericksonella attenuata* is found on the Gulf coast north to New England. *Saduria* has a long pointed pleotelson (Fig. 17–27). The first three pereopod pairs are subchelate, the last four much longer, setose and used for swimming. Also the branches of the anterior two pleopods and the exopodites of the third pair are swimmerets. *Saduria entomon* is the largest northern European isopod, with males to 5 cm long or 8 cm at lower salt concentrations, and females to 6 cm. It lives in brackish water, 3–18‰ salt, on sand and mudbanks in which it burrows. In experiments it can tolerate undiluted ocean water as well as dilute brackish water, and some races live in freshwater. Young animals swim with pleopods, venter up; animals over 5.5 cm swim only rarely. The female has about 122–380 eggs. *Chiridotea coeca*, about 1.5 cm long, is found in the tide zones from Nova Scotia to Florida on sand bottoms. *Edotea triloba*, up to 0.7 mm, is found in mud from Maine to New Jersey. *Glyptonotus antarcticus*, to 11 cm long, is a predator and scavenger in the Antarctic, from 1 to 600 m. Analysis of gut content for common foods reveals parts of ophiuroids, echinoids, gastropods, pelecypods, and isopods. Another genus is *Zenobiana*.

Arcturidae. The body is almost cylindrical (Fig. 17–18). The fourth pereomere is much longer than the others. The second antennae are long. The four anterior pereopods are thin and anteriorly directed; the posterior three walking legs have cleft claws. The European *Astacilla* is 12–27 mm long (Fig. 17–18). Its long fourth pereomere separates the spiny feeding basket formed by limbs 1–4 from walking legs 5–7. Pereomeres 5–7 are very movable, permitting the anterior of the body to be raised into vertical position. The flagellum of the long, raptorial, second antenna is short. The 2–5th pleomeres are fused. Oostegites 1–3 are very small, the 4th alone functional. Found mainly on Bryozoa and Hydrozoa, they climb by reaching and holding on with the long second antennae, pulling forward and then attaching the three posterior pereopods. They can swim on the back by beating pleopods 1 and 2. *Arctuella*, to 6 mm long, is of similar structure, the fourth pereomere shorter and wider. Another genus is *Arcturus*.

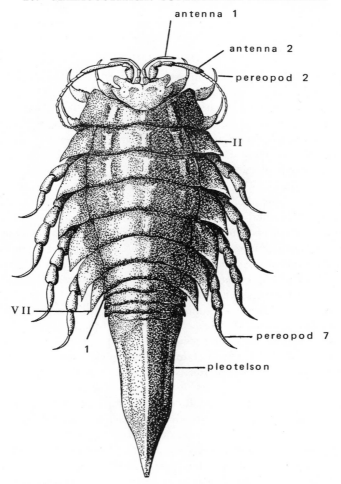

Fig. 17–27.   Valviferan isopod, *Saduria entomon*; 4 cm long. Roman numerals refer to pereomeres, Arabic to pleomere.

SUBORDER ASELLOTA

Pleomeres 2–3 to 6 and the telson have fused into a single large unit (Fig. 17–28). Rarely the pereomeres and anterior pleomeres are also fused with one another or with the rest of the abdomen. Mouthparts are adapted for chewing. The females lack the first pleopod pair. The uropods are cylindrical. Only few swim. The most posterior pleopods serve as gills and are housed in a chamber that is protected ventrally by 1 or 2 more anterior pleopods.

Infraorder Aselloida

This group contains only the family Asellidae, all freshwater animals. The exopodite of the third pleopod pair is very large, and acts as an operculum covering the following pairs. In the male the first pleopods are small, and the second are complicated gonopods. The coxae of the posterior six pereopods are usually movably articulated with the body and not modified as epimeres. *Asellus aquaticus* is the European fresh-water isopod (Fig. 17–28), with males about 12 mm, females 8 mm long. Found in standing water and slow streams, they move about by walking and feed on soft water plants or decaying leaves that have fallen into the water. Shallow-water populations can tolerate freezing into ice and populations found in caves lack pigment and have reduced eyes. *Asellus cavaticus,* with males to 6 mm, females to 8 mm, is found in deep wells and caves of Europe. The common eastern North American species is *A. militaris,* 15 mm long, found among aquatic vegetation. *Asellus intermedius* is found in the central United States. Numerous species are found in North American caves. Species of *Lirceus* are found in eastern North America; *L. lineatus* unlike other asellids is found in water as deep as 55 m.

Infraorder Paraselloida

The second pleopods of the female are fused into a large wide operculum that covers the posterior pairs. The many genera are very diverse and the division of the group is still controversial. There may be as many as 20 families. Most are marine and are an important and diverse part of the deep sea fauna. *Jaera marina,* 5 mm long, is found

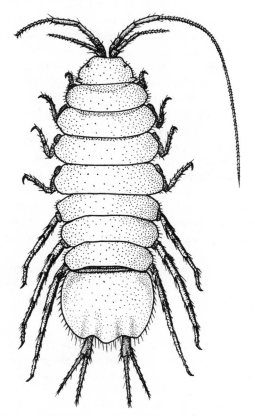

Fig. 17–28. Asellote isopod, freshwater *Asellus aquaticus*; female, 7.5 mm long. (From Govner.)

from Labrador to New England and on northern European coasts among weeds between tide lines. *Janira maculosa* of Europe, to 10 mm long, resembles *Asellus* in body
outline. They are wide and flat with laterally directed legs and run on the surface of
mud; *J. alta* is found on the North American Atlantic coast, *J. minuta* on the Gulf
coast. *Munna* has species in the northern Atlantic and on the North American Pacific
coast. *Pleurogonium rubicundum* of Europe grows to 1.5 mm long. *Munnopsis* is
adapted for life on a mud surface with its long third and fourth pereopods; *M. typica,*
to 18 mm long (Fig. 17–16), has second antennae four times the body length. The
first and second pereopods are short, third and fourth at least three times as long as the
body. *Eurycope,* 1.5–4 mm long, has long second antennae and second to fourth
pereopods as long as the body. Foraminifera and diatoms have been found in its
stomach.

The Microparasellidae contain blind, cylindrical dwarf forms, about 1 mm long
found in subterranean water or interstitial in the littoral zone of the ocean. They seem
derived from marine forms with little morphological specialization. Members have
only one free pleomere and a pleotelson; the 7 pereopods resemble each other. Most
species of *Microcharon* and *Microparasellus* are found in southern Europe.

SUBORDER PHREATOICIDEA

These are amphipod-like, the abdominal epimeres extending ventrally make the animals appear laterally flattened; also the pereopods are not positioned transversely (Fig.
17–6). There are chewing mouthparts. The abdomen is segmented, only the last
somite fused with the telson. The anterior four pereopods are directed anteriorly, the
posterior three directed posteriorly, just as in the amphipods. At least the first, rarely
the posterior pereopods are subchelate. Pereopods 1–4 have oostegites. Uropods,
attached on the sides, are cylindrical, biramous, and are used in locomotion. These are
freshwater animals but a few burrowers live in moist places on land, some in subterranean waters. They do not swim, but walk among plants, legs extended to the
sides. In matted plant material the uropods push the animal along and the legs are
carried folded back. Plant detritus is used as food. In South Africa they dig 2.5–5 cm
deep if the water dries up, then fold pleotelson and uropods over the gills. In this
condition they survive in the mud, and unroll as soon as thrown into free water. Most
are about 10 mm long, but species range from 5–45 mm. Found only in South Africa,
New Zealand, Australia and India, this ancient group apparently evolved on a former
southern continent that included India.

Most species of the families Amphisopidae and Phreatoicidae are from Australia and
New Zealand. *Mesamphisopus,* an amphisopid, is found in southern Africa in swamps.
Members of Nichollsiidae are found in ground water in India.

SUBORDER ONISCIDEA *

The usually dorsoventrally flattened trunk has wide epimeres on the pereon derived
from coxae, usually without sutures. The pleomeres are usually well separated, the last
one fused to the telson (Fig. 17–1). The first antennae are minute (Fig. 17–3). There

---

* The suborder is usually called Oniscoidea. But since the ending -oidea is recommended
for superfamilies in zoology, by the *International Code on Zoological Nomenclature,* the
suborder is here called Oniscidea.

are chewing mouthparts. The oniscids are amphibious and terrestrial, feeding on plants. The Oniscidea evolved in two directions, the two groups are secondarily convergent.

Infraorder Ligiomorpha

The head projection between eyes is weakly developed. The epimeres do not have a suture separating them from the tergites (Fig. 17–1).

GROUP DIPLOCHETA

This group is called Protoniscina in the older literature. Gills are used for respiration. The first antennae have several articles. The long, cylindrical branches of the uropods are parallel to each other and extend posteriorly. This group is transitional between marine isopods and the groups Synocheta and Crinocheta. This transition is indicated by the length of the flagella of the second antennae, endopodites of the maxillipeds, uropods, and number of ommatidia. The two diplochete families Ligiidae and Mesoniscidae have two genital papillae, each with an opening of a vas deferens.

Ligiidae. Rock slaters, *Ligia*° live only on ocean shores, hiding at day time in cracks of rocks or under stones. *Ligia oceanica,* with males to 28 mm and females to 20 mm, is the northern European sea slater which has also been reported several times from the New England coast. It was recently seen in large numbers at one point in Maine. The slater feeds at night and if disturbed runs rapidly toward water. It always finds the way to water and can remain in salt water for long periods. In the western Baltic it lives in up to 6‰ brackish water and has a life span of 15–18 months. It lives for about three years and has an average of 80 eggs in the marsupium. The old world warm water *L. exotica* to 3 cm is found on the coast of southeastern and Gulf states, south to southern Brazil and from California to Chile. It has probably been introduced by wooden ships; *L. occidentalis* is found on the Pacific coast of North America, *L. pallasii* from Alaska to California. *Ligidium,* a swamp isopod, has the inner branches of the uropods much thinner than the outer ones. The European *L. hypnorum,* 7–10 mm long, lives in moist places in deciduous forest litter, under bark and in moss. *Ligia gracile* is found from Alaska to California, *L. longicaudatum* has been reported from northeastern U.S. in moist places protected from flooding. It may remain active during the winter among ice crystals in moss.

GROUP SYNOCHETA

The Synocheta and Crinocheta used to be combined under the name Euoniscina. Both groups have several families, all containing many species which can respire in water as well as in air through differentiated lateral parts of the exopodites of some or all pleopods. In both groups the first antennae have only several articles, the flagellum of the second antennae has only few. The endopodites of the maxillipeds are short. In the Synocheta the vasa deferentia unite and a single duct opens on the single genital papilla.

Trichoniscidae. There is gill respiration and no water transport system. The flagella of the second antennae have few thin, long articles. Many populations are parthenogenetic. *Trichoniscus pusillus* is found in moist places in Europe, *T. pygmaeus* from

° The generic name *Ligia* is on the *Official List of Generic Names in Zoology.*

Europe has been imported to North America; *T. demivirgo*, to 4 mm, is common in wet habitats of northern North America; they are very active and gregarious animals, and males are unknown. *Trichoniscus halophilus* is found in New England salt marshes.

GROUP CRINOCHETA

The Crinocheta contain the common terrestrial wood lice or sow bugs. The cylindrical uropod branches are short. Usually there is a water transport system. The vasa deferentia remain separate and open separately on the single genital papilla. For that reason the group is believed derived separately from the Diplocheta (Vandel, 1960; Gruner, 1966). Of the many species found in America, many common ones have been introduced from Europe. Probably they have replaced much of the native fauna.

Squamiferidae. The flagellum of the second antennae appears to have only one article as the proximal article is very short. The body is densely covered with scales. The species, at most 6 mm long, usually lack eyes and have no tracheal system. The European *Platyarthrus hoffmannseggii*, 5 mm long, is an ant symbiont, found in ant hills, but it can survive away from ants. On meeting an ant the isopod presses to the substrate, remains motionless and turns up the uropod exopodites, which smear secretions on the mandibles of a biting ant. They feed on detritus, fecal matter of ants, fungal spores, and take up honeydew from plant louse colonies. They have been found several times in the eastern United States, once in a nest of *Formica sanguinea* and its slave *F. fusca.* *Trichorhina* has species in tropical America.

Oniscidae. The flagellum of the second antennae has three articles (Fig. 17–1). Besides the gills, the lateral edge of the pleopod exopodite, a thin grooved area, is used for respiration (Fig. 17–22). *Oniscus* asellus (Fig. 17–1), up to 18 mm, is found under stones, litter, bark, moist and mesic habitats, especially in deciduous forests, but also in cellars and greenhouses. It is widely distributed in Europe, and has been introduced and is common near human settlements in North America. *Philoscia muscorum*, up to 12 mm, is widely distributed in Europe and has been introduced locally in North America. *Philoscia variegata* is widespread in South America.

Cylisticidae. Members have five pairs of tracheal lungs. The trunk can roll up but does not form a complete ball, the second antennae cannot be hidden, also the long uropods extend between telson and cephalothorax. Otherwise the family resembles the Porcellionidae. *Cylisticus convexus* 13.5 mm long, is widely distributed in Europe and North America. It is found on humid, stony places in leaf litter and stones, but also in gardens and greenhouses. In central Europe it is found in gardens and courtyards of large cities. Despite occurring away from human habitations in America it has probably been introduced, as related species are found in Europe. Another genus is *Typhlisticus.*

Porcellionidae. The flagellum of the second antennae has two articles. On the lateral borders of two, three, or five pairs of pleopods are tracheal lungs. The genera with lungs on all pleopods are convergent with those having lungs only on the anterior two pairs of pleopods. The uropod exopodites are not flattened. *Trachelipus* and *Porcellium* have tracheal lungs on all pleopods. *Trachelipus rathkei*, to 15 mm long, found in dry and wet places, widespread in Europe, has been introduced and is wide-

---

* The generic name *Oniscus* is listed on the *Official List of Generic Names in Zoology.*

spread near human habitation in North America. *Porcellium conspersum,* to 9 mm long, can roll up slightly but does not form a closed sphere. It is found in moist shady places. *Porcellio* and *Hemilepistus* have tracheal lungs on only the two anterior pairs of pleopods. *Porcellio° scaber,* up to 18 mm long, is the commonest and most widespread European-sow bug found in cellars, greenhouses, rock debris, and gardens. They are active for about one hour during night time, then return to their previous hiding places. The distribution is cosmopolitan, including the tropics, due to accidental transport by man. *Porcellio spinicornis, P. dilatatus,* and *P. laevis* are other European species widespread by man. *Porcellio spinicornis* has been found in attics of abandoned farmhouses, a habitat otherwise not frequented by isopods. It may require lime. *Hemilepistus,* up to 20 mm long, is found in deserts. *Metoponorthus pruinosus* (= *Porcellionides pruinosus*) up to 11 mm, of southern Europe has become cosmopolitan near human habitations, it is found from northern United States to southern South America, including West Indian islands, far more widespread and common than in its native haunts.

Armadillididae. The exopodites of the two anterior pleopods are tracheal lungs. The body is arched, the exopodites of the uropods are flattened, the endopodites hidden by the telson. In the pill bug *Armadillidium vulgare,* up to 17 mm long, the uropods do not extend beyond the pleotelson, covering the space between epimera and pleotelson permitting *Armadillidium* to roll into a sphere, hiding even the antennae. It is found in dry sunny places, borders of woods, quarries, and leaf litter. European in origin, *A. vulgare* has become cosmopolitan in association with man. *Armadillidium nasatum* of southern Europe has become widespread by man and is found in North America.

Armadillidae. There are five pairs of tracheal lungs. There is no water transport system. The protopodites of the uropods are flattened, the exopodite is small. All can roll up completely. *Armadillo officinalis,* up to 19 mm, is native in Mediterranean countries and also is found in greenhouses in central Europe.

Infraorder Tylomorpha

The tracheal lungs of the exopodites of the pleopods open centrally on the ventral wall with one or several spiracles. The cephalothorax has a well developed projection between the eyes. Chromatophores are scattered in *Tylos.* The epimeres are separated by a suture from tergites. The uropods are flattened, inserting on the anterior edge of the pleotelson and, as in the Valvifera, have an anterior projection that extends beyond their articulation. Also as in Valvifera they can be folded ventromedially so that they cover the posterior pleopods (*Tylos*) or at least the anal tubercle (*Helleria*). The first antennae have only one article. This group has probably evolved convergently. The uropods and cephalic projection indicate ancestors belonging to the Valvifera, especially the Idoteidae (making the Oniscidea a polyphyletic group). The Tylomorpha should be a separate suborder.

Tylidae. This is the only family. *Tylos† latreilli* to 13 mm long, is found on ocean shores from the Atlantic Islands to the Mediterranean. It has been used for studies in orientation. Introduced, it is found on Bermuda, Puerto Rico, Florida, and South

---

° The name *Porcellio* is listed on the *Official List of Generic Names in Zoology.*
† The generic name *Tylos* is on the *Official Index of Generic Names in Zoology.*

America; *T. niveus* is found in Florida and Cuba. *Helleria brevicornis,* 18–20 mm long, occurs in forests far from the ocean of western Mediterranean countries.

SUBORDER EPICARIDEA

All members of this suborder parasitize crustaceans and feed on blood. The females often grow tremendously but with loss of appendages the body may transform (Figs. 17–14, 17–19). The males are small and remain isopod-like (Fig. 17–29). The mouthparts form a sucking cone containing one pair of converging, piercing stylets formed from the mandible (Fig. 17–4). The first and second maxillae are vestigial or lost. Sometimes all mouthparts are reduced and replaced by a long proboscis (Fig. 17–14). The oostegites are usually not lost and may be much enlarged (Fig. 17–29). The life cycle involves two hosts. The young have the appearance of isopods (Fig. 17–13) and attach to copepods. As adults, another crustacean is parasitized; many are host specific. The Epicaridea are mainly marine.

Infraorder Bopyrina

Sexes are separate. Females show some segmentation. The antennae are vestigial or have become specialized (Fig. 17–27). Mandibles and maxillipeds are present, usually two maxillae are rudimentary or absent. The oostegites are usually large, but very small in Dajidae, in which the marsupium is formed by a ventromedian fold of the body wall.

Bopyridae. The adult female shows segmentation (Fig. 17–27), and has appendages. But the trunk is more or less asymmetrical and the seven pereopods may only be present on one side. Oostegites are always present. The males are minute. They parasitize only decapod crustaceans. *Ione thompsoni* is found on *Callianassa* on the

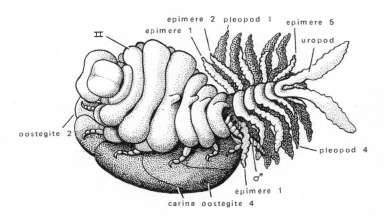

Fig. 17–29. Epicaridean isopod, the parasitic *Cancricepon elegans*, showing sexual dimorphism. Female in dorsolateral view. The large oostegites of the anterior 5 pereopods form a huge marsupium, at first red, later blue violet, for the many eggs of the parasite. The carina strengthens the support of the oostegites. The male (♂) has barely grown since the 2nd larval stage, and stays on the female. The epimeres are lateral extensions of the tergites of the pleon. Roman numeral refers to pereomere. Female, 9 mm; male, 1.4 mm long. (After Giard and Bonnier.)

eastern North American coast. *Cancricepon elegans,* with females 9.2 mm long and up to 5.3 mm wide, males 1.4 mm long, live in the branchial chambers of *Pilumnus hirtellus* (Fig. 17–29). *Stegophryxus hyptius* is found on the abdomen of hermit crabs on the New England coast. About 1.5% of the crabs are infected in the Woods Hole region. Little damage seems to be done. *Pseudoione,* to 8 mm long, is widespread on prawns and anomurans, *P. furcata* is found on the Gulf coast. *Probopyrus pandalicola* is found on all coasts of North America and northern Europe, parasitic in the gill chamber of *Palaemonetes vulgaris* and other prawns; *P. bithynis* is found in freshwater shrimp *Macrobrachium* and *Palaemonetes* in the Gulf of Mexico drainage. Other *Probopyrus* species are parasites of freshwater shrimp in America. *Bopyrus* is parasitic on species of *Leander.* Females of *Phryxus* live under the abdomen of shrimp losing all pereopods except one on one side and becoming asymmetrical.

Dajidae. Adult females are symmetrical, shield-shaped. Segmentation is lost or barely visible. They parasitize Mysidacea and eucarids. Genera are *Dajus* and *Prodajus,* the females only 3 mm long.

Entoniscidae. Females have a much deformed trunk, partly covered by dorsally attached oostegites (Fig. 17–19). Both antennae have only one article, distinctly visible on the cephalothorax as is the maxilliped. Only mandibles remain in the mouth cone. There are five pairs of oostegites. The pleopods are often drawn out into fringed lamellae. Individuals inhabiting the branchial chambers of Brachyura dent the inner wall of the gill chamber deep into the body cavity of the host (Fig. 17–19). The males are very small. *Entoniscoides okadai* hatches at an advanced stage and does not need the primary host. *Portunion* parasitizes *Portunus, Carcinus* and other crabs. *Portunion maenadis* has females up to 40 mm long. About 1% of the European shore crabs (green crabs), *Carcinus maenas,* are infected. The female, while growing, invaginates the wall of the gill cavity to the midline of the crab, clamping the soft wall between its thick, muscular antennae and piercing it with the mandibles. Aided by two large lateral ceca of the foregut, which widen the cephalothorax into a sphere, it then pumps blood from the host into its own midgut ceca. If the parasite is removed from the host's sac, it moves energetically and blows up the humps of its body. Aided by waving movements of the first oostegites, which are within the brood sac, and by contractions of the trunk wall, water is circulated from the host's gill chamber through the opening of the invagination into the parasite's marsupium. It enters through a crack between the second oostegites near the cephalothorax, circulates around the eggs and leaves the marsupium through a slit between the fifth oostegites posteriorly. It then flows through a channel made by the flat pleopods on one side, the venter of the abdomen on the other to the tip of the abdomen. The water then returns to the gill chamber of the host. The metamorphosis was discussed above under Development. *Paguritherium* attacks hermit crabs, but not those infested with *Stegophryxus.* The parasite causes the crab's gonads to atrophy. *Entionella fluviatilis* is the freshwater entoniscid that, parasitizes the Chinese mitten crab, *Eriocheir japonicus* in the Orient, apparently carried by its host from salt- to freshwater.

Infraorder Cryptoniscina

The Cryptoniscina are protandrous hermaphrodites, functional males when young and later changing into females. Females bearing many eggs lose their crustacean appearance and become a sac. The eggs are liberated by the bursting of the sac. No appendages can be seen, but segmentation is present at times. Lacking oostegites, the marsupium is formed by deep invaginations of the ventral body wall. They parasitize various crustacean groups: ostracods, barnacles, mysids, amphipods, isopods, and cumaceans. The metamorphosis was discussed above under development. The only family is Cryptoniscidae.

Danalia, with males 1.2 mm long, females up to 8 mm (Fig. 17–14), is a hyperparasite. Entering the crab *Inachus*, from the gill chamber it attacks *Sacculina* in the body cavity of the crab with its long proboscis. The mouthparts and all appendages have previously been lost. Within 5–6 weeks the female acquires its final form. It stays parasitic for 2–3 months. *Liriopsis pygmaea* male is 1.35 mm long, female up to 5 mm (Fig. 17–2). The pear-shaped female sucks on the mantle cavity wall of the rhizocephalan *Peltogaster*. Later its narrow body tip widens and the animal becomes almost spherical, the spherical body divided from the triangular posterior part by a narrow constriction. At this time the posterior part is in the mantle cavity while the anterior is free outside the *Peltogaster* and the parasite no longer feeds. Another genus is *Clypeoniscus*.

Referenecs

Alexandrowicz, J.S. 1952. Innervation of the heart of *Ligia oceanica*. *J. Marine Biol. Ass.* 31: 85–95.

Amar, R. 1951. Formations endocrines cérébrales des Isopodes marins et comportement chromatique d'*Idotea*. *Ann. Fac. Sci. Marseille* (2)20: 167–305.

Angelini, G. 1935. Struttura del mesointestino di *Armadillidium*. *Atti Soc. Toscana Sci. Natur.* 45: 128–140.

Arcangeli, A. 1952. La evoluzione del sistema respiratorio dell'exopodite dei pleopodi nelle famiglie dei Porcellionidi e degli Eubelidi. *Boll. Zool.* 19: 297–304.

Armitage, K.B. 1960. Chromatophore behavior in the isopod *Ligia occidentalis*. *Crustaceana* 1: 193–207.

Audoinot, R. 1956. Territoires de régénération des péréiopodes de l'Isopode *Sphaeroma serratum*. *Compt. Rend. Acad. Sci. Paris* 242: 1239–1240.

Baer, J.G. 1951. *Ecology of Animal Parasites*. Univ. Illinois Press, Urbana.

Barnard, K.H. 1927. The fresh water isopodan and amphipodan Crustacea of South Africa. *Trans. Roy. Soc. South Africa* 14: 139–215.

Barnes, T.C. 1932. Salt requirements and space orientation of the Isopod *Ligia* in Bermuda. *Biol. Bull.* 63: 496–504.

———— 1934. The salt requirements of *Ligia* in Bermuda. *Ibid.* 66: 124–132.

Barr, T.C. 1968. Cave ecology and evolution of troglobites. *Evol. Biol.* 2: 35–102.

Becker, F.D., and Mann, M.E. 1938. The reproductive system of the male isopod *Porcellio scaber*. *Trans. Amer. Microscop. Soc.* 57: 395–399.

Becker, G. and Kampf, W.D. 1955. Die Gattung *Limnoria* ihre Lebensweise, Entwicklung und Umweltabhängigkeit. *Z. Angew. Zool.* 42: 477–517.

Becker-Carus, C. 1967. Die Bedeutung der Tageslänge für die Ausbildung des Geschlechts bei *Armadillidium*. *Crustaceana* 13: 137–150.

Bock, A. 1942. Über das Lernvermögen bei Asseln. *Z. Vergl. Physiol.* 29: 595–637.

Bocquet, C. 1953. Le polymorphisme naturel des *Jaera marina*. *Arch. Zool. Expér. Gén.* 90: 187–450.

Bocquet, C. *et al.* 1951. Le polychromatisme de *Sphaeroma serratum. Ibid.* 87: 245–297.

Bourdillon, A. 1958. La dissémination des crustacés xylophage *Limnoria tripunctata* et *Chelura terebrans. Ann. Biol. Paris* (3) 34: 437–463.

Bowman, T.E. 1955. The isopod genus *Chiridotea. J. Washington Acad. Sci.* 45: 224–229.

Bursell, E. 1955. Transpiration of terrestrial isopods. *J. Exp. Biol.* 32: 238–255.

Cals, P. 1966. Adaption du complex stomo-appendiculaire à la vie parasitaire de Bopyridae. *Comp. Rend. Acad. Sci. Paris* (D) 263: 132–135.

Carlisle, D.B. 1956. The endocrinology of isopod crustaceans; moulting in *Ligia oceanica. J. Marine Biol. Ass.* 35: 515–520.

Chace, F.A. et al. 1959. Malacostraca *in* Edmondson, W.T. ed. *Ward and Whipple Freshwater Biology.* Wiley, New York.

Chappuis, P.A. and Delamare Debouteville, C.L. 1954. Les Isopodes psammiques de la Méditerranée. *Arch. Zool. Exp. Gén.* 91: 103–138.

Cicero, R. 1964. Fina struttura della cuticola intestinale di isopoda terrestri. *Arch. Zool. Italiano* 49: 75–91.

Cloudsley-Thompson, J.L. 1956. Bioclimatic observations in Tunisia in relation to the physiology of woodlice, centipedes, scorpions and beetles. *Ann. Mag. Natur. Hist.* (12) 9: 305–329.

Collinge, W.E. 1947. Abnormal reproduction in woodlice. *Nature* 160: 509.

Consiglio, C. and Argano, R. 1966. Polimorfismo in *Sphaeroma serratum. Arch. Zool. Italiano* 51: 47–96.

Dearborn, J.H. 1967. Food and reproduction of *Glyptonotus antarcticus. Trans. Roy. Soc. New Zealand* 8: 163–168.

Dresel, E.I.B. and Moyle, V. 1950. Nitrogenous excretion of amphipods and isopods. *J. Exp. Biol.* 27: 210–225.

Edney, E.B. 1954. Woodlice and the land habitat. *Biol. Rev.* 29: 185–219.

——————. 1964. Acclimation to temperature in terrestrial isopods. *Physiol. Zool.* 37: 364–394.

——————. 1968. Transition from water to land in isopod Crustacea. *Amer. Zool.* 8: 309–326.

—————— and Spencer, J.O. 1955. Cutaneous respiration in woodlice. *J. Exp. Biol.* 32: 256–269.

Fischbach, E. 1954. Licht, Schwere und Geruchssinn bei Isopoden. *Zool. Jahrb. Abt. Allg. Zool.* 65: 141–170.

Forsman, B. 1944. Beobachtungen über *Jaera albifrons* an der schwedischen Westküste. *Ark. Zool.* 35A: 1–33.

——————. 1956. Notes on the invertebrate fauna of the Baltic. *Ibid.* (2) 9: 389–419.

Frankenberg, D. and Burbanck, W.D. 1963. The physiology and ecology of the estuarine isopod *Cyathura polita* in Massachusetts and Georgia. *Biol. Bull.* 125: 81–95.

Franzl, W. 1940. Die Atmungsorte von *Asellus aquaticus. Zool. Anz.* 132: 44–48.

Goodrich, A.L. 1939. The origin and fate of the entoderm elements in the embryogeny of *Porcellio laevis* and *Armadillidium nasutum. J. Morphol.* 64: 401–429.

Gorvett, H. 1956. Tegumental glands and terrestrial life in woodlice. *Proc. Zool. Soc. London* 126: 291–314.

Gräber, H. 1933. Über die Gehirne der Amphipoden und Isopoden. *Z. Morphol. Ökol.* 26: 334–371.

Green, J. 1957. The feeding mechanism of *Mesidotea. Proc. Zool. Soc. London* 129: 245–254.

Gruner, H.-E. 1953. Der Rollmechanismus bei kugelnden Land - Isopoden und Diplopoden. *Mitt. Zool. Mus. Berlin* 29: 148–179.

——————. 1954. Über das Coxalglied der Pereiopoden der Isopoden. *Zool. Anz.* 152: 312–317.

——————. 1965, 1966, Isopoda *in* Dahl, F. *Die Tierwelt Deutschlands.* C. Fischer Verlag, Jena 51: 1–150; 53: 151–380.

Günther, K. 1931. Bau und Funktion der Mundwerkzeuge bei Crustaceen aus der Familie der Cymothoidae. *Z. Morphol. Ökol.* 23: 1–79.

Gupta, M. 1962. Seasonal variation in the humidity reaction of woodlice *Oniscus asellus* and *Procellio scaber. Proc. Nat. Inst. Sci. India* 29B: 203–206.

Haffner, K. von 1937. Die ursprüngliche und abgeleitete Stellung der Beine bei den Isopoden. *Z. Wiss. Zool.* 149: 513–536.

Hanström, B. 1947. The brain, the sense organs, and the incretory organs of the head in the Crustacea Malacostraca. *Lunds Univ. Arsskr.* 43: 1–44.

Hartenstein, R. 1964. Feeding, digestion, glycogen, and the environmental conditions of the digestive system and carbohydrate metabolism in *Oniscus. J. Insect Physiol.* 10: 611–631.

————. 1968. Nitrogen metabolism in *Oniscus asellus. Amer. Zool.* 8: 507–519.

Hatch, M.H. 1947. The Chelifera and Isopoda of Washington. *Publ. Univ. Washington Biol.* 10(5): 155–274.

Hatchett, S.P. 1947. Biology of the Isopoda of Michigan. *Ecol. Monogr.* 17: 47–79.

Howes, N.H. 1939. Biology and post-embryonic development of *Idotea viridis* from Southeast Essex. *J. Marine Biol. Ass.* 23: 279–310.

Hutchinson, G.E. 1967. *A Treatise on Limnology.* vol. 2. Wiley, New York.

Jans, D. and Ross, K.F.A. 1963. Histological study of peripheral receptors in the thorax of land isopods. *Quart. J. Microscop. Sci.* 104: 337–350.

Jensen, J.P. 1955. Biological observations on the isopod *Sphaeroma. Vidensk. Medd. Dansk Naturh. Foren.* 117: 305–339.

Jones, D. 1968. Functional morphology of the digestive system in *Eurydice. J. Zool. London* 156: 363–376.

Jöns, D. 1965. Biologie und Ökologie von *Ligia oceanica* in der westlichen Ostsee. *Kieler Meeresforsch.* 21: 203–207.

Juchault, P. 1967. La différention sexuelle male chez les isopodes. *Ann. Biol.* (4) 6: 191–212.

Kinne, O. 1954. Eidonomie, Anatomie und Lebenszyklus von *Sphaeroma. Kieler Meeresf.* 10: 100–120.

Koepcke, H.-W. 1948. Das Zeichnungsmuster einiger *Idothea*-Artem. Zool. Jahrb. Abt. All. *Zool.* 61: 413–460.

Kohlmeyer, J. *et al.* 1959. Ernährung der Holzbohrassel *Limnoria tripunctata* und ihre Beziehung zu holzzerstörenden Pilzen. *Z. Angew. Zool.* 46: 457–489.

Kosswig, C. and L. 1940. Die Variabilität bei *Asellus aquaticus. Rev. Fac. Sci. Univ. Istanbul* 5B: 78–132.

Kuenen, D.J. 1959. Excretion and water balance in some land isopods. *Entomol. Exp. Appl.* 2: 287–294.

————. and Nooteboom, H.P. 1963. Olfactory orientation in some land isopods. *Ibid.* 6: 133–142.

Lagarrigue, J.G. 1964. Changements cuticulaires au cours du cycle de mue chez un Isopode terrestre: *Armadillo officinalis. Bull. Soc. Zool. France* 89: 310–316.

Lang, K. 1961. *Microcerberus* with a description of a new species of central Californian coast. *Ark. Zool.* 13: 493–509.

Lattin, G. de 1939. Isopodenaugen. *Zool. Jahrb. Abt. Anat.* 65: 417–468.

Legrand, J.J. 1958. Comportement sexuel et modalités de la fécondation chez l'Oniscoide *Porcellio dilatatus. Compt. Rend. Acad. Sci. Paris* 246: 3120–3122.

Lockwood, A.P.M. and Croghan, P.C. 1957. Chloride regulation of the brackish and fresh water races of *Mesidotea entomon. J. Exp. Biol.* 34: 253–258.

Lueken, W. 1962. Geschlechtsbestimmung bei Landisopoden. *Z. Wiss. Zool.* 166: 251–351.

————. 1963. Zur Spermienspeicherung bei Armadillidien. *Crustaceana* 5: 27–34.

Maercks, H.H. 1930. Sexualbiologische Studien an *Asellus aquaticus. Zool. Jahrb. Abt. Allg. Zool.* 48: 399–508.

Manton, S.M. 1965. The evolution of arthropodan locomotory mechanisms. *J. Linnean Soc. London Zool.* 46: 251–483.

Mathes, I. and Strouhal, H. 1954. Ökologie und Biologie der Ameisenassel *Platyarthrus. Z. Morphol. Ökol.* 43: 82–93.

Matsakis, J. 1955. Contribution à l'étude du développement post-embryonnaire et de la croissance des Oniscoides. *Bull. Soc. Zool. France* 80: 52–65.

Mead, F. 1963. L'existence d'une cavité incubatrice complexe chez l'Isopode terrestre *Helleria. Compt. Rend. Acad. Sci. Paris* 257: 775–777.

———. 1964. Chevauchée nuptiale de longue durée chez l'Isopode terrestre *Helleria brevicornis. Ibid.* 258: 5268–5270.

Menzies, R.J. 1953. Chelifera of the eastern tropical and north temperate Pacific Ocean. *Bull. Mus. Comp. Zool.* 107: 443–496.

———. 1954. The comparative biology of reproduction in the wood-boring isopod crustacean *Limnoria. Bull. Mus. Comp. Zool.* 112: 364–388.

———. 1957. Marine borer family Limnoridae. *Bull. Marine Sci. Gulf Caribbean* 7: 101–200.

———. 1962. Reports of the Lund University Chile Expedition. *Lunds Univ. Årsskr.* (N.F.) 57(11): 1–162.

Menzies, R.J. and Frankenberg, D. 1966. *Handbook on the Common Marine Isopod Crustacea of Georgia*, Univ. Georgia Press, Athens.

Messner, B. 1965. Morphologisch-histologischer Beitrag zur Häutung von *Porcellio scaber* und *Oniscus asellus. Crustaceana* 9: 285–301.

Miller, M.A. and Burbanck, W.D. 1961. *Cyathura polita. Biol. Bull.* 120: 62–84.

Monod, T. 1926. Tanaidacés et Isopodes aquatiques de l'Afrique Occidentale et Septentrionale. *Bull. Soc. Sci. Nat. Maroc* 5: 61–77, 233–247.

Muchmore, W.B. 1957. Some exotic terrestrial isopods from New York state. *J. Washington Acad. Sci.* 47: 78–83.

Nair, S.G. 1956. Embryology of the isopod *Irona. J. Embryol. Exp. Morphol.* 4: 1–33.

Nambu, M. *et al.* 1960. Reducing sugar and glycogen of *Armadillidium vulgare* in relation to moulting. *Annot. Zool. Jap.* 33: 85–89.

Naylor, E. 1955. The diet and feeding mechanism of *Idotea. J. Marine Biol. Ass.* 34: 347–355.

———. 1955. The external morphology and revised taxonomy of the British species of *Idotea. Ibid.* 34: 467–493.

Needham, A.E. 1938. Abdominal appendages in the female and copulatory appendages in the male *Asellus. Quart. J. Microscop. Sci.* 81: 127–149.

Nicholls, A.G. 1931. Studies on *Ligia oceanica. J. Marine Biol. Ass.* 17: 655–706.

Nicolls, G.E. 1943, 1944. The Phreatoicoidea. *Pap. Proc. Roy. Soc. Tasmania* 1942: 1–45; 1943: 1–156.

Nierstrasz, F. et al. 1926. Epicaridea *in:* Grimpe, G. and Wagler, E. *Die Tierwelt der Nord- und Ostsee.* Akad. Verl., Geest & Portig, Leipzig 10: 1–56.

Nierstrasz, J.F. and Stekhoven, J.H.S. 1930. Isopoda. *Ibid.* 10: 57–132.

Pardi, L. 1954. Die Orientierung von *Tylos latreillii. Z. Tierpsychol.* 11: 175–181.

———. 1963. Orientamento astronomico vero in un isopodo marino *Idotea baltica. Monit. Zool. Ital.* 71: 491–495.

Paris, O.H. 1963. Ecology of *Armidillidium vulgare. Ecol. Monogr.* 33: 1–22.

——— and Pitelka, F.A. 1962. Population characteristics of *Armadillidium vulgare. Ecology* 43: 229–248.

Parry, G. 1953. Osmotic and ionic regulation in the isopod crustacean *Ligia oceanica. J. Exp. Biol.* 30: 567–574.

Patané, L. 1940. Sulla structura e le funzioni del marsupio di *Porcellio laevis. Arch. Zool. Ital.* 28: 271–296.

———— 1958. Biologia sessuale di *Porcellio laevis*. *Boll. Soc. Ital. Biol. Sper.* 34: 165–168.

Pennak, R.W. 1953. *Fresh-water Invertebrates of the United States.* Ronald, New York.

Ray, D.L. 1959. Nutritional physiology of *Limnoria*. *Friday Harbor Symp. Marine Biol.* 1:46–61.

———— and Stuntz, D.E. 1959. Marine fungi and *Limnoria* attack. *Science* 129: 93–94.

Reidenbach, J.M. 1965. Effets de l'ablation du complexe nuerosécréteur céphalique chez les femelles du Crustacé Isopode marin *Idotea. Compt. Rend. Acad. Sci. Paris* 261: 4237–4239.

———— 1965. La formation du marsupium et sur l'évolution d'un caractère neutre chez *Idotea. Ibid.* 261: 555–556.

Reinhard, E.G. 1949. The determination and differentiation of sex in the bopyrid *Stegophryxus hyptius. Biol. Bull.* 96: 17–31.

Reinders, P.E. 1933. Die Funktion der Corpora alba bei *Porcellio scaber. Z. Vergl. Physiol.* 20: 291–298.

Remmert, H. 1967. *Sphaeroma hookeri* ein semiterrestrischer Salzwiesenbewohner, *Naturwissenschaften* 54: 253.

Reverberi, G. and Pitotti, M. 1942. Distribuzione delle ossidasi e perossidasi lungo il "cell-lineage" di nova a mosaico. *Publ. Staz. Zool. Napoli* 19: 250–263.

Schmölzer, K. 1965. Ordnung Isopoda. *in* Aguilar, J. *et al.* ed. *Bestimmungsbücher zur Bodenfauna Europas,* Akad. Verl., Berlin 4–5: 1–468.

Schmidt, W.J. 1954. Die Integumentcuticula des Isopods *Idotea linearis. Z. Zellf.* 39: 537–549.

Schultz, G.A. 1969. *How to Know the Marine Isopod Crustaceans.* Brown, Dubuque, Iowa.

Siewing, R. 1951. Besteht eine engere Verwandtschaft swischen Isopoden und Amphipoden? *Zool. Anz.* 147: 166–180.

Silén, L. 1954. The circulatory system of the Isopoda Oniscoidea. *Acta Zool.* 35: 11–70.

Stevenson, J.R. 1961. Absence of chitinase in *Limnoria. Nature* 190: 463.

Stoll, A. 1962. Cycle évolutif de *Paragnathia formica. Cah. Biol. Marine* 3: 401–406.

Strömberg, J.-O. 1965. The embryology of the isopod *Idotea. Ark. Zool.* 17: 421–473.

———— 1968. Embryology of *Limnoria. Ark. Zool.* 20: 91–139.

Suneson, S. 1947. Colour change and chromatophore activities in *Idothea. Lunds Univ. Arsskr.* 43: 1–34.

Swan, E.E. 1956. *Ligia* on the New England coast. *Ecology* 37: 204–206.

Thienemann, A. 1950. Verbreitungsgeschichte der Süsswassertierwelt Europas. *Binnengewässer* 18: 1–809.

Tretzel, E. 1961. Biologie, Ökologie und Brutpflege von *Coelotes terrestris* (Araneae, Agelenidae). *Z. Morphol. Ökol.* 49: 658–745.

Unwin, E.E. 1932. The structure of the respiratory organs of the terrestrial Isopoda. *Pap. Proc. Roy. Soc. Tasmania* 1931: 37–104.

Vandel, A. 1938. Les mutations *alba* et *pallida* du *Trichoniscus elisabethae* et l'origine des formes cavernicoles. *Bull. Biol. France Belgique* 72: 121–146.

———— 1938. Le déterminisone du sexe et de la monogénie chez *Trichoniscus provisorius. Ibid.* 147–186.

———— 1940. Polyploide et distribution géographique. *Ibid.* 74: 94–100.

———— 1941. Monogénie chez les Oniscoides. *Ibid.* 75: 316–363.

———— 1943. L'origine, l'évolution et la classification des *Oniscoidea. Bull. Biol. France Belgique Suppl.* 30: 1–136.

———— 1960, 1962. Isopodes terrestres *in Faune de France.* Lechevalier, Paris, 64: 1–416, 66: 417–932.

Van Name, W.G. 1936. American land and fresh-water isopod Crustacea. *Bull. Amer. Mus. Natur. Hist.* 71: 1–535.

———— 1940, 1942. A supplement to the American isopod Crustacea. *Ibid.* 77: 109–142; 80: 299–329.

Verrier, L. 1932. La forme, de l'habitat et du comportement de quelques crustacés isopodes. *Bull. Biol. France Belgique* 66: 200–231.

Waloff, N. 1941. Humidity reactions of terrestrial isopods. *J. Exp. Biol.* 18: 115–135.

Warburg, M.R. 1965. Water relation and internal body temperature of isopods. *Physiol. Zool.* 38: 99–109.

———— 1967. Behavioral adaptations of terrestrial isopods. *Amer. Zool.* 8: 545–559.

Weygoldt, P. 1960. Die Keimblätterbildung bei *Asellus aquaticus. Z. Wiss. Zool.* 163: 342–354.

———— 1960. Mehrphasige Gastrulationen bei Arthropoden. *Zool. Anz.* 164: 381–385.

Wieser, W. 1963. Tageslänge für das Einsetzen der Fortpflanzungsperiode bei *Porcellio scaber. Z. Naturf.* 18b: 1090–1092.

———— 1965. Die Häutung von *Porcellio scaber. Verh. Deutschen Zool. Ges.* 28: 178–195.

———— 1965. Die Steuerung von Stoffwechselvorgängen bei *Porcellio scaber* durch Temperatur und Licht. *Ibid.* 359–364.

———— 1965. Ernährung und Gesamtstoffwechsel von *Porcellio scaber. Pedobiologia* 5: 304–331.

———— 1965. Blood proteins in an ecological series of isopod and amphipod species. *J. Marine Biol. Ass.* 45: 507–523.

———— 1966. Wachstumskosten von *Porcellio scaber. Helgoländer Wiss. Meeresunters* 14: 326–334.

———— 1966. Copper and the role of isopods in degradation of organic matter. *Science* 153: 67–69.

———— 1967. The copper problem from the isopod's point of view. *Helgoländer Wiss. Meeresunters.* 15: 282–293.

———— 1968. Nutrition and metabolism of copper in isopods. *Amer. Zool.* 8: 496–506.

Wolff, W. 1956. Isopoda from depths exceeding 6000 m. *Galathea Rep.* 2: 85–157.

Zimmer, C. 1927. Isopoda *in* Kükenthal, W. and Krumbach, T. *Handbuch der Zoologie,* de Gruyter, Berlin 3(1): 697–766.

18.

Order Amphipoda

There are 4,600 species of amphipods known. The largest of them are the bottom dwelling *Alicella gigantea,* and the pelagic *Thaumatops loveni,* both of which may reach a length of 14 cm.

Amphipods are peracarids in which the cephalothorax lacks a carapace and the whole head is fused to one, rarely two, thoracomeres. The anterior 3 pleopods are usually swimmerets with many articles, the posterior ones are saltatorial legs with few articles. The thoracopods bear the gills.

Anatomy

The majority of species has the first four pereopods directed anteriorly, the last three posteriorly (Figs. 1-3, 18-1). The trunk is laterally compressed, its coxal plates (epimeres) directed almost ventrally (Fig. 18-2). In the pereon these epimeres are always a plateshaped part of the leg coxa, while the epimeres of the abdomen are real folds of the trunk. This is similar to the arrangement seen in isopods. On the venter epimeres enclose a long channel, which is very deep in many bottom dwelling Gammaridea. This channel encloses the respiratory current, protecting the gills and marsupium within from getting plugged up by mud, sand or detritus (Fig. 18-2). In the pelagic Hyperiidea, however, the channel is very shallow (Fig. 18-15).

Besides the compressed species, there are numerous species with an almost spherical, domed cephalothorax and abdomen, for example, many pelagic Hyperiidea (Figs. 18-15, 18-16). Some have the trunk cylindrical, or lack epimeres (Ingolfiellidae, Caprellidae, Figs. 18-11, 18-12). Some genera even have the trunk dorsoventrally depressed, especially the isopod-like *Pereionotus* and the Cyamidae (Fig. 18-13). Less depressed but not compressed are *Corophium* and *Chelura* (Figs. 18-6, 18-10).

Usually only the first thoracomere is fused to the head, forming the cephalothorax. In the Caprellidea the second thoracomere is close to the head, but usually

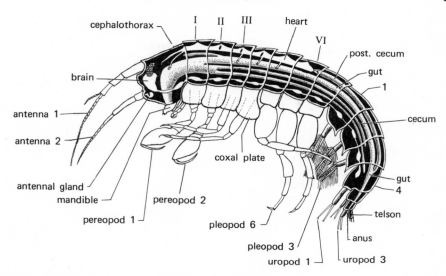

Fig. 18–1. Anatomy of a gammaridean amphipod. The body cavity is shown black, to emphasize the different origins of the thoracic coxal plates and abdominal epimeres. The first 2 pereopods are called gnathopods. The coxal plates of the first 4 pereomeres are large, the following 3 smaller. The mandibles are under the mouth, behind it are 2 pairs of maxillae and one maxilliped. The last three pleopods are termed uropods. In the cephalothorax the anterior aorta and the most anterior lateral arteries can be seen; under the posterior cecum the posterior aorta. Roman numerals indicate pereomeres, Arabic numerals, pleomeres. (After Klövekorn and other authors.)

with a suture showing. The pereomeres are freely moveable, only in some pelagic Hyperiidea where they are stiffly fused to each other. On the abdomen one can separate the three anterior somites (metasome or pleosome) from the smaller posterior (urosome) ones. The metasome bears the swimmerets and in some pelagic species is strongly developed (e.g., *Phronima* Fig. 18-4). In climbing Caprellidea, however, the abdomen is reduced to only a minute vestige (Fig. 18-12). The telson is short often with a posterior median cleft, sometimes divided into two.

The first thoracopod is transformed into a maxilliped. Its endopodite is short. Its basal article is fused with its opposite forming a kind of labium (not homologous to the labium of myriapods and insects). The pereopods are uniramous. They usually hang ventrally, their plane of movement parallel to the midline of the body. The claws of the first four limbs are directed posteriorly, those of the last three anteriorly. This position is a secondary one. Also secondary is the position of muscles and the blood channels within the legs. All are due to a rotation late in development. The first two pereopods are tipped by subchelae and are called gnathopods. True chelae are rare. The claws of the last three pereopods are strongly developed in parasites (Fig. 18-13).

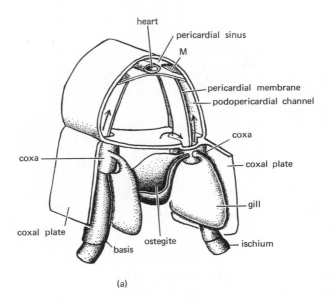

(a)

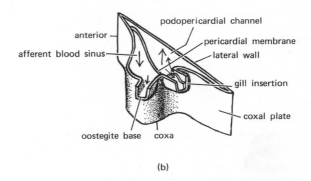

(b)

Fig. 18–2. (a) Cross section of the pereomere of a female gammarid, viewed from the posterior. The left oostegite is omitted from the illustration. (b) Cross section of the basal article of a pereopod to show its anterior and posterior widening into the coxal plate. Arrows indicate blood currents. From the domed middle part of the coxa the gills originate posteriorly, the oostegite toward the inside. The gill stalk contains a blood channel leading to the lateral pericardial sinus (arrows). M, opening of the podopericardial channel into the dorsal pericardial sinus. The pericardial sinus is bordered ventrally by the alar muscles. The pericardial membrane passes ventrally toward the leg attachment and enters the space within the leg, dividing it into an afferent and an efferent blood sinus. The blood supply to the leg comes from the ventral sinus (curved arrow at right). (After sections by Klövekorn and others.)

The pleopods are biramous. The anterior three pairs have 6–30 articles on each branch, bearing long dense setal fringes which function as swimmerets. The three posterior pairs, all considered uropods, have only short setae and have cylindrical branches with only one or two articles. The uropods function for pushing or jumping (Talitridae) and may also function in swimming (e.g., Hyperiidea and *Gammarus*), in mating and removal of exuviae (Talitridae) (Fig. 18-1). Sometimes one of the uropod pairs is vestigial. Caprellidea and Ingolfiellidea have vestigial pleopods or have lost them (Figs. 18-11, 18-12).

The surface of the integument is usually smooth, rarely are there platelets, spines or dorsal keels (as in *Gammarellus*). The exoskeleton is mineralized. While species living on the bottom are usually colored inconspicuously, others may be red, green or blue. The bright colors are contained in chromatophores, in pigments of the epidermis or the hemolymph. Many pelagic Hyperiidea are transparent.

The olive-brown coloration of *Gammarus pulex* is due to carotinoids. Nearly colorless individuals from a mine were cultured for 12 generations in permanent darkness or in the illumination normal for a colorless population from a mine. Coloration appeared independently of illumination where the food contained carotinoids. Presumably the carotinoids were unavailable to the amphipods in the mine. Crossings of wild brown and red inbred cultures from a mine in the Harz mountains indicate that color is controlled by three alleles. (R^2 dominant, brown, r+, olive, and r, brick red). There is complete dominance in the order given. At the same time the r has a strong, R^2 weaker, masculinization influence, while r+ does not influence sex. Sex chromosomes have not been found in the 11 species of amphipods examined.

NERVOUS SYSTEM. Ganglia of the three pairs of mouthparts and maxillipeds are fused into a subesophageal ganglion (Vol. 2, Fig. 4-10, p. 48) while the ganglia of the pereon and the three anterior pleomeres are separated from each other by connectives (Fig. 18-1). The last three ganglia have moved anteriorly and have fused in the fourth pleomere. Further fusions of ganglia are found in the Hyperiidea.

SENSE ORGANS. The compound eyes are never stalked and neither their cuticle nor hypodermis is divided into facets. Some pelagic amphipods (*Phronima*) have the eyes on each side divided into two independent parts, or three parts in the bottom dwelling Ampeliscidae. *Ampelisca* has a uniform lens covering each of the two pairs of anterior eyes and the retinulae converge toward the lens instead of diverging as in other compound eyes (Fig. 18-3). Thus a camera eye has evolved, which has the advantage, compared to the compound eye, of being more sensitive to light. The light coming in from the sides is not lost in the pigment surrounding the crystalline cone, but penetrates through the crystalline cone to the rhabdomeres. These species live at intermediate ocean depths. In subterranean *Niphargus* and Ingolfiellidea the eyes have been lost.

Esthetascs are found on the first antennae, sensory hairs on both pairs of antennae, and also on the legs and mouthparts. In a number of Gammaridea and Hyperiidea one pair of statocysts with one to five statoliths has been found dorsally in the head, innervated by the supraesophageal ganglion.

DIGESTIVE SYSTEM. The foregut forms a chewing stomach. From the anterior end of the midgut extend two or four posteriorly directed long tube-shaped ceca and a short median, anteriorly directed cecum (Gammaridea) or paired ceca (Caprellidae), lacking in the Hyperiidea (Fig. 18-1). In addition, at least in Gammaridea, anteriorly directed paired ceca from the posterior end of the midgut reach far anterior. Since these are enclosed by the posterior aorta (Fig. 18-1) they are believed to be excretory in function.

EXCRETORY SYSTEM. Besides the digestive ceca just mentioned there are antennal glands. The duct of the antennal gland is noticeably longer in freshwater

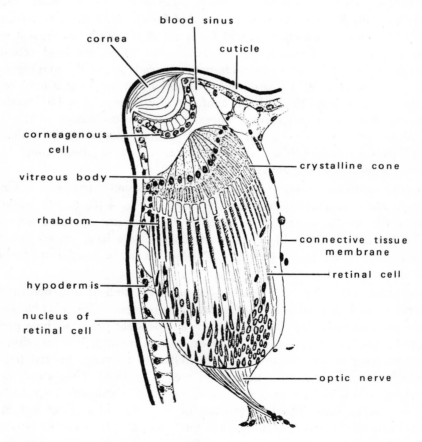

Fig. 18–3. Longitudinal section through a ventral eye of the amphipod *Ampelisca brevicornis*, 0.2 mm wide. (After Strauss from Plate.)

species than in their marine relatives. Nephrocytes are found in the blood sinus that leads from the pereopods to the pericardial sinus.

CIRCULATORY SYSTEM.   The heart, corresponding to the pereopod position of the gills, is in the pereon extending in *Gammarus* and *Phronima* from the first to the sixth pereomere (Fig. 18-1). The pericardial sinus is bordered ventrally by a pericardial septum containing transverse intersegmental alar muscles, the contraction of which contributes to the diastole of the heart. In *Gammarus* this septum extends to the telson, the alar muscles to the third pleomere. The lateral borders of the septum extend ventrally in each pereomere, forming, with the lateral body wall, a podopericardial channel (Fig. 18-2). Its membrane extends into each pereopod, dividing its lumen into two adjoining blood channels. The afferent sinus obtains its blood from the ventral sinus. The blood reaches the leg tip, where the membrane ends, and flows back in the adjoining channel, which leads dorsally to the pericardial sinus. The circulation of the gills attached to the leg bases is similar. The blood reaching the pericardial sinus from legs and gills passes ostia and is pumped back into the body by an anterior and a posterior aorta and paired lateral arteries. There is no subneural artery in amphipods.

Gammarus, Talitrus and *Phronima* have three pairs of ostia, one each in the second to fourth pereomeres. *Phronima* has two pairs of arteries that originate below the two posterior ostia and serve the stomach muscles only. *Gammarus* has a third pair of arteries that like the other two branch out on the midgut diverticula (Fig. 18-1). At each side of the origin of the anterior aorta a lateral artery branches off. Judging by comparison with other Peracarida (Figs. 17-10, 18-1), these must be lateral arteries rather than aortic branches. In *Gammarus* the end of the heart has two lateral openings into a wide posterior aorta which surrounds the posterior cecum. This might be the two hindmost arteries which have fused.

RESPIRATORY ORGANS.   The respiratory organs are leaf-shaped epipodites of at most six pereopods, of the second to seventh pereomeres (Fig. 18-9). But many genera have lost some; for example some Gammaridea have only five pairs (on pereopods 2–6), *Phronima* only three pairs (on pereopods 4–6), and *Cyamus* only two pairs (Fig. 18-13).   The location of the epipodites, on the inside of the leg or behind it rather than lateral, results from the shift of the coxa forming an epimere (Figs, 18-1, 18-2, 18-9).

REPRODUCTIVE SYSTEM.   The paired testes are cylindrical tubes that extend from the third to sixth pereomeres, limited to the last thoracomeres only in the caprellids. At the posterior end the testes continue as vasa deferentia, with glandular walls. The vasa deferentia open through short, muscular, ectodermal ejaculatory ducts at the tips of two long penis papillae. The papillae are on the midventer at the end of the thorax. The anterior pleopods do not function as gonopods except possibly in *Chiltonia* and a few other freshwater species.

The ovaries are tubes lying in the thorax. Their oviducts end as ectodermal vaginae which open in the sixth thoracomeres (fifth pereomere) into the marsupium. The marsupium is formed by spoonshaped oval or circular plates, the oostegites, attached to the coxae (Fig. 18-2). Gammaridea and Hyperiidea have four pairs of such oostegites (2nd–5th pereopods), Caprellidea only two pairs on the fourth or fifth thoracomeres. The Ingolfiellidea have very small oostegites.

The first minute oostegite primordia appear in *Gammarus* long before maturity. *Gammarus pulex*, 4.5 mm long, has one pair at first, the additional three pairs appearing after the next molt. The length of the oostegites increases at each molt, long setae appearing around the border at the parturial molt. After the young have left and the female has molted again, her oostegites are fully developed with setae. If this is the hibernating winter instar, the oostegites lack the setae.

The formation of oostegites and their setae depends on an ovarian hormone. Male differentiation is controlled by androgenic glands attached to the vas deferens. If the androgenic gland is transplanted into a female, the ovaries and secondary sexual characters transform toward maleness.

Reproduction

Sexes usually differ in size; in most gammarideans the male is larger, but in some families, e.g., most Lysianassidae, the Haustoriidae, some Gammaridae (*Crangonyx*), and many Phoxocephalidae, the male is smaller. In males the antennae are usually longer and the subchelae of the first two pereopods stronger.

The male usually rides on the back of the female for several days before the parturial molt, holding the anterior border of the first and posterior border of the fifth pereon tergite with his first and second pereopods. The observations are mostly on gammarideans, but differ in the terrestrial talitrids. *Gammarus pulex* thus waits 5–7 days during summer for the female to molt. After her parturial molt, the male turns the female venter up and by bending his abdomen, pushes the three anterior folded pleopods 8–14 times at half-second intervals between the posterior oostegites of the female marsupium. By these movements the sperm, issuing in long strings out of the male genital papillae, is transported to the female gonopores. About 1½–4 hr after molting, still during mating, the female deposits eggs, their number and size increasing with the age of the mother. The embryo arches concave ventrally, unlike isopod embryos.

Spawning in *Gammarus duebenii* depends on sperm transmission. If the male papillae are closed by searing with a hot needle, only half the mated (but not inseminated) females produce eggs, and the eggs are fewer in number.

Fertilization takes place in the marsupium. Movement of the pleopods and extension of the oostegites ventilate the marsupium. There is no nutrient fluid. Eggs of *G. duebenii* can develop outside the marsupium. In some species the young leave the marsupium a few days after hatching. It seems that many amphipods live only about a year.

The female of *Gammarus duebenii* molts 21 times and produces six or seven broods, totaling 220 eggs, during the life span of 13–14 months. The male molts 20 times during a life span of 14–16 months; in the laboratory they may live up to 3 years.

The young hatching out of the eggs resemble the adults. But recently hatched young of pelagic species do not have the specializations of the adult such as the swollen trunk or extreme elongation.

Young of parasitic hyperiids on coelenterates or tunicates have larval characteristics, the pleon having at most buds of the appendages. There are up to four such juvenile stages.

Relationships

The relationship of amphipods to other groups was discussed in Chapter 15.

Habits

Most species of amphipods live in the ocean between the tide zones and abyssal depths (10,000 m), either bottom-dwelling or pelagic. The Gammaridea are widely distributed bottom-dwellers in freshwater. Strong rheotaxis prevents them from being carried away by currents.

Lake Baikal contains almost 300 species, many of them brightly colored blue, red or yellow, up to 9 cm long, and including planktonic species. The plankter, *Macrohectopus branickii*, 33 mm long, is transparent. Apparently this ancient deep lake is a center of evolution and adaptive radiation within the family Gammaridae.

Many marine species tolerate brackish water and some are very euryhaline. The north American *Corophium lacustre* can live in salt concentrations of 6–0.37‰ and *Gammarus duebenii* can live even in freshwater that contains as little as 24 ppm NaCl; it cannot live in the interior of continents where the chloride ion is naturally absent or very low.

Few interstitial coastal amphipods are known. They have the usual elongated shape of interstitial animals, have short epimeres and most are barely 2.5 mm long (e.g., Ingolfiellidea and *Bogidiella*). The genus *Niphargus* (Fig. 18-9) widespread in groundwater of Central Europe, seems to be derived from blind interstitial ocean inhabitants. Many amphipods live in caves. A few genera are found along the upper tide line in wet sand (*Orchestia, Talitrus*).

A very few species (*Talitroides*) that live in moist soil away from the ocean are found in hot houses and now free-living throughout Florida and the Gulf coast states and in parts of California (e.g., Golden Gate Park in San Francisco). These Talitridae have few morphological adaptations to terrestrial life but have important physiological and behavioral adaptations. Loss of sexual dimorphism accompanies loss of the male carrying habit in copulation, so that the male does not have the large subchelate 2nd gnathopod, does not have the enlarged second antennal peduncle and has normal sized 7th pereopods. But for holding apart the female pleon plates during copulation, males have specialized 1st and 2nd uropods. The coxal gills, especially those of the second and sixth pereopods, are greatly enlarged and convoluted. Posterior processes that develop on the second to fourth lateral plates prevent the plates from separating when the body is strongly flexed. Enlargement of the first uropod gives greater leverage in jumping. Special spines on the uropod rami assist the removal of exuviae in air. The antennal gland cone has been lost and it is believed that excretion is through the digestive tract. Also as an adaptation to chewing dry food particles, the palpi and most fine setae of the mouthparts are lost—analogous to the trend in insects. As in terrestrial isopods, the first antennae have become reduced and the 2nd has become the principal chemoreceptor on land. There are hooked brood plate setae (in terrestrial *Talitroides*), presumably to hold eggs firmly in place when the female is jumping and landing hard on the ground.*

Within the tide zone the density of amphipods can be very large. *Corophium volutator* (Fig. 18-6) has been found in densities of 5,000–20,000/m², rarely to 40,000, in northern European shallows. Their abundance makes scuds of great importance as fish food.

Pelagic forms, where habits are otherwise barely known, make the daily vertical migration typical of many planktonic crustaceans. The pelagic forms have adaptations that lower the specific weight of the animals and increase their ability to float: swollen trunk, cephalothorax, and pereon enclosing tissues with a high water content (Figs. 18-4, 18-14, 18-16). Some Scinidae, especially *Rhabdosoma,* have increased resistance to sinking by extreme elongation of the trunk and the rostrum as in some pelagic Cladocera (*Bythotrephes*). Many marine bottom-dwellers can swim actively at night in open water (e.g., *Pontoporeia, Bathyporeia*).

The compressed shape facilitates passing through dense aquatic vegetation. Adaptations for concealment in vegetation have been developed by the caprellids. Their posterior two or three pereopods, used to hold onto plants or bryozoan colonies, insert on shortened metameres. The anterior somites are elongated and the other developed legs separated by some distance from those used to hold on. The first two leg pairs at the tip of the pereon are raptorial (compare Fig. 18-12 with that of the similarly adapted isopod *Astacilla*, Fig. 17-18). Parasites have the legshape specialized for external parasitism. In the Cyamidae they have large subchela on the five pairs of fully developed pereopods (Fig. 18-13). Interstitial amphipods are cylindrical (Fig. 18-11).

---

* Information on these terrestrial adaptations was kindly supplied by E.L. Bousfield in a letter.

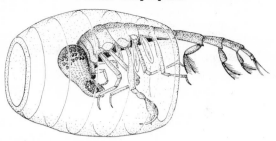

Fig. 18–4. Female *Phronima sedentaria* swimming within the tunic of a tunicate. The long abdomen with its large swimmerets is extended out of the tunic. The 5th pereopod is long and subchelate. *Phronima* is 3 cm long. (After Woltereck from Hesse-Doflein.)

PROTECTION. Many species are protectively colored, some have many color phases. Only *Caprella* (Fig. 18-12) is known to change color to match its background. The pigment of the hemolymph and the hypodermis after a molt resembles that of the substrate, green or red algae. While *Hyperia galba* is swimming the red-brown and black pigments of its chromatophores disperse, producing a red-brown appearance. The pigments contract as soon as the swimming animal settles down. As it usually attaches to transparent pelagic coelenterates the change from colored to transparent is an adaptation to the substrate.

The bottom dwelling amphipods seek protection of the substrate by crawling among fallen leaves, animal colonies, under stones, or into the ground, or they build tubes for themselves. Burrowing is always done head first, but differs in different genera. *Talitrus* pulls the anterior body with the first, third, and fourth pereopods into the sand. At the same time sand grains are torn loose which are pushed between the outspread posterior three pairs of pereopods to the dorsal surface of the uropods. Then the ventrally bent abdomen stretches and the sand is thrown back to 15 cm. The fifth to seventh pereopods, held laterally, support the body while burrowing and also push the animal head first into the burrow. *Haustorius* digs in rapidly using the 2nd antennae as a plow, the first and second pereopods as picks, the wide third and fourth pereopods as shovels to push the sand posteriorly, the pleopods to throw the sand and the fifth to seventh pereopods to push itself head first into the sand. Within the sand *Haustorius* seems to swim, metachronally beating the pleopods in the wide tube made by the large anterior coxal plates and the large proximal articles of the fifth to seventh pereopods. During these movements they throw the sand back and at the same time move the body anteriorly. Gills and marsupium are protected from the sand and space remains for the respiratory current in the high burrow lumen. *Neohela monstrosa* shovels mud with the stout gnathopods held together in front of it. By rapidly bending the thoracomeres dorsally it flings mud forward. A few such motions produce a depression 2 cm deep from which a horizontal tunnel is excavated. The animal presses its pleon toward the end of the tunnel then creeps forward shoveling material ahead of it to the opening of the burrow (Fig. 18-5). The material is left on the bottom or flung up by violent jerks of the body. It tunnels as much as 10 cm a day. The tunnel may have two openings but is only a shelter from which the animal emerges at night to swim slowly, soon to make a new burrow.

Some species make tubes in the sand, their inner walls solidified by secretions. Tubes of *Corophium volutator* are U-shaped (Fig. 18-6) extending 4–8 cm into the substrate. In wintertime one arm may reach as deep as 20 cm down. Species of many genera use a secretion from one celled glands that lie in the middle articles of the third and fourth pereopods and open at the tip of the leg. Holding the tip to an algal filament, the amphipod presses out liquid that solidifies into a thread and sticks to the alga. Then the leg tip moves and attaches the silk to another piece of alga. The web, strung among algae, may be a tube open at both ends, often with fecal material and detritus glued in (e.g., *Microdeutopus*). Other species burrow into the sand with the pereopods, then turn in all directions, up, down, to the sides, moving the third and fourth pereopods between the surrounding sand grains, spinning them together (e.g., *Microprotopus*). The tubes may be as long as the body or twice as long, to about 5 cm. *Ampelisca,* after digging itself in, builds a thin-walled pocket in sand or mud using secretions from integumental glands. From the pocket the cephalothorax with the antennae is extended. *Leptocheirus* builds with secretions of unicellular skin glands that open on the third and fourth pereopods and on the coxal plate, constructing a domed ceiling from mud and algal particles above vegetation or a *Cordylophora* stem.

LOCOMOTION.   Amphipods swim by metachronal beats of the three anterior pairs of pleopods. During their backward movement, the dense setal fringes of the pleopods spread out, increasing their surface. During forward movement, the water pressure pushes the setae backwards and the pleopod surface is reduced. From time to time the speed is jerkily increased by backward beats of the pleon.

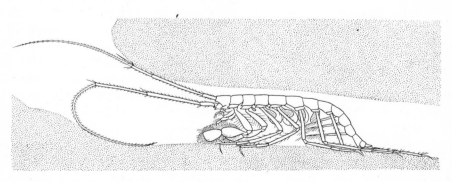

Fig. 18–5. *Neohela monstrosa* shovelling mud out of its burrow; total length 25 mm. (From Enequist.)

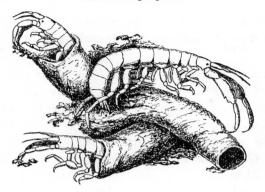

Fig. 18–6. **Three *Corophium* and their tubes; two individuals shown using their antennae to gather detritus for tube building; animals 1 cm long. (After Giesbrecht.)**

Not only pelagic species swim but also many bottom dwelling species swim at night, some haustoriids with the venter turned up, but most Pontoporeiinae swim with dorsum up. The *Gammarus* species of streams and ponds move only short distances by swimming, dorsum up, trying to keep the third and fourth pereopods in contact with the substrate, as they walk along. The pleon is used only to push off from the bottom and is then kept straight.

Different methods are used for running. Fresh water *Gammarus* walks with its side down, pushing forward with the fifth to seventh pereopods of that side while their partners on the other side are held tightly to the side, and all other pereopods against the venter. Once in a while the posterior three pairs of pleopods (uropods) help, hooking to the bottom and pushing the body forward with a jerk when the pleon beats backward. *Synurella* and *Corophium* walk dorsum up like most other crustaceans. The third and fourth pereopods pull the trunk forward. In the water the pleopods help by pushing the body forward. In areas exposed at low tide *Corophium* hooks the tips of the long 2nd antennae into the substrate and flexes them, pulling the body forward. The terrestrial *Talitrus* pulls with the third and fourth pereopods, and pushes with the sixth and seventh or uses the sixth and seventh also as lateral supports to prevent falling over. The fifth pereopods are used only for turning and backing up. When frightened, *Talitrus* can jump 10–30 cm, several cm high, in quick succession, by beating the abdomen back.

Caprella crawls like a geometrid caterpillar. It holds to an alga with both anterior pereopod pairs and bends the trunk dorsally. The posterior three pereopods reach forward and hold the algal filament. The body then stretches and the anterior reaches forward.

Bottom-dwelling amphipods of a few freshwater species have been observed to move temporarily away from their resting places. In central Europe *Niphargus aquilex* leaves the underwater springs it inhabits and searches for food in the current outside. Nets outside the springs caught numerous *Niphargus* in summer after sundown and (fewer) shortly before sunrise. By supplying food dyed with Congo red it was demonstrated that the animals return to the spaces between particles.

In the same area it was shown that in *Gammarus pulex*, especially large individuals, but not brood carrying females, drift downstream. It was estimated that 625,000 individuals a year moved down the brook to a place 600 m from its source. The maximum drift is in June, minimum in winter. During reproductive periods most of the emigration is by males. The populations were sampled by electric shocking. *Gammarus* marked with P[32] also move upstream for short periods so that in 24 hr they averaged a drift of only 2 m down. Most activity was after sundown and before sunrise, but return was often delayed.

Spread of distribution was observed in *Gammarus tigrinus* inhabiting brackish water on the North American Atlantic coast. A thousand individuals were transplanted into a branch of the Weser River polluted by potassium chloride (with average salt content 2.1 g Cl per liter, and maximum 6.4 g). Within 6 years the population spread along the Weser 300 km to Bremen.

SENSES.   Little is known about the mechanical senses. Studying the spontaneous activity of *Synchelidium* (Oedicerotidae) in the tidal zone it was noticed that the amphipod responded to differences of as little as 0.01 atm pressure by greater activity but receptors could not be found. On the second article of the first antenna *Caprella* has a typical chordotonal organ with a single scolopidium, but its biological function has not been studied. Well-developed statocysts, with statoliths 0.09 mm in diameter, give *Gammarus locusta* independence from the light coming from above. The species continues to swim normally if illuminated from below, while *G. pulex*, which has only a minute statocyst, with a statolith 0.02 mm in diameter turns over. Normally, however, *G. pulex* swims only short distances just above the bottom.

The eyes are much larger in pelagic and bathypelagic species than in species that live on the bottom (Figs. 18-4, 18-15) but their function has not been studied.

The role of vision in orientation has been studied in *Talitrus saltator*, which lives along the upper tide line. *Talitrus* and the related *Orchestia* and *Talorchestia* return to their narrow habitat of wet sand, even when blown inland by wind or washed out by a wave or if the sand dries up. The ability to return is lost if eyes are covered with black lacquer. If many *Talitrus* are placed in a concave vessel far from the coast they move in the direction that would have led them to the water in their original habitat, all gathering on one side. Individuals from the south side of a bay move in the opposite direction to those collected from the bay's north shore. Young hatched and reared in the laboratory retain the direction of movement of the parents, even though isolated immediately after birth. Each population thus is genetically adapted to a particular location.

Sun, moon and distribution of the polarized light in the sky serve as a compass. To prove orientation by sun, similar experiments were made with the wolf spider *Arctosa* (see Vol. 2, p. 166) with similar results. *Talitrus*, like *Arctosa*, can compensate for the movement of the sun during the course of the day. To reach the tide line early in the morning the animal takes a quite different angle to the sun than it would in the afternoon.

If *Talitrus* is transplanted to another longitude, from Italy by plane to Argentina, it will move left from the sun at 38° 30′ at the same time it moved left from the sun at 37° in Pisa, Italy, the direction that would have led to the coast at home. In Pisa it is 15 h (3 pm) in Argentina 11 am, the angle thus guiding the animal in a different direction. The compensation to the angle of the sun appears thus independent from the sensory perception, but is an internal clock. The internal clock can be influenced. If *Talitrus* is kept for 18 hr at an elevated temperature, 35–37°C, the clock is advanced and *Talitrus* orients as would have been correct several hours later. If *Talitrus* is kept cold, at 0–3°C for 5 hr, it takes an angle to the sun which would have been correct earlier; the internal clock has been retarded. The internal clock can also be changed by an artificial day and night change with a shift of 6 hr in a controlled light room. Astronomical orientation is disturbed when the sun passes the zenith or less than 7–8° to the zenith. Subequatorial populations of *Talorchestia mertensii* are disoriented at noon and avoid exposure during this time, or if disturbed they promptly dig in the sand.

Orientation by the moon has been demonstrated by watching the reaction to the reflection of moonlight in a mirror, the light now coming from the opposite direction. *Talitrus* displays an endogenous lunar periodicity, that, in contrast to the solar period, functions only at night following the lunar day which is ¾ hour longer. Animals which were brought into a dark room more than an hour before sunset orient less exactly relative to the moon than animals that have not been disturbed.

FEEDING. Some amphipods feed on coarse food, others are microphagous.

The predacious amphipods are mainly pelagic Hyperiidea. In aquaria at least *Parathemisto gracilipes* has been observed to catch living *Artemia*. The prey is captured with the posterior pereopods, the distal articles of which fold against the previous one. The third and fourth pereopods push the prey toward the first two thoracopods which tear pieces off and take them to the mouth. It took 7–15 min to consume an *Artemia*. *Parathemisto* almost 5 mm long overpowered fairy shrimp 7 mm long. *Hyperia galba* (Fig. 18-16) has been found with quantities of nematocysts in the stomach, and is therefore assumed to be parasitic on the jellyfish on which it rides. Phronimids are believed to take prey with their chelate fifth pereopods, on the tips of which one-celled glands open (Fig. 18-4), but there are no direct observations. Many predacious caprellids lie in wait, like preying mantids, capturing copepods, crustacean larvae, small amphipods and worms with the subchela of the first pereopods. The long setae of the second antennae, and more rarely the second subchela, bend over the catch, preventing escape of the prey. Some Caprellidae feed mainly on sessile diatoms. Lysianassidae feed on carrion and handicapped animals. Fishermen know that *Orchomonella* and *Scopelocheirus* in huge numbers attack fish kept in nets too long. Small fish carrion and plants washed ashore serve as food for *Orchestia* and *Talitrus*. *Talitrus* will also feed on living animals. *Gammarus locusta*, *G. duebenii*, and *G. chevreuxi* prefer animal food, but for their development fresh plant food is essential. Central European freshwater *Gammarus* and *Synurella* feed mainly on phanerogam leaves, even terrestrial ones that have fallen into the water. But they also take carrion, and may accumulate in large numbers on dead fish. *Gammarus tigrinus* feeds on algae, phanerogams, decaying vegetation, oligochaetes, cladocerans, and undecomposed carrion. Very hungry individuals may attack fishes up to 6 cm long but are shaken off. But in large numbers they may macerate a fish of this size if the fish is in a net, hampered in its movements. A *Daphnia* is grasped with the subchelae of anterior pereopods, and both pairs of antennae touch the prey which is bitten dorsally and killed, then turned so that the mouthparts move in between the carapace valves. A hungry individual may consume 3 *Daphnia*, one after another. A *Tubifex* is usually shared by several *Gammarus tigrinus* individuals.

North American freshwater scuds are considered omnivorous. *Chelura* (Fig. 18-10) feeds on wood: its gut is filled by wood particles although it also feeds on the feces of *Limnoria*. The feeding has not been described in detail in amphipods and almost nothing is known about the functioning of the generically different subchela.

There are four groups of microphagous amphipods:

1. Bathyporeia and other narrow-bodied haustoriids (Pontoporeiinae) are "sand-lickers," cleaning off sand particles.

2. Filter feeders may at times take algal filaments or larger detritus. There are two ways of filtering. In one method, the anterior pereopods are held under the pereon so that their long setae cross in the respiratory current flowing from head to the pleopods. *Cheirocratus* and *Dikerogammarus villosus*, living in Balaton Lake, Hungary, have long setae on the penultimate articles before the last of both the anterior pereopod pairs, and to these setae the filtrate clings. The maxilliped palps take it from the setae to the mouth. At the same time in *Dikerogammarus* floating detritus sticks to the setae of the third uropods which are then cleaned by the anterior pereopods and the detritus is brought to the maxilliped endopodites. The more complicated filter of *Leptocheirus* consists mainly of the merus of the second pereopods. The ventral side of the merus carries a row of long, densely feathered setae. The filtrate clinging to the setae is combed out by the shorter setae of the distal particles of the same leg and in the female is carried to the mouth by the pereopods and maxilliped.

In the other method, as in *Mysis,* the second maxilla serves as the filter, producing a current toward the mouth in *Haustorius* and *Ampelisca. Ampelisca* stirs up material from the bottom with the second antennae and sucks it up with the second maxillae. *Haploops tubicola* rests on top of its tube (Fig. 18-7). The antennae beat through the water with strong movements and the whirled up material is caught by the antennae or sucked to the mouthparts by the pleopod current. The cephalothorax faces away from the water currents, the antennae turn toward it.

3. *Dulichia* extends its densely setose long first and second antennae into the water where floating detritus and plankters settle. *Dulichia* uses its short posterior pereopods to climb on an algal filament and holds on in the same way as *Caprella. Melphidippella macra* (Melphidippidae, Gammaridea) can swim rapidly for short stretches, but usually stands still, its dorsum turned down resting on the posterior pereopods (Fig. 18-8a). Falling detritus clings to the rows of setae on the second antennae, third and fourth pereopods (Fig. 18-8b). While the appendages bend quickly they are combed out by the fifth and sixth articles of the first and second pereopods, transferring the detritus to the mouthparts (Enequist, 1949).

4. *Corophium* (Fig. 18-6) sweeps up the bottom. Extending the cephalothorax out of the opening of its tube, *C. volutator* uses the two antennae held next to each other to pull the thin surface layer of the surrounding mud into its tube. A fine groove remains as evidence, and as the activity is continued in all directions, the opening of the tube becomes surrounded by a rosette of grooves as well as by a wall of accumulated material. The prey, mainly the diatoms that cover the

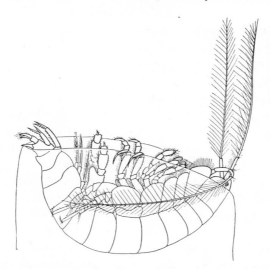

Fig. 18–7. *Haploops tubicola* **in its tube in feeding position; total length 10 mm. The tube is drawn as though transparent. (From Enequist.)**

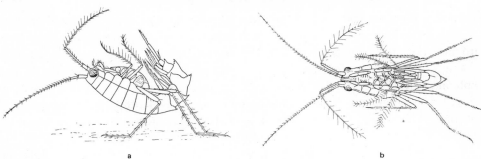

Fig. 18–8. *Melphidippella macra*: (a) from side; (b) from above; total length 6 mm. (From Enequist.)

mud in a thick layer, are swept up by the respiratory current and filtered by setae of both gnathopod pairs.

Parasites include the Cyamidae (Fig. 18-13), which feed on the soft skin of whales. Some Hyperiidae and *Leucothoe,* found in the gills of ascidians or water channels of sponges, are perhaps parasitic although nothing is known about their feeding habits. Lafystiidae and Laphystiopsidae, are exclusively external fish parasites, as are *Opisa* and *Trichizostoma* (Lysianassidae). Anamixidae and Colomastigidae include internal parasites of sponges and tunicates. (All these families belong in the Suborder Gammaridea.)

DIGESTION.   In *Orchestia* and *Marinogammarus* digestion starts in the cardiac stomach as in other malacostracans. Enzymes originate in the ventral midgut ceca. *Marinogammarus,* a vegetarian, and *Orchestia,* which tears off pieces from dead plants and animal carrion, can digest most carbohydrates. *Orchestia* has a weak cellulase that permits it to feed briefly even on filter paper, and strong proteases and lipases. Cells of the midgut ceca secrete and absorb, and the gut also absorbs digested food.

OSMOREGULATION.   Osmoregulation is of particular importance to amphipods as many live in littoral waters and some move into brackish water. Many, mainly Gammaridae and Talitridae, live in freshwater. *Gammarus duebenii* normally lives in brackish water of 3–12‰ but is also found in Scandinavia in a pool of 30‰ salinity and in freshwater in Canada, Ireland and England. In individuals from freshwater and brackish water, differences in osmoregulative competence have been reported, but no differences in morphology have been found. Usually *G. duebenii* lives in shallow water at depths of 5–100 cm and is exposed to changes of salinity during high evaporation or rain. Experiments show that the ability to adapt to these changes decreases at higher temperatures (about 19°C) and with increasing age of the animal. Adults adapted to salinity of 30‰ at 16°C, then suddenly changed to freshwater, die after 1–2½ days, unless moved back into brackish water. Young individuals just leaving the mother's marsupium are more resistant to changes in salinity. Adapted, as is their mother, to 2‰ salinity at 16°C, they die after an average of 105 days if changed to freshwater at 9°C, after 80 days at 19°C. Adapted to 30‰ salinity, they die after 60 days in freshwater at 9°C, after 40 days at 19°C. If returned to brackish water, their development is normal. The lethality of freshwater is not sudden but a long-term effect.

Osmoregulation has been examined quantitatively in only a few species. The loss of NaCl in water of lower salinity than the blood is caused by loss through urine and the body surface. Experiments with *Marinogammarus finmarchicus* have shown that in this species the integument has greater permeability than in the brackish water inhabitants *Gammarus tigrinus* and *G. zaddachi*. *Gammarus duebenii* adapts by reduction of body permeability, production of hypotonic urine, and uptake of NaCl through the body surface. On being changed to freshwater the NaCl loss is reduced and this brackish water species temporarily accomplishes a balance.

In *Gammarus pulex* and *G. lacustris* inhabiting fresh water, the lowest external concentration in which sodium balance was maintained was 0.06 mM/l NaCl. Restricting the loss of NaCl at such low concentrations (below 0.2 mM/l) is accomplished by reducing the NaCl concentration of the blood (lowering the diffusion rate). At the same time there is increased uptake of NaCl at the body surface and its resorption by the nephridium; the urine becomes diluted. The superior adaptation of *Gammarus pulex* to freshwater compared with the brackish water inhabitant, *G. duebenii*, results from its greater ability to take up Na at its body surface.

The physiological optima found in the laboratory (10–15‰ salt, 4–16°C, as expressed by greatest life span and highest rate of reproduction) do not necessarily correspond to conditions in the natural habitats of *G. duebenii*. The species has been found at 5–8‰ and even in freshwater. Probably biological factors such as competition with *Gammarus zaddachi* are responsible, and *G. duebenii* depends on its physiological adaptability to survive where there is no competition. In one place where pumps provided only an irregular supply of ocean water into freshwater *G. duebenii* was the dominant animal; 3–4 months after better pumps were installed, *G. duebenii* was replaced by *G. zaddachi*, previously rare.

CIRCULATION. The heart beat rate increases suddenly with rise of temperature. *Gammarus duebenii* has a heart beat rate of 112 times/min at 7°C and 10‰ salinity; with a rise in temperature to 9.5°C within a minute, the rate increases to 144 times/min. But there are also slow starting, long lasting adaptations that make the heart beat independent from the constant temperature as in winter. The heart of *Gammarus pulex* kept at 5°C in February beats faster when the water temperature drops to 2°C than that of individuals previously adapted to 20°C. With increased age of an individual, heart beat rate decreases.

RESPIRATION. Respiratory currents are produced by the movement of pleopods even if the abdomen is flexed. The pleopod movement is toward the posterior and in *Gammarus pulex* there are 12–140 beats/min at 14°C. The water current flows from anterior to posterior along the venter of the trunk between the gills. The channel, which is very high in many gammarids, is formed by the venter of the body as ceiling and the coxal plates of the pereopods and epimeres of the abdomen as walls. The marsupium is also flushed by the current.

Classification

SUBORDER GAMMARIDEA

The cephalothorax is made up of the head and first thoracomere. The eyes cover only a small part of the head, which is only rarely rounded. The trunk is laterally compressed, the abdomen well developed. The maxillipeds have seven articles, their endites almost always separated. The pereon has coxal plates, the anterior four pairs usually large. To this group belong the majority of the species, including the most primitive scuds. They are found in oceanic and freshwater and some moist terrestrial habitats, from the deep ocean to altitudes of 5,400 m. Many inhabit caves. Of over 3,100 species (Barnard, 1959), 232 are found in Lake Baikal, 450 in other freshwaters, including waves.

Gammaridae. The body is slender, the first antennae long, often longer than the second, and with an accessory flagellum. The gnathopods are strongly subchelate, especially in males, the other pereopods normal walking limbs. There are coxal gills on pereomeres 2–7 or 2–6. The pleon is usually setose, spinose, or carinate. The last uropod pair is usually foliate, laterally setose, and extends farther back than the others; it is a swimming limb. Most species are found in freshwater. *Gammarus** includes side-swimmers (Fig. 1–3); *G. locusta,* to 2 cm long, is found in northern Europe. It can tolerate brackish water of 5–6‰ and is often found at the mouth of rivers. The food is animal and vegetable matter. *Gammarus locusta* has often been confused with the following three similar species: *Gammarus zaddachi,* up to 1.7 cm long, found in brackish- and freshwater in northern and western Europe, in salt concentrations of 14–18‰, sometimes 0.26–1.8‰. *Gammarus oceanicus* is found on both sides of the northern Atlantic, Arctic, in brackish waters of the Baltic of 2.5‰. *Gammarus salinus,* a third species often confused with the preceding, occurs in brackish waters of the North Sea and Baltic. *Gammarus (Rivulogammarus) duebenii,* up to 1.8 cm long, is euryhaline and found on both sides of the north Atlantic in coastal North America from Labrador to Massachusetts, often with *G. zaddachi* in dilute water. *Gammarus (R.) pulex* is the European freshwater scud, 2.4 cm long, found in running water north of the Danube, and rarely in freshwater lakes in aquatic vegetation. After 10 molts, at 3–4 months old, it is mature. Females spawn 8–30 eggs which are carried for 3–4 weeks or longer during cold seasons. The life span is about 10 months. A female has six to nine broods, molting and mating between. The young stay 1–2 days in the marsupium after hatching. *Gammarus (R.) lacustris,* to 2.4 cm long, is found in northern European and American lakes, and a form is found in springs and spring-fed brooks of northeastern North America. *Gammarus fasciatus,* to 2.0 cm, is common in

---

* The name *Gammarus* is listed on the *Official List of Generic Names in Zoology.*

ponds and larger rivers of the North Atlantic drainages. *Gammarus annulatus* is found from Nova Scotia to Long Island Sound. The American Atlantic brackish water *G. tigrinus* was introduced to brackish water of the German river Werra in 1967 where it occurs in large numbers in brackish water and is fed on by fishes. *Gammarus mucronatus*, up to 15 mm long, is found along the Atlantic and Gulf coasts of North America. A few North American *Gammarus* species are known from caves. *Dikerogammarus*, 1–1.8 cm, is found in freshwater in Europe. *Gammarellus*, up to 4.4 cm long with a keeled back, occurs along rocky surf shores of northern Europe and North America. Another genus is *Marinogammarus*. *Pallasea quadrispinosa*, up to 2.7 cm long, carries 15–100 eggs in the marsupium; it is found in northern European freshwater lakes that, during glacial periods, were connected with the Baltic. *Cheirocratus*, up to 1 cm long, is marine. The many species of *Niphargus* (Fig. 18–9), up to 3 cm long, are found in wells, caves, and springs of Europe and on the bottoms of Alpine lakes. Colorless and eyeless, they have the trunk elongated like many troglobites. Two species are known to construct U-shaped burrows which branch to the sides as resting chambers. For digging, the large subchelae of the first two pereopods are used. After the animal has submerged in the bottom, the anterior pereopods 3–5 push large particles back, the pleopods small ones. *Niphargus* sits at the entrance of the burrow, antennae extended. The species observed swimming do so on one side. *Niphargus puteanus*, up to 3 cm long, lives in crevices of rapidly running subterranean waters and feeds on obligochaetes (*Dorydrilus wiardi*) several cm long. *Niphargus aquilex*, up to 1.5 cm long, lives in slow moving subterranean water, wells, and trickling springs. The body is less deep but more elongate than that of the troglobite *N. puteanus*. *Crangonyx*, 0.6 cm long, is found in subterranean waters of Europe and North America. *Crangonyx gracilis* is found in ponds and lakes in the northeastern United States and eastern

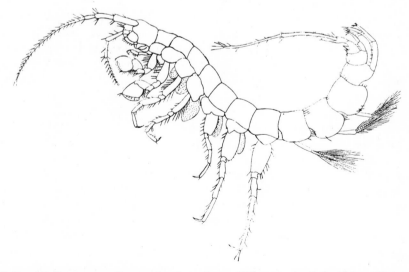

Fig. 18–9. *Niphargus aquilex*, male; 1 cm long. Gills are stippled. (After Wrzesniowski from Schellenberg.)

Canada; a number of other species are known. *Crangonyx gracilis* populations survive in temporary ponds as well as in permanent waters. *Crangonyx forbesi* of the Ozark region, migrates upstream in spring, downstream in fall; it breeds during winter. *Stygonectes* includes almost 30 species that inhabit caves of the eastern United States. Males of North American cave *Stygobromus* are rare; the females may reproduce parthenogenetically, producing only one to a few eggs per brood. *Stygodytes balcanicus* from Yugoslavian caves is 5 cm long. *Synurella*, 7 mm long and strikingly green in color, lives among plants of slowly flowing streams and ditches, forcing reeds and willow leaves between left and right gnathopods and pressing them against the mouth. Four species are known from North America. *Melita nitida*, of the North American Atlantic, is found in seaweed or under stones along the New England coast and is also reported from the Pacific coast. Another genus is *Elasmopus* with representatives on both coasts of the Americas.

Bogidiellidae. Bogidiella, up to 2.4 mm long, is a very slender European interstitial form. A coastal species is known from Brazil. The related *Pseudingolfiella*, found in interstitial waters of Chile, resembles ingolfiellids in shape.

Calliopiidae. Members of the family have first antennae shorter than the second with the secondary flagellum vestigial. Unlike many gammarids the telson is entire. *Calliopius laeviusculus*, up to 1.8 cm long and light green in color, is found in the tide zones of the North Sea and Baltic, and from Greenland to Cape Hatteras, and in the North Pacific, often on pilings but also on shells and plants. It is most active at night, moving about in swarms, and can be collected with lights. *Paraleptamphopus* is found in freshwater in New Zealand.

Lysianassidae. Unlike other Gammaridea the mandibles lack molar processes. The first antennae have a stout base and a secondary flagellum. The subchela of the second pereopods is weak and covered with setae and the third article is elongate. *Orchomenella*, up to 1.4 cm long, and *Scopelocheirus*, to 0.7 cm long, are found in the North Sea, *O. pinguis* along the New England shores. Another genus is *Alicella*. *Lysianopsis alba* is found abundantly in shelly sand below the tide level along southern New England to Florida. Other North American genera are *Anonyx* and *Hippomedon*.

Leucothoidae. The first pereopods have large subchela. *Leucothoe*, up to 1.5 cm long, is found within sponges or the gills of ascidians but also between plants and animals. *Leucothoe spinicarpa* has a wide distribution in the American Atlantic and Pacific, but several species may be confused under this name (Bousfield, in letter).

Talitridae. The first antennae are much shorter than the second and lack a secondary flagellum. The mandibles lack a palp. The second gnathopod is larger than the first. The third uropod is uniramous. This is the only family that contains terrestrial amphipods as well as marine and freshwater scuds. *Talitrus saltator*, beach flea or sand hopper of Europe, up to 1.5 cm long, inhabits the upper tide line in sand, where the waves have deposited a line of algae, carrion and debris. It has lost its ability to swim well but survives submerged in water for considerable periods. It depends on saturated air and succumbs at 75% relative humidity in about 5 hr. Thus during the day it remains buried in sand, protected against desiccation, to emerge at sunset and run about on the wet sand. It respires mainly through gills, and if the first four pairs are lacquered, the animal soon dies as the fifth pair together with the integument cannot supply sufficient oxygen. *Talitrus* survives longer in unsaturated air than *Orchestia gammarella* and *Talorchestia deshayesii*. Wet food forestalls desiccation. In laboratory choice experiments at 92% relative humidity at 15°C, *Talitrus* still reacts to small differences in humidity; at higher humidities it does not. The humidity receptors are still unknown. Females carry up to 17 eggs in the marsupium and live to be 1½ years old. The male does not ride on the female before her final molt, but mates with a freshly molted female. They overwinter to 50 cm deep far above the tide lines. *Orchestia gammarella*, up to 2.2 cm long, lives as does *Talitrus*, at the upper tide line on European and Canadian Atlantic coasts but prefers stones and rocks covered by spray. During daytime it lives under stones, often in large numbers. At the southern part of its distribution it inhabits freshwater shores, in Sicily to 800 m altitude. *Orchestia cavimana*, up to 2 cm long, is found on the bank of brackish and freshwater under stones with algae, plant fragments, etc. in Europe. It overwinters in masses of hundreds in cavities above the water. *Orchestia platensis* (=*O. agilis*), beach flea, to 14 mm, brown, is found in the line of decaying algae near the high water mark along the Atlantic from Newfoundland to Patagonia and on European coasts. It feeds on vegetation and animal matter. *Orchestia grillus* (=*O. palustris*) is found in Atlantic salt marshes away from the beach; *O. traskiana*, up to 15 mm and brown, is found on the Pacific coast from Alaska to Mexico. *Orchestoidea californiana*, up to 3 cm and orange-brown, is found burrowing in sand in California. *Talorchestia deshayesii*, up to 1.5 cm long, is found in the upper tide zone on flat sand in northern Europe. *Talorchestia longicornis* is the common large beach flea up to 2.5 cm, along the high tide mark from Gulf of St. Lawrence to northern Florida. Other species are found on the Atlantic and Gulf coasts. Terrestrial *Talitroides* is found in the tropics of the new and old world and is found also in hothouses. It has only vestiges of pleopods and walks upright, dorsum up, as does *Talitrus*. *Talitriator* found in forests of South Africa has only plant remains in the gut. It can stay submerged for days in freshwater but dies rapidly in dry air. *"Talitrus" sylvaticus* to 1.5 cm long occurs in southeastern Australia, and has been introduced in California's Golden Gate Park. *Talitroides topitotum* (=*T. pacificus*) up to 12 mm in length and *T. alluaudi*, up to 6 mm, have spread widely in Florida and the Gulf States following introduction from native India and Madagascar, respectively.

Hyalidae. Like Talitridae, but antenna 1 is longer than peduncle of antenna 2, telson is cleft; uropod 3 is occasionally biramous; gnathopods are normally subchelate in female, and maxillipeds have dactylate 4th palp article. *Hyale* has species in the intertidal zone in the Pacific and Atlantic. *Hyale plumulosa* (=*A. littoralis*) to 10 mm is found intertidally from Cape Cod to New Jersey and on the American Pacific coast. But two species live in freshwater wells in Zanzibar. *Hyachelia* has been found as symbiont in the buccal cavity of sea turtles where it feeds on food residues. *Allorchestes angustus* is found on the North American Pacific coast.

Hyalellidae. Like Hyalidae, but the telson is entire, and accessory gills are on pereopod 7. The third uropods are uniramous. *Parhyalella* is marine. *Hyalella azteca*, up to 8 mm, is common and widespread in North and Central America among submerged vegetation in freshwater springs, ponds, brooks, and lakes. It may have as many as 15 broods per season and mates after releasing each batch of young. Each

brood averages about 18 eggs. The young pass through at least 15 instars before mating. Lake Titicaca has numerous endemic species of *Hyalella*. *Chiltonia* from New
Zealand belongs to this family.

Dogielinotidae. Like Hyalidae but has walking limbs broadened and modified for
burrowing. *Dogielinotus loquax*, up to 13 mm, occurs in upper midtide level of surf
beaches of northern California to Washington.

Ampeliscidae. The first antennae are half the length of the second without accessory flagellum. The second antennae are some distance from the first. The gnathopods
are weak. The lateral eyes are divided into two or three pairs usually two pairs of
which have a simple lens (Fig. 18–3). The last two pleomeres are fused and the telson
is cleft. The animals burrow. *Ampelisca* lives in a tube under the bottom surface from
the intertidal to more than 600 m depth. Glands of all pereopods, uropods and coxal
plates take part in formation of the tube. Species are found on both sides of the
Atlantic. *Haploops tubicola* is found in the Baltic (Fig. 18–7). *Byblis serrata* is found
on sandy substrates along New England coastlines.

Haustoriidae. Body appendages are very spinose and setose, their articles expanded
and modified for burrowing; the gnathopods are weak. Mandibles each have a strong
molar process; mouthparts are setose, filter-feeding in function; pereopod 6 longer than
7, bases very broadly expanded. The subfamily Pontoporeiinae contains slender-bodied
"sand-licking" feeders. *Pontoporeia femorata*, to 1.7 cm long, survives in water to 6‰
salt, burrowing in mud. Its distribution is circumpolar, but a Pleistocene relict population is found in the Baltic. *Pontoporeia affinis*, 0.95 cm long, occurs at depths of 35–36
m in brackish bays of the Baltic and also has Pleistocene relict populations in Scandi-
navian and north German lakes and is found in the Great Lakes and other deep North
American lakes, to 300 m in Lake Superior. They live for 3 years, swimming about at
night, and breeding between December and April. Densities of 4,553/m² have been
reported from depth of 50–60 m from Green Lake, Wisconsin.

The subfamily Haustoriinae contains broad-bodied, filter-feeders. The European
Bathyporeia, to 0.8 cm long, excavates its burrows in the tide zone and in shallow
water, actively swimming at night and cleaning off sand grains for food. They can
rapidly submerge in sand, the 2–4 pereopods digging, the posterior pushing the trunk.
The fourth and fifth pleopods push sand back. *Haustorius* has very large first to fourth
coxal plates and the merus and carpus of the 5–7 pereopods are widened. *Haustorius
arenarius*, 1.3 cm long, lives in the wave packed sand of sandbanks of European Atlantic beaches; *H. canadensis* is its American counterpart. At night they swim, venter
up. A large and diverse haustoriid fauna of 11 genera and about 30 species inhabits
the American Atlantic and Gulf coast beaches; genera are: *Haustorius, Lepidactylus,
Protohaustorius, Parahaustorius, Acanthohaustorius, Neohaustorius, Pseudohaustorius,
Bathyporeia, Pontoporeia, Priscillina* as well as the American endemic genus *Amphiporeia. Eohaustorius* occurs on American Pacific beaches.

Phliasidae. Most members, except *Phlias*, are dorsoventrally flattened. *Pereionotus*,
about 0.5 cm long, has the last two pleomeres folded under the anterior two. The fifth
pleomere with its appendages is missing. Both pairs of antennae are short. They live
on algae.

Aoridae. The body is slender. The first antennae have a secondary flagellum and
are longer than the second pair. The seventh pereopod is longest. *Microdeutopus*

gryllotalpa, up to 1 cm long, found on both sides of the North Atlantic, builds a tube between branched algae. Its first pereopods have a large complexly subchelate hand.

Photidae. The first antennae are sometimes larger than the second and sometimes lack a secondary flagellum. The second pair of gnathopods is larger than the first. The fourth and fifth pairs of pereopods are longer than the others. The third uropods are biramous with simple apical spines. *Leptocheirus pilosus,* 0.5 cm long, found in the brackish water of northern Europe and at the mouths of streams entering the Mediterranean, weaves a domed cover over the sea floor that retains water during low tide. The water is driven to the gills by beating pleopods. *Leptocheirus pinguis* is very common from eastern Canada south to New Jersey, living on mud out to deep water. The North Sea *Microprotopus maculatus,* up to 0.3 cm long, climbs on algae and runs or swims dorsum up on the bottom. It builds tubes of sand. Its American counterpart is *M. ranei* found from Cape Cod to Georgia.

Ampithoidae. The first antennae, with or without a secondary flagellum, are the same length as the second. The second gnathopods are larger and different in shape from the first. The lower lip has deeply notched outer lobes; on uropod 3 the outer ramus terminates in hooked spines. *Ampithoe rubricata,* to 2 cm long, on both sides of the North Atlantic among plants or gravel, inhabits spun tubes to 5 cm long. From the tube opening, using the subchelae of the two pairs of anterior pereopods, it takes pieces of algae for food or for building extensions to the tube. Numerous species are found on the North American Pacific coast. *Cymadusa compta,* 12 mm, is common in eelgrass from Cape Cod to the Carolinas.

Corophiidae. The body is more or less dorsoventrally flattened. The coxal plates, epimeres, and pleon are small (Figs. 18–5, 18–6). Members are found in marine, brackish, and freshwater. *Ericthonius* species are found in all oceans except the Arctic and Antarctic. *Corophium volutator,* up to 1 cm long, live in huge numbers on muddy shallows, and below the low tide mark in the North Sea, and provide a food supply for flounder and other fish, and for *Crangon.* Along the Finnish coast they are known to live 1 year. *Corophium curvispinum,* up to 0.9 cm long, has moved from the Caspian Sea into freshwaters, probably transported by ships carrying attached tubes. It was first found in freshwater near Berlin in 1912 and has since been widely dispersed, but has not gone into the Baltic Sea. It lives exclusively in slowly flowing waters where it spins sandgrains together with threads to form tubes on embankments, stones, boats, and plants. Many tubes may form clumps. As many as 1,680 individuals have been found per m², remaining in their tubes during daytime but becoming active at night. A brackish-water species *C. lacustre,* up to 0.5 cm long, is found in the mouths of rivers of the North Sea, the Baltic and the American Atlantic coast. A number of species of *Corophium* are found on both North American coasts; *C. acherusicum* extends to the Gulf coast. *Unciola irrorata,* 17 mm, mottled red and white, is common along the New England coast, inhabiting the tubes of annelid worms. *Kamaka* has brackish and one freshwater species in Japan. *Neohela monstrosa* (Fig. 18–5) has been studied from mud of the Skagerak.

Cheluridae. The posterior three pleomeres are fused. The coxal plates are short. The male has a remarkably long paddlelike outer ramus of the third uropod (Fig. 18–10). The wood borer *Chelura terebrans,* (Fig. 18–10), to 0.6 cm long, bores in submerged timbers as does the isopod *Limnoria.* Always found with the isopod, which

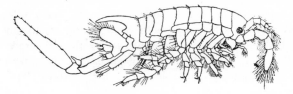

Fig. 18–10.   The wood borer, *Chelura terebrans*; 0.5 cm long. (After Schellenberg.)

first invades the wood and does the main damage, it burrows in oaks as well as in softer woods. *Chelura* swims dorsum down and can hop on dry wood by flexing the abdomen. The gut is filled with wood particles, but *Chelura* also feeds on the fecal material of *Limnoria*.

Podoceridae. The body is more or less dorsoventrally flattened. The third uropod and its somite are vestigial. Species of *Podocerus* are found on the North American Pacific coast. *Dulichia porrecta*, 5–8 mm long, is found on both sides of the North Atlantic.

Hadziidae. This family, closely related to Gammaridae, contains peculiar, minute amphipods from southern European caves. None has thus far been found in the Americas.

Phoxocephalidae. Members have the body fusiform, and have burrowing append-ages. The rostrum on the head is a broad hood overhanging bases of antennae. Deep coxal plates are setose below. The sixth pereopod is longest, the seventh shortest. Mandibular incisor is strong, molar weak. Genera are: *Paraphoxus, Phoxocephalus,* and *Harpinia.*

Liljeborgiidae. Small to medium sized tube dwelling amphipods having deep coxal plates, the first expanded anteriorly. The seventh pereopod is the longest and strongest. The telson is deeply cleft. Mandibular incisor is strong, molar weak, the palp genic-ulate at the second article. The North American *Listriella* has several species com-mensal in burrows of polychaete worms.

SUBORDER INGOLFIELLIDEA

The body is elongate, almost wormshape, circular in cross section. There are no epimera or coxal plates (Fig. 18–11). The cephalothorax is made up of the head and one thoracomere. The pleon is almost as long as the pereon. The maxillipeds have six or seven articles. The first and second pereopods have large subchelae, with the fifth article enlarged; the third to fifth pereopods have short coxal gills (Fig. 18–11). The anterior three pairs of pleopods are vestigial triangular plates. The first and second uropods are biramous, the third reduced to a stump (Fig. 18–11). The oostegites are smaller than the epipodites but with two long terminal setae. There are no eyes. Proto-, deuto-, and tritocerebrum have paired, posteriorly directed extensions contain-ing ganglionic cells. The chewing stomach is tubeshaped and unusually long, reaching to the posterior border of the 4th pereomere. Large species have a heart with three pairs of ostia; small species have only one pair of ostia, in the fourth pereomere. The peculiar shape of the brain and the stomach can be explained by lack of space in the body cavity in these dwarf forms, but it persists in the large *Ingolfiella leleupi.*

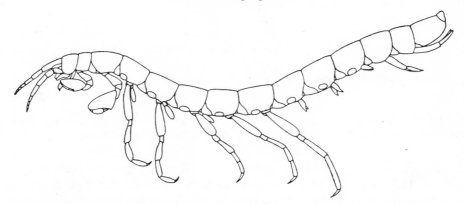

Fig. 18–11. *Ingolfiella leleupi*; 1.4 cm long. (After Ruffo from Siewing.)

The only genus is *Ingolfiella*. Geographically widespread, the small species live interstitially in sand of coasts, coastal ground water and in fine silt and flocculent ooze at great depths. One species is known from the English Channel, Davis Strait, others from southeast Asian coast, one from interstitial waters in Peru, from Yugoslavian and Congo caves. One can watch them slither rapidly between sand grains. The marine species known are 1–2.5 mm long. *Ingolfiella leleupi,* up to 14 mm long (Fig. 18–11), is found in cave waters of central Africa.

SUBORDER CAPRELLIDEA

The suborder is also called Laemodipodea. The cephalothorax consists of the head and two thoracomeres. The trunk is not laterally compressed. The maxillipeds often have seven articles but are reduced in Cyamidae. Coxal plates or epimeres are usually lacking or vestigial. The first antennae are longer than the second. The fourth and fifth thoracopods are often vestigial, only the gills and oostegites remaining (Fig. 18–12). Only the fourth and fifth thoracomeres have oostegites. The pleon is very short; females usually lack appendages but males sometimes have one or two pairs of vestigial pleopods (Fig. 18–12). All species are marine.

Caprellidae. The body is elongated, cylindrical, almost circular in cross section (Fig. 18–12). The second and third pairs of thoracopods have raptorial subchelae. The posterior thoracopods have long dactyls that are used in holding on to plants or colonial animals. At times they also are tipped by subchelae. In most species the fourth and fifth thoracopods have been lost or reduced to stumps. Rarely the sixth thoracopod is also reduced. Gills are present on the third to fifth, usually only on the fourth or fifth thoracomere. Oostegites are present on only the 4th and 5th thoracomeres, so the marsupium is very short. The pleon is very short, pleopods have been lost or are vestigial. Most species are predators, but some feed on diatoms and organic detritus. Caprellids resemble the substrate and are hard to find. *Pariambus typicus* lives on *Asterias rubens* and *Crossaster papposus, Caprella grahami* on *Asterias forbesi. Caprella* is found on all parts of the starfish, between tubefeet and, in contrast to other caprellids, is only rarely held by the pedicellaria. The animal moves like a geometrid caterpillar and feeds mainly on detritus particles that stick to the mucus of the starfish's

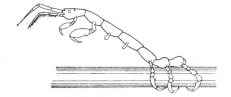

Fig. 18–12. *Caprella* **waiting in ambush. The 4th and 5th thoracomeres have lost the appendages except gills; 2 cm long. (After Wetzel from Schellenberg.)**

tubefeet, spines and papulae. Of 29 *Asterias forbesi* examined, 28 averaged 8 *C. grahami*. They swim by flexing the anterior body including the third thoracomere posterioventrally and then stretching. *Caprogammarus* of the North Pacific region is the most primitive genus; it has functional pleopods and uropods. *Proto* and *Phthisica* are also considered primitive genera with eight pairs of thoracopods and two pairs of uropod stumps. *Proto*, up to 1 cm long, has three to five thoracopods and two pairs of two to three articled pleopods. *Phthisica* of the North Sea is up to 2 cm long. *Caprella* lacks vestigial fourth and fifth thoracopods and pleopods. *Caprella septentrionalis*, up to 3.2 cm long, is found in the North Sea and along the Greenland and Labrador coast to New England. Several other species of *Caprella* are found on the Pacific and Gulf coasts of North America. *Pariambus* has a stumpy sixth thoracopod. *Pariambus typicus*, up to 0.7 cm long, is found in the North Sea on vegetation, starfish, and in other places on *Maja*. Other genera found along American coasts are *Metacaprella*, *Paracaprella*, *Deutella*, and *Tritella*. *Aeginella longicornis* is found on both sides of the North Atlantic.

Cyamidae. These amphipods, called whale lice, are parasitic on the skin of whales. All are dorsoventrally flattened, wide, and only a stump remains of the pleon. The females lack pleopods, the males have a vestigial pair (Fig. 18–13). The first antennae are longer than the short second antennae. There are small piercing mouthparts. Thoracopods 4 and 5 have become vestigial except for the long gills. All other thoracopods have strong subchelae. Only the 4th and 5th thoracomeres have oostegites. Since there are no swimming stages, they must be transmitted by direct contact of the hosts. *Cyamus boopis*, up to 1.3 cm long (Fig. 18–13), lives on humpbacked whales (*Megaptera novaeangliae*) by the hundred thousand causing fist-sized cavities that reach to the blubber. The epidermis and corium are 5 mm thick. *Cyamus* species are mostly host specific. Feeding and other habits, as in other species of the family, are completely unknown. Other genera are *Paracyamus* and *Platycyamus*.

SUBORDER HYPERIIDEA

The integument is more or less transparent. The cephalothorax, consisting of the head and the first thoracomere, is swollen, as is usually also the pereon (Figs. 18–14, 18–15). The eyes are very large and may cover the whole side of the cephalothorax (Fig. 18–4). The maxillipeds have only a coxa, basis and ischium. The other four articles are lost or fused into a single article. The coxal plates and epimeres are short. All forms are pelagic and marine. Their habits are virtually unknown.

The families Lanceolidae and Euminonectidae (Cohort Gammaridea) have the maxillipeds with paired median endites. There are only a few genera: *Lanceola* and

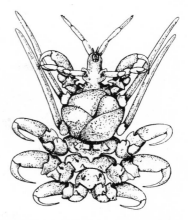

Fig. 18–13. Whalelouse, *Cyamus erraticus*: female with brood pouch, in ventral view; 1 cm long. The narrow cephalothorax, small mouthparts, minute abdomen and long gills on the 3rd and 4th pereomeres are adaptations to its parasitic habits. (After Iwasa.)

Mimonectus. The pelagic *Mimonectus* are spherical animals which, except for the little tail, resemble jellyfish (Fig. 18–14).

The other families of Hyperiidea (Cohort Genuina) have median endites of the maxillipeds fused to form an undivided plate. Most Hyperiidea belong here.

Scinidae. The body is not swollen but dorsoventrally flattened. The eyes are small. *Scina* is one genus.

Thaumatopsidae. The anterior body is swollen, the cephalothorax very large. The eyes take up the whole dorsum (Fig. 18–15). *Thaumatops* is found in the deep ocean.

Hyperiidae. The eyes are large and the large cephalothorax is almost spherical. *Hyperia galba* (Fig. 18–16), up to 2 cm long, occurs from northern Europe, and the Arctic to New England. It may be free-swimming or may attach to the subumbrella or oval tentacles of scyphomedusae, at times penetrating into the gastral cavities. It has also been collected on the ctenophore *Beroe*. It hooks with its last three pereopods posteriorly bent into the jellyfish so that its venter is held away from the jellyfish body. *Euthemisto* and *Parathemisto* contain free-swimming and carnivorous species. Another genus is *Hyperoche*.

Phronimidae. The cephalothorax is large and conical. The eyes are divided, one on each side facing up, the other to the side. The fifth pereopod is armed with a large

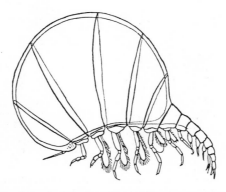

Fig. 18–14.   *Mimonectes loveni*, female, about 25 mm long. (From Schellenberg after Bovallius.)

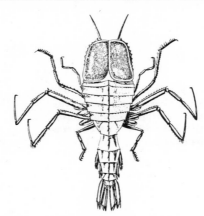

Fig. 18–15.   *Thaumatops neptuni*, dorsal view of transparent bathypelagic amphipod; 10 cm long. (After Thomson from Murray and Hjort.)

reversed subchela or chela. The females often swim in the shells of tunicates or diphyids (Fig. 18–4). *Phronima* is 0.5–4 cm long.

Lycaeidae. The cephalothorax is large, spherical, with large eyes. *Lycaea* is found mainly in *Salpa.*

Oxycephalidae. The cephalothorax is long, drawn out in a rostrum. *Rhabdosoma*, including rostrum and telson measures about 7 cm long, the body elongated, the cephalothorax with its narrow rostrum being 2 cm long, the telson 1–9 cm long. The legs are short, the eyes large.

Typhidae (=Platyscelidae). Members have the trunk wide and the cephalothorax very large, the eyes completely covering its sides. The bases of the fifth or sixth pereopods are widened to large, long plates that can be folded like doors below over the gills and oostegites. Genera are *Platyscelus* and *Tetrathyrus.*

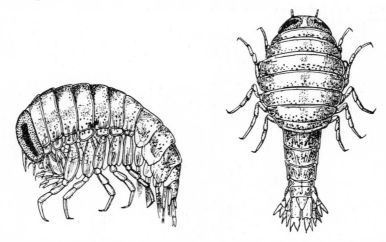

Fig. 18–16.   *Hyperia galba*, female from the side and in dorsal view; 2 cm long. This pelagic species has cephalothorax and pereon swollen and has small coxal plates. (After Sars from Schellenberg.)

References

Agrawal, V.P. 1964. Optimum pH for the activity of caecal carbohydrases of the amphipod *Orchestia*. *Proc. Zool. Soc. London* 143: 545–567.

Alexandrowicz, J.S. 1954. Innervation of an amphipod heart. *J. Marine Biol. Ass.* 33: 709–719.

Anders, A. 1959. Das Augenpigmentsystem von *Gammarus pulex*. *Z. Vererbungslehre* 90: 94–115.

Anders, F. 1953. Die carotinoidean Körperpigmente und die Geschlechtsbestimmung von *Gammarus pulex*. *Naturwissenschaften* 40: 27.

———— 1956. Ausbildung und Vererbung der Köperfarbe bei *Gammarus pulex ssp. subterraneus*. *Z. Indukt. Abstamm. Vererbungslehre* 87: 567–579.

———— 1957. Modifikation und erbliche Variabilität der Pigmentierung bei einer Höhlenform von *Gammarus pulex*. *Verhandl. Deutschen Zool. Ges.* 20: 197–203.

———— 1957. Die Geschlechtsbeeinflussende Wirkung con Farballelen bei *Gammarus pulex*. *Z. Indukt. Abstamm. Vererbungslehre* 88: 291–332.

Barnard, J.L. 1958. Index to families, genera and species of gammaridean Amphipoda. *Occas. Pap. Allan Hancock Found. Publ.* 19: 1–145.

———— 1966. Galapagan amphipod inhabiting the buccal cavity of the sea turtle *Chelonia mydas*. *Proc. Symp. Crustacea Ernakulum* 1: 119–125.

———— 1967. Dogielinotid marine Amphipoda. *Crustaceana* 13: 281–291.

Bock, K. 1967. Experimente zur Ökologie von *Orchestia platensis*. *Z. Morphol. Ökol.* 58: 405–428.

Bousfield, E.L. 1958. Fresh water amphipod crustaceans of glaciated North America. *Canadian Field Natur.* 72: 55–113.

———— 1958. Ecology of the Terrestrial Talitridae (Crustacea: Amphipoda) of Canada. *Proc. Tenth Int. Congr. Entomol.* 1956. 1: 883–898.

———— 1965. Haustoriidae of New England. *Proc. U.S. Nat. Mus.* 117: 159–240.

———— 1969. Records of *Gammarus* from the Middle Atlantic Region. *Chesapeake Science* 10. in press.

Chace, F.A. *et al.* 1959. Malacostraca *in* Edmondson, W.T. ed. 1959 *Ward and Whipple Freshwater Biology*, Wiley, New York.

Charniaux-Cotton, H. 1953. Existence, chez un crustacé amphipode, de l'action hormonale des ovaires sur un caractère sexuel secondaire. *Intern. Congr. Zool. Copenhagen*: 305–306.

———— 1956. Déterminisme hormonal de la différenciation sexuelle chez les Crustacés. *Ann. Biol.* (3) 32: 371–399.

Dahl, E. 1960. *Hyperia galba*, a true ectoparasite on jellyfish. *Arb. Univ. Bergen*, 1959, 1–8.

Debenedetti, E.T. 1963. Observations on the orientation of *Talitrus saltator* in fresh and sea water. *Naturwissenschaften* 50: 25–26.

Dennell, R.B. 1933. Habits and feeding mechanism of the amphipod *Haustorius*. *J. Linnean Soc. London* 38: 363–388.

Dudich, E. 1931. Die Wiederherstellung des Mosaikpanzers bei den Gammariden nach ihrer Häutung. *Vehr. Int. Ver. Limnol.* 5: 600–617.

Enequist, P. 1949. Soft bottom amphipods of the Skagerak. *Zool. Bidr. Uppsala* 28: 297–492.

Enright, J.T. 1961. Pressure sensitivity of an amphipod. *Science* 133: 758–760.

———— 1963. Tidal rhythm of a sand beach amphipod. *Z. Vergl. Physiol.* 46: 276–313.

———— 1967. Temperature compensation in short duration time measurement by an intertidal amphipod. *Science* 156: 1510–1512.

Ercolini, A. 1963. Orientamento solare degli Anfipodi. *Arch. Zool. Ital.* 48: 147–180.

———— 1964, 1965. Orientamento astronomico di anfipodi litorali della zona equatoriale. *Z. Vergl. Physiol.* 49: 138–171; *Redia*, Firenze 49: 119–128.

Fries, G. and Tesch, F. 1965. Massenvorkommen von *Gammarus tigrinus* auf Fische und niedere Tierwelt in der Weser. *Arch. Fischereiwiss.* 16: 133–150.

Geppetti, L. and Tongiorgi, P. 1967. Nocturnal migrations of *Talitrus saltator*. *Monit. Zool. Italiano* (N.S.) 1: 37–40.

———— 1967. Le Migrazione di *Talitrus saltator*. *Redia Staz. Entomol. Agrar. Firenze* 50: 309–336.

Ginet, R. 1956. Études sur la biologie d'Amphipodes troglodytes du genre *Niphargus*. *Bull. Soc. Zool. France* 80: 332–349.

Gräber, H. 1933. Die Gehirne der Amphipoden und Isopoden. *Z. Morphol. Ökol.* 26: 234–371.

Graf, F. 1967. Réabsorption intensive par l'épithélium caecal du calcium stocké dans les caecums postérieurs d'*Orchestia*. *Compt. Rend. Acad. Sci. Paris* (D) 264: 2211–2213.

Günzler, E. 1967. Verlust der endogenen Tagesrhythmik bei *Niphargus puteanus*. *Biol. Zentralbl.* 83: 677–694.

Haffner, K. von. 1935. Die ursprüngliche Stellung der Beine bei den Amphipoden. *Zool. Anz.* 111: 177–189.

———— 1935. Der Blutkreislauf von *Phronima sedentaria*, mit besonderer Berücksichtigung des lacunären Systems. *Z. Wiss. Zool.* 146: 283–328.

Hanström, B. 1947. The brain, the sense organs, and the incretory organs of the head in the Crustacea Malacostraca. *Lunds Univ. Arsskr.* 43: 1–44.

Heinze, K. 1932. Fortpflanzung und Brutpflege bei *Gammarus pulex* und *Carinogammarus roeselli*. *Zool. Jahrb. Abt. Allg. Zool.* 51: 397–440.

Holsinger, J.R. 1967. Subterranean amphipod *Stygonectes*. *Bull. U.S. Nat. Mus.* 259: 1–176.

———— 1969. Subterranean amphipod genus *Apocrangonyx*. *Amer. Midland Natur.* 81: 1–28.

Hubricht, L. 1959. Amphipoda *in* Edmondson, W.T. ed. *Ward and Whipple Freshwater Biology*. Wiley, New York.

Hutchinson, G.E. 1967. *A Treatise on Limnology*, Vol. 2. Wiley, New York.

Ingle, R. 1966. Burrowing behavior of the amphipod *Corophium arenarium*. *Ann. Mag. Natur. Hist.* (13) 9: 309–317.

Kane, J.E. 1963. Observations on the moulting and feeding of a hyperiid amphipod. *Crustaceana* 6: 129–132.

Kinne, O. 1952. Untersuchungen über Blutkonzentration, Herzfrequenz und Atmung bei *Gammarus*. *Kieler Meeresf.* 9: 134–149.

———— 1953. Biologie und Physiologie von *Gammarus duebeni*. *Z. Wiss. Zool.* 157: 427–491.

———— 1953. Sperma als Stimulus für die Oviposition bei *Gammarus duebeni*. *Naturwissenschaften* 40: 417.

———— 1953. Wird die Häutungsfolge der Amphipoden *Gammarus zaddachi* durch die lunare Periodizität beeinflusst? *Kieler Meeresf.* 9: 271–279.

———— 1954. Die Bedeutung der Kopulation für Eiablage und Häutungsfrequenz bei *Gammarus*. *Biol. Zentralbl.* 73: 190–200.

———— 1960. *Gammarus salinus*, über den Umwelteinfluss auf Wachstum, Häutungsfolge, Herzfrenquenz und Eintwicklungsdauer. *Crustaceana* 1: 208–217.

———— 1961. Growth, moulting frequency, heart beat, number of eggs and incubation time in *Gammarus zaddachi* exposed to different environments. *Ibid.* 2: 26–36.

———— 1964. Physiologische und Ökologische Aspekte des Lebens in Ästuarien. *Biol. Anst. Helgoländ Jahresber.* pp. 131–156.

Klövekorn, J. 1935. Das Organsystem der Blutbewegung bei *Gammarus*. *Z. Wiss. Zool.* 146: 153–192.

Kühne, H. and Becker, G. 1964. Der Holz-flohkrebs *Chelura terebrans*. *Z. Angew. Zool.* (Beihefte) 1: 1–141.

Kureck, A. 1957. Tagesperiodische Ausdrift von *Niphargus aquilex schellenbergi* aus Quellen. *Z. Morphol. Ökol.* 58: 247–262.

Langenbuch, R. 1928. Über die Statocysten einiger Crustaceen. *Zool. Jahrb. Abt. Allg. Zool.* 44: 575–622.

Laval, P. 1964. Présence d'une période larvaire au début du développement de certains Hypérides parasites. *Compt. Rend. Acad. Sci. Paris* 260: 6195–6198.

————. 1966. *Bougisia ornata*, Hyperiidea. *Crustaceana* 10: 210–218.

Lehmann, U. 1967. Drift und Populationsdynamik von *Gammarus pulex fossarum*. *Z. Morphol. Ökol.* 60: 227–274.

Lockwood, A.P.M. 1965. Losses of sodium in the urine and across the body surface in the amphipod, *Gammarus duebeni*. *J. Exp. Biol.* 42: 59–69.

Mabillot, S. 1955. Histophysiologique de l'appareil digestif de *Gammarus*. *Arch. Zool. Exp. Gén.* 92: 20–38.

McCain, J.C. 1968. The Caprellidae of the Western North Atlantic. *Bull. U.S. Nat. Mus.* 278: 1–147.

Martin, A.L. 1964. The alimentary canal of *Marinogammarus*. *Proc. Zool. Soc. London* 143: 525–544.

————. 1965. The histochemistry of the moulting cycle in *Gammarus*. *J. Zool.* 147: 185–200.

————. 1966. Feeding and digestion of two intertidal gammarids. *J. Zool. London* 148: 515–525.

Meadows, P.S. 1964. Experiments on substrata selection by *Corophium* species. *J. Exp. Biol.* 41: 499–511; 677–687.

————. and Reid, A. 1966. Behaviour of *Corophium*. *J. Zool. London* 150: 387–399.

Mills, E.L. 1967. Deep-Sea Amphipods from the western North Atlantic Ocean; Ingolfiellidea and Pardaliscidae. *Canadian J. Zool.* 45: 347–355.

Morgan, E. 1965. The activity rhythm of the amphipod *Corophium volutator* and hydrostatic pressure associated with the tides. *J. Anim. Ecol.* 34: 731–746.

Palluault, M. 1954. Notes écologiques sur le *Talitrus saltator*. *Arch. Zool. Exp. Gén.* 91: 105–129.

Papi, E. 1955. The sense of time in *Talitrus saltator*. *Experientia* 11: 201–202.

————. 1960. Orientation by night. *Cold Spring Harbor Symp. Quant. Biol.* 25: 475–480.

Papi, F. and Pardi, L. 1953. Orientamento di *Talitrus saltator*. *Z. Vergl. Physiol.* 35: 490–518.

————. 1959. Orientamento lunare di *Talitrus saltator*. *Ibid.* 41: 583–596.

————. 1963. On the lunar orientation of sandhoppers. *Biol. Bull.* 124: 97–105.

Pardi, L. 1957. Modificazione sperimentale della direzione di fuga negli anfipodi ad orientamento solare. *Z. Tierpsychol.* 14: 261–275.

————. 1960. Innate components in the solar orientation of littoral arthropods. *Cold Spring Harbor Symp. Quant. Biol.* 25: 395–401.

————. et al. 1958. Orientamento degli Anfipodi del litorale. *Atti. Accad. Sci. Torino* 92: 308–315.

Patton, W. 1967. Habits, behavior and host specificity of *Caprella grahami* commensal on *Asterias forbesi*. *Biol. Bull.* 134: 148–153.

Pennak, R.W. 1953. *Fresh water Invertebrates of the United States*. Ronald, New York.

Platzman, S.J. 1960. Comparative ecology of two species of intertidal amphipods: *Talorchestia* and *Orchestia*. *Biol. Bull.* 119: 333.

Ponyi, E. 1956. Ökologische, ernährungsbiologische und systematische Untersuchungen an verschiedenen *Gammarus*-Arten. *Arch. Hydrobiol.* 52: 367–387.

Ruoff, K. 1968. Experimentelle Untersuchungen über den in der Weser eingebürgerten amerikanischen *Gammarus tigrinus*. *Arch. Fischereiwiss.* 19: 134–158.

Saunders, C.G. 1966. Dietary analysis of caprellids. *Crustaceana* 10: 314–315.

Schellenberg, A. 1942. Flohkrebse oder Amphipoda, in Dahl, F. *Die Tierwelt Deutschlands,* G. Fischer, Jena 40: 1–252.

Schlick, W. 1943. *Haustorius arenarius* als "Schwimmgräber" im feuchten Sandstrande der Nordsee. *Zool. Anz.* 142: 160–172.

Schmitz, E. 1967. Visceral anatomy of *Gammarus lacustris. Amer. Midland Natur.* 78: 1–54.

Schumann, F. 1928. Salze, insbesondere des kohlensauren Kalkes, für Gammariden und ihren Einfluss auf deren Häutungs-Physiologie und Lebensmöglichkeit. *Zool. Jahrb. Abt. Allg. Zool.* 44: 623–704.

Schwartzkopff, J. 1955. Herzfrequenz bei Krebsen. *Biol. Zentralbl.* 74: 480–497.

_____. 1955. Herzfrenquenz von Krebsen. *Experientia* 11: 323–325.

Segerstråle, S.G. 1947. Distribution and morphology of *Gammarus zaddachi. J. Marine Biol. Ass.* 27: 219–244.

Siewing, R. 1951. Verwandtschaft zwischen Isopoden und Amphipoden. *Zool. Anz.* 147: 166–180.

_____. 1963. Zur Morphologie der aberranten Amphipodengruppe Ingolfiellidae. *Ibid.* 171: 76–91.

Spooner, G.M. 1947. Distribution of *Gammarus* species in estuaries. *J. Marine Biol. Ass.* 27: 1–52.

Ståhl, F. 1938. Inkretorische Organe und Farbwechselhormone im Kopf eininger Crustaceen. *Kungl. Fysiogr. Sällsk. Handl.* (N.F.) 49: 3–20.

Sutcliffe, D. 1966. Sodium regulation in *Gammarus pulex. J. Exp. Biol.* 46: 499–518.

_____. 1968. Sodium regulation and adaptation to fresh water in gammarid crustaceans. *J. Exp. Biol.* 48: 359–380.

_____ and Shaw, J. 1968. Sodium regulation in *Gammarus duebeni. Ibid.* 48: 339–358.

Thiem, E. 1942. Darmkanal und die Nahrungsaufnahme von *Synurella ambulans. Z. Morphol. Ökol.* 38: 63–79.

Thienemann, A. 1950. Verbreitungsgeschichte der Süsswassertierwelt Europas. *Binnengewässer* 18: 1–809.

Thore, S. 1932. Statocysten und Frontalorgane bei *Gammarus. Zool. Jahrb. Abt. Anat.* 55: 489–504.

Traut, W. 1962. Zur Geschlechtsbestimmung bei *Gammarus. Z. Wiss. Zool.* 167: 1–72.

Watkin, E.E. 1939. Swimming and burrowing habits of some species of the Amphipod genus *Bathyporeia. J. Marine Biol. Ass.* 23: 457–465.

Werntz, H.O. 1963. Osmotic regulation in marine and fresh-water gammarids. *Biol. Bull.* 124: 225–239.

Wetzel, A. 1932. Caprelliden, Bewegung, Nahrungserwerb, Aufenthaltsort. *Z. Wiss. Zool.* 141: 347–398.

_____. 1933. Caprelliden, Raumorientierung, Farbanpassung und Farbwechsel. *Ibid.* 143: 77–125.

Weygoldt, P. 1958. Die Embryonalentwicklung des Amphipoden *Gammarus pulex. Zool. Jahrb. Abt. Anat.* 77: 51–110.

Williamson, D.I. 1951. Talitridae: Effects of Atmospheric Humidity. *J. Marine Biol. Ass.* 30: 73–90.

_____. 1956. Visual Orientation in *Talitrus saltator. Ibid.* 30: 91–99.

Subject Index

A

abbreviated development, 290
abdomen, 2, 4, 6
abdominal somites, 2
abductor muscles, **14***
Acanthephyra, 316
Acanthodiaptomus, 112
Acanthohaustorius, 492
Acanthomysis, 384
Acanthosquilla, 255
Acartia, 168, 169, 171
Actheres, **163,** 164, 180, 181
acorn barnacle, 200, **201**
acron, 2, 4, 6
Acroscalpellum, **216**
Acrothoracica, 195, 201, 218
actinian, 296
acyclic, 109
adductur muscle, 8, **14,** 126, 196, 204, **237**
adhesive gland, **208**
Aega, 429, **443,** 444, **452,** 453
Aegidae, 435, 439, 443, 444, 453
Aeginae, 420
Aeginella, 496
Aegla, 333, **334**
Aeglidae, 292, 333, **334**
aesthetasc, *see* esthetase
aesthete, 152
Akentrogonida, 222, 226
Akessonia, 155, 173
alar muscles, 31, 475
Albunea, 298, **336**
Albuneidae, 335, **336**
Alepas, 216
Alicella, 470, 490

alima, **250**
Allorchestes, 491
Alona, 107, 112, 119
Alpheidae, 284, **314,** 317
Alpheodiea, 316
Alpheus, 276, 280, 296, 306, **314,** 317
Alteutha, 171
Amblyops, 373, 383
amictic, 106, 108
ammonia, 29, 309
Ampelisca, 473, **474,** 480, 485, 492
Ampeliscidae, 473, 492
Amphiascus, 172
Amphipoda, 370, 371, 470
Amphiporeia, 492
Amphisopidae, 458
Ampithoe, 493
Ampithoidae, 493
Anamixidae, 486
anamorphic development, 49, 53, **56,** 57, 58, 290
Anapagurus, 288
Anaspidacea, **33,** 257
Anaspides, 231, 257, **258,** 259, 260, 261
Anaspididae, 261
anatomy, 2
Anchistropus, 112, 114, 115, 119
Anchistus, 299, 318
androgenic gland, 36, 37, 424, 476
anecdysis, 67, 68
Anelasma, 215, 216
Anilocra, 416, 423, 453
annulus, 287
annulus ventralis, 325
Anomalocera, 169, 171
Anomopoda, 118

*Boldface numbers refer to illustrations.

503